Grundlehren der mathematischen Wissenschaften 325

A Series of Comprehensive Studies in Mathematics

Springer
Berlin
Heidelberg
New York
Barcelona
Hong Kong
London
Milan
Paris
Singapore
Tokyo

Constantine M. Dafermos

Hyperbolic Conservation Laws in Continuum Physics

With 38 Figures

 Springer

Phys

Constantine M. Dafermos
Division of Applied Mathematics
Brown University
Providence, RI 02912
USA

e-mail: dafermos@cfm.brown.edu

Library of Congress Cataloging-in-Publication Data applied for

Die Deutsche Bibliothek – CIP-Einheitsaufnahme

Dafermos, Constantine M.: Hyperbolic conservation laws in continuum physics /
Constantine M. Dafermos. – Berlin; Heidelberg; New York; Barcelona; Hong Kong;
London; Milan; Paris; Singapore; Tokyo: Springer, 2000
(Grundlehren der mathematischen Wissenschaften; 325)
ISBN 3-540-64914-X

Mathematics Subject Classification (1991):
35L65, 35L67, 73Bxx, 76L05

ISSN 0072-7830

ISBN 3-540-64914-X Springer-Verlag Berlin Heidelberg New York

© Springer-Verlag Berlin Heidelberg 2000
Printed in Germany

Cover design: MetaDesign plus GmbH, Berlin
Photocomposed from the author's TEX files after editing and reformatting by Kurt
Mattes, Heidelberg, using the MathTime fonts and a Springer TEX macro-package
Printed on acid-free paper   SPIN: 10690815   41/3143Ko-5 4 3 2 1 0

For Mihalis and Thalia

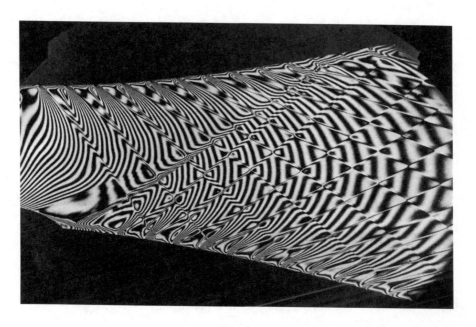

The universe is a gigantic system of reflexes produced by shocks.

Bernard Shaw ("The black girl in search of God")

Preface

The seeds of Continuum Physics were planted with the works of the natural philosophers of the eighteenth century, most notably Euler; by the mid-nineteenth century, the trees were fully grown and ready to yield fruit. It was in this environment that the study of gas dynamics gave birth to the theory of quasilinear hyperbolic systems in divergence form, commonly called "hyperbolic conservation laws"; and these two subjects have been traveling hand-in-hand over the past one hundred and fifty years. This book aims at presenting the theory of hyperbolic conservation laws from the standpoint of its genetic relation to Continuum Physics. Even though research is still marching at a brisk pace, both fields have attained by now the degree of maturity that would warrant the writing of such an exposition.

In the realm of Continuum Physics, material bodies are realized as continuous media, and so-called "extensive quantities", such as mass, momentum and energy, are monitored through the fields of their densities, which are related by balance laws and constitutive equations. A self-contained, though skeletal, introduction to this branch of classical physics is presented in Chapter II. The reader may flesh it out with the help of a specialized text on the subject.

In its primal formulation, the typical balance law stipulates that the time rate of change in the amount of an extensive quantity stored inside any subdomain of the body is balanced by the rate of flux of this quantity through the boundary of the subdomain together with the rate of its production inside the subdomain. In the absence of production, a balanced extensive quantity is conserved. The special feature that renders Continuum Physics amenable to analytical treatment is that, under quite natural assumptions, statements of gross balance, as above, reduce to field equations, i.e., partial differential equations in divergence form.

The collection of balance laws in force demarcates and identifies particular continuum theories, such as Mechanics, Thermomechanics, Electrodynamics and so on. In the context of a continuum theory, constitutive equations specify the nature of the medium, for example viscous fluid, elastic solid, elastic dielectric, etc. In conjunction with these constitutive relations, the field equations yield closed systems of partial differential equations, dubbed "balance laws" or "conservation laws", from which the equilibrium state or motion of the continuous medium is to be determined. Historically, the vast majority of noteworthy partial differential equations were generated through that process. The central thesis of this book is that the umbilical cord joining Continuum Physics with the theory of partial differential equations should not be severed, as it is still carrying nourishment in both directions.

Systems of balance laws may be elliptic, typically in statics; hyperbolic, in dynamics, for media with "elastic" response; mixed elliptic-hyperbolic, in statics or dynamics, when the medium undergoes phase transitions; parabolic or mixed parabolic-hyperbolic, in the presence of viscosity, heat conductivity or other diffusive mechanisms. Accordingly, the basic notions shall be introduced, in Chapter I, at a level of generality that would encompass all of the above possibilities. Nevertheless, since the subject of this work is hyperbolic conservation laws, the discussion will eventually focus on such systems, beginning with Chapter III.

Solutions to hyperbolic conservation laws may be visualized as propagating waves. When the system is nonlinear, the profiles of compression waves get progressively steeper and eventually break, generating jump discontinuities which propagate on as shocks. Hence, inevitably, the theory must deal with weak solutions. This difficulty is compounded further by the fact that, in the context of weak solutions, uniqueness is lost. It thus becomes necessary to devise proper criteria for singling out admissible weak solutions. Continuum Physics naturally induces such admissibility criteria through the Second Law of thermodynamics. These may be incorporated in the analytical theory, either directly, by stipulating outright that admissible solutions should satisfy "entropy" inequalities, or indirectly, by equipping the system with a minute amount of diffusion, which has negligible effect on smooth solutions but reacts stiffly in the presence of shocks, weeding out those that are not thermodynamically admissible. The notions of "entropy" and "vanishing diffusion", which will play a central role throughout the book, are first introduced in Chapters III and IV.

From the standpoint of analysis, a very elegant, definitive theory is available for the case of scalar conservation laws, in one or several space dimensions, which is presented in detail in Chapter VI. By contrast, systems of conservation laws in several space dimensions are still terra incognita, as the analysis is currently facing insurmountable obstacles. The relatively modest results derived thus far, pertaining to local existence and stability of smooth or piecewise smooth solutions, underscore the importance of the special structure of the field equations of Continuum Physics and the stabilizing role of the Second Law of thermodynamics. These issues are discussed in Chapter V.

Beginning with Chapter VII, the focus of the investigation is fixed on systems of conservation laws in one-space dimension. In that setting, the theory has a number of special features, which are of great help to the analyst, so major progress has been achieved.

Chapter VIII provides a systematic exposition of the properties of shocks. In particular, various shock admissibility criteria are introduced, compared and contrasted. Admissible shocks are then combined, in Chapter IX, with another class of particular solutions, called centered rarefaction waves, to synthesize wave fans that solve the classical Riemann problem. Solutions of the Riemann problem may in turn be employed as building blocks for constructing solutions to the Cauchy problem, in the class BV of functions of bounded variation. For that purpose, two construction methods will be presented here: The random choice scheme, in Chapter XIII, and a front tracking algorithm, in Chapter XIV. Uniqueness and stability of these solutions will also be established. The main limitation of this approach is that it generally applies only when the initial data have sufficiently small total variation. This restriction seems to be generally necessary, as it turns out that, in certain systems, when the initial data are "large" even weak solutions to the Cauchy problem may blow up in finite time. However, whether such catastrophes may occur to solutions of the field equations of Continuum Physics is at present a major open problem.

There are other interesting properties of weak solutions, beyond existence and uniqueness. In Chapter X, the notion of characteristic is extended from classical to weak solutions and is employed for obtaining a very precise description of regularity and long time behavior of solutions to scalar conservation laws, in Chapter XI, as well as to systems of two conservation laws, in Chapter XII.

Finally, Chapter XV introduces the concept of measure-valued solution and outlines the functional analytic method of compensated compactness, which determines solutions to hyperbolic systems of conservation laws as weak limits of sequences of approximate solutions, constructed via a variety of approximating schemes.

In order to highlight the fundamental ideas, the discussion proceeds from the general to the particular, notwithstanding the clear pedagogical advantage of the reverse course. Moreover, the pace of the proofs is purposely uneven: slow for the basic, elementary propositions that may provide material for an introductory course; faster for the more advanced technical results that are addressed to the experienced analyst. Even though the various parts of this work fit together to form an integral entity, readers may select a number of independent itineraries through the book. Thus, those principally interested in the conceptual foundations of the theory of hyperbolic conservation laws, in connection to Continuum Physics, need only go through Chapters I–V. Chapter VI, on the scalar conservation law, may be read virtually independently of the rest. Students intending to study solutions as compositions of interacting elementary waves may begin with Chapters VII–IX and then either continue on to Chapters X–XII or else pass directly to Chapter XIII and/or Chapter XIV. Finally, only Chapter VII is needed as a prerequisite for the functional analytic approach expounded in Chapter XV.

Twenty-five years ago, it might have been feasible to write a treatise surveying the entire area; however, the explosive development of the subject over the past several years has rendered such a goal unattainable. Thus, even though this work strives to present a panoramic view of the terrain, certain noteworthy features had to be left out. The most conspicuous absence is a discussion of numerics. This is regrettable, because, beyond its potential practical applications, the numerical analysis of hyperbolic conservation laws provides valuable insight to the theory. Fortunately, a number of specialized texts on that subject are currently available. Several other important topics receive only superficial treatment here, so the reader may have to resort to the cited references for a more thorough investigation. On the other hand, certain topics are perhaps discussed in excessive detail, as they are of special interest to the author. A number of results are published here for the first time. Though extensive, the bibliography is far from exhaustive. In any case, the whole subject is in a state of active development, and significant new publications appear with considerable frequency.

My teachers, Jerry Ericksen and Clifford Truesdell, initiated me to Continuum Physics, as living scientific subject and as formal mathematical structure with fascinating history. I trust that both views are somehow reflected in this work.

I am grateful to many scientists – teachers, colleagues and students alike – who have helped me, over the past thirty years, to learn Continuum Physics and the theory of hyperbolic conservation laws. Since it would be impossible to list them all here by name, let me single out Stu Antman, John Ball, Alberto Bressan, Gui-Qiang Chen, Bernie Coleman, Ron DiPerna, Jim Glimm, Jim Greenberg, Mort Gurtin, Ling Hsiao, Barbara Keyfitz, Peter Lax, Philippe LeFloch, Tai-Ping Liu, Andy Majda, Piero Marcati, Walter Noll, Denis Serre, Marshal Slemrod, Luc Tartar, Konstantina Trivisa, Thanos Tzavaras and Zhouping Xin, who have also honored me with their friendship. In particular, Denis Serre's persistent encouragement helped me to carry this arduous project to completion.

The frontispiece figure depicts the intricate wave pattern generated by shock reflections in the supersonic gas flow through a Laval nozzle with wall disturbances. This beautiful interferogram, brought to my attention by John Ockendon, was produced by W.J. Hiller and G.E.A. Meier at the Max-Planck-Institut für Strömungsforschung, in Göttingen. It is reprinted here, by kind permission of the authors, from *An Album of Fluid Motion*, assembled by Milton Van Dyke and published by Parabolic Press in 1982.

I am indebted to Janice D'Amico for her skilful typing of the manuscript, while suffering cheerfully through innumerable revisions. I also thank Changqing (Peter) Hu for drawing the figures from my rough sketches. I am equally indebted to Karl-Friedrich Koch, of the Springer book production department, for his friendly cooperation. Finally, I gratefully acknowledge the continuous support from the National Science Foundation and the Office of Naval Research.

Constantine M. Dafermos

Table of Contents

Chapter I. Balance Laws

The ambient space for the balance law will be $\mathbb{R}^k$, with typical point X. In the applications to Continuum Physics, $\mathbb{R}^k$ will stand for physical space, of dimension one, two or three, in the context of statics; and for space-time, of dimension two, three or four, in the context of dynamics.

The generic balance law in a domain of $\mathbb{R}^k$ will be introduced through its primal formulation, as a postulate that the production of an extensive quantity in any subdomain is balanced by a flux through the boundary; it will then be reduced to a field equation. It is this reduction that renders Continuum Physics mathematically tractable. It will be shown that the divergence form of the field equation is preserved under change of coordinates.

The field equation for the general balance law will be combined with constitutive equations, relating the flux and production density with a state vector, to yield a quasilinear first order system of partial differential equations in divergence form.

It will be shown that symmetrizable systems of balance laws are endowed with companion balance laws which are automatically satisfied by smooth solutions, though not necessarily by weak solutions. The issue of admissibility of weak solutions will be raised.

Solutions will be considered with shock fronts or weak fronts, in which the state vector field or its derivatives experience jump discontinuities across a manifold of codimension one.

The theory of BV functions, which provide the natural setting for solutions with shock fronts, will be surveyed and the geometric structure of BV solutions will be described.

Highly oscillatory weak solutions will be constructed, and a first indication of the stabilizing role of admissibility conditions will be presented.

The setting being Euclidean space, it will be expedient to employ matrix notation, which may be deficient in elegance but is efficient for calculation. The symbol $\mathcal{M}^{r,s}$ will generally denote the space of $r \times s$ matrices and $\mathbb{R}^r$ will be identified with $\mathcal{M}^{r,1}$. Certain objects that are naturally rank $(0,2)$ tensors shall be here represented by matrices. Consequently, standard conventions notwithstanding, in order to retain consistency with matrix operations, gradients must be realized as row vectors and the divergence operator will be acting on row vectors. The unit sphere in $\mathbb{R}^r$ will be denoted throughout by $\mathcal{S}^{r-1}$.

1.1 Formulation of the Balance Law

A balance law on an open subset $\mathscr{X}$ of $\mathbb{R}^k$ postulates that the *production* of a (generally vector-valued) "extensive" quantity in any bounded measurable subset $\mathscr{D}$ of $\mathscr{X}$ with finite perimeter is balanced by the *flux* of this quantity through the measure-theoretic boundary $\partial\mathscr{D}$ of $\mathscr{D}$. Note that $\partial\mathscr{D}$ is defined as the set of points whose density relative to both $\mathscr{D}$ and $\mathbb{R}^k \setminus \mathscr{D}$ is nonzero; and $\mathscr{D}$ has finite perimeter when $\partial\mathscr{D}$ has finite $(k-1)$-dimensional Hausdorff measure: $\mathscr{H}^{k-1}(\partial\mathscr{D}) < \infty$. With almost all (with respect to $\mathscr{H}^{k-1}$) points X of $\partial\mathscr{D}$ is associated a vector $N(X) \in \mathscr{S}^{k-1}$ which may be naturally interpreted as the measure-theoretic exterior normal to $\mathscr{D}$ at X. A Borel subset $\mathscr{C}$ of $\partial\mathscr{D}$, oriented through the exterior normal N, constitutes an *oriented surface*. The reader unfamiliar with the above concepts may consult the brief survey in Section 1.7 and the references on geometric measure theory cited in Section 1.10 or may assume, without much loss, that we are dealing here with open bounded subsets of $\mathscr{X}$ whose topological boundary is a Lipschitz $(k-1)$-dimensional manifold.

The production is introduced through a functional $\mathscr{P}$, defined on bounded measurable subsets $\mathscr{D}$ of $\mathscr{X}$ with finite perimeter, taking values in $\mathbb{R}^n$, and satisfying the conditions

$$(1.1.1) \qquad \mathscr{P}(\mathscr{D}_1 \cup \mathscr{D}_2) = \mathscr{P}(\mathscr{D}_1) + \mathscr{P}(\mathscr{D}_2) , \quad \text{if } \mathscr{D}_1 \cap \mathscr{D}_2 = \emptyset ,$$

$$(1.1.2) \qquad |\mathscr{P}(\mathscr{D})| \leq c|\mathscr{D}| ,$$

for some constant $c \geq 0$, where $|\mathscr{D}|$ denotes the Lebesgue measure of $\mathscr{D}$.

The flux through $\partial\mathscr{D}$ is induced by a functional Q, defined on the set of oriented surfaces $\mathscr{C}$, which takes values in $\mathbb{R}^n$, and satisfies the conditions

$$(1.1.3) \qquad |Q(\mathscr{C})| \leq c\mathscr{H}^{k-1}(\mathscr{C}) ,$$

for some constant $c \geq 0$, and

$$(1.1.4) \qquad Q(\mathscr{C}_1 \cup \mathscr{C}_2) = Q(\mathscr{C}_1) + Q(\mathscr{C}_2) ,$$

for all disjoint Borel subsets $\mathscr{C}_1$, $\mathscr{C}_2$ of $\partial\mathscr{D}$.

Consequently, the balance law states

$$(1.1.5) \qquad Q(\partial\mathscr{D}) = \mathscr{P}(\mathscr{D})$$

for any bounded measurable subset $\mathscr{D}$ of $\mathscr{X}$ with finite perimeter.

1.2 Reduction to Field Equations

Due to (1.1.1) and (1.1.2), there is a *production density* $P \in L^\infty(\mathscr{X}; \mathbb{R}^n)$ such that

$$(1.2.1) \qquad \mathscr{P}(\mathscr{D}) = \int_{\mathscr{D}} P(X)dX .$$

Similarly, by virtue of (1.1.3) and (1.1.4), with any bounded measurable sub-set $\mathscr{D}$ of $\mathscr{X}$, with finite perimenter, is associated a bounded Borel *flux density* function $Q_{\partial\mathscr{D}} : \partial\mathscr{D} \to \mathbb{R}^n$ such that

$$(1.2.2) \qquad Q(\mathscr{C}) = \int_{\mathscr{C}} Q_{\partial\mathscr{D}}(X)d.\mathscr{H}^{k-1}(X)$$

holds for any oriented surface $\mathscr{C} \subset \partial\mathscr{D}$. Clearly, if $\mathscr{C} \subset \partial\mathscr{D}_1$ and $\mathscr{C} \subset \partial\mathscr{D}_2$, then $Q_{\partial\mathscr{D}_1}$ and $Q_{\partial\mathscr{D}_2}$ restricted to $\mathscr{C}$ must coincide, a.e. with respect to $\mathscr{H}^{k-1}$.

It is remarkable that the seemingly mild assumptions (1.1.3) and (1.1.4) in conjunction with (1.1.5) imply severe restrictions on the density flux function:

Theorem 1.2.1 *Under the assumptions* (1.1.3), (1.1.4), (1.1.5), (1.2.1), *and* (1.2.2), *the value of* $Q_{\partial\mathscr{D}}$ *at* $X \in \partial\mathscr{D}$ *depends on* $\partial\mathscr{D}$ *solely through the exterior normal* $N(X)$ *to* $\mathscr{D}$ *at* X, *namely, there is a bounded measurable function* $Q : \mathscr{X} \times \mathscr{S}^{k-1} \to \mathbb{R}^n$ *such that*

$$(1.2.3) \qquad Q_{\partial\mathscr{D}}(X) = Q(X, N(X)) , \quad \text{a.e. on } \partial\mathscr{D} , \text{ with respect to } \mathscr{H}^{k-1} .$$

Furthermore, Q *depends "linearly" on* N, *i.e., there is a flux density field* $A \in L^\infty(\mathscr{X}; \mathscr{M}^{n,k})$ *such that*

$$(1.2.4) \qquad Q(X, N) = A(X)N , \quad \text{a.e. on } \mathscr{X} ,$$

and

$$(1.2.5) \qquad \text{div} A = P ,$$

in the sense of distributions.

Proof. To establish (1.2.3), fix $\overline{X} \in \mathscr{X}, \overline{N} \in \mathscr{S}^{k-1}$ and consider any two bounded measurable subsets $\mathscr{D}_1$ and $\mathscr{D}_2$ of $\mathscr{X}$, with finite perimeter, such that $\overline{X} \in \partial\mathscr{D}_1, \overline{X} \in \partial\mathscr{D}_2$, and $N_{\mathscr{D}_1}(\overline{X}) = N_{\mathscr{D}_2}(\overline{X}) = \overline{N}$; see Fig. 1.2.1. The aim is to show that $Q_{\partial\mathscr{D}_1}(\overline{X}) = Q_{\partial\mathscr{D}_2}(\overline{X})$. Let $\mathscr{B}_r$ denote the ball in $\mathbb{R}^k$ of (small) radius r centered at $\overline{X}$. We write the balance law (1.1.5), first for $\mathscr{D} = \mathscr{D}_1 \cap \mathscr{B}_r$ then for $\mathscr{D} = \mathscr{D}_2 \cap \mathscr{B}_r$ and subtract the resulting equations to get

$$(1.2.6) \qquad \begin{aligned} & Q(\mathscr{B}_r \cap \partial\mathscr{D}_1) - Q(\mathscr{B}_r \cap \partial\mathscr{D}_2) \\ & = \mathscr{P}((\mathscr{D}_1\backslash\mathscr{D}_2) \cap \mathscr{B}_r) - \mathscr{P}((\mathscr{D}_2\backslash\mathscr{D}_1) \cap \mathscr{B}_r) \\ & \quad - Q((\mathscr{D}_1\backslash\mathscr{D}_2) \cap \partial\mathscr{B}_r) + Q((\mathscr{D}_2\backslash\mathscr{D}_1) \cap \partial\mathscr{B}_r) . \end{aligned}$$

As $r \downarrow 0$, the first two terms on the right-hand side of (1.2.6) are $O(r^k)$, by virtue of (1.1.2); the last two terms are $o(r^{k-1})$, except possibly on a set of r for which the origin is a point of rarefaction, on account of (1.1.3), since $\mathscr{D}_1$ and $\mathscr{D}_2$ are tangential to each other at $\overline{X}$. Consequently, (1.2.2) and (1.2.6) yield

$$(1.2.7) \quad \int_{B_r \cap \partial\mathscr{D}_1} Q_{\partial\mathscr{D}_1}(X)d.\mathscr{H}^{k-1}(X) - \int_{B_r \cap \partial\mathscr{D}_2} Q_{\partial\mathscr{D}_2}(X)d.\mathscr{H}^{k-1}(X) = o(r^{k-1}) .$$

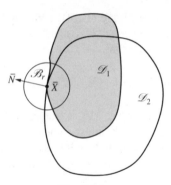

Fig. 1.2.1.

Thus, if $\overline{X}$ is a Lebesgue point of both $Q_{\partial\mathscr{D}_1}$ and $Q_{\partial\mathscr{D}_2}$ then $Q_{\partial\mathscr{D}_1}(\overline{X}) = Q_{\partial\mathscr{D}_2}(\overline{X})$.

The proof of (1.2.4) will be attained by means of the celebrated *Cauchy tetrahedron argument*. Consider the standard orthonormal basis $\{E_\alpha : \alpha = 1, \cdots, k\}$ in $\mathbb{R}^k$.

For fixed α and $\overline{X}$, let us apply the balance law to the rectangle $\mathscr{D} = \{X : -\delta < X_\alpha - \overline{X}_\alpha < \varepsilon, |X_\beta - \overline{X}_\beta| < r, \beta \neq \alpha\}$ with δ, ε and r positive small. Letting $\varepsilon \downarrow 0$ and $\delta \downarrow 0$, one easily deduces $Q(X, -E_\alpha) = -Q(X, E_\alpha)$, a.e. on $\mathscr{X}$.

Now fix $N \in \mathscr{S}^{k-1}$ with nonzero components N_α, $\alpha = 1, \cdots, k$, and $\overline{X} \in \mathscr{X}$ which is a Lebesgue point of $Q(\cdot, N)$ as well as of $Q(\cdot, \pm E_\alpha)$, $\alpha = 1, \cdots, k$. Consider the simplex $\mathscr{D} = \{X : (X_\alpha - \overline{X}_\alpha)N_\alpha > -r, \alpha = 1, \cdots, k, (X - \overline{X}) \cdot N < r\}$ with r positive and small. Notice that $\partial\mathscr{D}$ contains a face $\mathscr{C}$ with exterior normal N and faces $\mathscr{C}_\alpha$, $\alpha = 1, \cdots, k$, with exterior normal $-(\operatorname{sgn} N_\alpha)E_\alpha$. Moreover, we have $\mathscr{H}^{k-1}(\mathscr{C}_\alpha) = |N_\alpha|\mathscr{H}^{k-1}(\mathscr{C})$, $\alpha = 1, \cdots, k$. Applying the balance law to this $\mathscr{D}$, dividing through by $\mathscr{H}^{k-1}(\mathscr{C})$ and letting $r \downarrow 0$ yields

$$(1.2.8) \qquad Q(\overline{X}, N) = \sum_{\alpha=1}^{k} Q(\overline{X}, E_\alpha)N_\alpha \ ,$$

which establishes (1.2.4).

It remains to show (1.2.5). When A is Lipschitz, the balance law takes the form

$$(1.2.9) \qquad \int_{\partial\mathscr{D}} A(X)N(X)d\mathscr{H}^{k-1}(X) = \int_{\mathscr{D}} P(X)dX$$

so that (1.2.5) follows directly from Green's theorem. In the general case, when A is merely in L^∞, even though (1.2.9) may no longer make sense for arbitrary $\mathscr{D}$, it will still hold for translates $\mathscr{D}_Y = \{X \in \mathbb{R}^k : X - Y \in \mathscr{D}\}$ of any fixed hypercube $\mathscr{D}$ by almost all Y in a ball $\{Y \in \mathbb{R}^k : |Y| < \varepsilon\}$, with ε sufficiently small to retain $\overline{\mathscr{D}_Y} \subset \mathscr{X}$. Accordingly, we fix any test function $\psi \in C_0^\infty(\mathbb{R}^k)$ with total mass 1, supported in the unit ball, we rescale it by ε,

(1.2.10)
$$\psi_\varepsilon(X) = \varepsilon^{-k}\psi(\varepsilon^{-1}X) \ ,$$

and use it to mollify, in the customary fashion, the fields P and A on the set $\mathscr{X}_\varepsilon \subset \mathscr{X}$ of points whose distance from $\mathscr{X}^c$ exceeds ε:

(1.2.11)
$$P_\varepsilon = \psi_\varepsilon * P \ , \qquad A_\varepsilon = \psi_\varepsilon * A \ .$$

For any hypercube $\mathscr{D} \subset \mathscr{X}_\varepsilon$, we apply Green's theorem to the smooth field A_ε and use Fubini's theorem to get

$$
\begin{aligned}
\int_{\mathscr{D}} \operatorname{div} A_\varepsilon(X)dX &= \int_{\partial\mathscr{D}} A_\varepsilon(X)N(X)d.\mathscr{H}^{k-1}(X) \\
&= \int_{\partial\mathscr{D}} \int_{\mathbb{R}^k} \psi_\varepsilon(Y)A(X-Y)N(X)dY d.\mathscr{H}^{k-1}(X) \\
&= \int_{\mathbb{R}^k} \psi_\varepsilon(Y) \int_{\partial\mathscr{D}_Y} A(Z)N(Z)d.\mathscr{H}^{k-1}(Z)dY \\
&= \int_{\mathbb{R}^k} \psi_\varepsilon(Y) \int_{\mathscr{D}_Y} P(Z)dZdY \\
&= \int_{\mathscr{D}} \int_{\mathbb{R}^k} \psi_\varepsilon(Y)P(X-Y)dYdX \\
&= \int_{\mathscr{D}} P_\varepsilon(X)dX \ ,
\end{aligned}
$$

(1.2.12)

whence we conclude $\operatorname{div} A_\varepsilon = P_\varepsilon$ on $\mathscr{X}_\varepsilon$. Letting $\varepsilon \downarrow 0$, yields (1.2.5) on $\mathscr{X}$, in the sense of distributions. This completes the proof.

Conversely, a field equation (1.2.5), with $A \in L^\infty(\mathscr{X}; \mathscr{M}^{n,k})$ and $P \in L^\infty(\mathscr{X}; \mathbb{R}^n)$, induces a balance law (1.1.5), where $\mathscr{P}$ is defined by (1.2.1), and Q is obtained from (1.2.2), for some function $Q_{\partial\mathscr{D}} \in L^\infty(\partial\mathscr{D}; \mathbb{R}^n)$ identified through its action on test functions $\phi \in C^\infty(\mathbb{R}^k)$:

(1.2.13)
$$\int_{\partial\mathscr{D}} \phi(X)Q_{\partial\mathscr{D}}(X)d.\mathscr{H}^{k-1}(X) = \int_{\mathscr{D}} \phi(X)P(X)dX + \int_{\mathscr{D}} A(X)(\operatorname{grad}\phi)^T(X)dX \ .$$

Clearly, (1.2.13) is derived formally upon multiplying (1.2.5) by ϕ, integrating over $\mathscr{D}$ and applying Green's theorem.

In fact, the function $Q_{\partial\mathscr{D}}$ may be constructed, through (1.2.13), even in the more general case where $A \in L^\infty(\mathscr{X}; \mathscr{M}^{n,k})$ satisfies a field equation (1.2.5) with P a measure on $\mathscr{X}$. Of course in that case it is no longer generally true that the value of $Q_{\partial\mathscr{D}}$ at $X \in \partial\mathscr{D}$ depends on $\partial\mathscr{D}$ solely through the exterior normal $N(X)$ to $\partial\mathscr{D}$ at X. Details may be found in the references cited in Section 1.10.

1.3 Change of Coordinates

The divergence form of the field equations of balance laws is preserved under coordinate changes, so long as the fields transform according to appropriate rules.

Theorem 1.3.1 *Let $\mathscr{X}$ be an open subset of $\mathbb{R}^k$ and assume that functions $A \in L^1_{loc}(\mathscr{X}; \mathscr{M}^{n,k})$ and $P \in L^1_{loc}(\mathscr{X}; \mathbb{R}^n)$ satisfy the field equation*

$$(1.3.1) \qquad \qquad \operatorname{div} A = P \ ,$$

in the sense of distributions on $\mathscr{X}$. Consider any bilipschitz homeomorphism X^* of $\mathscr{X}$ to a subset $\mathscr{X}^*$ of $\mathbb{R}^k$, with Jacobian matrix

$$(1.3.2) \qquad \qquad J = \frac{\partial X^*}{\partial X}$$

such that

$$(1.3.3) \qquad \qquad \det J \geq a > 0 \ , \quad a.e. \ on \ \mathscr{X} \ .$$

Then $A^* \in L^1_{loc}(\mathscr{X}^*; \mathscr{M}^{n,k})$, $P^* \in L^1_{loc}(\mathscr{X}^*; \mathbb{R}^n)$ defined by

$$(1.3.4) \qquad A^* \circ X^* = (\det J)^{-1} A J^T \ , \quad P^* \circ X^* = (\det J)^{-1} P$$

satisfy the field equation

$$(1.3.5) \qquad \qquad \operatorname{div} A^* = P^* \ ,$$

in the sense of distributions on $\mathscr{X}^*$.

Proof. From (1.3.1) it follows that

$$(1.3.6) \qquad \qquad \int_{\mathscr{X}} [A(\operatorname{grad}\phi)^T + P\phi] dX = 0$$

holds for any test function $\phi \in C_0^\infty(\mathscr{X})$ and thereby, by completion in $W^{1,\infty}$, for any Lipschitz function ϕ with compact support in $\mathscr{X}$.

Given any test function $\phi^* \in C_0^\infty(\mathscr{X}^*)$, consider the Lipschitz function $\phi = \phi^* \circ X^*$, with compact support in $\mathscr{X}$. Notice that $\operatorname{grad}\phi = (\operatorname{grad}\phi^*)J$. Furthermore, $dX^* = (\det J)dX$. By virtue of these and (1.3.4), (1.3.6) yields

$$(1.3.7) \qquad \qquad \int_{\mathscr{X}^*} [A^*(\operatorname{grad}\phi^*)^T + P^*\phi^*] dX^* = 0 \ ,$$

which establishes (1.3.5). The proof is complete.

1.4 Systems of Balance Laws

On an open subset $\mathscr{X}$ of $\mathbb{R}^k$, we consider the field equation (1.2.5) of a balance law, which we now write in components form:

$$(1.4.1) \qquad \sum_{\alpha=1}^{k} \partial_\alpha A_\alpha(X) = P(X) \ .$$

The symbol ∂_α stands for $\partial/\partial X_\alpha$, $\alpha = 1, \cdots, k$. Recall that P and A_α, $\alpha = 1, \cdots, k$, take values in $\mathbb{R}^n$.

We assume that the state of the medium is described by a *state vector* field U, taking values in an open subset $\mathscr{O}$ of $\mathbb{R}^n$, which determines the flux density field A and the production density field P at the point $X \in \mathscr{X}$ by *constitutive equations*

$$(1.4.2) \qquad A(X) = G(U(X), X) \ , \qquad P(X) = \Pi(U(X), X) \ ,$$

where G and Π are given smooth functions defined on $\mathscr{O} \times \mathscr{X}$ and taking values in $\mathscr{M}^{n,k}$ and $\mathbb{R}^n$, respectively.

Notation. Throughout this work, the symbol D will denote the gradient operator $[\partial/\partial U_1, \cdots, \partial/\partial U_n]$ with respect to the state vector U, visualized as a n-row. Also, D^2 will denote the operation of forming the ($n \times n$ matrix-valued) Hessian with respect to U.

Combining (1.4.1) with (1.4.2), yields the quasilinear first order system of partial differential equations

$$(1.4.3) \qquad \sum_{\alpha=1}^{k} \partial_\alpha G_\alpha(U(X), X) = \Pi(U(X), X) \ ,$$

from which the state vector field U is to be determined. Any equation of the form (1.4.3) will henceforth be called a *system of balance laws*, if $n \geq 2$, or a *scalar balance law* when $n = 1$. In the special case where there is no production, $\Pi \equiv 0$, (1.4.3) will be called a *system of conservation laws*, if $n \geq 2$, or a *scalar conservation law* when $n = 1$. This terminology is not quite standard: In lieu of "system of balance laws" certain authors favor the term "system of conservation laws with source". When G and Π do not depend explicitly on X, the system of balance laws is called *homogeneous*.

Notice that when coordinates are stretched in the vicinity of some fixed point $\overline{X} \in \mathscr{X}$, i.e., $X = \overline{X} + \varepsilon Y$, then, as $\varepsilon \downarrow 0$, the system of balance laws (1.4.3) reduces to a homogeneous system of conservation laws with respect to the Y variable. It is for this reason that local properties of solutions of general systems of balance laws may be investigated, without loss of generality, in the simpler setting of homogeneous systems of conservation laws.

A Lipschitz continuous field U which satisfies (1.4.3) almost everwhere on $\mathscr{X}$ will be called a *classical solution*. A bounded measurable field U which satisfies (1.4.3) in the sense of distributions, i.e.,

$$(1.4.4) \qquad \int_{\mathscr{X}} \left[\sum_{\alpha=1}^{k} \partial_\alpha \varphi(X) G_\alpha(U(X), X) + \varphi(X) \Pi(U(X), X) \right] dX = 0 ,$$

for any test function $\varphi \in C_0^\infty(\mathscr{X})$, is a *weak solution*. Any weak solution which is Lipschitz continuous is necessarily a classical solution.

1.5 Companion Systems of Balance Laws

Consider a system (1.4.3) of balance laws on an open subset $\mathscr{X}$ of $\mathbb{R}^k$, resulting from combining the field equation (1.4.1) with constitutive relations (1.4.2). A smooth function q, defined on $\mathscr{O} \times \mathscr{X}$ and taking values in $\mathscr{M}^{1,k}$, is called a *companion* of G if there is a smooth function B, defined on $\mathscr{O} \times \mathscr{X}$ and taking values in $\mathbb{R}^n$, such that, for all $U \in \mathscr{O}$, $X \in \mathscr{X}$,

$$(1.5.1) \qquad Dq_\alpha(U, X) = B(U, X)^T DG_\alpha(U, X) , \quad \alpha = 1, \cdots, k .$$

The relevance of (1.5.1) stems from the observation that any classical solution U of the system of balance laws (1.4.3) is automatically also a (classical) solution of the *companion balance law*

$$(1.5.2) \qquad \sum_{\alpha=1}^{k} \partial_\alpha q_\alpha(U(X), X) = h(U(X), X)$$

with

(1.5.3)
$$h(U, X) = B(U, X)^T \Pi(U, X) + \sum_{\alpha=1}^{k} \left[\frac{\partial q_\alpha(U, X)}{\partial X_\alpha} - B(U, X)^T \frac{\partial G_\alpha(U, X)}{\partial X_\alpha} \right] .$$

In (1.5.3) $\partial/\partial X_\alpha$ denotes partial derivative with respect to X_α, holding U fixed.

The task of determining the companion balance laws (1.5.2) of a given system of balance laws (1.4.3) may be accomplished by identifying the integrating factors B that render the right-hand side of (1.5.1) a gradient of a function of U. The relevant integrability condition is

$$(1.5.4) \quad DB(U, X)^T DG_\alpha(U, X) = DG_\alpha(U, X)^T DB(U, X) , \quad \alpha = 1, \cdots, k ,$$

for all $U \in \mathscr{O}$ and $X \in \mathscr{X}$. Clearly, one can satisfy (1.5.4) by employing any B which may vary with X but not with U; in that case, however, the resulting companion balance law (1.5.2) is just a trivial linear combination of the equations of the original system (1.4.3). For determining nontrivial B, which vary with U, (1.5.4) imposes $\frac{1}{2}n(n-1)k$ conditions on the n unknown components of B. Thus, when $n = 1$ and k is arbitrary one may use any (scalar-valued) function B. When $n = 2$ and $k = 2$, (1.5.4) reduces to a system of two equations in two unknowns from which a family of B may presumably be determined. In all other cases, however, (1.5.4) is formally overdetermined and the existence of nontrivial companion balance laws should not be generally expected. Nevertheless, as we shall see in Chapter III, the systems of balance laws of continuum thermomechanics are endowed with natural companion balance laws.

The system of balance laws (1.4.3) is called *symmetric* when the $n \times n$ matrices $DG_\alpha(U, X)$, $\alpha = 1, \cdots, k$, are symmetric, for any $U \in \mathcal{O}, X \in \mathcal{X}$, say $\mathcal{O}$ is simply connected and

$$(1.5.5) \qquad G_\alpha(U, X) = Dg_\alpha(U, X)^T , \qquad \alpha = 1, \cdots, k ,$$

for some smooth function g, defined on $\mathcal{O} \times \mathcal{X}$ and taking values in $\mathcal{M}^{1,k}$. In that case one may satisfy (1.5.4) by taking $B(U, X) \equiv U$, which induces the companion

$$(1.5.6) \qquad q(U, X) = U^T G(U, X) - g(U, X) .$$

Conversely, if (1.5.1) holds for some B with the property that for every fixed $X \in \mathcal{X}$, $B(\cdot, X)$ maps diffeomorphically $\mathcal{O}$ to some open subset $\mathcal{O}^*$ of $\mathbb{R}^n$, then the change $U^* = B(U, X)$ of state vector reduces (1.4.3) to the equivalent system of balance laws

$$(1.5.7) \qquad \sum_{\alpha=1}^{k} \partial_\alpha G_\alpha^*(U^*(X), X) = \Pi^*(U^*(X), X) , \qquad \bullet$$

with
$$(1.5.8)$$
$$G^*(U^*, X) = G(B^{-1}(U^*, X), X) , \qquad \Pi^*(U^*, X) = \Pi(B^{-1}(U^*, X), X) ,$$

which is symmetric. Indeed, upon setting

$$(1.5.9) \qquad q^*(U^*, X) = q(B^{-1}(U^*, X), X) ,$$

$$(1.5.10) \qquad g^*(U^*, X) = U^{*T} G^*(U^*, X) - q^*(U^*, X) ,$$

it follows from (1.5.1) that

$$(1.5.11) \qquad G_\alpha^*(U^*, X) = Dg_\alpha^*(U^*, X)^T , \qquad \alpha = 1, \cdots, k .$$

We have thus demonstrated that a system of balance laws is endowed with nontrivial companion balance laws if and only if it is *symmetrizable*.

Despite (1.5.1), and in contrast to the behavior of classical solutions, weak solutions of (1.4.3) need not satisfy (1.5.2). Nevertheless, one of the tenets of the theory of systems of balance laws is that *admissible weak solutions* should at least satisfy the inequality

$$(1.5.12) \qquad \sum_{\alpha=1}^{k} \partial_\alpha q_\alpha(U(X), X) \leq h(U(X), X) \ ,$$

in the sense of distributions, for a designated family of companions. Relating this postulate to the Second Law of thermodynamics and investigating its implications on stability of weak solutions are among the principal objectives of this book.

Notice that when the inequality (1.5.12) holds, its left-hand side is necessarily a measure and, therefore, following the discussion at the end of Section 1.2, may be associated with a balance law.

1.6 Weak and Shock Fronts

The regularity of solutions of a system of balance laws will depend on the nature of the constitutive functions. The focus will be on solutions with "fronts", that is, singularities assembled on manifolds of codimension one. To get acquainted with this sort of solutions, we consider here two kinds of fronts in a particularly simple setting.

In what follows, $\mathscr{F}$ will be a smooth $(k-1)$-dimensional manifold, embedded in the open subset $\mathscr{X}$ of $\mathbb{R}^k$, with orientation induced by the unit normal field N. U will be a solution of the system of balance laws (1.4.3) on $\mathscr{X}$ which is continuously differentiable on $\mathscr{X}\backslash\overline{\mathscr{F}}$, but is allowed to be singular on $\mathscr{F}$. In particular, (1.4.3) holds for any $X \in \mathscr{X}\backslash\overline{\mathscr{F}}$.

First we consider the case where $\mathscr{F}$ is a *weak front*, that is, U is Lipschitz continuous on $\mathscr{X}$ and as one approaches $\mathscr{F}$ from either side the partial derivatives $\partial_\alpha U$ of U attain distinct limits $\partial_\alpha^- U$, $\partial_\alpha^+ U$. Thus $\partial_\alpha U$ experiences a jump $[\partial_\alpha U] = \partial_\alpha^+ U - \partial_\alpha^- U$ across $\mathscr{F}$. Since U is continuous, tangential derivatives of U cannot jump across $\mathscr{F}$ and hence $[\partial_\alpha U] = N_\alpha[\partial U/\partial N]$, $\alpha = 1, \cdots, k$, where $[\partial U/\partial N]$ denotes the jump of the normal derivative $\partial U/\partial N$ across $\mathscr{F}$. Therefore, taking the jump of (1.4.3) across $\mathscr{F}$ at any point $X \in \mathscr{F}$ yields the following condition on $[\partial U/\partial N]$:

$$(1.6.1) \qquad \sum_{\alpha=1}^{k} N_\alpha DG_\alpha(U(X), X) \left[\frac{\partial U}{\partial N}\right] = 0 \ .$$

Next we assume $\mathscr{F}$ is a *shock front*, that is, as one approaches $\mathscr{F}$ from either side, U attains distinct limits U_-, U_+ and thus experiences a jump $[U] = U_+ - U_-$ across $\mathscr{F}$. Since U is a (weak) solution of (1.4.3), we may write (1.4.4) for any $\varphi \in C_0^\infty(\mathscr{X})$. In (1.4.4) integration over $\mathscr{X}$ may be replaced with integration over $\mathscr{X}\backslash\overline{\mathscr{F}}$. Since U is C^1 on $\mathscr{X}\backslash\overline{\mathscr{F}}$, we may integrate by parts in (1.4.4). Using that φ has compact support in $\mathscr{X}$ and that (1.4.3) holds for any $X \in \mathscr{X}\backslash\overline{\mathscr{F}}$, we get

$$(1.6.2) \qquad \int_{\mathscr{F}} \varphi(X) \sum_{\alpha=1}^{k} N_\alpha[G_\alpha(U_+, X) - G_\alpha(U_-, X)]d\mathscr{H}^{k-1}(X) = 0 \ ,$$

whence we deduce that the following *jump condition* must be satisfied at every point X of the shock front $\mathscr{F}$:

$$(1.6.3) \qquad \sum_{\alpha=1}^{k} N_\alpha [G_\alpha(U_+, X) - G_\alpha(U_-, X)] = 0 .$$

Notice that (1.6.3) may be rewritten in the form

$$(1.6.4) \qquad \sum_{\alpha=1}^{k} N_\alpha \left\{ \int_0^1 DG_\alpha(\tau U_+ + (1-\tau)U_-, X)d\tau \right\} [U] = 0 .$$

Comparing (1.6.4) with (1.6.1) we conclude that weak fronts may be regarded as shock fronts with "infinitesimal" strength: $|[U]|$ vanishingly small.

With each $U \in \mathcal{O}$ and $X \in \mathscr{X}$ we associate the variety

$$(1.6.5) \qquad \mathscr{V}(U, X) = \left\{ (N, V) \in \mathscr{S}^{k-1} \times \mathbb{R}^n : \sum_{\alpha=1}^{k} N_\alpha DG_\alpha(U, X)V = 0 \right\} .$$

The number of weak fronts and shock fronts of small amplitude that may be sustained by solutions of (1.4.3) will depend on the size of $\mathscr{V}$. In the extreme case where, for all (U, X), the projection of $\mathscr{V}(U, X)$ onto $\mathbb{R}^n$ contains only the vector $V = 0$, (1.4.3) is called *elliptic*. Thus a system of balance laws is elliptic if and only if it cannot sustain any weak fronts or shock fronts of small amplitude. In Continuum Physics, ellipticity manifests itself as a condition of stability in elastostatics. Ellipticity may fail in elastostatics when the constitutive equation allows for instabilities, like phase transitions. In that case, shock fronts are interpreted as *phase boundaries*, separating different phases of the material. The opposite extreme to ellipticity, where $\mathscr{V}$ attains the maximal possible size, is typically encountered in elastodynamics and will be discussed in Chapter III.

1.7 Survey of the Theory of *BV* Functions

In this section we shall get acquainted with *BV* functions, in which discontinuities assemble on manifolds of codimension one, and thus provide the natural setting for solutions of systems of balance laws with shock fronts. Comprehensive treatment of the theory of *BV* functions can be found in the references cited in Section 1.10, so only properties relevant to our purposes will be listed here, mostly without proofs.

Definition 1.7.1 A function U defined on an open subset $\mathscr{X}$ of $\mathbb{R}^k$ and taking values in $\mathbb{R}^n$ is of *locally bounded variation* if $U \in L^1_{\text{loc}}(\mathscr{X})$ and for $\alpha = 1, \cdots, k$ the distributional derivative $\partial_\alpha U$ is a locally finite (vector-valued) Radon measure μ_α on $\mathscr{X}$, i.e.,

$$(1.7.1) \qquad - \int_{\mathscr{X}} \partial_\alpha \phi(X) U(X) dX = \int_{\mathscr{X}} \phi(X) d\mu_\alpha(X) , \quad \alpha = 1, \cdots, k ,$$

holds for any test function $\phi \in C_0^\infty(\mathscr{X})$. When $U \in L^1(\mathscr{X})$ and the μ_α are finite, U is a function of *bounded variation*, with *total variation*

$$(1.7.2) \qquad\qquad TV_{\mathscr{X}} U = \sum_{\alpha=1}^{k} |\mu_\alpha|(\mathscr{X}) ,$$

where $|\mu_\alpha|$ denotes the total variation of the measure μ_α. The set of functions of bounded variation and locally bounded variation over $\mathscr{X}$ will be denoted by $BV(\mathscr{X})$ and $BV_{\text{loc}}(\mathscr{X})$, respectively.

Clearly, the Sobolev space $W^{1,1}(\mathscr{X})$ of functions $U \in L^1(\mathscr{X})$ with distributional derivatives $\partial_\alpha U \in L^1(\mathscr{X})$ is contained in $BV(\mathscr{X})$ and $W_{\text{loc}}^{1,1}(\mathscr{X})$ is contained in $BV_{\text{loc}}(\mathscr{X})$.

The following proposition may be used to test whether some function has locally bounded variation.

Theorem 1.7.1 *Let* $\{E_\alpha, \alpha = 1, \cdots, k\}$ *denote the standard orthonormal basis of* $\mathbb{R}^k$. *If* $U \in BV_{\text{loc}}(\mathscr{X})$, *then*

$$(1.7.3) \quad \limsup_{h\downarrow 0} \frac{1}{h} \int_{\mathscr{Y}} |U(X + hE_\alpha) - U(X)| dX = |\mu_\alpha|(\mathscr{Y}) , \qquad \alpha = 1, \cdots, k ,$$

for any open bounded set $\mathscr{Y}$ *with* $\overline{\mathscr{Y}} \subset \mathscr{X}$. *Conversely, if* $U \in L_{\text{loc}}^1(\mathscr{X})$ *and the left-hand side of* (1.7.3) *is finite for any* $\mathscr{Y}$ *as above, then* $U \in BV_{\text{loc}}(\mathscr{X})$.

Proof. Fix any test function $\phi \in C_0^\infty(\mathscr{X})$ supported in $\mathscr{Y}$ and notice that

$$(1.7.4) \quad \frac{1}{h} \int_{\mathscr{Y}} \phi(X)[U(X + hE_\alpha) - U(X)] dX = - \int_{\mathscr{Y}} \partial_\alpha \psi_\alpha(X; h) U(X) dX ,$$

where

$$(1.7.5) \qquad\qquad \psi_\alpha(X; h) = \int_0^1 \phi(X - h\tau E_\alpha) d\tau .$$

Clearly, $\|\psi_\alpha\|_{L_\infty} \leq \|\phi\|_{L^\infty}$ and so, recalling (1.7.1),

$$(1.7.6) \qquad\qquad \frac{1}{h} \int_{\mathscr{Y}} |U(X + hE_\alpha) - U(X)| dX \leq |\mu_\alpha|(\mathscr{Y}) .$$

On the other hand,

$$(1.7.7) \quad \lim_{h\downarrow 0} \frac{1}{h} \int_{\mathscr{Y}} \phi(X)[U(X + hE_\alpha) - U(X)] dX = - \int_{\mathscr{Y}} \partial_\alpha \phi(X) U(X) dX ,$$

whence

$$(1.7.8) \qquad\qquad |\mu_\alpha|(\mathscr{Y}) \leq \limsup_{h\downarrow 0} \frac{1}{h} \int_{\mathscr{Y}} |U(X + hE_\alpha) - U(X)| dX .$$

This completes the proof.

Theorem 1.7.2 *Any sequence* $\{U_\ell\}$ *in* $BV_{loc}(\mathscr{X})$ *such that* $\|U_\ell\|_{L^1(\mathscr{Y})}$ *and* $TV_{\mathscr{Y}}U_\ell$
are uniformly bounded on every open bounded $\mathscr{Y}$ *with* $\overline{\mathscr{Y}} \subset \mathscr{X}$ *contains a*
subsequence which converges in $L^1_{loc}(\mathscr{X})$, *as well as almost everywhere, to some*
function U *in* $BV_{loc}(\mathscr{X})$, *with* $TV_{\mathscr{Y}}U \leq \liminf_{\ell \to \infty} TV_{\mathscr{Y}}U_\ell$.

Proof. A bounded subset of L^1 is relatively compact when its members are L^1-
equicontinuous. The assertion then follows directly from estimate (1.7.6). This
completes the proof.

   The relevance of BV functions to our purposes stems from their local properties described in

Theorem 1.7.3 *The domain* $\mathscr{X}$ *of any* $U \in BV_{loc}(\mathscr{X})$ *is the union of three,*
pairwise disjoint, subsets $\mathscr{C}$, $\mathscr{J}$, *and* $\mathscr{I}$ *with the following properties:*
(a) $\mathscr{C}$ *is the set of points of approximate continuity of* U, *i.e., with each* $\overline{X} \in \mathscr{C}$ *is*
associated U_0 *in* $\mathbb{R}^n$ *such that, for any* $\varepsilon > 0$, *as* $r \downarrow 0$,

(1.7.9) $meas\{X \in \mathscr{X} : |X - \overline{X}| < r, |U(X) - U_0| > \varepsilon\} = o(r^k)$.

(b) $\mathscr{J}$ *is the set of points of approximate jump discontinuity of* U, *i.e., with each*
$\overline{X} \in \mathscr{J}$ *are associated* N *in* $\mathscr{S}^{k-1}$ *and distinct* U_-, U_+ *in* $\mathbb{R}^n$ *such that, for any*
$\varepsilon > 0$, *as* $r \downarrow 0$,

(1.7.10)
  $meas\{X \in \mathscr{X} : |X - \overline{X}| < r, (X - \overline{X}) \cdot N \gtrless 0, |U(X) - U_\pm| > \varepsilon\} = o(r^k)$.

$\mathscr{J}$ *is essentially covered by the countable union of* C^1 $(k-1)$-*dimensional*
manifolds $\{\mathscr{F}_i\}$ *embedded in* $\mathbb{R}^k$: $\mathscr{H}^{k-1}(\mathscr{J} \backslash \bigcup \mathscr{F}_i) = 0$. *Furthermore, when*
$\overline{X} \in \mathscr{J} \cap \mathscr{F}_i$ *then* N *is normal on* $\mathscr{F}_i$ *at* $\overline{X}$.
(c) $\mathscr{I}$ *is the set of irregular points of* U; *its* $(k-1)$- *dimensional Hausdorff measure*
is zero: $\mathscr{H}^{k-1}(\mathscr{I}) = 0$.

   Up to this point, the identity of a BV function is unaffected by modifying
its values on any set of (k-dimensional Lebesgue) measure zero, i.e., $BV_{loc}(\mathscr{X})$
is actually a space of equivalence classes of functions, specified only up to a
set of measure zero. However, when dealing with the finer, local behavior of
these functions, it is expedient to designate a particular representative of each
equivalence class, with values specified up to a set of $(k-1)$-dimensional Hausdorff
measure zero. This will be effected in the following way.
   Suppose A is a continuous function from $\mathbb{R}^n$ to $\mathscr{M}^{r,s}$ and let $U \in BV_{loc}(\mathscr{X})$,
with values in $\mathbb{R}^n$. With reference to the notation of Theorem 1.7.3, the *normalized*
composition $\widehat{A \circ U}$ of A and U is defined by

(1.7.11) $\widehat{A \circ U}(X) = \begin{cases} A(U_0), & \text{if } X \in \mathscr{C} \\ \int_0^1 A(\tau U_- + (1-\tau)U_+)d\tau, & \text{if } X \in \mathscr{J} \end{cases}$

and arbitrarily on the set $\mathscr{I}$ of irregular points, whose $(k-1)$-dimensional Hausdorff measure is zero. In particular, we may employ as A the identity map to normalize U itself:

$$(1.7.12) \qquad \tilde{U}(X) = \begin{cases} U_0, & \text{if } X \in \mathscr{C} \\ \frac{1}{2}(U_- + U_+), & \text{if } X \in \mathscr{J}. \end{cases}$$

The appropriateness of the above normalization is justified by the following generalization of the classical chain rule:

Theorem 1.7.4 *Assume H is a continuously differentiable map from $\mathbb{R}^n$ to $\mathbb{R}^r$ and let $U \in BV_{\text{loc}}(\mathscr{X}) \cap L^\infty(\mathscr{X})$. Then $H \circ U \in BV_{\text{loc}}(\mathscr{X}) \cap L^\infty(\mathscr{X})$. The normalized function $\widetilde{DH \circ U}$ is locally integrable with respect to the measures $\mu_\alpha = \partial_\alpha U, \alpha = 1, \cdots, k$, and*

$$(1.7.13) \qquad \partial_\alpha(H \circ U) = \widetilde{DH \circ U}\partial_\alpha U, \qquad \alpha = 1, \cdots, k,$$

in the sense

$$(1.7.14) \qquad -\int_{\mathscr{X}} \partial_\alpha \phi(X) H(U(X)) dX = \int_{\mathscr{X}} \phi(X) \widetilde{DH \circ U}(X) d\mu_\alpha(X),$$

for any test function $\phi \in C_0^\infty(\mathscr{X})$.

Next we review certain geometric aspects of the theory of BV functions.

Definition 1.7.2 A subset $\mathscr{D}$ of $\mathbb{R}^k$ has *(locally) finite perimeter* when its indicator function $\chi_\mathscr{D}$ has (locally) bounded variation on $\mathbb{R}^k$.

Let us apply Theorem 1.7.3 to the indicator function $\chi_\mathscr{D}$ of a set $\mathscr{D}$ with locally finite perimeter. Clearly, the set $\mathscr{C}$ of points of approximate continuity of $\chi_\mathscr{D}$ is the union of the sets of density points of $\mathscr{D}$ and $R^k \setminus \mathscr{D}$. The complement of $\mathscr{C}$, i.e., the set of X in $\mathbb{R}^k$ that are not points of density of either $\mathscr{D}$ or $R^k \setminus \mathscr{D}$, constitutes the *measure theoretic boundary* $\partial \mathscr{D}$ of $\mathscr{D}$, which we already encountered in Section 1.1. It can be shown that $\mathscr{D}$ has finite perimeter if and only if $\mathscr{H}^{k-1}(\partial \mathscr{D}) < \infty$, and its perimeter may be measured equivalently by $TV_{\mathbb{R}^k}\chi_\mathscr{D}$ or by $\mathscr{H}^{k-1}(\partial \mathscr{D})$. The set of points of approximate jump discontinuity of $\chi_\mathscr{D}$ is called the *reduced boundary* of $\mathscr{D}$ and is denoted by $\partial^* \mathscr{D}$. By Theorem 1.7.3, $\partial^* \mathscr{D} \subset \partial \mathscr{D}, \mathscr{H}^{k-1}(\partial \mathscr{D} \setminus \partial^* \mathscr{D}) = 0$, and $\partial^* \mathscr{D}$ is covered by the countable union of C^1 $(k-1)$-dimensional manifolds. Moreover, the vector $N \in \mathscr{S}^{k-1}$ associated with each point X of $\partial^* \mathscr{D}$ may naturally be interpreted as the *measure theoretic exterior normal* to $\mathscr{D}$ at X. This was also noted in Section 1.1.

Definition 1.7.3 Assume $\mathscr{D}$ has finite perimeter and let $V \in BV_{\text{loc}}(\mathbb{R}^k)$. V has *inward* (or *outward*) *trace* V_+ (or V_-) at the point $\overline{X}$ of the reduced boundary $\partial^* \mathscr{D}$ of $\mathscr{D}$, where the exterior normal is N, if for any $\varepsilon > 0$, as $r \downarrow 0$,

(1.7.15)

$$meas\{X \in \mathbb{R}^k : |X - \overline{X}| < r, (X - \overline{X}) \cdot N \lessgtr 0, |V(X) - V_\pm| > \varepsilon\} = o(r^k) .$$

It can be shown that the traces $V_\pm$ are defined for almost all (with respect to $\mathcal{H}^{k-1}$) points of $\partial^* \mathcal{D}$ and are locally integrable on $\partial^* \mathcal{D}$. Furthermore, the following version of the Gauss-Green theorem holds:

Theorem 1.7.5 *Assume* $V \in BV(\mathbb{R}^k)$ *so* $\partial_\alpha V$ *are finite measures* μ_α, $\alpha = 1, \cdots, k$. *Consider any bounded set* $\mathcal{D}$ *of finite perimeter, with set of density points* $\mathcal{D}^*$ *and reduced boundary* $\partial^* \mathcal{D}$. *Then*

(1.7.16)
$$\mu_\alpha(\mathcal{D}^*) = \int_{\partial^* \mathcal{D}} N_\alpha V_+ d.\mathcal{H}^{k-1} , \quad \alpha = 1, \cdots, k .$$

Furthermore, for any Borel subset $\mathcal{F}$ *of* $\partial \mathcal{D}$ *(an "oriented surface" in the terminology of Section 1.1),*

(1.7.17)
$$\mu_\alpha(\mathcal{F}) = \int_{\mathcal{F}} N_\alpha(V_- - V_+) d.\mathcal{H}^{k-1} , \quad \alpha = 1, \cdots, k .$$

In particular, the set $\mathcal{J}$ of points of approximate jump discontinuity of any $U \in BV_{\text{loc}}(\mathbb{R}^k)$ may be covered by the countable union of oriented surfaces and so (1.7.17) will hold for any measurable subset $\mathcal{F}$ of $\mathcal{J}$.

For $k = 1$, the theory of BV functions is intimately related with the classical theory of functions of bounded variation. Assume U is a BV function on a (bounded or unbounded) interval $(a, b) \subset (-\infty, \infty)$. Let $\tilde{U}$ be the normalized form of U. Then

(1.7.18)
$$TV_{(a,b)}U = \sup \sum_{j=1}^{\ell-1} |\tilde{U}(x_{j+1}) - \tilde{U}(x_j)| ,$$

where the supremum is taken over all (finite) meshes $a < x_1 < x_2 < \cdots < x_\ell < b$. Furthermore, (classical) one-sided limits $\tilde{U}(x\pm)$ exist at every $x \in (a, b)$ and are both equal to $\tilde{U}(x)$, except possibly on a countable set of points. When $k = 1$, the compactness Theorem 1.7.2 reduces to the classical *Helly theorem*.

As we shall see in the following section, the above results have significant implications to the theory of solutions of systems of balance laws.

1.8 *BV* Solutions of Systems of Balance Laws

We consider here weak solutions $U \in L^\infty(\mathcal{X})$ of the system (1.4.3) of balance laws, which are in $BV_{\text{loc}}(\mathcal{X})$. In that case, by virtue of Theorem 1.7.4, the functions $G_\alpha \circ U$, $\alpha = 1, \cdots, k$, are also in $BV_{\text{loc}}(\mathcal{X}) \cap L^\infty(\mathcal{X})$ and (1.4.3) is satisfied as an equality of measures. The first task is to examine the local form of (1.4.3), in the light of Theorems 1.7.3, 1.7.4, and 1.7.5.

Theorem 1.8.1 *A function* $U \in BV_{loc}(\mathscr{X}) \cap L^\infty(\mathscr{X})$ *is a weak solution of the system* (1.4.3) *of balance laws if and only if* (a) *the measure equality*

$$(1.8.1) \qquad \sum_{\alpha=1}^{k} DG_\alpha(\tilde{U}(X), X)\partial_\alpha U + \sum_{\alpha=1}^{k} \frac{\partial G_\alpha}{\partial X_\alpha}(\tilde{U}(X), X) = \Pi(\tilde{U}(X), X)$$

holds on the set $\mathscr{C}$ *of points of approximate continuity of* U*; and* (b) *the jump condition*

$$(1.8.2) \qquad \sum_{\alpha=1}^{k} N_\alpha[G_\alpha(U_+, X) - G_\alpha(U_-, X)] = 0$$

is satisfied for almost all (with respect to $\mathscr{H}^{k-1}$*)* X *on the set* $\mathscr{J}$ *of points of approximate jump discontinuity of* U*, with normal vector* N *and one-sided limits* U_-, U_+.

Proof. Let μ denote the measure defined by the left-hand side of (1.4.3). On $\mathscr{C}$, μ reduces to the measure on the left-hand side of (1.8.1), by virtue of Theorem 1.7.4, (1.7.13), (1.7.11) and (1.7.12). Recalling the Definition 1.7.3 of trace and the characterization of one-sided limits in Theorem 1.7.3, we deduce $(G \circ U)_\pm = G \circ U_\pm$ at every point of $\mathscr{J}$. Thus, if $\mathscr{F}$ is any Borel subset of $\mathscr{J}$, then by account of the remark following the proof of Theorem 1.7.5 and (1.7.17),

$$(1.8.3) \qquad \mu(\mathscr{F}) = \int_{\mathscr{F}} \sum_{\alpha=1}^{k} N_\alpha[G_\alpha(U_-, X) - G_\alpha(U_+, X)] d\mathscr{H}^{k-1} .$$

Therefore, $\mu = \Pi$ in the sense of measures if and only if (1.8.1) and (1.8.2) hold. This completes the proof.

Consequently, the set of points of approximate jump discontinuity of a *BV* solution is the countable union of *shock fronts*.

As we saw in Section 1.5, when G has a companion q, the companion balance law (1.5.2) is automatically satisfied by any classical solution of (1.4.3). The following proposition describes the situation in the context of *BV* weak solutions.

Theorem 1.8.2 *Assume the system of balance laws* (1.4.3) *is endowed with a companion balance law* (1.5.2). *Let* $U \in BV_{loc}(\mathscr{X}) \cap L^\infty(\mathscr{X})$ *be a weak solution of* (1.4.3). *Then the measure*

$$(1.8.4) \qquad v = \sum_{\alpha=1}^{k} \partial_\alpha q_\alpha(U(X), X) - h(U(X), X)$$

is concentrated on the set $\mathscr{J}$ *of points of approximate jump discontinuity of* U *and the inequality* (1.5.12) *will be satisfied in the sense of measures if and only if*

$$(1.8.5) \qquad \sum_{\alpha=1}^{k} N_\alpha[q_\alpha(U_+, X) - q_\alpha(U_-, X)] \geq 0$$

holds for almost all (with respect to $\mathscr{H}^{k-1}$*)* $X \in \mathscr{J}$.

Proof. By virtue of Theorem 1.7.4, we may write (1.4.3) and (1.8.4) as

$$(1.8.6) \qquad \sum_{\alpha=1}^{k} \widetilde{DG_\alpha \circ U} \partial_\alpha U + \sum_{\alpha=1}^{k} \frac{\partial G_\alpha}{\partial X_\alpha} - \Pi = 0 \; ,$$

$$(1.8.7) \qquad v = \sum_{\alpha=1}^{k} \widetilde{Dq_\alpha \circ U} \partial_\alpha U + \sum_{\alpha=1}^{k} \frac{\partial q_\alpha}{\partial X_\alpha} - h \; .$$

By account of (1.7.11) and (1.7.12), if X is in the set $\mathscr{C}$ of points of approximate continuity of U,

$$(1.8.8) \quad \widetilde{DG_\alpha \circ U}(X) = DG_\alpha(\tilde{U}(X), X) \; , \quad \widetilde{Dq_\alpha \circ U}(X) = Dq_\alpha(\tilde{U}(X), X) \; .$$

Combining (1.8.6), (1.8.7), (1.8.8) and using (1.5.1), (1.5.3), we deduce that v vanishes on $\mathscr{C}$.

From the Definition 1.7.3 of trace and the characterization of one-sided limits in Theorem 1.7.3, we infer $(q \circ U)_\pm = q \circ U_\pm$. If $\mathscr{F}$ is a bounded Borel subset of $\mathscr{J}$, we apply (1.7.17), keeping in mind the remark following the proof of Theorem 1.7.5. This yields

$$(1.8.9) \qquad v(\mathscr{F}) = \int_{\mathscr{F}} \sum_{\alpha=1}^{k} N_\alpha [q_\alpha(U_-, X) - q_\alpha(U_+, X)] d\mathscr{H}^{k-1} \; .$$

Therefore, $v \leq 0$ if and only if (1.8.5) holds. This completes the proof.

1.9 Rapid Oscillations and the Stabilizing Effect of Companion Balance Laws

Consider a homogeneous system of conservation laws

$$(1.9.1) \qquad \sum_{\alpha=1}^{k} \partial_\alpha G_\alpha(U(X)) = 0$$

and assume that

$$(1.9.2) \qquad \sum_{\alpha=1}^{k} N_\alpha [G_\alpha(W) - G_\alpha(V)] = 0$$

holds for some states V, W in $\mathscr{O}$ and $N \in \mathscr{S}^{k-1}$. Then one may construct highly oscillatory weak solutions of (1.9.1) on $\mathbb{R}^k$ by the following procedure: Consider any finite family of parallel $(k-1)$-dimensional hyperplanes, all of them orthogonal to N, and define a function U on $\mathbb{R}^k$ which is constant on each slab confined between two adjacent hyperplanes, taking the values V and W in alternating order. It is clear that U is a weak solution of (1.9.1), by virtue of (1.9.2) and Theorem 1.8.1.

We may thus construct a sequence of solutions which converges in L^∞ weak* to some U of the form $U(X) = \rho(X \cdot N)V + [1 - \rho(X \cdot N)]W$, where ρ is any measurable function from $\mathbb{R}$ to $[0, 1]$. It is clear that, in general, such U will not be a solution of (1.9.1), unless $G(\cdot)N$ happens to be affine along the straight line segment in $\mathbb{R}^n$ that joins V with W. This type of instability distinguishes systems that may support shock fronts from elliptic systems that can not.

Assume now G is equipped with a companion q having the property $\sum N_\alpha[q_\alpha(W) - q_\alpha(V)] \neq 0$. Notice that imposing the admissibility condition $\sum \partial_\alpha q_\alpha(U) \leq 0$ would rule out the oscillating solutions constructed above, because, by virtue of Theorem 1.8.2, it would be prohibited to have jumps both from V to W and from W to V, in the direction N. Consequently, inequalities (1.5.12) seem to play a stabilizing role. To what extent this stabilizing is effective will be a major issue for discussion in the book.

1.10 Notes

The principles of the theory of balance laws were conceived in the process of laying down the foundations of elasticity, in the 1820's. Theorem 1.2.1 has a long and celebrated history. The crucial discovery that the flux density is necessarily a linear function of the exterior normal was made by Cauchy [1,2]. The argument that the flux density through a surface may depend on the surface solely through its exterior normal is attributed to Hamel and to Noll [2]. The formulation of the balance law and the proof of Theorem 1.2.1 at the level of generality presented here is adapted from Ziemer [1]. The recovery of the balance law from its field equation is described in Anzellotti [1]. An alternative, more explicit, construction of the flux function is due to Chen and Frid [1,6].

The observation that systems of balance laws are endowed with nontrivial companions if and only if they are symmetrizable is due to Godunov [1,2,3], Friedrichs and Lax [1] and Boillat [1]. See also Ruggeri and Strumia [1].

In one space dimension, weak fronts are first encountered in the acoustic research of Euler while shock fronts were introduced by Stokes [1]. Fronts in several space dimensions were first studied by Christoffel [1]. The classical reference is Hadamard [1]. For a historical account of the early development of the subject, with emphasis on the contributions of Riemann and Christoffel, see Hölder [1]. The connection between shock fronts and phase transitions will not be pursued here. For references to this active area of research see Section 8.7.

Comprehensive expositions of the theory of BV functions can be found in the treatise of Federer [1], the monograph of Giusti [1], and the texts of Evans and Gariepy [1] and Ziemer [2]. Theorems 1.7.4 and 1.7.5 are taken from Volpert [1].

An insightful discussion of the issues raised in Section 1.9 is found in DiPerna [8]. These questions will be elucidated by the presentation of the method of compensated compactness, in Chapter XV.

Chapter II. Introduction to Continuum Physics

In Continuum Physics, material bodies are modelled as continuous media whose motion and equilibrium is governed by balance laws and constitutive relations.

The list of balance laws in force identifies the theory, for example mechanics, thermomechanics, electrodynamics, etc. The referential (Lagrangian) and the spatial (Eulerian) formulation of the typical balance law will be presented. The balance laws of mass, momentum, energy, and the Clausius-Duhem inequality, which demarcate continuum thermomechanics, will be recorded.

The type of constitutive relation characterizes the nature of material response. The constitutive equations of thermoelasticity and thermoviscoelasticity will be introduced. Restrictions imposed by the Second Law of thermodynamics, the principle of material frame indifference, and material symmetry will be discussed.

The matrix notational conventions of Chapter I will be used here as well, with $\mathcal{M}^{r,s}$ denoting the space of $r \times s$ matrices and $\mathbb{R}^r$ identified with $\mathcal{M}^{r,1}$.

2.1 Bodies and Motions

The ambient physical space is $\mathbb{R}^m$, of dimension one, two or three. A *body* is identified by a *reference configuration*, namely an open subset $\mathcal{B}$ of $\mathbb{R}^m$. Points of $\mathcal{B}$ will be called *particles*. The typical particle will be denoted by x and time will be denoted by t.

A *placement* of the body is a bilipschitz homeomorphism of its reference configuration $\mathcal{B}$ to some open subset of $\mathbb{R}^m$. A *motion* of the body over the time interval (t_1, t_2) is a Lipschitz map χ of $\mathcal{B} \times (t_1, t_2)$ to $\mathbb{R}^m$ whose restriction to each fixed t in (t_1, t_2) is a placement. Thus, for fixed $x \in \mathcal{B}$ and $t \in (t_1, t_2)$, $\chi(x, t)$ specifies the position in space of the particle x at time t; for fixed $t \in (t_1, t_2)$, the map $\chi(\cdot, t) : \mathcal{B} \to \mathbb{R}^m$ yields the placement of the body at time t; finally, for fixed $x \in \mathcal{B}$, the curve $\chi(x, \cdot) : (t_1, t_2) \to \mathbb{R}^m$ describes the trajectory of the particle x. See Fig. 2.1.1.

The aim of Continuum Physics is to monitor the evolution of various fields associated with the body, such as density, stress, temperature, etc. In the *referential* approach, one follows the evolution of fields along particle trajectories, while in the *spatial* approach one monitors the evolution of fields at fixed position in space. The motion allows to pass from one formulation to the other. For example, considering some illustrative field ω, we write $\omega = f(x, t)$ for its referential

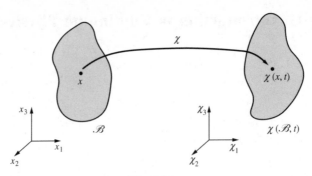

Fig. 2.1.1.

description and $\omega = \phi(\chi, t)$ for its spatial description. The motion relates f and ϕ by $\phi(\chi(x, t), t) = f(x, t)$, for $x \in \mathscr{B}, t \in (t_1, t_2)$.

Hydrodynamicists commonly use the terms *Lagrangian* for referential and *Eulerian* for spatial. This terminology has become standard, and will be adopted here, notwithstanding that the "Lagrangian" description was actually introduced by Euler and the "Eulerian" description was first employed by Daniel Bernoulli and D'Alembert.

Either formulation has its relative advantages, so both will be used here. Thus, in order to keep proper accounting, three symbols should be needed for each field, one to identify it, one for its referential description, and one for its spatial description (ω, f, and ϕ in the example, above). However, in order to control the proliferation of symbols, the standard notational convention is to employ the single identifying symbol of the field for all three purposes. To prevent ambiguity in the notation of derivatives, the following rules will apply: Partial differentiation with respect to t will be denoted by an overdot in the referential description and by a t-subscript in the spatial description. Gradient and divergence will be denoted by Grad and Div, with respect to the material variable x, and by grad and div, with respect to the space variable χ. Thus, referring again to the typical field ω with referential description $\omega = f(x, t)$ and spatial description $\omega = \phi(\chi, t)$, $\dot{\omega}$ will denote $\partial f / \partial t$, ω_t will denote $\partial \phi / \partial t$, Grad ω will denote $\text{grad}_x f$, and grad ω will denote $\text{grad}_\chi \phi$. This notation may appear confusing at first but the student of the subject soon learns to use it efficiently and correctly.

The motion χ induces two important kinematical fields, the *velocity*

$$(2.1.1) \qquad\qquad v = \dot{\chi} \, ,$$

in $L^\infty(\mathscr{B} \times (t_1, t_2); \mathbb{R}^m)$, and the *deformation gradient*

$$(2.1.2) \qquad\qquad F = \text{Grad} \, \chi \, ,$$

in $L^\infty(\mathscr{B} \times (t_1, t_2); \mathscr{M}^{m,m})$. In accordance with the definition of placement, we shall be assuming

$$(2.1.3) \qquad\qquad \det F \geq a > 0 \qquad \text{a.e.}$$

These fields allow one to pass from spatial to material derivatives, for example, assuming ω is a Lipschitz field,

$$(2.1.4) \qquad \dot{\omega} = \omega_t + (\operatorname{grad}\omega)v \ ,$$

$$(2.1.5) \qquad \operatorname{Grad}\omega = (\operatorname{grad}\omega)F \ .$$

By virtue of the polar decomposition theorem, the local deformation of the medium, determined by the deformation gradient F, may be realized as the composition of a pure stretching and a rotation:

$$(2.1.6) \qquad F = RU \ ,$$

where the symmetric, positive definite matrix

$$(2.1.7) \qquad U = (F^T F)^{1/2}$$

is called the *right stretch tensor* and the proper orthogonal matrix R is called the *rotation tensor*.

Turning to the rate of change of deformation, we introduce the referential and spatial velocity gradients:

$$(2.1.8) \qquad \dot{F} = \operatorname{Grad}v \ , \quad L = \operatorname{grad}v \ .$$

L is decomposed into the sum of the symmetric *stretching tensor* D and the skew-symmetric *spin tensor* W:

$$(2.1.9) \qquad L = D + W \ , \quad D = \tfrac{1}{2}(L + L^T) \ , \quad W = \tfrac{1}{2}(L - L^T) \ .$$

The class of Lipschitz continuous motions allows for shocks but is not sufficiently broad to also encompass motions involving cavitation in elasticity, vortices in hydrodynamics, vacuum in gas dynamics, etc. Even so, we shall continue to develop the theory under the assumption that motions are Lipschitz continuous, deferring considerations of generalization until such need arises.

2.2 Balance Laws in Continuum Physics

Consider a motion χ of a body with reference configuration $\mathscr{B} \subset \mathbb{R}^m$, over a time interval (t_1, t_2). The typical balance law of Continuum Physics postulates that the change over any time interval in the amount of a certain extensive quantity stored in any part of the body is balanced by a flux through the boundary and a production in the interior during that time interval. With space and time fused into space-time, the above statement yields a balance law of the type considered in Chapter I, ultimately reducing to a field equation of the form (1.2.5).

To adapt to the present setting the notation of Chapter I, we take space-time $\mathbb{R}^{m+1}$ as the ambient space $\mathbb{R}^k$, and set $\mathscr{X} = \mathscr{B} \times (t_1, t_2)$, with typical point $X = (x, t)$. With reference to (1.2.5), we decompose the flux density field A into a spatial part and a temporal part, namely $A = [-\Psi, \Theta]$, with $\Psi \in L^1_{loc}(\mathscr{X}; \mathscr{M}^{n,m})$, $\Theta \in L^1_{loc}(\mathscr{X}; \mathbb{R}^n)$. In the notation of the previous section, (1.2.5) now takes the form

$$(2.2.1) \qquad \dot{\Theta} = \text{Div}\,\Psi + P \ .$$

This is the referential field equation for the typical balance law of Continuum Physics. The field Θ is the density of the balanced quantity; Ψ is the flux density field through material surfaces; and P is the production density.

The corresponding spatial field equation may be derived by appealing to Theorem 1.3.1. The map X^* that sends (x, t) to the point $(\chi(x, t), t)$ is a bilipschitz homeomorphism of $\mathscr{X}$ to some subset $\mathscr{X}^*$ of $\mathbb{R}^{m+1}$, with Jacobian matrix (cf. (1.3.2), (2.1.1), and (2.1.2)):

$$(2.2.2) \qquad J = \left[\begin{array}{c|c} F & v \\ \hline 0 & 1 \end{array} \right] \ .$$

Notice that (1.3.3) is satisfied by virtue of (2.1.3). Theorem 1.3.1 now implies that (2.2.1) holds in the sense of distributions on $\mathscr{X}$ if and only if

$$(2.2.3) \qquad \Theta^*_t + \text{div}\,(\Theta^* v^T) = \text{div}\,\Psi^* + P^*$$

is satisfied in the sense of distributions on $\mathscr{X}^*$, with $\Theta^* \in L^1_{loc}(\mathscr{X}^*; \mathbb{R}^n)$, $\Psi^* \in L^1_{loc}(\mathscr{X}^*; \mathscr{M}^{n,m})$ and $P^* \in L^1_{loc}(\mathscr{X}^*; \mathbb{R}^n)$ defined by

$$(2.2.4) \qquad \Theta^* = (\det F)^{-1}\Theta \ , \quad \Psi^* = (\det F)^{-1}\Psi F^T \ , \quad P^* = (\det F)^{-1} P \ .$$

It has thus been established that the referential field equations (2.2.1) and the spatial field equations (2.2.3) of the balance laws of Continuum Physics are related through (2.2.4) and are equivalent whenever the fields are in L^1_{loc}.

In anticipation of the forthcoming discussion of material symmetry, it is useful to investigate how the field equations (2.2.1) and (2.2.3) transform under *isochoric* changes of the reference configuration of the body, induced by a bilipschitz homeomorphism $\bar{x}$ of $\mathscr{B}$ to some subset $\bar{\mathscr{B}}$ of $\mathbb{R}^m$, with Jacobian matrix

$$(2.2.5) \qquad H = \frac{\partial \bar{x}}{\partial x} \ , \quad \det H = 1 \ .$$

By virtue of Theorem 1.3.1, the referential field equation (2.2.1) on $\mathscr{B}$ will transform into an equation of exactly the same form on $\bar{\mathscr{B}}$, with fields $\bar{\Theta}, \bar{\Psi}, \bar{P}$ related to Θ, Ψ, P by

$$(2.2.6) \qquad \bar{\Theta} = \Theta \ , \quad \bar{\Psi} = \Psi H^T \ , \quad \bar{P} = P \ .$$

In the corresponding spatial field equation, the fields $\bar{\Theta}^*, \bar{\Psi}^*, \bar{P}^*$ are obtained through (2.2.4) : $\bar{\Theta}^* = (\det \bar{F})^{-1}\bar{\Theta}$, $\bar{\Psi}^* = (\det \bar{F})^{-1}\bar{\Psi}\,\bar{F}^T, \bar{P}^* = (\det \bar{F})^{-1}\bar{P}$, where $\bar{F}$ denotes the deformation gradient relative to the new reference configuration $\bar{\mathscr{B}}$. Since $\bar{F} = FH^{-1}$, using (2.2.6), (2.2.4) and (2.2.5) yields

(2.2.7) $$\overline{\Theta}^* = \Theta^* , \quad \overline{\Psi}^* = \Psi^* , \quad \overline{P}^* = P^* ,$$

i.e., as was to be expected, the spatial field equations are unaffected by changes of the reference configuration of the body.

In Continuum Physics, theories are identified by means of the list of balance laws that apply in their context. The illustrative example of thermomechanics will be considered in the following section.

2.3 The Balance Laws of Continuum Thermomechanics

Continuum thermomechanics, which will serve as a representative model through-out this work, is demarcated by the balance laws of mass, linear momentum, angular momentum, and energy, whose referential and spatial field equations will now be introduced.

In the *balance law of mass* there is neither flux nor production so the referential and spatial field equations read

(2.3.1) $$\dot{\rho}_0 = 0 ,$$

(2.3.2) $$\rho_t + \mathrm{div}\,(\rho v^T) = 0 ,$$

where ρ_0 is the *reference density* and ρ is the *density* associated with the motion, related through

(2.3.3) $$\rho = \rho_0 (\det F)^{-1} .$$

Note that (2.3.1) implies that the value of the reference density associated with a particle does not vary with time: $\rho_0 = \rho_0(x)$.

In the *balance law of linear momentum*, the production is induced by the *body force* (per unit mass) vector b, with values in $\mathbb{R}^m$, while the flux is represented by a *stress tensor* taking values in $\mathcal{M}^{m,m}$. The referential and spatial field equations read

(2.3.4) $$(\rho_0 v)^{\cdot} = \mathrm{Div}\,S + \rho_0 b ,$$

(2.3.5) $$(\rho v)_t + \mathrm{div}\,(\rho v v^T) = \mathrm{div}\,T + \rho b ,$$

where S denotes the *Piola-Kirchhoff stress* and T denotes the *Cauchy stress*, related by

(2.3.6) $$T = (\det F)^{-1} S F^T .$$

For $v \in \mathscr{S}^{m-1}$, the value of Sv at (x, t) yields the stress (force per unit area) vector transmitted at the particle x and time t across a material surface with normal v; while the value of Tv at (χ, t) gives the stress vector transmitted at the point χ in space and time t across a spatial surface with normal v.

In the *balance law of angular momentum*, production and flux are the moments about the origin of the production and flux involved in the balance of linear momentum. Consequently, the referential field equation is

$$(2.3.7) \qquad (\chi \wedge \rho_0 v)^{\cdot} = \mathrm{Div}\,(\chi \wedge S) + \chi \wedge \rho_0 b\,,$$

where $\wedge$ denotes vectorial exterior product. Under the assumption that $\rho_0 v$, S and $\rho_0 b$ are in L^1_{loc} while the motion χ is Lipschitz continuous, we may use (2.3.4), (2.1.1) and (2.1.2) to reduce (2.3.7) into

$$(2.3.8) \qquad SF^T = FS^T\,.$$

Similarly, the spatial field equation of the balance of angular momentum reduces, by virtue of (2.3.5), to the statement that the Cauchy stress tensor is symmetric:

$$(2.3.9) \qquad T^T = T\,.$$

There is no need to perform that calculation since (2.3.9) also follows directly from (2.3.6) and (2.3.8).

In the *balance law of energy*, the energy density is the sum of the (specific) *internal energy* (per unit mass) ε and kinetic energy. The production is the sum of the rate of work of the body force and the *heat supply* (per unit mass) r. Finally, the flux is the sum of the rate of work of the stress tensor and the *heat flux*. The referential and spatial field equations thus read

$$(2.3.10) \qquad (\rho_0 \varepsilon + \tfrac{1}{2}\rho_0 v^T v)^{\cdot} = \mathrm{Div}\,(v^T S + Q) + \rho_0 v^T b + \rho_0 r\,,$$

$$(2.3.11) \quad (\rho\varepsilon + \tfrac{1}{2}\rho v^T v)_t + \mathrm{div}\,[(\rho\varepsilon + \tfrac{1}{2}\rho v^T v)v^T] = \mathrm{div}\,(v^T T + q) + \rho v^T b + \rho r\,,$$

where the referential and spatial heat flux vectors Q and q, with values in $\mathscr{M}^{1,m}$, are related by

$$(2.3.12) \qquad q = (\det F)^{-1} Q F^T\,.$$

Note that when the velocity field v is Lipschitz continuous, so that the standard product rule of differentiation applies to $v^T v$ and $v^T S$, the field equation (2.3.10) reduces, by account of (2.3.4), to

$$(2.3.13) \qquad \rho_0 \dot{\varepsilon} = \mathrm{tr}\,(S\dot{F}^T) + \mathrm{Div}\,Q + \rho_0 r\,.$$

Continuum Physics assigns different roles to the balance laws: The field equations (2.3.2), (2.3.4), (2.3.5), (2.3.10) and (2.3.11), for the balance of mass, linear momentum and energy, are viewed as conditions on the evolution of the various fields that serve to determine the motion of the body. By contrast, the balance of angular momentum (2.3.8), (2.3.9) is regarded as a condition on material response, to be satisfied identically by the constitutive relations.

The list of balance laws is complemented by the Second Law of thermodynamics, which postulates that the growth, over any time interval, in the entropy stored in any part of the body exceeds the sum of the entropy flux through the boundary and the entropy production in the interior during that time interval. It is assumed that the entropy flux is heat flux divided by temperature and the entropy production density is heat supply divided by temperature. Thus, if s denotes the (specific) *entropy* (per unit mass) and θ is the absolute *temperature*, $\theta > 0$, the above "imbalance" law leads to the *Clausius-Duhem inequality*, in referential and spatial form:

$$(2.3.14) \qquad (\rho_0 s)^{\cdot} \geq \text{Div}\left(\frac{1}{\theta}Q\right) + \rho_0 \frac{r}{\theta} ,$$

$$(2.3.15) \qquad (\rho s)_t + \text{div}\,(\rho s v^T) \geq \text{div}\left(\frac{1}{\theta}q\right) + \rho\frac{r}{\theta} .$$

A motion together with an entropy field constitute a *thermodynamic process* of the body; they determine all the fields involved in the balance laws through constitutive relations, depending on the nature of the material. A fundamental premise of Continuum Physics is that the Clausius-Duhem inequality should hold identically for any smooth process that balances mass, momentum and energy. Consequently, in the context of smooth processes the inequality is regarded as a condition on material response, inducing restrictions on constitutive relations. Introducing the referential and spatial temperature gradient fields

$$(2.3.16) \qquad G = \text{Grad}\,\theta , \quad g = \text{grad}\,\theta , \quad g = GF^{-1} ,$$

we note that when the velocity field v and the temperature field θ are Lipschitz continuous, the reduced form (2.3.13) of the energy equation combines with the Clausius-Duhem inequality (2.3.14) to yield the dissipation inequality

$$(2.3.17) \qquad \rho_0\dot{\varepsilon} - \rho_0\theta\dot{s} - \text{tr}\,(S\dot{F}^T) - \frac{1}{\theta}QG^T \leq 0 .$$

After appropriate reduction of the constitutive equations, the Clausius-Duhem inequality will hold identically for smooth thermodynamic processes but not necessarily for processes that are not smooth. Consequently, in addition to being employed as a condition on material response, the inequality is charged with the responsibility of testing the *thermodynamic admissibility* of nonsmooth processes.

To prepare for forthcoming investigation of material symmetry, it is necessary to discuss the law of transformation of the fields involved in the balance laws when the reference configuration undergoes a change induced by an isochoric bilipshitz homeomorphism $\bar{x}$, with unimodular Jacobian matrix H (2.2.5). The deformation gradient F and the stretching tensor D (cf. (2.1.9)) will transform into new fields $\overline{F}$ and $\overline{D}$:

$$(2.3.18) \qquad \overline{F} = FH^{-1} , \quad \overline{D} = D .$$

The reference density ρ_0, internal energy ε, Piola-Kirchhoff stress S, entropy s, temperature θ, referential heat flux vector Q, density ρ, Cauchy stress T, and spatial heat flux vector q, involved in the balance laws, will also transform into new fields $\overline{\rho_0}, \overline{\varepsilon}, \overline{S}, \overline{s}, \overline{\theta}, \overline{Q}, \overline{\rho}, \overline{T}$, and $\overline{q}$ according to the rule (2.2.6) or (2.2.7), namely,

(2.3.19) $\overline{\rho_0} = \rho_0$, $\overline{\varepsilon} = \varepsilon$, $\overline{S} = SH^T$, $\overline{s} = s$, $\overline{\theta} = \theta$, $\overline{Q} = QH^T$,

(2.3.20) $\overline{\rho} = \rho$, $\overline{T} = T$, $\overline{q} = q$.

Also the referential and spatial temperature gradients G and g will transform into $\overline{G}$ and $\overline{g}$ with

(2.3.21) $\overline{G} = GH^{-1}$, $\overline{g} = g$.

2.4 Material Frame Indifference

The body force and heat supply are usually induced by external factors and are assigned in advance, while the fields of internal energy, stress, entropy and heat flux are determined by the thermodynamic process. Motions may influence these fields in as much as they deform the body: Rigid motions, which do not change the distance between particles, should have no effect on internal energy, temperature or referential heat flux and should affect the stress tensor in a manner that the resulting stress vector, observed from a frame attached to the moving body, looks fixed. This requirement is postulated by the fundamental *principle of material frame indifference* which will now be stated with precision.

Consider any two thermodynamic processes (χ, s) and $(\chi^\#, s^\#)$ of the body such that the entropy fields coincide, $s^\# = s$, while the motions differ by a rigid (time dependent) rotation:

(2.4.1) $\chi^\#(x, t) = O(t)\chi(x, t)$, $x \in \mathcal{B}$, $t \in (t_1, t_2)$,

(2.4.2) $O^T(t)O(t) = O(t)O^T(t) = I$, $\det O(t) = 1$, $t \in (t_1, t_2)$.

Note that the fields of deformation gradient $F, F^\#$, spatial velocity gradient $L, L^\#$ and stretching tensor $D, D^\#$ (cf. (2.1.8), (2.1.9)) of the two processes (χ, s), $(\chi^\#, s^\#)$ are related by

(2.4.3) $F^\# = OF$, $L^\# = OLO^T + \dot{O}O^T$, $D^\# = ODO^T$.

Let $(\varepsilon, S, \theta, Q)$ and $(\varepsilon^\#, S^\#, \theta^\#, Q^\#)$ denote the fields for internal energy, Piola-Kirchhoff stress, temperature and referential heat flux associated with the processes (χ, s) and $(\chi^\#, s^\#)$. The principle of material frame indifference postulates:

(2.4.4) $\varepsilon^\# = \varepsilon$, $S^\# = OS$, $\theta^\# = \theta$, $Q^\# = Q$.

From (2.4.4), (2.3.16) and (2.4.3) it follows that the referential and spatial temperature gradients $G, G^\#$ and $g, g^\#$ of the two processes are related by

(2.4.5) $G^\# = G$, $g^\# = gO^T$.

Furthermore, from (2.3.6), (2.3.12) and (2.4.3) we deduce the following relations between the Cauchy stress tensors T, $T^\#$ and the spatial heat flux vectors q, $q^\#$ of the two processes:

$$(2.4.6) \qquad T^\# = O T O^T , \quad q^\# = q O^T .$$

The principle of material frame indifference should be reflected in the constitutive relations of continuous media, irrespectively of the nature of material response. Illustrative examples will be considered in the following two sections.

2.5 Thermoelasticity

In the framework of continuum thermomechanics, a *thermoelastic* medium is characterized by the constitutive assumption that, for any fixed particle x and any motion, the value of the internal energy ε, the Piola-Kirchhoff stress S, the temperature θ, and the referential heat flux vector Q, at x and time t, is determined solely by the value at (x, t) of the deformation gradient F, the entropy s, and the temperature gradient G, through constitutive equations

$$(2.5.1) \qquad \begin{cases} \varepsilon = \hat{\varepsilon}(F, s, G) \\[2mm] S = \hat{S}(F, s, G) \\[2mm] \theta = \hat{\theta}(F, s, G) \\[2mm] Q = \hat{Q}(F, s, G) \end{cases}$$

where $\hat{\varepsilon}, \hat{S}, \hat{\theta}, \hat{Q}$ are smooth functions defined on the subset of $\mathcal{M}^{m,m} \times \mathbb{R} \times \mathcal{M}^{1,m}$ with $\det F > 0$. Moreover, $\hat{\theta}(F, s, G) > 0$. When the thermoelastic medium is *homogeneous*, the same functions $\hat{\varepsilon}, \hat{S}, \hat{\theta}, \hat{Q}$ and the same value ρ_0 of the reference density apply at all particles $x \in \mathcal{B}$.

The Cauchy stress T and the spatial heat flux q are also determined by constitutive equations of the same form, which may be derived from (2.5.1) and (2.3.6), (2.3.12). When employing the spatial description of the motion, it is natural to substitute on the list (2.5.1) the constitutive equations of T and q for the constitutive equations of S and Q; also on the list (F, s, G) of the state variables to replace the referential temperature gradient G with the spatial temperature gradient g (cf. (2.3.16)).

The above constitutive equations will have to comply with the conditions stipulated earlier. To begin with, as postulated in Section 2.3, every smooth thermodynamic process which balances mass, momentum and energy must satisfy identically the Clausius-Duhem inequality (2.3.14) or, equivalently, the dissipation inequality (2.3.17). Substituting from (2.5.1) into (2.3.17) yields

$$(2.5.2) \qquad \mathrm{tr}\left[(\rho_0 \partial_F \hat{\varepsilon} - \hat{S})\dot{F}^T\right] + \rho_0(\partial_s \hat{\varepsilon} - \hat{\theta})\dot{s} + \rho_0 \partial_G \hat{\varepsilon} \dot{G}^T - \frac{1}{\hat{\theta}} \hat{Q} G^T \leq 0 .$$

It is clear that by suitably controlling the body force b and the heat supply r one may construct smooth processes which balance mass, momentum and energy and attain at some point (x, t) arbitrarily prescribed values for $F, s, G, \dot{F}, \dot{s}, \dot{G}$, subject only to the constraint $\det F > 0$. Hence (2.5.2) cannot hold identically unless the constitutive relations (2.5.1) are of the following special form:

(2.5.3)
$$\begin{cases} \varepsilon = \hat{\varepsilon}(F, s) \\ S = \rho_0 \partial_F \hat{\varepsilon}(F, s) \\ \theta = \partial_s \hat{\varepsilon}(F, s) \\ Q = \hat{Q}(F, s, G) \ , \end{cases}$$

(2.5.4)
$$\hat{Q}(F, s, G)G^T \geq 0 \ .$$

Thus the internal energy may depend on the deformation gradient and on the entropy but not on the temperature gradient. The constitutive equations for stress and temperature are induced by the constitutive equation of internal energy, through caloric relations, and are likewise independent of the temperature gradient. Only the heat flux may depend on the temperature gradient, subject to the condition (2.5.4) which implies that heat always flows from the hotter to the colder part of the body.

Another requirement on constitutive relations is that they observe the principle of material frame indifference, formulated in Section 2.4. By combining (2.4.4) and (2.4.3)$_1$ with (2.5.3), we deduce that the functions $\hat{\varepsilon}$ and $\hat{Q}$ must satisfy the conditions

(2.5.5) $\hat{\varepsilon}(OF, s) = \hat{\varepsilon}(F, s) \ , \quad \hat{Q}(OF, s, G) = \hat{Q}(F, s, G) \ ,$

for all proper orthogonal matrices O. A simple calculation verifies that when (2.5.5) hold then the remaining conditions in (2.4.4) will be automatically satisfied, by virtue of (2.5.3)$_2$ and (2.5.3)$_3$.

To see the implications of (2.5.5), we apply it with $O = R^T$, where R is the rotation tensor in (2.1.6), to deduce

(2.5.6) $\hat{\varepsilon}(F, s) = \hat{\varepsilon}(U, s) \ , \quad \hat{Q}(F, s, G) = \hat{Q}(U, s, G) \ .$

It is clear that, conversely, if (2.5.6) hold then (2.5.5) will be satisfied for any proper orthogonal matrix O. Consequently, the principle of material frame indifference is completely encoded in the statement (2.5.6) that the internal energy and the referential heat flux vector may depend on the deformation gradient F solely through the right stretch tensor U.

When the spatial description of motion is to be employed, the constitutive equation for the Cauchy stress

(2.5.7)
$$T = \rho \partial_F \hat{\varepsilon}(F, s)F^T \ ,$$

which follows from (2.3.6), (2.3.3) and (2.5.3)$_2$, will satisfy the principle of material frame indifference (2.4.6)$_1$ so long as (2.5.6) hold. For the constitutive equation of the spatial heat flux vector

(2.5.8) $$q = \hat{q}(F, s, g)$$

the principle of material frame indifference requires (recall (2.4.6)$_2$, (2.4.3)$_1$ and (2.4.5)$_2$):

(2.5.9) $$\hat{q}(OF, s, gO^T) = \hat{q}(F, s, g)O^T ,$$

for all proper orthogonal matrices O.

The final general requirement on constitutive relations is that the Piola-Kirchhoff stress satisfy (2.3.8), for the balance of angular momentum. This imposes no additional restrictions, however, because a simple calculation reveals that once (2.5.5)$_1$ holds, S computed through (2.5.3)$_2$ will automatically satisfy (2.3.8). Thus in thermoelasticity material frame indifference implies balance of angular momentum.

The constitutive equations undergo further reduction when the medium is endowed with *material symmetry*. Recall from Section 2.3 that when the reference configuration of the body is changed by means of an isochoric bilipschitz homeomorphism $\bar{x}$ with unimodular Jacobian matrix H (2.2.5), then the fields transform according to the rules (2.3.18), (2.3.19), (2.3.20) and (2.3.21). It follows, in particular, that any medium which is thermoelastic relative to the original reference configuration, will stay so relative to the new one, as well, even though the constitutive functions will generally change. Any isochoric transformation of the reference configuration that leaves invariant all constitutive functions signals material symmetry of the medium. Consider any such transformation and let H be its Jacobian matrix. By virtue of (2.3.18)$_1$, (2.3.19)$_2$ and (2.5.3)$_1$, the constitutive function $\hat{\varepsilon}$ of the internal energy will remain invariant, provided

(2.5.10) $$\hat{\varepsilon}(FH^{-1}, s) = \hat{\varepsilon}(F, s) .$$

A simple calculation verifies that when (2.5.10) holds the constitutive functions for S and θ determined through (2.5.3)$_2$ and (2.5.3)$_3$ satisfy automatically the invariance requirements for that same H. The remaining constitutive equation, for the heat flux vector, will be treated for convenience in its spatial description (2.5.8). By account of (2.3.18)$_1$, (2.3.21)$_2$ and (2.3.20)$_3$, $\hat{q}$ will remain invariant if

(2.5.11) $$\hat{q}(FH^{-1}, s, g) = \hat{q}(F, s, g) .$$

It is clear that the set of matrices H with determinant one for which (2.5.10) and (2.5.11) hold forms a subgroup $\mathscr{G}$ of the special linear group $\mathscr{SL}_m$, called the *symmetry group* of the medium. In certain media, $\mathscr{G}$ may contain only the identity matrix I in which case material symmetry is minimal. When $\mathscr{G}$ is nontrivial, it dictates through (2.5.10) and (2.5.11) conditions on the constitutive functions of the medium.

Maximal material symmetry is attained when $\mathscr{G} \equiv \mathscr{SL}_m$. In that case the medium is a *thermoelastic fluid*. Applying (2.5.10) and (2.5.11) with $H = (\det F)^{-1/m} F \in \mathscr{SL}_m$, we deduce that $\hat{\varepsilon}$ and $\hat{q}$ may depend on F solely through its determinant or, equivalently by virtue of (2.3.3), through the density ρ:

$$(2.5.12) \qquad \varepsilon = \tilde{\varepsilon}(\rho, s) , \quad q = \tilde{q}(\rho, s, g) .$$

The Cauchy stress may then be obtained from (2.5.7) and the temperature from (2.5.3)$_3$. The calculation gives

$$(2.5.13) \qquad T = -\tilde{p}(\rho, s)I , \quad \theta = \partial_s \tilde{\varepsilon}(\rho, s) ,$$

where

$$(2.5.14) \qquad \tilde{p}(\rho, s) = \rho^2 \partial_\rho \tilde{\varepsilon}(\rho, s) .$$

The constitutive function $\tilde{q}$ in (2.5.12) must also satisfy the requirement (2.5.9) of material frame indifference which now assumes the simple form

$$(2.5.15) \qquad \tilde{q}(\rho, s, gO^T) = \tilde{q}(\rho, s, g)O^T ,$$

for all proper orthogonal matrices O. In three space dimensions ($m = 3$), the final reduction of $\tilde{q}$ that satisfies (2.5.15) is

$$(2.5.16) \qquad q = \tilde{\kappa}(\rho, s, |g|)g ,$$

where $\tilde{\kappa}$ is a scalar-valued function. We have thus shown that in a thermoelastic fluid the internal energy depends solely on density and entropy. The Cauchy stress is a *hydrostatic pressure*, likewise depending only on density and entropy. The heat flux obeys Fourier's law with *thermal conductivity* κ that may vary with density, entropy and the magnitude of the heat flux.

An *isotropic thermoelastic solid* is a thermoelastic material with symmetry group $\mathscr{G}$ the proper orthogonal group $\mathscr{SO}_m$. In that case, to obtain the reduced form of the internal energy function $\hat{\varepsilon}$ we combine (2.5.10) with (2.5.6)$_1$. Recalling (2.1.7) we conclude that

$$(2.5.17) \qquad \hat{\varepsilon}(OUO^T, s) = \hat{\varepsilon}(U, s)$$

for any proper orthogonal matrix O. In particular, we may apply (2.5.17) for the proper orthogonal matrices O that diagonalize the symmetric matrix $U : OUO^T = \Lambda$. This establishes that in consequence of material frame indifference and material symmetry the internal energy of an isotropic thermoelastic solid may depend on F solely as a symmetric function of the eigenvalues of the right stretch tensor U. Equivalently,

$$(2.5.18) \qquad \varepsilon = \tilde{\varepsilon}(I_1, \cdots, I_m, s) ,$$

where $(I_1, \cdots, I_m)$ are the principal invariants of U; in particular, $I_1 = \text{tr}\, U$, $I_m = \det U = \det F$. The reduced form of the Cauchy stress, computed from

(2.5.18) and (2.5.7), and also of the heat flux vector, for the isotopic thermoelastic solid, are recorded in the references cited in Section 2.7.

We conclude the discussion of thermoelasticity with remarks on special thermodynamic processes. A process is called *adiabatic* if the heat flux Q vanishes identically; it is called *isothermal* when the temperature field θ is constant; and it is called *isentropic* if the entropy field s is constant. Note that (2.5.4) implies $\hat{Q}(F, s, 0) = 0$ so, in particular, all isothermal processes are adiabatic. Materials that are poor conductors of heat are commonly modeled as *nonconductors of heat*, characterized by the constitutive assumption $\hat{Q} \equiv 0$. Thus every thermodynamic process of a nonconductor is adiabatic.

In an isentropic process, the entropy is set equal to a constant, $s \equiv \bar{s}$; the constitutive relations for the temperature and the heat flux are discarded and those for the internal energy and the stress are restricted to $s = \bar{s}$:

(2.5.19)
$$\begin{cases} \varepsilon = \hat{\varepsilon}(F, \bar{s}) \\ S = \rho_0 \partial_F \hat{\varepsilon}(F, \bar{s}). \end{cases}$$

The motion is determined solely by the balance laws of mass and momentum. In practice this simplifying assumption is made when it is judged that entropy fluctuations have insignificant effect. Later on we shall encounter situations where this is indeed the case. We should keep in mind, however, that an isentropic process cannot be sustained unless the heat supply r is regulated in such a manner that the ensuing motion together with the constant entropy field satisfy the balance law of energy.

Isentropic thermoelasticity rests solely on the balance laws of mass and momentum and this may leave the impression that it is a mechanical, rather than a thermomechanical, theory. In fact the constitutive relations (2.5.19) suggest that isentropic thermoelasticity is isomorphic to a mechanical theory called *hyperelasticity*. It should be noted, however, that isentropic thermoelasticity inherits from thermodynamics the Second Law under the following guise: Assuming that the process is adiabatic as well as isentropic and combining the balance law of energy (2.3.10) with the Clausius-Duhem inequality (2.3.14) yields

(2.5.20)
$$(\rho_0 \varepsilon + \tfrac{1}{2}\rho_0 v^T v)^{\cdot} \leq \mathrm{Div}\,(v^T S) + \rho_0 v^T b \ .$$

The spatial description of this inequality is

(2.5.21)
$$(\rho\varepsilon + \tfrac{1}{2}\rho v^T v)_t + \mathrm{div}\,[(\rho\varepsilon + \tfrac{1}{2}\rho v^T v)v^T] \leq \mathrm{div}\,(v^T T) + \rho v^T b \ .$$

The above inequalities play in isentropic thermoelasticity the role played by the Clausius-Duhem inequality (2.3.14), (2.3.15) in general thermoelasticity: For smooth motions, they hold identically, as equalities[1], by virtue of (2.3.4) and (2.5.19). By contrast, in the context of motions that are merely Lipschitz continuous, they are extra conditions serving as the test of *thermodynamic admissibility* of the motion.

---

[1] In particular, this implies that smooth isentropic processes may be sustained with $r = 0$, that is without supplying or extracting any amount of heat.

2.6 Thermoviscoelasticity

We now consider an extension of thermoelasticity, which encompasses materials with *internal dissipation* induced by *viscosity of the rate type*. The internal energy ε, the Piola-Kirchhoff stress S, the temperature θ, and the referential heat flux vector Q may now depend not only on the deformation gradient F, the entropy s and the temperature gradient G, as in (2.5.1), but also on the time rate $\dot{F}$ of the deformation gradient:

(2.6.1)
$$\begin{cases} \varepsilon = \hat{\varepsilon}(F, \dot{F}, s, G) \\ S = \hat{S}(F, \dot{F}, s, G) \\ \theta = \hat{\theta}(F, \dot{F}, s, G) \\ Q = \hat{Q}(F, \dot{F}, s, G) \,. \end{cases}$$

As stipulated in Section 2.3, every smooth thermodynamic process which balances mass, momentum and energy must satisfy identically the dissipation inequality (2.3.17). Substituting from (2.6.1) into (2.3.17) yields

(2.6.2)
$$\mathrm{tr}\,[(\rho_0 \partial_F \hat{\varepsilon} - \hat{S})\dot{F}^T] + \mathrm{tr}\,(\rho_0 \partial_{\dot{F}} \hat{\varepsilon}\ddot{F}^T) + \rho_0 (\partial_s \hat{\varepsilon} - \hat{\theta})\dot{s}$$
$$+ \rho_0 \partial_G \hat{\varepsilon}\dot{G}^T - \frac{1}{\theta}\hat{Q}G^T \le 0\,.$$

By suitably controlling the body force b and heat supply r, one may construct smooth processes which balance mass, momentum and energy and attain at some point (x, t) arbitrarily prescribed values for F, $\dot{F}$, s, G, $\ddot{F}$, $\dot{s}$, $\dot{G}$, subject only to the constraint $\det F > 0$. Consequently, the inequality (2.6.2) cannot hold identically unless the constitutive function in (2.6.1) have the following special form:

(2.6.3)
$$\begin{cases} \varepsilon = \hat{\varepsilon}(F, s) \\ S = \rho_0 \partial_F \hat{\varepsilon}(F, s) + Z(F, \dot{F}, s, G) \\ \theta = \partial_s \hat{\varepsilon}(F, s) \\ Q = \hat{Q}(F, \dot{F}, s, G)\,, \end{cases}$$

(2.6.4)
$$\mathrm{tr}\,[Z(F, \dot{F}, s, G)\dot{F}^T] + \frac{1}{\hat{\theta}(F, s)}\hat{Q}(F, \dot{F}, s, G)G^T \ge 0\,.$$

Comparing (2.6.3) with (2.5.3) we observe that, again, the internal energy, which may depend solely on the deformation gradient and the entropy, determines the constitutive equation for the temperature by the same caloric equation of state. On the other hand, the constitutive equation for the stress now includes the additional term Z which contributes the viscous effect and induces internal dissipation manifested in (2.6.4).

The constitutive functions will have to be reduced further to comply with the principle of material frame indifference, postulated in Section 3.4. In particular, frame indifference imposes to internal energy the same condition $(2.5.5)_1$ as in thermoelasticity and the resulting reduction is, of course, the same:

$$(2.6.5) \qquad \hat{\varepsilon}(F, s) = \hat{\varepsilon}(U, s) \, ,$$

where U denotes the right stretch tensor (2.1.7). Furthermore, when (2.6.5) holds the constitutive equation for the temperature, derived through $(2.6.3)_3$, and the term $\rho_0 \partial_F \hat{\varepsilon}(F, s)$, in the constitutive equation for the stress, will be automatically frame indifferent. It remains to investigate the implications of frame indifference on Z and the heat flux. Since the analysis will focus eventually on thermoviscoelastic fluids, it will be expedient to switch at this point from S and Q to T and q; also to replace, on the list $(F, \dot{F}, s, G)$ of state variables, $\dot{F}$ with L (cf. (2.1.8)) and G with g (cf. (2.3.16)). We thus write

$$(2.6.6) \qquad T = \rho \partial_F \hat{\varepsilon}(F, s) F^T + \hat{Z}(F, L, s, g) \, ,$$

$$(2.6.7) \qquad q = \hat{q}(F, L, s, g) \, .$$

Recalling (2.4.3) and (2.4.5), we deduce that the principle of material frame indifference requires

$$(2.6.8) \qquad \begin{cases} \hat{Z}(OF, OLO^T + \dot{O}O^T, s, gO^T) = O\hat{Z}(F, L, s, g)O^T, \\ \hat{q}(OF, OLO^T + \dot{O}O^T, s, gO^T) = \hat{q}(F, L, s, g)O^T, \end{cases}$$

for any proper orthogonal matrix O. In particular, for any fixed state (F, L, s, g) with spin W (cf. (2.1.9)), we may pick $O(t) = \exp(-tW)$, in which case $O(0) = I$, $\dot{O}(0) = -W$. It then follows from (2.6.8) that $\hat{Z}$ and $\hat{q}$ may depend on L solely through its symmetric part D and hence (2.6.6) and (2.6.7) may be written as

$$(2.6.9) \qquad T = \rho \partial_F \hat{\varepsilon}(F, s) F^T + \hat{Z}(F, D, s, g)$$

$$(2.6.10) \qquad q = \hat{q}(F, D, s, g) \, ,$$

with $\hat{Z}$ and $\hat{q}$ such that

$$(2.6.11) \qquad \begin{cases} \hat{Z}(OF, ODO^T, s, gO^T) = O\hat{Z}(F, D, s, g)O^T, \\ \hat{q}(OF, ODO^T, s, gO^T) = \hat{q}(F, D, s, g)O^T, \end{cases}$$

for all proper orthogonal matrices O.

For the balance law of angular momentum (2.3.9) to be satisfied, $\hat{Z}$ must also be symmetric: $\hat{Z}^T = \hat{Z}$. Notice that in that case the dissipation inequality (2.6.4) may be rewritten in the form

$$(2.6.12) \qquad \mathrm{tr}\,[\hat{Z}(F, D, s, g)D] + \frac{1}{\hat{\theta}(F, s)} \hat{q}(F, D, s, g)g^T \geq 0 \, .$$

Further reduction of the constitutive functions obtains when the medium is endowed with material symmetry. As in Section 2.5, we introduce here the *symmetry group* $\mathscr{G}$ of the material, namely the subgroup of $\mathscr{SL}_m$ formed by the

Jacobian matrices H of those isochoric transformations $\bar{x}$ of the reference configuration that leave all constitutive functions invariant. The rules of transformation
of the fields under change of the reference configuration are recorded in (2.3.18),
(2.3.19), (2.3.20) and (2.3.21). Thus, $\mathscr{G}$ is the set of all $H \in \mathscr{SL}_m$ with the
property

(2.6.13)
$$\begin{cases} \hat{\varepsilon}(FH^{-1}, s) = \hat{\varepsilon}(F, s) \ , \\ \hat{Z}(FH^{-1}, D, s, g) = \hat{Z}(F, D, s, g) \ , \\ \hat{q}(FH^{-1}, D, s, g) = \hat{q}(F, D, s, g) \ . \end{cases}$$

The material will be called a *thermoviscoelastic fluid* when $\mathscr{G} \equiv \mathscr{SL}_m$.
In that case, applying (2.6.13) with $H = (\det F)^{-1/m} F \in \mathscr{SL}_m$, we conclude
that $\hat{\varepsilon}$, $\hat{Z}$, and $\hat{q}$ may depend on F solely through its determinant or, equivalently,
through the density ρ. Therefore, the constitutive equations of the thermoviscoelas-
tic fluid reduce to

(2.6.14)
$$\begin{cases} \varepsilon = \tilde{\varepsilon}(\rho, s) \\ T = -\tilde{p}(\rho, s)I + \tilde{Z}(\rho, D, s, g) \\ \theta = \partial_s \tilde{\varepsilon}(\rho, s) \\ q = \tilde{q}(\rho, D, s, g) \end{cases}$$

where $\tilde{p}$ is given by (2.5.14). For frame indifference, $\tilde{Z}$ and $\tilde{q}$ should still satisfy,
for any proper orthogonal matrix O,

(2.6.15)
$$\begin{cases} \tilde{Z}(\rho, ODO^T, s, gO^T) = O\tilde{Z}(\rho, D, s, g)O^T, \\ \tilde{q}(\rho, ODO^T, s, gO^T) = \tilde{q}(\rho, D, s, g)O^T \ , \end{cases}$$

which follow from (2.6.11). It is possible to write down explicitly the form of the
most general functions $\tilde{Z}$ and $\tilde{q}$ that conform with (2.6.15). Here, it will suffice
to record the most general linear constitutive relations that are compatible with
(2.6.15),

(2.6.16)
$$T = -\tilde{p}(\rho, s)I + \tilde{\lambda}(\rho, s)(\operatorname{tr} D)I + 2\tilde{\mu}(\rho, s)D \ ,$$

(2.6.17)
$$q = \tilde{\kappa}(\rho, s)g \ ,$$

which identify the (compressible) *Newtonian fluid*.

2.7 Notes

The venerable field of Continuum Physics has been enjoying a revival, concomitant
with the rise of interest in the behavior of materials with nonlinear response.
The encyclopedic works of Truesdell and Toupin [1] and Truesdell and Noll [1]
contain reliable historical information as well as massive bibliographies and may

serve as excellent guides for following the development of the subject from its inception, in the 18th century, to the mid 1960's. The text by Gurtin [1] provides a clear, elementary introduction to the area. A more advanced treatment, with copious references, is found in the book of Silhavy [1]. Other good sources, emphasizing elasticity theory, are the books of Ciarlet [1], Hanyga [1], Marsden and Hughes [1] and Wang and Truesdell [1]. The recent monograph by Antman [3] contains a wealth of material on the theory of elastic strings, rods, shells and three-dimensional bodies, with emphasis on the qualitative analysis of the governing balance laws.

On the equivalence of the referential (Lagrangian) and spatial (Eulerian) description of the field equations for the balance laws of Continuum Physics, see Dafermos [17] and Wagner [3]. It would be useful to know whether this holds under more general assumptions on the motion than Lipschitz continuity. For instance, when the medium is a thermoelastic gas, it is natural to allow regions of vacuum in the placement of the body. In such a region the density vanishes and the specific volume (determinant of the deformation gradient) becomes infinitely large. For particular results in that direction, see Wagner [2].

The field equations for the balance laws considered here were originally derived by Euler [1,2], for mass, Cauchy [3,4], for linear and angular momentum, and Kirchhoff [1], for energy. The Clausius-Duhem inequality was postulated by Clausius [1], for the adiabatic case; the entropy flux term was introduced by Duhem [1] and the entropy production term was added by Truesdell and Toupin [1]. This last reference also contains an exhaustive treatment of the balance laws of Continuum Physics.

The postulate that constitutive equations should be reduced so that the Clausius-Duhem inequality be satisfied automatically by smooth thermodynamic processes which balance mass, momentum and energy was first stated as a general principle by Coleman and Noll [1]. The examples presented here were adapted from Coleman and Noll [1], for thermoelasticity, and Coleman and Mizel [1], for thermoviscoelasticity.

The use of frame indifference and material symmetry to reduce constitutive equations originated in the works of Cauchy [4] and Poisson [2]. In the ensuing century, this program was implemented (mostly correctly but occasionally incorrectly) by many authors, for a host of special constitutive equations. In particular, the work of the Cosserats [1], Rivlin and Ericksen [1] and others in the 1940's and 1950's contributed to the clarification of the concepts. The principle of material frame indifference and the definition of the symmetry group were ultimately postulated with generality and mathematical precision by Noll [1].

Chapter III. Hyperbolic Systems of Balance Laws

The ambient space for the system of balance laws, introduced in Chapter I, will be visualized here as space-time, and the central notion of hyperbolicity in the time direction will be motivated and defined. Companions to the flux, considered in Section 1.5, will now be realized as entropy-entropy flux pairs.

Numerous examples will be presented of hyperbolic systems of balance laws arising in Continuum Physics.

3.1 Hyperbolicity

Returning to the setting of Chapter I, let us visualize $\mathbb{R}^k$ as $\mathbb{R}^m \times \mathbb{R}$, where $\mathbb{R}^m$, $m = k - 1$, is "space" with typical point x, and $\mathbb{R}$ is "time" with typical value t, so $X = (x, t)$. We write ∂_t for ∂_k, denote G_k by H and thus rewrite the system of balance laws (1.4.3) in the equivalent form

$$(3.1.1) \qquad \partial_t H(U(x,t), x, t) + \sum_{\alpha=1}^{m} \partial_\alpha G_\alpha(U(x,t), x, t) = \Pi(U(x,t), x, t) .$$

Definition 3.1.1 The system of balance laws (3.1.1) is called *hyperbolic* in the t-direction if, for any fixed $U \in \mathscr{O}$, $(x, t) \in \mathscr{X}$ and $\nu \in \mathscr{S}^{m-1}$, the $n \times n$ matrix $DH(U, x, t)$ is nonsingular and the eigenvalue problem

$$(3.1.2) \qquad \left[\sum_{\alpha=1}^{m} \nu_\alpha DG_\alpha(U, x, t) - \lambda DH(U, x, t) \right] R = 0$$

has real eigenvalues $\lambda_1(\nu; U, x, t), \cdots, \lambda_n(\nu; U, x, t)$, called *characteristic speeds*, and n linearly independent eigenvectors $R_1(\nu; U, x, t), \cdots, R_n(\nu; U, x, t)$.

A class of great importance are the *symmetric hyperbolic* systems of balance laws (3.1.1), in which, for any $U \in \mathscr{O}$ and $(x, t) \in \mathscr{X}$, the $n \times n$ matrices $DG_\alpha(U, x, t), \alpha = 1, \cdots, m$, are symmetric and $DH(U, x, t)$ is symmetric positive definite.

The definition of hyperbolicity may be naturally interpreted in terms of the notion of fronts, introduced in Section 1.6. A front $\mathscr{F}$ of the system of balance laws (3.1.1) may be visualized as a one-parameter family of $m - 1$ dimensional manifolds in $\mathbb{R}^m$, parametrized by t, i.e., as a surface propagating in space. In that context, if we renormalize the normal N on $\mathscr{F}$ so that $N = (\nu, -s)$ with $\nu \in \mathscr{S}^{m-1}$, then the wave will be propagating in the direction ν with speed s. Therefore, comparing (3.1.2) with (1.6.1) we conclude that a system of n balance laws is hyperbolic if and only if n distinct weak waves can propagate in any space direction. The eigenvalues of (3.1.2) will determine the speed of propagation of these waves while the corresponding eigenvectors will specify the direction of the amplitude.

When $\mathscr{F}$ is a shock front, (1.6.3) may be written in the current notation as

(3.1.3)

$$-s[H(U_+, x, t) - H(U_-, x, t)] + \sum_{\alpha=1}^{m} \nu_\alpha [G_\alpha(U_+, x, t) - G_\alpha(U_-, x, t)] = 0 ,$$

which is called the *Rankine-Hugoniot jump condition*. By virtue of Theorem 1.8.1, this condition should hold at every point of approximate jump discontinuity of any function U of class BV_{loc} that satisfies the system (3.1.1) in the sense of measures.

It is clear that hyperbolicity is preserved under any change $U^* = U^*(U, x, t)$ of state vector with $U^*(\cdot, x, t)$ a diffeomorphism for every fixed $(x, t) \in \mathscr{X}$. In particular, since $DH(U, x, t)$ is nonsingular, we may employ, locally at least, H as the new state vector. Thus, without much loss of generality, one may limit the investigation to hyperbolic systems of balance laws that have the special form

(3.1.4) $$\partial_t U(x, t) + \sum_{\alpha=1}^{m} \partial_\alpha G_\alpha(U(x, t), x, t) = \Pi(U(x, t), x, t) .$$

For simplicity and convenience, we shall regard henceforth the special form (3.1.4) as *canonical*. The reader should keep in mind, however, that when dealing with systems of balance laws arising in Continuum Physics it may be advantageous to keep the state vector naturally provided, even at the expense of having to face the more complicated form (3.1.1) rather than the canonical form (3.1.4).

3.2 Entropy-Entropy Flux Pairs

Assume that the system of balance laws (1.4.3), which we now write in the form (3.1.1), is endowed with a companion balance law (1.5.2). To recast (1.5.2) in the new notation, we rewrite it in the form

(3.2.1) $$\partial_t \eta(U(x, t), x, t) + \sum_{\alpha=1}^{m} \partial_\alpha q_\alpha(U(x, t), x, t) = h(U(x, t), x, t) ,$$

by setting $q_k \equiv \eta$. As we shall see in Section 3.3, in the applications to Continuum Physics companion balance laws of the form (3.2.1) are intimately related with the Second Law of thermodynamics. For that reason, η is called an *entropy* for the system (3.1.1) of balance laws and $(q_1, \cdots, q_m)$ is called the *entropy flux* associated with η.

Equation (1.5.1), for $\alpha = k$, should now be written as

(3.2.2) $$D\eta(U, x, t) = B(U, x, t)^T DH(U, x, t) .$$

Assume the system is in canonical form (3.1.4) so that (3.2.2) reduces to $D\eta = B^T$. Then (1.5.1) and the integrability condition (1.5.4) become

(3.2.3) $$Dq_\alpha(U, x, t) = D\eta(U, x, t)DG_\alpha(U, x, t) , \quad \alpha = 1, \cdots, m ,$$

(3.2.4)

$$D^2\eta(U, x, t)DG_\alpha(U, x, t) = DG_\alpha(U, x, t)^T D^2\eta(U, x, t) , \quad \alpha = 1, \cdots, m .$$

Notice that (3.2.4) imposes $\frac{1}{2}n(n-1)m$ conditions on the single unknown function η. Therefore, as already noted in Section 1.5, the problem of determining a nontrivial entropy-entropy flux pair for (3.1.1) is formally overdetermined, unless either $n = 1$ and m is arbitrary, or $n = 2$ and $m = 1$. However, when the system is symmetric, we may satisfy (3.2.4) with $\eta = \frac{1}{2}|U|^2$. Conversely, if (3.2.4) holds and $\eta(U, x, t)$ is uniformly convex in U, then the change $U^* = D\eta^T$ of state vector renders the system symmetric hyperbolic.

3.3 Examples of Hyperbolic Systems of Balance Laws

Out of a host of hyperbolic systems of balance laws in Continuum Physics, only a small sample will be presented here. They will serve as beacons for guiding the development of the general theory.

(a) The Scalar Balance Law. The single balance law ($n = 1$)

(3.3.1) $$\partial_t u(x, t) + \sum_{\alpha=1}^{m} \partial_\alpha g_\alpha(u(x, t), x, t) = \varpi(u(x, t), x, t)$$

is always hyperbolic. Any function $\eta(u, x, t)$ may serve as entropy, with associated entropy flux and entropy production computed by

(3.3.2) $$q_\alpha = \int^u \frac{\partial \eta}{\partial u} \frac{\partial g_\alpha}{\partial u} du , \quad \alpha = 1, \cdots, m ,$$

(3.3.3) $$h = \sum_{\alpha=1}^{m} \left[\frac{\partial \eta}{\partial u} \frac{\partial g_\alpha}{\partial x_\alpha} - \frac{\partial q_\alpha}{\partial x_\alpha} \right] + \varpi \frac{\partial \eta}{\partial u} + \frac{\partial \eta}{\partial t} .$$

Equation (3.3.1), the corresponding homogeneous scalar conservation law, and especially their one-space dimensional ($m = 1$) versions will serve extensively as models for developing the theory of general systems.

(b) Thermoelastic Nonconductors of Heat. The theory of thermoelastic media was discussed in Chapter II. Here we shall employ the referential (Lagrangian) description so the fields will be functions of (x, t). For consistency with the notation of the present chapter, we shall use ∂_t to denote material time derivative (in lieu of the overdot employed in Chapter II) and ∂_α to denote partial derivative with respect to the α-component x_α of x.

The constitutive equations are recorded in Section 2.5. Since there is no longer danger of confusion, we may simplify the notation by dropping the "hat" from the symbols of the constitutive functions. Also for simplicity we assume that the medium is homogeneous, with reference density $\rho_0 \equiv 1$.

As explained in Chapter II, a thermodynamic process is determined by a motion χ and an entropy field s. In order to cast the field equations of the balance laws into a first order system of the form (3.1.1), we monitor χ through its derivatives (2.1.1), (2.1.2) and thus work with the state vector $U = (F, v, s)$, taking values in $\mathbb{R}^{m^2+m+1}$. In that case we must append to the balance laws of linear momentum (2.3.4) and energy (2.3.10) the compatibility condition (2.1.8)$_1$. Consequently, our system of balance laws reads

(3.3.4)
$$
\begin{cases}
\partial_t F_{i\alpha} - \partial_\alpha v_i = 0 , & i, \alpha = 1, \cdots, m , \\
\partial_t v_i - \sum_{\alpha=1}^{m} \partial_\alpha S_{i\alpha}(F, s) = b_i , & i = 1, \cdots, m , \\
\partial_t \left[\varepsilon(F, s) + \frac{1}{2}|v|^2 \right] - \sum_{\alpha=1}^{m} \partial_\alpha \left[\sum_{i=1}^{m} v_i S_{i\alpha}(F, s) \right] = \sum_{i=1}^{m} b_i v_i + r ,
\end{cases}
$$

with (cf. (2.5.3))

(3.3.5)
$$
S_{i\alpha}(F, s) = \frac{\partial \varepsilon(F, s)}{\partial F_{i\alpha}} , \qquad \theta(F, s) = \frac{\partial \varepsilon(F, s)}{\partial s} .
$$

A lengthy calculation verifies that the system (3.3.4) is hyperbolic on a certain region of the state space if for every (F, s) lying in that region

(3.3.6)
$$
\frac{\partial \varepsilon(F, s)}{\partial s} > 0 ,
$$

(3.3.7)
$$
\sum_{i,j=1}^{m} \sum_{\alpha,\beta=1}^{m} \frac{\partial^2 \varepsilon(F, s)}{\partial F_{i\alpha} \partial F_{j\beta}} v_\alpha v_\beta \xi_i \xi_j > 0 , \quad \text{for all } v \text{ and } \xi \text{ in } \mathscr{S}^{m-1} .
$$

By account of (3.3.5)$_2$, condition (3.3.6) simply states that the absolute temperature must be positive. (3.3.7), called the *Legendre-Hadamard condition*, means that ε is *rank-one convex* in F, i.e., it is convex along any direction $\xi \otimes v$ with rank one. An alternative way of expressing (3.3.7) is to state that for any $v \in \mathscr{S}^{m-1}$ the *acoustic tensor* $N(v, F, s)$, defined by

(3.3.8)
$$
N_{ij}(v, F, s) = \sum_{\alpha,\beta=1}^{m} \frac{\partial^2 \varepsilon(F, s)}{\partial F_{i\alpha} \partial F_{j\beta}} v_\alpha v_\beta , \quad i, j = 1, \cdots, m
$$

is positive definite. In fact, for the system (3.3.4), the characteristic speeds are the $2m$ square roots of the m eigenvalues of the acoustic tensor, and zero with multiplicity $m^2 - m + 1$.

Recall from Chapter II that, in addition to the system of balance laws (3.3.4), thermodynamically admissible processes should also satisfy the Clausius-Duhem inequality (2.3.14) which here takes the form

$$(3.3.9) \qquad\qquad -\partial_t s \leq -\frac{r}{\theta(F, s)} \ .$$

By virtue of (3.3.5), every classical solution of (3.3.4) will satisfy (3.3.9) identically as an equality. Thus, in the terminology of Section 3.2, $-s$ is an entropy for the system (3.3.4) with associated entropy flux zero.[1] Weak solutions of (3.3.4) will not necessarily satisfy (3.3.9). Therefore, the role of (3.3.9) is to weed out undesirable weak solutions. The extension of a companion balance law from an identity for classical solutions into an inequality for weak solutions will play a crucial role in the general theory of hyperbolic systems of balance laws.

(c) Isentropic Process of Thermoelastic Nonconductors of Heat.

The physical background of isentropic process was discussed in Section 2.5. The entropy is fixed at a constant value $\bar{s}$ and, for simplicity, is dropped from the notation. The state vector reduces to $U = (F, v)$ with values in $\mathbb{R}^{m^2+m}$. The system of balance laws results from (3.3.4) by discarding the balance of energy:

$$(3.3.10) \qquad \begin{cases} \partial_t F_{i\alpha} - \partial_\alpha v_i = 0 , & i, \alpha = 1, \cdots, m \\ \partial_t v_i - \sum_{\alpha=1}^{m} \partial_\alpha S_{i\alpha}(F) = b_i , & i = 1, \cdots, m \end{cases}$$

and we still have

$$(3.3.11) \qquad\qquad S_{i\alpha}(F) = \frac{\partial \varepsilon(F)}{\partial F_{i\alpha}} , \qquad i, \alpha = 1, \cdots, m \ .$$

The system (3.3.10) is hyperbolic if ε is rank-one convex, i.e., (3.3.7) holds at $s = \bar{s}$.

As explained in Section 2.5, in addition to (3.3.10) thermodynamically admissible isentropic motions must also satisfy the inequality (2.5.20), which in the current notation reads

$$(3.3.12) \qquad \partial_t [\varepsilon(F) + \tfrac{1}{2}|v|^2] - \sum_{\alpha=1}^{m} \partial_\alpha \left[\sum_{i=1}^{m} v_i S_{i\alpha}(F) \right] \leq \sum_{i=1}^{m} b_i v_i \ .$$

By virtue of (3.3.11), any classical solution of (3.3.10) satisfies identically (3.3.12) as an equality. Thus, in the terminology of Section 3.2, $\eta = \varepsilon(F) + \tfrac{1}{2}|v|^2$ is an entropy for the system (3.3.10). Note that (3.3.10) is in canonical form (3.1.4) and that $D\eta = (S(F), v)$. Therefore, as shown in Section 3.2, if the internal energy $\varepsilon(F)$ is uniformly convex, then changing the state vector from $U = (F, v)$ to $U^* = (S, v)$ will render the system (3.3.10) symmetric hyperbolic.

---

[1] Identifying $-s$ as the "entropy" here, rather than s which is the physical entropy, may look strange. This convention is made because it is more convenient to deal with functionals of the solution that are nonincreasing with time.

Weak solutions of (3.3.10) will not necessarily satisfy (3.3.12). We thus encounter again the situation in which a companion balance law is extended from an identity for classical solutions into an inequality serving as admissibility condition on weak solutions.

The passing from (3.3.4) to (3.3.10) provides an example of a truncation process which is commonly employed in Continuum Physics for simplifying systems of balance laws by dropping a number of the equations while simultaneously reducing proportionally the size of the state vector. In a canonical truncation, which preserves the entropy structure and does not increase wave speeds, the elimination of any equation should be paired with freezing the corresponding component of the special state vector that symmetrizes the system. Thus, for instance, one may canonically truncate the system (3.3.10) by dropping the i-th of the last m equations while freezing the i-th component v_i of velocity, or else by dropping the (i, α)-th of the first m^2 equations while freezing the (i, α)-th component $S_{i\alpha}(F)$ of the Piola-Kirchhoff stress.

(d) Thermoelastic Fluid Nonconductors of Heat. The system of balance laws (3.3.4) governs the adiabatic thermodynamic processes of all thermoelastic media, including, in particular, thermoelastic fluids. In the latter case, however, it is advantageous to employ spatial (Eulerian) description. The reason is that, as shown in Section 2.5, the internal energy, the temperature, and the Cauchy stress in a thermoelastic fluid depend on the deformation gradient F solely through the density ρ. We may thus dispense altogether with F and describe the state of the medium through the state vector $U = (\rho, v, s)$ which takes values in the (much smaller) space $\mathbb{R}^{m+2}$.

The fields will now be functions of (χ, t). For consistency with the notational conventions of this chapter, we will be using ∂_t (rather than a t-subscript as in Chapter II) to denote partial derivative with respect to t. Also, the typical components of the position vector χ will be labeled with Latin indices, e.g. χ_i, χ_j, and the corresponding partial derivatives $\partial/\partial\chi_i$, $\partial/\partial\chi_j$ will be denoted by ∂_i, ∂_j.

The balance laws in force are for mass (2.3.2), linear momentum (2.3.5) and energy (2.3.11). The constitutive relations are (2.5.12), with $\tilde{q} \equiv 0$, (2.5.13) and (2.5.14). To simplify the notation, we drop the "tilde" from the symbols of the constitutive functions. Therefore, the system of balance laws takes the form

$$(3.3.13) \quad \begin{cases} \partial_t\rho + \sum_{j=1}^m \partial_j(\rho v_j) = 0 \\ \partial_t(\rho v_i) + \sum_{j=1}^m \partial_j(\rho v_i v_j) + \partial_i p(\rho, s) = \rho b_i \quad , \quad i = 1, \cdots, m \\ \partial_t[\rho\varepsilon(\rho, s) + \frac{1}{2}\rho|v|^2] + \sum_{j=1}^m \partial_j[(\rho\varepsilon(\rho, s) + \frac{1}{2}\rho|v|^2 + p(\rho, s))v_j] \\ \qquad\qquad = \sum_{j=1}^m \rho b_j v_j + \rho r, \end{cases}$$

with

$$(3.3.14) \qquad p(\rho, s) = \rho^2 \varepsilon_\rho(\rho, s) , \quad \theta(\rho, s) = \varepsilon_s(\rho, s) .$$

The system (3.3.13) will be hyperbolic if

(3.3.15) $$\varepsilon_s(\rho, s) > 0 , \quad p_\rho(\rho, s) > 0 .$$

In addition to (3.3.13), thermodynamically admissible processes must also satisfy the Clausius-Duhem inequality (2.3.15), which here reduces to

(3.3.16) $$\partial_t(-\rho s) + \sum_{j=1}^{m} \partial_j(-\rho s v_j) \leq -\rho \frac{r}{\theta(\rho, s)} .$$

When the process is smooth, it follows from (3.3.13) and (3.3.14) that (3.3.16) holds identically, as an equality. Consequently, $\eta = -\rho s$ is an entropy for the system (3.3.13) with associated entropy flux $-\rho s v$. Once more we see that a companion balance law is extended from an identify for classical solutions into an inequality serving as a test for the physical admissibility of weak solutions.

(e) Isentropic Process of Thermoelastic Fluids. The entropy is fixed at a constant value and is dropped from the notation. The state vector is $U = (\rho, v)$, with values in $\mathbb{R}^{m+1}$. The system of balance laws results from (3.3.13) by discarding the balance of energy:

(3.3.17) $$\begin{cases} \partial_t \rho + \sum_{j=1}^{m} \partial_j(\rho v_j) = 0 \\ \partial_t(\rho v_i) + \sum_{j=1}^{m} \partial_j(\rho v_i v_j) + \partial_i p(\rho) = \rho b_i , \quad i = 1, \cdots, m \end{cases}$$

with

(3.3.18) $$p(\rho) = \rho^2 \varepsilon'(\rho) .$$

The system (3.3.17) is hyperbolic if

(3.3.19) $$p'(\rho) > 0 .$$

A classical example is the *polytropic gas*:

(3.3.20) $$p = \kappa \rho^\gamma , \quad \kappa > 0 , \quad \gamma > 1 .$$

Thermodynamically admissible isentropic motions must satisfy the inequality (2.5.21), which here reduces to

(3.3.21) $$\partial_t[\rho\varepsilon(\rho) + \tfrac{1}{2}\rho|v|^2] + \sum_{j=1}^{m} \partial_j[(\rho\varepsilon(\rho) + \tfrac{1}{2}\rho|v|^2 + p(\rho))v_j] \leq \sum_{j=1}^{m} \rho b_j v_j .$$

The pattern has become by now familiar: By virtue of (3.3.18), any classical solution of (3.3.17) satisfies identically (3.3.21), as an equality, so that the function $\eta = \rho\varepsilon(\rho) + \tfrac{1}{2}\rho|v|^2$ is an entropy for the system (3.3.17). At the same time, the inequality (3.3.21) is employed to weed out physically inadmissible weak solutions.

(f) Maxwell's Equations in Nonlinear Dielectrics. Another rich source of interesting systems of hyperbolic balance laws is electromagnetism. The underlying system is Maxwell's equations

$$(3.3.22) \qquad \begin{cases} \partial_t B = -\mathrm{curl}\, E \\ \partial_t D = \mathrm{curl}\, H - J \end{cases}$$

on $\mathbb{R}^3$, relating the *electric field* E, the *magnetic field* H, the *electric displacement* D, the *magnetic induction* B and the *current* J, all of them taking values in $\mathbb{R}^3$.

Constitutive relations determine E, H, and J from the state vector $U = (B, D)$. For example, when the medium is a homogeneous electric conductor, with linear dielectric response, at rest relative to the inertial frame, then $D = \varepsilon E, B = \mu H$, and $J = \sigma E$, where ε is the dielectric constant, μ is the magnetic permeability and σ is the electric conductivity. In order to account for (possibly) moving media with nonlinear dielectric response and cross coupling of electromagnetic fields, one postulates general constitutive equations

$$(3.3.23) \qquad E = E(B, D) , \quad H = H(B, D) , \quad J = J(B, D) ,$$

where the functions E and H satisfy the *lossless condition*

$$(3.3.24) \qquad \frac{\partial H}{\partial D} = \frac{\partial E}{\partial B} .$$

Physically admissible fields must also satisfy the dissipation inequality

$$(3.3.25) \qquad \partial_t \eta(B, D) + \mathrm{div}\, q(B, D) \le h(B, D)$$

where (recall (3.3.24))

$$(3.3.26) \qquad \eta = \int [H \cdot dB + E \cdot dD] , \quad q = E \wedge H , \quad h = -J \cdot E .$$

Thus η is the *electromagnetic field energy* and q is the *Poynting vector*. A straightforward calculation shows that smooth solutions of (3.3.22), (3.3.23) satisfy (3.3.25) identically, as an equality. Therefore, (η, q) constitutes an entropy-entropy flux pair for the system of balance laws (3.3.22), (3.3.23). Since $D\eta = (H, E)$, it follows from the discussion in Section 3.2 that when the electromagnetic field energy function is uniformly convex, then the change of state vector from $U = (B, D)$ to $U^* = (H, E)$ renders the system symmetric hyperbolic.

The dielectric is *isotropic* when the electromagnetic field energy is invariant under rigid rotations of the vectors B and D, in which case η may depend on (B, D) solely through the three scalar products $B \cdot B$, $D \cdot D$, and $B \cdot D$.

(g) Lundquist's Equations of Magnetohydrodynamics. Interesting systems of hyperbolic balance laws arise in the context of electromechanical phenomena, where the balance laws of mass, momentum and energy of continuum thermomechanics are coupled with Maxwell's equations. As an illustrative example, we consider here the theory of *magnetohydrodynamics* which describes the interaction of a magnetic field with an electrically conducting thermoelastic fluid. The equations follow from a number of simplifying assumptions, which will now be outlined.

Beginning with Maxwell's equations, the electric displacement D is considered negligible so (3.3.22) yields $J = \text{curl } H$. The magnetic induction B is related to the magnetic field H by the classical relation $B = \mu H$. The electric field is totally generated by the motion of the fluid in the magnetic field and so is given by $E = B \wedge v = \mu H \wedge v$.

The fluid is a thermoelastic nonconductor of heat whose thermomechanical properties are still described by the constitutive relations (3.3.14). The balance of mass (3.3.13)$_1$ remains unaffected by the presence of the electromagnetic field. On the other hand, the electromagnetic field exerts a force on the fluid which should be accounted as body force in the balance of momentum (3.3.13)$_2$. The contribution of the electric field E to this force is assumed negligible while the contribution of the magnetic field is $J \wedge B = -\mu H \wedge \text{curl } H$. By account of the identity

$$(3.3.27) \qquad -H \wedge \text{curl } H = \text{div}\left[H \otimes H - \tfrac{1}{2}(H \cdot H)I\right],$$

this body force may be realized as the divergence of the *Maxwell stress tensor*. We assume there is no external body force. The electromagnetic effects on the energy equation (3.3.13)$_3$ are derived by virtue of (3.3.26): The internal energy should be augmented by the electromagnetic field energy $\tfrac{1}{2}\mu|H|^2$. The Poynting vector $\mu(H \wedge v) \wedge H = \mu|H|^2 v - \mu(H \cdot v)H$ should be added to the flux. Finally, the electromagnetic energy production $-J \cdot E = -\mu(H \wedge v) \cdot \text{curl } H$ and the rate of work $(J \wedge B) \cdot v = -\mu(H \wedge \text{curl } H) \cdot v$ of the electromagnetic body force cancel each other out.

We thus derive *Lundquist's equations*:

$$(3.3.28)$$
$$\begin{cases} \partial_t \rho + \text{div}(\rho v) = 0 \\[2mm] \partial_t(\rho v) + \text{div}\left[\rho v \otimes v - \mu H \otimes H\right] + \text{grad}\left[p(\rho, s) + \tfrac{1}{2}\mu|H|^2\right] = 0 \\[2mm] \partial_t\left[\rho\varepsilon(\rho, s) + \tfrac{1}{2}\rho|v|^2 + \tfrac{1}{2}\mu|H|^2\right] \\[1mm] \qquad + \text{div}\left[(\rho\varepsilon(\rho, s) + \tfrac{1}{2}\rho|v|^2 + p(\rho, s) + \mu|H|^2)v - \mu(H \cdot v)H\right] = \rho r \\[2mm] \partial_t H - \text{curl}(v \wedge H) = 0. \end{cases}$$

The above system of balance laws, with state vector $U = (\rho, v, s, H)$, will be hyperbolic if (3.3.15) hold. Thermodynamically admissible solutions of (3.3.28) should also satisfy the Clausius-Duhem inequality (3.3.16). By virtue of (3.3.14), it is easily seen that any classical solution of (3.3.28) satisfies identically (3.3.16) as an equality. Thus $-\rho s$ is an entropy for the system (3.3.28), with associated entropy flux $-\rho s v$.

3.4 Notes

The theory of nonlinear hyperbolic systems of balance laws traces its origins to the mid 19th century and has developed over the years conjointly with gas dynamics.

English translations of the seminal papers, with commentaries, are collected in the book of Johnson and Chéret [1]. Research was particularly intense during the Second World War; see the papers of Bethe [1], von Neumann [1,2,3] and Weyl [1]. The state of the art in the late 1940's is vividly presented in the classic monograph by Courant and Friedrichs [1]. It is the distillation of this material that has laid the foundations of the formalized mathematical theory in its present form.

A number of synthetic works, with different scopes, are available to the student of the field. The survey articles by Lax [5] and Dafermos [6] provide quick, elementary introduction to the area. The book of Smoller [1] is a comprehensive text at an introductory level, while the more recent, two-volume treatise by Serre [9] combines a general introduction to the subject with advanced, deeper investigation of certain topics. The earlier monograph by Rozdestvenskii and Janenko [1] emphasizes the contributions of the Russian school, which was very active in the 1950's and 1960's. An elementary text, with more narrow scope, is Jeffrey [2]. The lecture notes of Bressan [4] and Majda [3] provide clear, insightful accounts of the theory in the one-space and multi-space dimensional setting, respectively. The theory is also covered, to a certain extent, in the textbook of Hörmander [1] and the treatise of Taylor [1]. From the standpoint of numerical analysis, LeVeque [1] is an introductory text, while the books of Godlewski and Raviart [1,2] provide a more comprehensive and technical coverage together with a voluminous list of references. The lecture notes of Tadmor [1] is also a good source of insightful information and references. Another valuable resource is the text of Whitham [2] which presents a panorama of connections of the theory with a host of diverse applications as well as a survey of ideas and techniques devised over the years by applied mathematicians studying wave propagation, of which many are ready for more rigorous analytical development. The book of Zeldovich and Raizer [1] is an excellent introduction to gas dynamics from the perspective of physicists and may be consulted for building intuition. Monographs with more specialized scope will be referenced later, as they become relevant to the discussion.

The term "entropy" in the sense employed here was introduced by Lax [4].

Simple physical models that lead to scalar conservation laws (or simple systems thereof) are described in Whitham [2]. The study of weak fronts in isentropic thermoelasticity has exerted seminal influence in the development of the notions of hyperbolicity, stability, etc. An account, together with an extensive list of references, can be found in Truesdell and Noll [1].

The systems (3.3.13) and (3.3.17) are commonly called *Euler equations*. There is voluminous literature on various aspects of their theory, some of which will be cited in subsequent chapters. For a detailed analytical study in several space dimensions together with an extensive bibliography, see the monograph of Lions [2].

The Euler equations for ideal gases may also be derived from Boltzmann's equation, in the kinetic theory, upon identifying density, velocity, pressure, temperature, etc. with appropriate moments of the molecular velocity distribution function. By monitoring additional moments, the theory of *Extended Thermodynamics* (cf. Müller and Ruggeri [1]) embeds the Euler equations into a hierarchy of, progressively larger, hyperbolic systems of balance laws. One may pass from more complex to simpler systems in this hierarchy via canonical truncation (cf. Boillat and Ruggeri [1]).

The one-space dimensional versions of (3.3.4), (3.3.10), and (3.3.17) will be recorded in Section 7.1. These systems provide the governing equations, in Lagrangian or Eulerian coordinates, for gas flow in ducts and have been studied extensively in that context.

For a systematic development of electrothermomechanics, along the lines of the development of continuum thermomechanics in Chapter II, see Coleman and Dill [1] and Grot [1]. Numerous examples of electrodynamical problems involving hyperbolic systems of balance laws are presented in Bloom [1]. For magnetohydrodynamics see for example the texts of Cabannes [1] and Jeffrey [1].

The theory of relativity is a rich source of interesting hyperbolic systems of balance laws, which will not be tapped in this work. When the fluid velocity is comparable to the speed of light, the Euler equations should be modified to account for special relativistic effects; cf. Taub [1] and Friedrichs [2]. For hyperbolic systems of balance laws arising in general relativity, see Ruggeri [1,2], Smoller and Temple [1,2], Pant [1] and J. Chen [1].

Hyperbolic systems of balance laws, with special structure, govern separation processes, like chromatography and electrophoresis, employed in chemical engineering. In that connection the reader may consult the monograph by Rhee, Aris and Amundson [1]. The system of electrophoresis will be recorded later, in Section 7.3, and some of its special properties will be discussed in Chapters VII and VIII.

Chapter IV. The Initial-Value Problem: Admissibility of Solutions

The initial-value problem for a hyperbolic system of conservation laws will be formulated, and classical as well as weak solutions will be considered. Nonuniqueness of weak solutions will be demonstrated in the context of the simplest nonlinear scalar conservation law, the well-known Burgers equation. This raises the need to devise conditions that will weed out unstable, physically irrelevant, or otherwise undesirable solutions, hopefully singling out a unique admissible solution of the initial-value problem. Two admissibility criteria will be introduced in this chapter: The requirement that admissible solutions satisfy a designated entropy inequality; and the principle that the hyperbolic system should be viewed as the "vanishing viscosity" limit of a family of systems with diffusive terms. A preliminary investigation of the compatibility of the above two criteria will be conducted. The chapter will close with remarks on the interpretation of boundary conditions in the context of weak solutions.

4.1 The Initial-Value Problem

To avoid trivial complications, we focus the investigation on homogeneous hyperbolic systems of conservation laws in canonical form,

$$(4.1.1) \qquad \partial_t U(x,t) + \sum_{\alpha=1}^{m} \partial_\alpha G_\alpha(U(x,t)) = 0 \ ,$$

even though everything that will be done in this chapter can be extended in a routine manner to general hyperbolic systems of balance laws (3.1.1). Here, x takes values in $\mathbb{R}^m$ and t in $[0, \infty)$; U takes values in some open subset $\mathcal{O}$ of $\mathbb{R}^n$ and G_α, $\alpha = 1, \cdots, m$, are given smooth functions from $\mathcal{O}$ to $\mathbb{R}^n$. Hyperbolicity means that for every fixed $U \in \mathcal{O}$ and $\nu \in \mathscr{S}^{m-1}$ the $n \times n$ matrix

$$(4.1.2) \qquad \Lambda(\nu; U) = \sum_{\alpha=1}^{m} \nu_\alpha DG_\alpha(U)$$

has real eigenvalues $\lambda_1(\nu; U), \cdots, \lambda_n(\nu; U)$ and n linearly independent eigenvectors $R_1(\nu; U), \cdots, R_n(\nu; U)$.

With (4.1.1) we associate initial conditions

(4.1.3) $U(x, 0) = U_0(x) , \quad x \in \mathbb{R}^m ,$

where U_0 is a given bounded measurable function from $\mathbb{R}^m$ to $\mathscr{O}$.

A *classical solution* of (4.1.1) on a time interval $[0, T)$ is a bounded, locally Lipschitz, function U, defined on $\mathbb{R}^m \times [0, T)$ and taking values in $\mathscr{O}$, which satisfies (4.1.1) almost everywhere. This function solves the initial-value problem (4.1.1), (4.1.3) when it also satisfies (4.1.3), for all $x \in \mathbb{R}^m$.

A *weak solution* of (4.1.1) on the time interval $[0, T)$ is a bounded measurable function U, defined on $\mathbb{R}^m \times [0, T)$ and taking values in $\mathscr{O}$, which satisfies (4.1.1) in the sense of distributions. Any weak solution which is locally Lipschitz is necessarily a classical solution.

The type of the system (4.1.1) induces a certain degree of regularity in the time behavior of weak solutions:

Theorem 4.1.1 *Any weak solution U of (4.1.1) on $[0, T)$ may be normalized so that*

(4.1.4) $$U(\cdot, \tau) = \lim_{\varepsilon \to 0} \frac{1}{\varepsilon} \int_\tau^{\tau+\varepsilon} U(\cdot, t) dt , \quad \tau \in [0, T) ,$$

where the limit is taken in L^∞ weak*.

Proof. Recalling the definition of L^∞ weak* and using that almost all points in the domain of any measurable real-valued function are Lebesgue points, we conclude that (4.1.4) holds almost everywhere on $[0, T)$.

We now fix $0 \leq \tau < \sigma < T$ and take any $\chi \in C_0^\infty(\mathbb{R}^m)$. For positive small ε and δ, we define the Lipschitz function $\phi(x, t) = \chi(x)\theta(t)$, where $\theta(t) = 0$ for $0 \leq t \leq \tau$, $\theta(t) = \varepsilon^{-1}(t - \tau)$ for $\tau < t < \tau + \varepsilon$, $\theta(t) = 1$ for $\tau + \varepsilon \leq t \leq \sigma - \delta$, $\theta(t) = \delta^{-1}(\sigma - t)$ for $\sigma - \delta < t < \sigma$, and $\theta(t) = 0$ for $\sigma \leq t < T$. We then write that U satisfies (4.1.1), in the sense of distributions, with test function ϕ:

(4.1.5) $$\frac{1}{\varepsilon} \int_\tau^{\tau+\varepsilon} \int_{\mathbb{R}^m} \chi(x) U(x, t) dx dt - \frac{1}{\delta} \int_{\sigma-\delta}^\sigma \int_{\mathbb{R}^m} \chi(x) U(x, t) dx dt$$
$$+ \int_\sigma^\tau \int_{\mathbb{R}^m} \sum_{\alpha=1}^m \partial_\alpha \chi(x) G_\alpha(U(x, t)) dx dt = O(\varepsilon) + O(\delta) .$$

By letting δ and ε, separately, go to zero, we infer that the limit on the right-hand side of (4.1.4), as $\varepsilon \to 0$, exists for any $\tau \in [0, T)$. Hence, by modifying U, if necessary, on a set of τ of measure zero, we attain a normalized representative which satisfies (4.1.4), for all $\tau \in [0, T)$. The proof is complete.

From now on, we shall always consider weak solutions in their normalized form. In particular, the initial conditions (4.1.3) will now be satisfied in the classical sense. Equivalently,

$$(4.1.6) \qquad \int_0^T \int_{\mathbb{R}^m} \left[\partial_t \phi U + \sum_{\alpha=1}^m \partial_\alpha \phi G_\alpha(U) \right] dx dt + \int_{\mathbb{R}^m} \phi(x, 0) U_0(x) dx = 0 \,,$$

for every Lipschitz test function ϕ with compact support in $\mathbb{R}^m \times [0, T)$. It should also be noted that normalization renders evolutionarity, that is if $U(x, t)$ is a weak solution on $[0, T)$ with initial value $U(x, 0)$, then, for any $\tau \in (0, T)$, the function $U_\tau(x, t) = U(x, t + \tau)$ is also a weak solution on $[0, T - \tau)$, with initial value $U_\tau(x, 0) = U(x, \tau)$.

The system converts spatial to temporal regularity as seen in the following proposition.

Theorem 4.1.2 *Let U be a weak solution of* (4.1.1) *on* $[0, T)$ *such that, for any fixed* $t \in [0, T)$, $U(\cdot, t) \in BV(\mathbb{R}^m)$ *and* $TV_{\mathbb{R}^m}(U(\cdot, t)) \leq V$, *for all* $t \in [0, T)$. *Then* $t \mapsto U(\cdot, t)$ *is Lipschitz continuous in* $L^1(\mathbb{R}^m)$ *on* $[0, T)$,

$$(4.1.7) \qquad \| U(\cdot, \sigma) - U(\cdot, \tau) \|_{L^1(\mathbb{R}^m)} \leq aV|\sigma - \tau| \,, \quad 0 \leq \tau < \sigma < T \,,$$

where a depends solely on the Lipschitz constant of G. In particular, U is in BV_{loc} on $\mathbb{R}^m \times [0, T)$.

Proof. Fix $0 \leq \tau < \sigma < T$ and any $\Psi \in C_0^\infty(\mathbb{R}^m; \mathscr{M}^{1,n})$ with $|\Psi(x)| \leq 1$, $x \in \mathbb{R}^m$. We write the i-th component of (4.1.5) with $\chi = \Psi_i$, we sum over $i = 1, \cdots, n$, and then let $\delta \downarrow 0$ and $\varepsilon \downarrow 0$. Upon using (4.1.4),

$$(4.1.8) \qquad \int_{\mathbb{R}^m} \Psi(x)[U(x, \tau) - U(x, \sigma)] dx$$
$$= - \int_\tau^\sigma \int_{\mathbb{R}^m} \sum_{\alpha=1}^m \partial_\alpha \Psi(x) G_\alpha(U(x, t)) dx dt \,.$$

The integrand on the right-hand side is estimated in terms of the total variation of $G(U(\cdot, t))$, which in turn is bounded by aV. Taking the supremum of (4.1.8) over all Ψ with $|\Psi(x)| \leq 1$, we arrive at (4.1.7).

Theorem 1.7.1 together with (4.1.7) imply that U is in BV_{loc} on $\mathbb{R}^m \times [0, T)$. The proof is complete.

4.2 The Burgers Equation and Nonuniqueness of Weak Solutions

A fundamental difficulty in the theory of weak solutions to the initial-value problem for nonlinear hyperbolic systems of balance laws is that uniqueness generally fails. This may be demonstrated through the simplest, scalar conservation law in one space variable

$$(4.2.1) \qquad \partial_t u(x, t) + \partial_x \left(\frac{1}{2} u^2(x, t) \right) = 0 \,,$$

called the *Burgers equation*. Indeed, notice that the initial-value problem for (4.2.1) with data

(4.2.2)
$$u(x, 0) = \begin{cases} -1, & x < 0 \\ 1, & x > 0 \end{cases}$$

admits the family of BV weak solutions

(4.2.3)
$$u_\alpha(x, t) = \begin{cases} -1, & -\infty < x \le -t \\ \dfrac{x}{t}, & -t < x \le -\alpha t \\ -\alpha, & -\alpha t < x \le 0 \\ \alpha, & 0 < x \le \alpha t \\ \dfrac{x}{t}, & \alpha t < x \le t \\ 1, & t < x < \infty, \end{cases}$$

where α is any number in $[0, 1]$. In particular, the solution obtained for $\alpha = 0$ is locally Lipschitz away from the origin, while the rest contain a stationary shock at $x = 0$.

The above example highlights the need to devise criteria that will screen weak solutions for admissibility.

4.3 Entropies and Admissible Solutions

We assume that our system of conservation laws (4.1.1) is endowed with a designated entropy η with associated entropy flux $(q_1, \cdots, q_m)$. As we saw in Section 3.2, for any $U \in \mathcal{O}$,

(4.3.1) $Dq_\alpha(U) = D\eta(U)DG_\alpha(U)$, $\alpha = 1, \cdots, m$,

(4.3.2) $D^2\eta(U)DG_\alpha(U) = DG_\alpha(U)^T D^2\eta(U)$, $\alpha = 1, \cdots, m$.

We shall employ this entropy-entropy flux pair to weed out undesirable weak solutions. We shall call a weak solution U of (4.1.1), (4.1.3), on $[0, T)$, *admissible* if it satisfies the inequality

(4.3.3) $$\partial_t \eta(U(x, t)) + \sum_{\alpha=1}^{m} \partial_\alpha q_\alpha(U(x, t)) \le 0 ,$$

in the sense of distributions. The motivation for this notion of admissibility is provided by the observation that all the systems of balance laws from Continuum Physics encountered in Chapter II are indeed accompanied by some inequality in the form (4.3.3) which expresses, explicitly or implicitly, the Second Law of thermodynamics.

Turning to weak solutions U of the initial-value problem (4.1.1), (4.1.3), on $[0, T)$, one may define admissibility in analogy to (4.1.6):

(4.3.4)

$$\int_0^T \int_{\mathbb{R}^m} \left[\partial_t \psi \eta(U) + \sum_{\alpha=1}^m \partial_\alpha \psi q_\alpha(U) \right] dx dt + \int_{\mathbb{R}^m} \psi(x, 0) \eta(U_0(x)) dx \geq 0 ,$$

for all nonnegative Lipschitz continuous test functions ψ, with compact support in $\mathbb{R}^m \times [0, T)$. It should be noted, however, that (4.3.4) is not generally compatible with the principle of evolutionarity, as it does not necessarily guarantee that, for each $\tau \in [0, T)$, the solution $U_\tau(x, t) = U(x, t + \tau)$, with initial data $U_\tau(x, 0) = U(x, \tau)$, will also be admissible. One may thus consider replacing (4.3.4) with the stricter condition

(4.3.5)

$$\int_\tau^T \int_{\mathbb{R}^m} \left[\partial_t \psi \eta(U) + \sum_{\alpha=1}^m \partial_\alpha \psi q_\alpha(U) \right] dx dt + \int_{\mathbb{R}^m} \psi(x, \tau) \eta(U(x, \tau)) dx \geq 0 ,$$

which is to hold for all $\tau \in [0, T)$.

Clearly, all classical solutions of (4.1.1), (4.1.3) do satisfy (4.3.3), (4.3.5), and a fortiori (4.3.4), as equalities, and are therefore admissible.

Convex entropies play a very important role in the theory of hyperbolic systems of conservation laws. As we saw in Section 3.2, the presence of a uniformly convex entropy implies that the system is symmetrizable. The following proposition shows that entropy inequalities with convex entropies induce stability by strengthening the time regularity of solutions.

Theorem 4.3.1 *Assume $D^2 \eta(U)$ is positive definite, uniformly on compact subsets of $\mathcal{O}$. If U is any weak solution of (4.1.1), (4.1.3) on $[0, T)$, taking values in some convex compact subset of $\mathcal{O}$, which satisfies the entropy admissibility condition (4.3.5), then, for any $\tau \in [0, T)$ and $R > 0$,*

(4.3.6)
$$\lim_{\varepsilon \downarrow 0} \frac{1}{\varepsilon} \int_\tau^{\tau+\varepsilon} \int_{|x|<R} |U(x, t) - U(x, \tau)|^2 dx dt = 0 .$$

Proof. Fix $\tau \in [0, T)$, $R > 0$ and $\varepsilon > 0$. Write (4.3.5) for the test function $\psi(x, t) = \chi(|x|)\theta(t)$, where $\chi(r) = 1$ for $0 \leq r < R$, $\chi(r) = 1 - \varepsilon^{-1}(r - R)$ for $R \leq r < R + \varepsilon$, $\chi(r) = 0$ for $R + \varepsilon \leq r < \infty$, $\theta(t) = 1 - \varepsilon^{-1}(t - \tau)$ for $0 \leq t < \tau + \varepsilon$ and $\theta(t) = 0$ for $\tau + \varepsilon \leq t < T$. This gives

(4.3.7) $$-\frac{1}{\varepsilon} \int_\tau^{\tau+\varepsilon} \int_{|x|<R} \eta(U(x, t)) dx dt + \int_{|x|<R} \eta(U(x, \tau)) dx + O(\varepsilon) \geq 0.$$

Combining (4.3.7) with Theorem 4.1.1 yields

$$(4.3.8) \quad \limsup_{\varepsilon \downarrow 0} \frac{1}{\varepsilon} \int_\tau^{\tau+\varepsilon} \int_{|x|<R} \{\eta(U(x,t)) - \eta(U(x,\tau))$$

$$- D\eta(U(x,\tau))[U(x,t) - U(x,\tau)]\}dxdt \le 0$$

whence (4.3.6) follows. This completes the proof.

The reader should bear in mind that convexity is a relevant property of the entropy only when the system is in canonical form. In the general case, convexity of η should be replaced with the condition that the ($n \times n$ matrix-valued) derivative $DB(U, x, t)$ of the (n-vector-valued) function $B(U, x, t)$ in (3.2.2) is positive definite.

A review of the examples considered in Section 3.3 reveals that the entropy, as a function of the state vector which renders the system of balance laws in canonical form, is indeed convex in the case of the thermoelastic fluid (example (d)), the isentropic thermoelastic fluid (example (e)) and magnetohydrodynamics (example (g)). This may raise expectations that in the equations of Continuum Physics entropy is generally convex. However, as we shall see in Section 5.3, this is not always the case; hence the necessity to introduce a broader class of entropies.

Whenever the admissible solution U is of class BV_{loc}, the inequality (4.3.3) will be satisfied in the sense of measures. As already noted in Section 3.1, the normal N at any point of approximate jump discontinuity may be renormalized as $N = (v, -s)$, where $v \in \mathscr{S}^{m-1}$ is the direction and s is the speed of propagation of the shock. It then follows from Theorem 1.8.2 that (4.3.3) will hold if and only if

$$(4.3.9) \quad -s[\eta(U_+) - \eta(U_-)] + \sum_{\alpha=1}^m v_\alpha[q_\alpha(U_+) - q_\alpha(U_-)] \le 0$$

at every point of the shock set.

4.4 The Vanishing Viscosity Approach

The premise here is that a solution U of (4.1.1) is admissible provided it is the $\mu \downarrow 0$ limit of solutions U_μ to a system of conservation laws with diffusive terms:

$$(4.4.1) \quad \partial_t U(x,t) + \sum_{\alpha=1}^m \partial_\alpha G_\alpha(U(x,t)) = \mu \sum_{\alpha,\beta=1}^m \partial_\alpha[B_{\alpha\beta}(U(x,t))\partial_\beta U(x,t)] .$$

The motivation and terminology for this approach derive from Continuum Physics: As we saw in earlier chapters, the balance laws for thermoelastic materials under adiabatic conditions induce first order systems of hyperbolic type. By contrast, the balance laws for thermoviscoelastic, heat conducting materials, introduced in Section 2.6, generate systems of the second order in the spatial variables, containing diffusive terms like (4.4.1). In nature, every material has viscous response and conducts heat, to some degree. Classifying a certain material as an elastic nonconductor simply means that viscosity and heat conductivity are considered negligible, though not totally absent. Consequently, the theory of adiabatic thermoelasticity may be physically meaningful only as a limiting case of thermoviscoelasticity, with viscosity and heat conductivity tending to zero. It is this general philosophy that underlies the vanishing viscosity approach.

In writing down (4.4.1), the first question is how does one select the $n \times n$ matrices $B_{\alpha\beta}(U), \alpha, \beta = 1, \cdots, m$. When dealing with specific systems (4.1.1) of conservation laws from Continuum Physics, there are natural, physically motivated choices for (4.4.1). For example, when (4.1.1) is the system (3.3.13) of conservation laws of mass, momentum and energy for a thermoelastic fluid nonconductor (with zero body force and heat supply), which we visualize as a Newtonian fluid with constitutive relations (2.6.16), (2.6.17) having vanishingly small viscosity and heat conductivity, the natural choice would be

(4.4.2)
$$\begin{cases} \partial_t \rho + \sum_{j=1}^{m} \partial_j (\rho v_j) = 0 \ , \\[2mm] \partial_t (\rho v_i) + \sum_{j=1}^{m} \partial_j (\rho v_i v_j) + \partial_i p(\rho, s) \\[1mm] \qquad = \lambda \sum_{j=1}^{m} \partial_i \partial_j v_j + \mu \sum_{j=1}^{m} \partial_j (\partial_i v_j + \partial_j v_i) \ , \\[2mm] \partial_t \left[\rho\varepsilon(\rho, s) + \frac{1}{2}\rho|v|^2 \right] + \sum_{j=1}^{m} \partial_j \left[[\rho\varepsilon(\rho, s) + \frac{1}{2}\rho|v|^2 + p(\rho, s)]v_j \right] \\[1mm] \qquad = \lambda \sum_{i,j=1}^{m} \partial_i [v_i \partial_j v_j] + \mu \sum_{i,j=1}^{m} \partial_j [v_i (\partial_i v_j + \partial_j v_i)] + \kappa \sum_{i=1}^{m} \partial_i^2 \theta \ . \end{cases}$$

The reader should note that (4.4.2) contain three independent physical parameters, namely the bulk viscosity λ, the shear viscosity μ and the heat conductivity κ, which might all be very small albeit of different orders of magnitude. Thus, one should be prepared to consider formulations of the vanishing viscosity principle, more general than (4.4.1), containing several independent small parameters. However, such a generalization will not be pursued here.

Physics suggests the natural form for (4.4.1) in every example considered in Section 3.3, including electromagnetism, magnetohydrodynamics, etc. On the other hand, in the absense of guidelines from Physics, or for mere analytical and computational convenience, one may experiment with *artificial viscosity* added to the right-hand side of (4.1.1). For example, one may add artificial viscosity to (4.2.1) to derive the *Burgers equation with viscosity*:

(4.4.3)
$$\partial_t u(x, t) + \partial_x \left(\frac{1}{2}u^2(x, t) \right) = \mu \partial_x^2 u(x, t) \ .$$

It is clear that artificial viscosity should be selected in a way that the $B_{\alpha\beta}$ induce dissipation and thus render the initial-value problem (4.4.1) well-posed. The temptation is to use for $B_{\alpha\beta}$ matrices that would render (4.4.1) parabolic, and in particular 0 if $\alpha \neq \beta$ and I if $\alpha = \beta$, which would reduce the right-hand side of (4.4.1) to $\mu \Delta U$. The physically motivated example (4.4.2), however, sounds a warning that confining attention to the parabolic case would be ill-advised. In general, we shall have to deal with systems (4.4.1) that are of intermediate parabolic-hyperbolic type, in which case establishing the well-posedness of the initial-value problem may require considerable effort.

Assuming a vanishing viscosity mechanism has been selected, rendering the initial-value problem (4.4.1), (4.1.3) well-posed, the question arises in what sense should one expect the family $\{U_\mu\}$ of solutions to converge, as $\mu \downarrow 0$. This is of course a serious issue: If the sense of convergence is too weak, it is not clear that $\lim U_\mu$ will be a weak solution of (4.1.1), (4.1.3). On the other hand, requiring very strong convergence may raise unreasonable expectations for compactness of the family $\{U_\mu\}$. Various aspects of this problem will be discussed later in the book.

Another important task is to compare admissibility of solutions in the sense of the vanishing viscosity approach and admissibility in the sense of a designated entropy inequality (4.3.3), as discussed in Section 4.3. Reviewing the survey of continuum thermodynamics in Chapter II, reveals that whenever (4.4.1) results from actual constitutive equations and (4.3.3) derives from the Clausius-Duhem inequality, solutions of (4.1.1) obtained by the vanishing viscosity approach will automatically satisfy (4.3.3). For example, solutions of (3.3.13) obtained as the $(\lambda, \mu, \kappa) \downarrow 0$ limit of solutions of (4.4.2) will satisfy automatically the inequality (3.3.16). A necessary condition for compatibility of (4.4.1) with (4.3.3) is that the matrices $D^2 \eta B_{\alpha\beta}$ be positive semidefinite. Under the popular choice $B_{\alpha\beta} = 0$ if $\alpha \neq \beta$ and $B_{\alpha\beta} = I$ if $\alpha = \beta$, this requirement reduces to the familiar condition that η is convex. The following proposition exhibits a set of sufficient conditions for compatibility:

Theorem 4.4.1 *Consider the system of conservation laws (4.1.1) which is endowed with a designated entropy η and is equipped with artificial viscosity (4.4.1), such that*

$$(4.4.4) \qquad \sum_{\alpha,\beta=1}^{m} \xi_\alpha^T D^2 \eta(U) B_{\alpha\beta}(U) \xi_\beta \geq \sum_{\alpha=1}^{m} |\sum_{\beta=1}^{m} B_{\alpha\beta}(U) \xi_\beta|^2 \,,$$

for any $U \in \mathcal{O}$ and all $\xi_\alpha \in \mathbb{R}^n$, $\alpha = 1, \cdots, m$. Assume the initial data U_0 take values in a compact subset of $\mathcal{O}$ and also $U_0 - \bar{U} \in L^2(\mathbb{R}^m)$, with $\bar{U}$ a state where η attains its minimum on $\mathcal{O}$. Suppose that for any $\mu > 0$ the initial-value problem (4.4.1), (4.1.3) admits a solution U_μ on $[0, T)$, which takes values in a compact subset of $\mathcal{O}$, independent of μ, is locally Lipschitz on $\mathbb{R}^m \times (0, T)$, assumes the initial data in a strong sense, tends to $\bar{U}$ as $|x| \to \infty$, and satisfies $\sum_\beta B_{\alpha\beta}(U_\mu) \partial_\beta U_\mu \in L^2(\mathbb{R}^m \times (0, T))$, for $\alpha = 1 \cdots, m$. Suppose, further, that $U_{\mu_k} \to U$, a.e. on $\mathbb{R}^m \times (0, T)$, for some sequence $\{\mu_k\}$ with $\mu_k \downarrow 0$ as $k \to \infty$. Then U is a weak solution of (4.1.1), (4.1.3) on $[0, T)$ which satisfies the entropy admissibility condition (4.3.5).

Proof. It will suffice to check (4.3.4), namely the particular case of (4.3.5) for $\tau = 0$. We fix any Lipschitz continuous test function ϕ on $\mathbb{R}^m \times [0, T)$, with compact support. Multiplying (4.4.1) by ϕ and integrating by parts over $\mathbb{R}^m \times [0, T)$ yields

$$
(4.4.5) \quad \int_0^T \int_{\mathbb{R}^m} \left[\partial_t \phi U_\mu + \sum_{\alpha=1}^m \partial_\alpha \phi G_\alpha(U_\mu) \right] dx dt + \int_{\mathbb{R}^m} \phi(x, 0) U_0(x) dx
$$

$$
= \mu \int_0^T \int_{\mathbb{R}^m} \sum_{\alpha,\beta=1}^m \partial_\alpha \phi B_{\alpha\beta}(U_\mu) \partial_\beta U_\mu dx dt \ .
$$

We normalize the designated entropy-entropy flux pair so that $\eta(\overline{U}) = 0, q_\alpha(\overline{U}) = 0, \alpha = 1, \cdots, m$. We fix any nonnegative Lipschitz continuous test function ψ on $\mathbb{R}^m \times [0, T)$, with compact support, we multiply (4.4.1) by $\psi D\eta(U_\mu)$ and integrate over $\mathbb{R}^m \times [0, T)$. Using (4.3.1) and after an integration by parts, we deduce

$$
(4.4.6) \quad \int_0^T \int_{\mathbb{R}^m} \left[\partial_t \psi \eta(U_\mu) + \sum_{\alpha=1}^m \partial_\alpha \psi q_\alpha(U_\mu) \right] dx dt + \int_{\mathbb{R}^m} \psi(x, 0) \eta(U_0(x)) dx
$$

$$
= \mu \int_0^T \int_{\mathbb{R}^m} \sum_{\alpha,\beta=1}^m \partial_\alpha \psi D\eta(U_\mu) B_{\alpha\beta}(U_\mu) \partial_\beta U_\mu dx dt
$$

$$
+ \mu \int_0^T \int_{\mathbb{R}^m} \psi \sum_{\alpha,\beta=1}^m (\partial_\alpha U_\mu)^T D^2 \eta(U_\mu) B_{\alpha\beta}(U_\mu) \partial_\beta U_\mu dx dt \ .
$$

We introduce a large positive parameter r and apply (4.4.6) for the special test function $\psi = \chi(x)\theta(t)$, where

$$
(4.4.7) \quad \theta(t) = \begin{cases} 1 & 0 \le t < T - 2r^{-1} \\ r(T - t) - 1 & T - 2r^{-1} \le t < T - r^{-1} \\ 0 & T - r^{-1} \le t < T \ , \end{cases}
$$

$$
(4.4.8) \quad \chi(x) = \begin{cases} 1 & 0 \le |x| < r \\ -|x| + r + 1 & r \le |x| < r + 1 \\ 0 & r + 1 \le |x| < \infty \ . \end{cases}
$$

Then we let $r \to \infty$. Noting that $\partial_t \psi \eta(U_\mu) \le 0$, recalling the assumed behavior of U_μ as $|x| \to \infty$, and using (4.4.4), we conclude

$$
(4.4.9) \quad \mu \sum_{\alpha=1}^m \int_0^T \int_{\mathbb{R}^m} \left| \sum_{\beta=1}^m B_{\alpha\beta}(U_\mu) \partial_\beta U_\mu \right|^2 dx dt \le \int_{\mathbb{R}^m} \eta(U_0(x)) dx \ .
$$

By virtue of (4.4.9), the right-hand side of (4.4.5) tends to zero, as $\mu \downarrow 0$, and thus $U = \lim U_{\mu_k}$ satisfies (4.1.6), i.e., it is a weak solution of (4.1.1), (4.1.3).

Moreover, U also satisfies the inequality (4.3.4), since, as $\mu \downarrow 0$, the first term on the right-hand side of (4.4.6) tends to zero, by account of (4.4.9), while the second term is nonnegative, because of (4.4.4). This completes the proof.

4.5 Initial-Boundary-Value Problems

Suppose the system (4.1.1) of hyperbolic conservation laws is posed not on the whole of $\mathbb{R}^m$ but just on a proper, open subset $\mathscr{D}$ of it, say with finite perimeter. Let $\partial^*\mathscr{D}$ denote the reduced boundary of $\mathscr{D}$, on which the exterior normal field ν to $\mathscr{D}$ is defined (cf. Section 1.7). In order to formulate a well-posed problem for (4.1.1) on the cylinder $\mathscr{X} = \mathscr{D} \times (0, T)$, in addition to assigning initial data $U(x, 0) = U_0(x)$ on its base $\mathscr{D}$, we must also impose boundary conditions on its lateral boundary $\mathscr{B} = \partial^*\mathscr{D} \times (0, T)$. We shall not address here the issue of what would constitute appropriate boundary conditions for a particular system (4.1.1). We will only discuss the preliminary question, namely how boundary conditions of any kind may be realized on the "thin" set $\mathscr{B}$, within the class of weak solutions.

When the solution U is a BV function on $\mathscr{X}$, its inner trace U_- is well-defined on $\mathscr{B}$ (cf. Section 1.7). Consequently, within the BV framework, boundary conditions may be formulated in an almost classical, pointwise sense.

By contrast, when the solution U is just a bounded measurable function, there is no proper way of defining its trace on a manifold of codimension 1, like $\mathscr{B}$. On the other hand, recalling the remark following the proof of Theorem 1.2.1, one may define in a natural sense normal components of vector fields whose space-time divergence is a bounded measure on $\mathscr{X}$. The obvious example that comes to mind is the field $(G_1(U), \cdots, G_m(U), U)$ whose space-time divergence vanishes by virtue of (4.1.1). We may thus define a bounded measurable function $G_{\mathscr{B}}$ on $\mathscr{B}$ which is naturally interpreted as the trace

$$(4.5.1) \qquad\qquad G_{\mathscr{B}} = \sum_{\alpha=1}^{m} \nu_\alpha G_\alpha(U) \ .$$

Furthermore, if the solution satisfies an entropy admissibility condition (4.3.3), the space-time divergence of the field $(q_1(U), \cdots, q_m(U), \eta(U))$ will be a bounded measure on $\mathscr{X}$ and thus we may also define on $\mathscr{B}$ a bounded measurable function $q_{\mathscr{B}}$ to be interpreted as the trace

$$(4.5.2) \qquad\qquad q_{\mathscr{B}} = \sum_{\alpha=1}^{m} \nu_\alpha q_\alpha(U) \ .$$

From this standpoint, natural boundary conditions for (4.1.1) should involve special functions of U, like (4.5.1) and (4.5.2), whose trace on $\mathscr{B}$ may properly be defined.

An alternative viewpoint is to follow the vanishing viscosity approach, described in Section 4.4, one step further, by imposing appropriate, classical boundary conditions on the system (4.4.1) and then let the limiting process $\mu \downarrow 0$ pick natural boundary conditions for (4.1.1). What these boundary conditions may be is not clear *a priori*, because boundary layers generally form near $\mathscr{B}$, when μ is small. Assume (4.1.1) is endowed with an entropy-entropy flux pair (η, q) which satisfies (4.4.4). Let U_μ be a solution of (4.4.1) in the cylinder $\mathscr{X}$ satisfying Dirichlet boundary conditions $U_\mu = \hat{U}$ on $\mathscr{B}$. Suppose some sequence $\{U_{\mu_k}\}$ converges a.e. to a solution U of (4.1.1) on $\mathscr{X}$. In particular, as explained above, the traces $G_{\mathscr{B}}$ and $q_{\mathscr{B}}$ are defined on $\mathscr{B}$. One may then show, at least formally, that

$$(4.5.3) \qquad q_{\mathscr{B}} - \hat{q}_{\mathscr{B}} - D\eta(\hat{U})[G_{\mathscr{B}} - \hat{G}_{\mathscr{B}}] \le 0 \, ,$$

where

$$(4.5.4) \qquad \hat{G}_{\mathscr{B}} = \sum_{\alpha=1}^{m} \nu_\alpha G_\alpha(\hat{U}) \, , \quad \hat{q}_{\mathscr{B}} = \sum_{\alpha=1}^{m} \nu_\alpha q_\alpha(\hat{U}) \, .$$

It is conceivable that (4.5.3) should be interpreted as a boundary entropy admissibility condition on solutions of (4.1.1) on $\mathscr{X}$, with boundary conditions $U = \hat{U}$ on $\mathscr{B}$. The issue of boundary conditions is currently under active investigation.

4.6 Notes

Apparently, it was Bateman [1] who first suggested, in a little noticed paper, that (4.2.1) and (4.4.3) should be employed as models for the system of conservation laws of inviscid and viscous gases. The commonly used name of Burgers [1] was attached to these equations by Hopf [1] (cf. Section 6.9).

The issue of admissibility of weak solutions of hyperbolic systems of conservation laws stirred up a debate quite early in the development of the subject. Responding to the introduction of shock fronts in gas dynamics by Stokes [1] (cf. Section 1.10), Kelvin (in private correspondence) and Rayleigh [1] raised the objection that, in the presence of shocks, (isentropic) flows that conserve mass and momentum fail to conserve (mechanical) energy; in other words, weak solutions of the system (3.3.17) do not generally satisfy (3.3.21) as an equality. Intimidated by this criticism, Stokes [2] revised his paper, renouncing the idea of a shock. By the turn of the century, following the development of thermodynamics, weak solutions had been reinstated in Physics, albeit under conditions of admissibility, in the form of inequalities derived from the Second Law, as we saw in Section 3.3 (cf. Burton [1], Weber [1], Rayleigh [2]). The jump conditions associated with entropy inequalities were first written down by Jouguet [1], for the equations of gas dynamics. In the framework of the general theory of hyperbolic systems of conservation laws, the use of entropy inequalities to characterize admissible solutions was proposed by Kruzkov [1] and elaborated by Lax [4]. We shall return to this topic on several occasions.

The idea of regarding inviscid gases as viscous gases with vanishingly small viscosity is quite old; there are hints even in the aforementioned, seminal paper by Stokes [1]. The important contributions of Rankine [1], Hugoniot [1] and Rayleigh

[3] helped to clarify the issue. In later chapters, we shall have frequent encounters with the vanishing viscosity approach, as a method for constructing admissible solutions or as a means of identifying admissible shocks. References to relevant papers will be provided in the proper context.

An exposition of the theory of systems of intermediate parabolic-hyperbolic type is given in the monographs of S. Zheng [1] and Hsiao [2], where the reader will find an extensive list of references.

Boundary conditions for L^∞ solutions as traces on the boundary are considered in Heidrich [1] and Kan, Santos and Xin [1]. A systematic study of boundary-value problems is contained in Serre [9]. In the linear case, well-posedness is governed by the Lopatinski condition. On the issue of boundary layers and condition (4.5.3) see Bardos, Leroux and Nédélec [1], Benabdallah and Serre [1], DuBois and LeFloch [1], Gisclon [1], Gisclon and Serre [1], Grenier [1], Joseph and LeFloch [1], Otto [1] and Xin [4].

Chapter V. Entropy and the Stability of Classical Solutions

It is a tenet of Continuum Physics that the Second Law of thermodynamics is essentially a statement of stability. In the examples discussed in the previous chapters, the Second Law manifests itself in the presence of companion balance laws, to be satisfied identically, as equalities, by classical solutions, and to be imposed as inequality thermodynamic admissibility constraints on weak solutions of the systems of balance laws. A recurring theme in the exposition of the theory of hyperbolic systems of balance laws in this book will be that companion balance laws induce stability under various guises. The reader will get here a glimpse of the implications of entropy inequalities on the stability of classical solutions.

It will be shown that when the system of balance laws is endowed with a companion balance law induced by a convex entropy, the initial-value problem is locally well-posed in the context of classical solutions: Sufficiently smooth initial data generate a classical solution defined on a maximal time interval, of finite or infinite duration. Moreover, this solution is unique and depends continuously on the initial data, not only within the class of classical solutions but even within the broader class of weak solutions that satisfy the companion balance law as an inequality admissibility constraint. It will further be demonstrated that the same conclusion holds even when the entropy is convex only in the direction of a certain cone in state space, provided that the system of balance laws is equipped with special companion balance laws, called involutions, whose presence compensates for the lack of convexity in complementary directions.

From the standpoint of analytical technique, this chapter presents the aspects of the theory of quasilinear hyperbolic systems of balance laws that can be tackled by the methodology of the linear theory, namely energy estimates and Fourier analysis.

5.1 Convex Entropy and the Existence of Classical Solutions

As in Chapter IV, we consider here the initial-value problem

$$(5.1.1) \qquad \partial_t U(x,t) + \sum_{\alpha=1}^{m} \partial_\alpha G_\alpha(U(x,t)) = 0 \; ,$$

(5.1.2) $$U(x, 0) = U_0(x) , \quad x \in \mathbb{R}^m$$

for a homogeneous hyperbolic system of conservation laws in canonical form. The results may be extended to general hyperbolic systems of balance laws (3.1.1) at the expense of trivial technical complications.

It will be shown that when the system of conservation laws is equipped with a uniformly convex entropy, a classical solution of the initial-value problem exists on a maximal time interval, provided the initial data are sufficiently smooth.

In what follows, a *multi-index* r will stand for a m-tuple of nonnegative integers: $r = (r_1, \cdots, r_m)$. We put $|r| = r_1 + \cdots + r_m$ and $\partial^r = \partial_1^{r_1} \cdots \partial_m^{r_m}$. Thus ∂^r is a differential operator of order $|r|$. For the gradient operator $(\partial_1, \cdots, \partial_m)$, we shall be using the symbol ∇.

For $\ell = 0, 1, 2, \cdots$, H^ℓ will be the Sobolev space $W^{\ell,2}(\mathbb{R}^m; \mathcal{M}^{n,m})$ of $n \times m$ matrix-valued functions. The norm of H^ℓ will be denoted by $\| \cdot \|_\ell$. By the Sobolev embedding theorem, for $\ell > m/2$, H^ℓ is continuously embedded in the space of continuous functions on $\mathbb{R}^m$.

Theorem 5.1.1 *Assume the system of conservation laws* (5.1.1) *is endowed with an entropy η with $D^2\eta(U)$ positive definite, uniformly on compact subsets of $\mathcal{O}$. Suppose the initial data U_0 are continuously differentiable on $\mathbb{R}^m$, take values in some compact subset of $\mathcal{O}$ and $\nabla U_0 \in H^\ell$ for some $\ell > m/2$. Then there exists $T_\infty, 0 < T_\infty \leq \infty$, and a unique continuously differentiable function U on $\mathbb{R}^m \times [0, T_\infty)$, taking values in $\mathcal{O}$, which is a classical solution of the initial-value problem* (5.1.1), (5.1.2) *on $[0, T_\infty)$. Furthermore,*

(5.1.3) $$\nabla U(\cdot, t) \in C^0([0, T_\infty); H^\ell) .$$

The interval $[0, T_\infty)$ is maximal, in the sense that whenever $T_\infty < \infty$

(5.1.4) $$\limsup_{t \uparrow T_\infty} \|\nabla U(\cdot, t)\|_{L^\infty} = \infty$$

and/or the range of $U(\cdot, t)$ escapes from every compact subset of $\mathcal{O}$ as $t \uparrow T_\infty$.

Proof. It is lengthy and tedious. Just an outline will be presented here, so as to illustrate the role of the convex entropy. For the details the reader may consult the references cited in Section 5.4.

Fix any open subset $\mathscr{B}$ of $\mathbb{R}^m$ which contains the closure of the range of U_0 and whose closure $\bar{\mathscr{B}}$ is in turn contained in $\mathcal{O}$. With positive constants ω and T, to be fixed later, we associate the class $\mathscr{F}$ of Lipschitz continuous functions V, defined on $\mathbb{R}^m \times [0, T]$, taking values in $\mathscr{B}$, satisfying the initial condition (5.1.2) and

(5.1.5) $$\nabla V(\cdot, t) \in L^\infty([0, T]; H^\ell) , \quad \partial_t V(\cdot, t) \in L^\infty([0, T]; L^2 \cap L^\infty)$$

with

(5.1.6) $$\sup_{[0,T]} \|\nabla V(\cdot, t)\|_\ell \leq \omega ,$$

(5.1.7) $\sup\limits_{[0,T]} \|\partial_t V(\cdot,t)\|_{L^\infty} \le b\omega$, $\sup\limits_{[0,T]} \|\partial_t V(\cdot,t)\|_{L^2} \le b\omega$,

where

(5.1.8) $$b^2 = \max_{V \in \overline{\mathscr{B}}} \sum_{\alpha=1}^m |DG_\alpha(V)|^2 .$$

For ω sufficiently large, $\mathscr{F}$ is nonempty; for instance, $V(x,t) \equiv U_0(x)$ is a member of it.

By standard weak lower semicontinuity of norms, $\mathscr{F}$ is a complete metric space under the metric

(5.1.9) $$\rho(V,\overline{V}) = \sup_{[0,T]} \|V(\cdot,t) - \overline{V}(\cdot,t)\|_{L^2} .$$

Notice that, even though $V(\cdot,t)$ and $\overline{V}(\cdot,t)$ are not necessarily in L^2, $\rho(V,\overline{V}) \le 2b\omega T < \infty$, by virtue of $V(\cdot,0) - \overline{V}(\cdot,0) = 0$ and (5.1.7).

We now linearize (5.1.1) about any fixed $V \in \mathscr{F}$:

(5.1.10) $$\partial_t U(x,t) + \sum_{\alpha=1}^m DG_\alpha(V(x,t))\partial_\alpha U(x,t) = 0 .$$

The existence of a solution to (5.1.1), (5.1.2) on $[0,T]$ will be established by showing that

(a) When ω is sufficiently large and T is sufficiently small, the initial-value problem (5.1.10), (5.1.2) admits a solution $U \in \mathscr{F}$ on $[0,T]$.
(b) The aforementioned solution U is endowed with regularity (5.1.3), slightly better than (5.1.5) that mere membership in $\mathscr{F}$ would guarantee.
(c) For T sufficiently small, the map that carries $V \in \mathscr{F}$ to the solution $U \in \mathscr{F}$ of (5.1.10), (5.1.2) is a contraction in the metric (5.1.9) and thus pocesses a unique fixed point in $\mathscr{F}$, which is the desired solution of (5.1.1), (5.1.2).

In the following sketch of proof of assertion (a), above, we shall take for granted that the solution U of (5.1.10), (5.1.2), with the requisite regularity, exists and will proceed to establish that it belongs to $\mathscr{F}$. In a complete proof, one should first mollify V and the initial data, then employ the classical theory of symmetrizable linear hyperbolic systems, and finally pass to the limit.

We fix any multi-index r of order $1 \le |r| \le \ell + 1$, set $\partial^r U = U_r$, and apply ∂^r to equation (5.1.10) to get

(5.1.11) $\partial_t U_r + \sum\limits_{\alpha=1}^m DG_\alpha(V)\partial_\alpha U_r = \sum\limits_{\alpha=1}^m \{DG_\alpha(V)\partial^r\partial_\alpha U - \partial^r[DG_\alpha(V)\partial_\alpha U]\}$.

The L^2 norm of the right-hand side of (5.1.11) may be majorized with the help of Moser-type inequalities combined with (5.1.6):

(5.1.12) $$\left\| \sum_{\alpha=1}^m \{DG_\alpha(V)\partial^r\partial_\alpha U - \partial^r[DG_\alpha(V)\partial_\alpha U]\} \right\|_{L^2}$$

$$\le c\|\nabla V\|_{L^\infty}\|\nabla U\|_\ell + c\|\nabla U\|_{L^\infty}\|\nabla V\|_\ell \le 2ac\omega\|\nabla U\|_\ell .$$

Here and below c will stand for a generic positive constant which may depend on bounds of derivatives of the G_α over $\mathscr{B}$ but is independent of ω and T.

Let us now multiply (5.1.11), from the left, by $2U_r^T D^2\eta(V)$, sum over all multi-indices r with $1 \leq |r| \leq \ell + 1$ and integrate the resulting equation over $\mathbb{R}^m \times [0, t]$. Note that

$$(5.1.13) \qquad 2U_r^T D^2\eta(V)\partial_t U_r = \partial_t[U_r^T D^2\eta(V)U_r] - 2U_r^T\partial_t D^2\eta(V)U_r .$$

Moreover, by virtue of (4.3.2),

$$(5.1.14) \qquad 2U_r^T D^2\eta(V)DG_\alpha(V)\partial_\alpha U_r = \partial_\alpha[U_r^T D^2\eta(V)DG_\alpha(V)U_r]$$
$$- 2U_r^T\partial_\alpha[D^2\eta(V)DG_\alpha(V)]U_r .$$

Recall that $D^2\eta(V)$ is positive definite, uniformly on compact sets, so that

$$(5.1.15) \qquad U_r^T D^2\eta(V)U_r \geq \delta|U_r|^2 , \qquad V \in \mathscr{B} ,$$

for some $\delta > 0$. Therefore, combining the above we end up with an estimate

$$(5.1.16) \qquad \|\nabla U(\cdot, t)\|_\ell^2 \leq c\|\nabla U_0(\cdot)\|_\ell^2 + c\omega \int_0^t \|\nabla U(\cdot, \tau)\|_\ell^2 d\tau$$

whence, by Gronwall's inequality,

$$(5.1.17) \qquad \sup_{[0,T]} \|\nabla U(\cdot, t)\|_\ell^2 \leq ce^{c\omega T}\|\nabla U_0(\cdot)\|_\ell^2 .$$

From (5.1.17) follows that if ω is selected sufficiently large and T is sufficiently small, $\sup_{[0,T]} \|\nabla U(\cdot, t)\|_\ell \leq \omega$. Then (5.1.10) implies $\sup_{[0,T]} \|\partial_t U(\cdot, t)\|_{L^\infty} \leq b\omega$, $\sup_{[0,T]} \|\partial_t U(\cdot, T)\|_{L^2} \leq b\omega$, with b given by (5.1.8). Finally, for T sufficiently small, U will take values in $\mathscr{B}$ on $\mathbb{R}^m \times [0, T]$. Thus $U \in \mathscr{F}$.

Assertion (b), namely that U is regular as in (5.1.3), may be established by carefully monitoring the mode of convergence of solutions of (5.1.10), with V mollified, to those with V in $\mathscr{F}$. This, less central, issue will not be addressed here.

Turning now to assertion (c), let us fix V and $\overline{V}$ in $\mathscr{F}$ which induce solutions U and $\overline{U}$ of (5.1.10), (5.1.2), also in $\mathscr{F}$. Thus

$$(5.1.18) \quad \partial_t[U-\overline{U}]+\sum_{\alpha=1}^m DG_\alpha(V)\partial_\alpha[U-\overline{U}] = -\sum_{\alpha=1}^m[DG_\alpha(V)-DG_\alpha(\overline{V})]\partial_\alpha\overline{U} .$$

Multiply (5.1.18), from the left, by $2(U - \overline{U})^T D^2\eta(V)$ and integrate the resulting equation over $\mathbb{R}^m \times [0, t]$, $0 \leq t \leq T$. Notice that

$$(5.1.19) \qquad 2(U - \overline{U})^T D^2\eta(V)\partial_t(U - \overline{U}) = \partial_t[(U - \overline{U})^T D^2\eta(V)(U - \overline{U})]$$
$$- 2(U - \overline{U})^T\partial_t D^2\eta(V)(U - \overline{U})$$

and also, by virtue of (4.3.2),

(5.1.20)
$$2(U - \overline{U})^T D^2\eta(V)DG_\alpha(V)\partial_\alpha(U - \overline{U}) = \partial_\alpha[(U - \overline{U})^T D^2\eta(V)DG_\alpha(V)(U - \overline{U})]$$
$$- 2(U - \overline{U})^T \partial_\alpha[D^2\eta(V)DG_\alpha(V)](U - \overline{U}).$$

Since $D^2\eta(V)$ is positive definite,

(5.1.21)
$$(U - \overline{U})^T D^2\eta(V)(U - \overline{U}) \geq \delta|U - \overline{U}|^2.$$

Therefore, combining the above with (5.1.6), (5.1.7) and the Sobolev embedding theorem we arrive at the estimate

(5.1.22)
$$\|(U - \overline{U})(\cdot, t)\|_{L^2}^2 \leq c\omega \int_0^t \|(U - \overline{U})(\cdot, \tau)\|_{L^2}^2 d\tau$$
$$+ c\omega \int_0^t \|(V - \overline{V})(\cdot, \tau)\|_{L^2}\|(U - \overline{U})(\cdot, \tau)\|_{L^2} d\tau .$$

Using (5.1.9) and Gronwall's inequality, we infer from (5.1.22) that

(5.1.23)
$$\rho(U, \overline{U}) \leq c\omega T e^{c\omega T} \rho(V, \overline{V}) .$$

Consequently, for T sufficiently small, the map that carries V in $\mathscr{F}$ to the solution U of (5.1.10), (5.1.2) is a contraction on $\mathscr{F}$ and thus possesses a unique fixed point U which is the unique solution of (5.1.1), (5.1.2) on $[0, T]$, in the function class $\mathscr{F}$.

Since the restriction $U(\cdot, T)$ of the constructed solution to $t = T$ belongs to the same function class as $U_0(\cdot)$, we may repeat the above construction and extend U to a larger time interval $[0, T']$. Continuing the process, we end up with a solution U defined on a maximal interval $[0, T_\infty)$ with $T_\infty \leq \infty$. Furthermore, if $T_\infty < \infty$ then the range of $U(\cdot, t)$ must escape from every compact subset of $\mathscr{O}$ as $t \uparrow T_\infty$ and/or

(5.1.24)
$$\|\nabla U(\cdot, t)\|_\ell \to \infty , \quad \text{as } t \uparrow T_\infty .$$

In order to see the implications of (5.1.24), we retrace the steps that led to (5.1.16). We use again (5.1.12), (5.1.13), (5.1.14), and (5.1.15), setting $V \equiv U$, but we no longer majorize $\|\nabla U\|_{L^\infty}$ by $a\omega$. Thus, in the place of (5.1.16) we now get

(5.1.25) $$\|\nabla U(\cdot, t)\|_\ell^2 \leq c\|\nabla U_0(\cdot)\|_\ell^2 + c \int_0^t \|\nabla U(\cdot, \tau)\|_{L^\infty}\|\nabla U(\cdot, \tau)\|_\ell^2 d\tau .$$

Gronwall's inequality then implies that (5.1.24) cannot occur unless (5.1.4) does. This completes the sketch of the proof.

It should be noted that the possibility of a finite life span of the solution, raised by Theorem 5.1.1, is not an artifact of the proof. As we shall see later, in consequence of nonlinearity classical solutions generally break down in a finite time and the "catastrophe" (5.1.4) triggers the development of a shock front.

5.2 Convex Entropy and the Stability of Classical Solutions

The aim here is to show that the presence of a convex entropy guarantees that classical solutions of the initial-value problem depend continuously on the initial data, even within the broader class of admissible bounded weak solutions.

Theorem 5.2.1 *Assume the system of conservation laws* (5.1.1) *is endowed with an entropy* η *with* $D^2\eta(U)$ *positive definite, uniformly on compact subsets of* $\mathcal{O}$. *Suppose* $\overline{U}$ *is a classical solution of* (5.1.1) *on* $[0, T)$, *taking values in a convex compact subset* $\mathscr{D}$ *of* $\mathcal{O}$, *with initial data* $\overline{U}_0$. *Let* U *be any admissible weak solution of* (5.1.1) *on* $[0, T)$, *taking values in* $\mathscr{D}$, *with initial data* U_0. *Then*

$$(5.2.1) \qquad \int_{|x|<R} |U(x,t) - \overline{U}(x,t)|^2 dx \le ae^{bt} \int_{|x|<R+st} |U_0(x) - \overline{U}_0(x)|^2 dx$$

holds for any $R > 0$ *and* $t \in [0, T)$, *with positive constants* s, a, *depending solely on* $\mathscr{D}$, *and* b *that also depends on the Lipschitz constant of* $\overline{U}$. *In particular,* $\overline{U}$ *is the unique admissible weak solution of* (5.1.1) *with initial data* $\overline{U}_0$ *and values in* $\mathscr{D}$.

Proof. On $\mathscr{D} \times \mathscr{D}$ we define the functions

$$(5.2.2) \qquad h(U, \overline{U}) = \eta(U) - \eta(\overline{U}) - D\eta(\overline{U})[U - \overline{U}] \ ,$$

$$(5.2.3) \qquad f_\alpha(U, \overline{U}) = q_\alpha(U) - q_\alpha(\overline{U}) - D\eta(\overline{U})[G_\alpha(U) - G_\alpha(\overline{U})] \ ,$$

$$(5.2.4) \qquad Z_\alpha(U, \overline{U}) = G_\alpha(U) - G_\alpha(\overline{U}) - DG_\alpha(\overline{U})[U - \overline{U}] \ ,$$

all of quadratic order in $U - \overline{U}$ (recall (4.3.1)). Consequently, since $D^2\eta(U)$ is positive definite, uniformly on $\mathscr{D}$, there is a positive constant s such that

$$(5.2.5) \qquad \left[\sum_{\alpha=1}^{m} |f_\alpha(U, \overline{U})|^2 \right]^{1/2} \le sh(U, \overline{U}) \ .$$

Let us fix any nonnegative, Lipschitz continuous test function ψ on $\mathbb{R}^m \times [0, T)$, with compact support, and evaluate h, f_α and Z_α along the two solutions $U(x, t), \overline{U}(x, t)$. Recalling that U, as an admissible weak solution, must satisfy inequality (4.3.4), while $\overline{U}$, being a classical solution, will identically satisfy (4.3.4) as an equality, we deduce

$(5.2.6)$

$$\int_0^T \int_{\mathbb{R}^m} \left\{ \partial_t \psi h(U, \overline{U}) + \sum_{\alpha=1}^m \partial_\alpha \psi f_\alpha(U, \overline{U}) \right\} dx dt + \int_{\mathbb{R}^m} \psi(x, 0) h(U_0(x), \overline{U}_0(x)) dx$$

$$\ge - \int_0^T \int_{\mathbb{R}^m} \left\{ \partial_t \psi D\eta(\overline{U})[U - \overline{U}] + \sum_{\alpha=1}^m \partial_\alpha \psi D\eta(\overline{U})[G_\alpha(U) - G_\alpha(\overline{U})] \right\} dx dt$$

$$- \int_{\mathbb{R}^m} \psi(x, 0) D\eta(\overline{U}_0(x))[U_0(x) - \overline{U}_0(x)] dx \ .$$

Next we write (4.1.6) for both solutions U and $\overline{U}$, using components of the Lipschitz continuous vector field $\psi D\eta(\overline{U})$ as test function ϕ, to get

(5.2.7)
$$\int_0^T \int_{\mathbb{R}^m} \left\{ \partial_t [\psi D\eta(\overline{U})][U - \overline{U}] + \sum_{\alpha=1}^m \partial_\alpha [\psi D\eta(\overline{U})][G_\alpha(U) - G_\alpha(\overline{U})] \right\} dxdt$$

$$+ \int_{\mathbb{R}^m} \psi(x,0) D\eta(\overline{U}_0(x))[U_0(x) - \overline{U}_0(x)]dx = 0 .$$

Since $\overline{U}$ is a classical solution of (5.1.1) and by virtue of (4.3.2),

(5.2.8)
$$\partial_t D\eta(\overline{U}) = \partial_t \overline{U}^T D^2\eta(\overline{U}) = -\sum_{\alpha=1}^m \partial_\alpha \overline{U}^T DG_\alpha(\overline{U}) D^2\eta(\overline{U})$$

$$= -\sum_{\alpha=1}^m \partial_\alpha \overline{U}^T D^2\eta(\overline{U}) DG_\alpha(\overline{U})$$

so that, recalling (5.2.4),

(5.2.9)
$$\partial_t D\eta(\overline{U})[U - \overline{U}] + \sum_{\alpha=1}^m \partial_\alpha D\eta(\overline{U})[G_\alpha(U) - G_\alpha(\overline{U})]$$

$$= \sum_{\alpha=1}^m \partial_\alpha \overline{U}^T D^2\eta(\overline{U}) Z_\alpha(U, \overline{U}) .$$

Combining (5.2.6), (5.2.7) and (5.2.9) yields

(5.2.10)
$$\int_0^T \int_{\mathbb{R}^m} \left[\partial_t \psi h(U, \overline{U}) + \sum_{\alpha=1}^m \partial_\alpha \psi f_\alpha(U, \overline{U}) \right] dxdt + \int_{\mathbb{R}^m} \psi(x,0) h(U_0(x), \overline{U}_0(x)) dx$$

$$\geq \int_0^T \int_{\mathbb{R}^m} \psi \sum_{\alpha=1}^m \partial_\alpha \overline{U}^T D^2\eta(\overline{U}) Z_\alpha(U, \overline{U}) dxdt .$$

We now fix $R > 0$, $t \in (0, T)$ and ε positive small, and write (5.2.10) for the test function $\psi(x, \tau) = \chi(x, \tau)\theta(\tau)$, with

(5.2.11)
$$\theta(\tau) = \begin{cases} 1 & 0 \leq \tau < t \\ \dfrac{1}{\varepsilon}(t - \tau) + 1 & t \leq \tau < t + \varepsilon \\ 0 & t + \varepsilon \leq \tau < T \end{cases}$$

(5.2.12)

$$\chi(x,\tau) = \begin{cases} 1 & \begin{aligned} & 0 \leq \tau < T \,, \\ & 0 \leq |x| < R + s(t-\tau) \end{aligned} \\[2ex] \dfrac{1}{\varepsilon}[R + s(t-\tau) - |x|] + 1 & \begin{aligned} & 0 \leq \tau < T \,, \\ & R + s(t-\tau) \leq |x| < R + s(t-\tau) + \varepsilon \end{aligned} \\[2ex] 0 & \begin{aligned} & 0 \leq \tau < T \,, \\ & R + s(t-\tau) + \varepsilon \leq |x| < \infty \end{aligned} \end{cases}$$

where s is the constant appearing in (5.2.5). The calculation gives

(5.2.13)
$$\frac{1}{\varepsilon}\int_t^{t+\varepsilon}\int_{|x|<R} h(U(x,\tau),\overline{U}(x,\tau))dxd\tau \leq \int_{|x|<R+st} h(U_0(x),\overline{U}_0(x))dx$$
$$-\frac{1}{\varepsilon}\int_0^t\int_{R+s(t-\tau)<|x|<R+s(t-\tau)+\varepsilon} \left\{ sh(U,\overline{U}) + \sum_{\alpha=1}^m \frac{x_\alpha}{|x|} f_\alpha(U,\overline{U}) \right\} dxd\tau$$
$$-\int_0^t\int_{|x|<R+s(t-\tau)} \sum_{\alpha=1}^m \partial_\alpha \overline{U}^T D^2\eta(\overline{U})Z_\alpha(U,\overline{U})dxd\tau + O(\varepsilon) \,.$$

We let $\varepsilon \downarrow 0$. The second integral on the right-hand side of (5.2.13) is nonnegative by account of (5.2.5). Using that $\overline{U}$ is Lipschitz continuous, η is convex, and Theorem 4.1.1, we deduce

(5.2.14)
$$\int_{|x|<R} h(U(x,t),\overline{U}(x,t))dx \leq \int_{|x|<R+st} h(U_0(x),\overline{U}_0(x))dx$$
$$-\int_0^t\int_{|x|<R+s(t-\tau)} \sum_{\alpha=1}^m \partial_\alpha \overline{U}^T D^2\eta(\overline{U})Z_\alpha(U,\overline{U})dxd\tau \,.$$

As noted above, $h(U,\overline{U})$ and the $Z_\alpha(U,\overline{U})$ are of quadratic order in $U - \overline{U}$ and, in addition, $h(U,\overline{U})$ is positive definite, due to the convexity of η. Therefore, (5.2.14) in conjunction with Gronwall's inequality imply (5.2.1). Notice that a and s depend solely on $\mathscr{D}$ while b depends also on the Lipschitz constant of $\overline{U}$. This completes the proof.

It is remarkable that a single entropy inequality, with convex entropy, manages to weed out all but one solution of the initial-value problem, so long as a classical solution exists. As we shall see, however, when no classical solution exists, just one entropy inequality is no longer generally sufficient to single out any particular weak solution. The issue of uniqueness of weak solutions is knotty and will be a major issue for discussion in subsequent chapters.

5.3 Partially Convex Entropies and Involutions

The previous two sections have illustrated the beneficent role of convex entropies. Nevertheless, the entropy associated with systems of balance laws in Continuum

Physics is not always convex. An illustrative case is example (c) of Section 3.3, namely isentropic, adiabatic thermoelasticity, with system of balance laws (3.3.10), and entropy function $\eta = \varepsilon(F) + \frac{1}{2}|v|^2$ which would be convex if ε were convex. Even though ε may indeed be convex on certain regions of state space, global convexity of it is incompatible with experience and, in particular, would violate the principle of material frame indifference, which requires $\varepsilon(OF) = \varepsilon(F)$ for all proper orthogonal matrices O (cf. (2.5.5)). Of course, ε may still be convex in certain directions. In fact, recall that the system (3.3.10) is hyperbolic when ε is rank-one convex (cf. (3.3.7)). In contrast to convexity, rank-one convexity does not violate any laws of physics and is supported by measurements in materials like rubber.

It will be shown here that the failure of ε to be convex in certain directions is compensated by the property that solutions of the system (3.3.10) satisfy identically the additional conservation laws

$$(5.3.1) \qquad \partial_\beta F_{i\alpha} - \partial_\alpha F_{i\beta} = 0 \; , \quad i = 1, \cdots, m \; ; \quad \alpha, \beta = 1, \cdots, m \; .$$

The extra conservation laws (5.3.1) also apply to the system (3.3.4) of balance laws of adiabatic thermoelasticity (example (b) of Section 3.3).

Systems exhibiting such behavior arise quite commonly in Continuum Physics. For example, solutions of Maxwell's equations (3.3.22), with current $J \equiv 0$, satisfy identically the additional conservation laws

$$(5.3.2) \qquad \operatorname{div} B = 0 \; , \quad \operatorname{div} D = 0 \; ,$$

whenever the initial data do so. Similarly, solutions to Lundquist's equations (3.3.28) of magnetohydrodynamics satisfy

$$(5.3.3) \qquad \operatorname{div} H = 0 \; .$$

Similar cases are encountered in the general theory of relativity; see references in Section 5.4. In view of the above, it is warranted to investigate systems of balance laws with this special structure in a general framework:

Definition 5.3.1 The first order system

$$(5.3.4) \qquad \sum_{\alpha=1}^{m} M_\alpha \partial_\alpha U = 0$$

of differential equations, with M_α constant $k \times n$ matrices, $\alpha = 1, \cdots, m$, is called an *involution* of the system (5.1.1) of conservation laws if any (generally weak) solution of the initial-value problem (5.1.1), (5.1.2) satisfies (5.3.4) identically, whenever the initial data do so.

Thus (5.3.1) is an involution of (3.3.10) as well as of (3.3.4); (5.3.2) is an involution of (3.3.22); and (5.3.3) is an involution of (3.3.28). The reader should exercise caution to distinguish involutions (5.3.4) which must be satisfied by all, even weak, solutions of (5.1.1) from conditions like (5.3.4) that need only hold for classical solutions. An example of the latter case is the vanishing of vorticity in smooth *irrotational flows* of Newtonian fluids: A standard calculation, that may be found in every text on hydrodynamics, shows that in any classical solution on $\mathbb{R}^3 \times (0, T)$ of the Euler equations (3.3.17) with body force derived from a potential, $b = \text{grad}\,\phi$, when the *vorticity* curl v vanishes at $t = 0$ then it vanishes everywhere in space-time:

$$(5.3.5) \qquad\qquad\qquad \text{curl}\, v = 0 .$$

However, (5.3.5) is not an involution of (3.3.17) because it does not necessarily hold for weak solutions.

A sufficient condition for (5.3.4) to be an involution of (5.1.1) is that

$$(5.3.6) \qquad M_\alpha G_\beta(U) + M_\beta G_\alpha(U) = 0 , \quad \alpha, \beta = 1, \cdots, m ,$$

for any $U \in \mathcal{O}$. We shall focus our investigation here to this special case which covers, in particular, the prototypical examples (5.3.1), (5.3.2) and (5.3.3). Alternative, more general, sufficient conditions are exhibited in the references cited in Section 5.4.

With the involution (5.3.4) and any $v \in \mathscr{S}^{m-1}$ we associate the $k \times n$ matrix

$$(5.3.7) \qquad\qquad\qquad N(v) = \sum_{\alpha=1}^{m} v_\alpha M_\alpha .$$

Recalling the notation (4.1.2), it follows from (5.3.6) that

$$(5.3.8) \qquad\qquad\qquad N(v)\Lambda(v; U) = 0$$

so, in particular, any eigenvector $R(v; U)$ of $\Lambda(v; U)$ with nonzero eigenvalue $\lambda(v; U)$ must lie in the kernel of $N(v)$. We make the simplifying assumption, valid in the prototypical examples, that for any $v \in \mathscr{S}^{m-1}$ the rank of $N(v)$ equals the dimension of the kernel of $\Lambda(v; U)$.

The premise is that in the presence of involutions the entropy need only be convex in the direction of a cone defined by

Definition 5.3.2 The *involution cone* in $\mathbb{R}^n$ of the involution (5.3.4) is

$$(5.3.9) \qquad\qquad\qquad \mathscr{C} = \bigcup_{v \in \mathscr{S}^{m-1}} \ker N(v)$$

with $N(v)$ given by (5.3.7).

In what follows, for $p > 0$, functions on $\mathbb{R}^m$ will be called $2p$-periodic when they are periodic, with period $2p$, in each variable x_α, $\alpha = 1, \cdots, m$; and $\mathscr{H}$ will denote the standard hypercube $\{x \in \mathbb{R}^m : |x_\alpha| < p, \alpha = 1, \cdots, m\}$ with edge length $2p$.

Lemma 5.3.1 *Assume the system of conservation laws (5.1.1) is endowed with an involution (5.3.4). Fix $\hat{U} \in \mathcal{O}$ and consider the differential operator*

$$(5.3.10) \qquad \mathscr{L} = \sum_{\beta=1}^{m} DG_\beta(\hat{U}) \partial_\beta .$$

A 2p-periodic L_{loc}^2 function S from $\mathbb{R}^m$ to $\mathbb{R}^n$ with

$$(5.3.11) \qquad \int_{\mathscr{H}} S(x)dx = 0$$

satisfies

$$(5.3.12) \qquad \sum_{\alpha=1}^{m} M_\alpha \partial_\alpha S = 0$$

in the sense of distributions if and only if there is a 2p-periodic $W_{\mathrm{loc}}^{1,2}$ function χ from $\mathbb{R}^m$ to $\mathbb{R}^n$ such that

$$(5.3.13) \qquad S = \mathscr{L}\chi ,$$

$$(5.3.14) \qquad \|\chi\|_{L^2(\mathscr{H})} \le ap\|S\|_{W^{-1,2}(\mathscr{H})} , \quad \|\chi\|_{W^{1,2}(\mathscr{H})} \le ap\|S\|_{L^2(\mathscr{H})} ,$$

where a is independent of p and S.

Proof. That (5.3.13) implies (5.3.12) follows immediately from (5.3.10) and (5.3.6).

To show necessity, expand S in Fourier series

$$(5.3.15) \qquad S(x) = \sum_{\xi} \exp\left\{\frac{i\pi}{p}(\xi \cdot x)\right\} X(\xi) ,$$

where the summation runs over all vectors $\xi = (\xi_1, \cdots, \xi_m)$ in $\mathbb{R}^m$ with integer components. By (5.3.11), $X(0) = 0$. Note that

$$(5.3.16)$$
$$\|S\|_{L^2(\mathscr{H})}^2 = (2p)^m \sum_{\xi} |X(\xi)|^2 , \quad \|S\|_{W^{-1,2}(\mathscr{H})}^2 = (2p)^m \sum_{\xi} (1 + |\xi|^2)^{-1} |X(\xi)|^2 .$$

By virtue of (5.3.12) and (5.3.7), for $\xi \ne 0$,

$$(5.3.17) \qquad |\xi| N(|\xi|^{-1}\xi) X(\xi) = \left(\sum_{\alpha=1}^{m} \xi_\alpha M_\alpha \right) X(\xi) = 0$$

so that $X(\xi)$ lies in the kernel of $N(|\xi|^{-1}\xi)$. By assumption, the rank of $N(|\xi|^{-1}\xi)$ equals the dimension of the kernel of $\Lambda(|\xi|^{-1}\xi; \hat{U})$. Therefore, for any $\xi \ne 0$ we may determine $Y(\xi)$ in $\mathbb{R}^n$ such that

$$(5.3.18) \qquad \Lambda(|\xi|^{-1}\xi; \hat{U}) Y(\xi) = -\frac{ip}{\pi}\frac{1}{|\xi|} X(\xi) ,$$

$$(5.3.19) \qquad\qquad |Y(\xi)| \le \frac{ap}{2|\xi|}|X(\xi)| \; ,$$

for some constant a independent of p and S. It follows that the Fourier series

$$(5.3.20) \qquad\qquad \chi(x) = \sum_{\xi \ne 0} \exp\left\{ \frac{i\pi}{p}(\xi \cdot x) \right\} Y(\xi)$$

defines a $2p$-periodic $W_{\mathrm{loc}}^{1,2}$ function χ from $\mathbb{R}^m$ to $\mathbb{R}^n$, which satisfies (5.3.13), by virtue of (5.3.10), (4.1.2) and (5.3.18), as well as (5.3.14), by account of (5.3.16), (5.3.19) and

(5.3.21)

$$\|\chi\|_{L^2(\mathscr{H})}^2 = (2p)^m \sum_{\xi} |Y(\xi)|^2 \; , \quad \|\chi\|_{W^{1,2}(\mathscr{H})}^2 = (2p)^m \sum_{\xi}(1 + |\xi|^2)|Y(\xi)|^2 \; .$$

This completes the proof.

Lemma 5.3.2 *Assume the system of conservation laws (5.1.1) is endowed with an involution (5.3.4), with involution cone $\mathscr{C}$. Suppose P is a symmetric $n \times n$ matrix-valued L^∞ function on $\mathbb{R}^m$ which is uniformly positive definite in the direction of $\mathscr{C}$, i.e.,*

$$(5.3.22) \qquad\qquad Z^T P(x) Z \ge \mu |Z|^2 \; , \quad Z \in \mathscr{C} \; , \quad x \in \mathbb{R}^m \; ,$$

for some $\mu > 0$, and its local oscillation is less than μ, i.e.,

$$(5.3.23) \qquad\qquad \limsup_{\varepsilon \downarrow 0} \; \sup_{|y-x|<\varepsilon} \; |P(y) - P(x)| < \mu - 2\delta \; ,$$

for some $\delta > 0$. It W is any L^2 function from $\mathbb{R}^m$ to $\mathbb{R}^n$ which is compactly supported in the hypercube $\mathscr{H}$ and satisfies

$$(5.3.24) \qquad\qquad \sum_{\alpha=1}^{m} M_\alpha \partial_\alpha W = Q$$

in the sense of distributions, for some Q in $W^{-1,2}$, then

$$(5.3.25) \qquad \int_{\mathbb{R}^m} W(x)^T P(x) W(x) dx \ge \delta \|W\|_{L^2}^2 - b\|W\|_{W^{-1,2}}^2 - b\|Q\|_{W^{-1,2}}^2 \; ,$$

where b does not depend on W or Q.

Proof. Expand W and Q in Fourier series over $\mathscr{H}$:

$$(5.3.26) \qquad\qquad W(x) = \sum_{\xi} \exp\left\{ \frac{i\pi}{p}(\xi \cdot x) \right\} X(\xi) \; , \quad x \in \mathscr{H} \; ,$$

$$(5.3.27) \qquad\qquad Q(x) = \sum_{\xi} \exp\left\{ \frac{i\pi}{p}(\xi \cdot x) \right\} Y(\xi) \; , \quad x \in \mathscr{H} \; .$$

Note that $Y(0) = 0$ and

(5.3.28)
$$\|W\|_{L^2}^2 = (2p)^m \sum_\xi |X(\xi)|^2 \ , \quad \|W\|_{W^{-1,2}}^2 = (2p)^m \sum_\xi (1 + |\xi|^2)^{-1} |X(\xi)|^2 \ ,$$

(5.3.29)
$$\|Q\|_{W^{-1,2}}^2 = (2p)^m \sum_\xi (1 + |\xi|^2)^{-1} |Y(\xi)|^2 \ .$$

Furthermore, by virtue of (5.3.24) and (5.3.7), for any $\xi \neq 0$,

(5.3.30)
$$|\xi| N(|\xi|^{-1}\xi) X(\xi) = \left(\sum_{\alpha=1}^m \xi_\alpha M_\alpha \right) X(\xi) = Y(\xi) \ .$$

We may thus split $X(\xi)$ into

(5.3.31)
$$X(\xi) = \Phi(\xi) + \Psi(\xi) \ , \quad \xi \neq 0 \ ,$$

where $\Phi(\xi)$ lies in the kernel of $N(|\xi|^{-1}\xi)$ while $\Psi(\xi)$ satisfies

(5.3.32)
$$|\Psi(\xi)| \leq \frac{c}{2|\xi|} |Y(\xi)| \ .$$

Here and below, c will stand for a generic constant, independent of W. In turn, (5.3.31) induces a splitting of W into

(5.3.33)
$$W(x) = X(0) + S(x) + T(x) \ , \quad x \in \mathcal{H} \ ,$$

with
(5.3.34)
$$S(x) = \sum_{\xi \neq 0} \exp \left\{ \frac{i\pi}{p} (\xi \cdot x) \right\} \Phi(\xi) \ , \quad T(x) = \sum_{\xi \neq 0} \exp \left\{ \frac{i\pi}{p} (\xi \cdot x) \right\} \Psi(\xi) \ .$$

Notice that S satisfies (5.3.11) and (5.3.12) while

(5.3.35)
$$\|T\|_{L^2(\mathcal{H})} \leq c \|Q\|_{W^{-1,2}} \ .$$

We now cover $\overline{\mathcal{H}}$ by the union of a finite collection $\mathcal{H}_1, \cdots, \mathcal{H}_J$ of open hypercubes, centered at points $y^1, \cdots, y^J$, such that

(5.3.36)
$$\sup_{x \in \mathcal{H}_I} |P(x) - P(y^I)| \leq \mu - 2\delta \ , \quad I = 1, \cdots, J \ .$$

With the above covering we associate a partition of unity induced by C^∞ functions $\theta_1, \cdots, \theta_J$ on $\mathbb{R}^m$ such that $spt\ \theta_I \subset \mathcal{H}_I \cap \mathcal{H}$, $I = 1, \cdots, J$, and

(5.3.37)
$$\sum_{I=1}^J \theta_I^2(x) = 1 \ , \quad x \in spt\ W \ .$$

Then

$$(5.3.38) \qquad \int_{\mathbb{R}^m} W(x)^T P(x) W(x) dx = \sum_{I=1}^{J} \int_{\mathscr{H}_I} \theta_I^2(x) W(x)^T P(x) W(x) dx$$

$$= \sum_{I=1}^{J} \int_{\mathscr{H}_I} \theta_I^2(x) W(x)^T P(y^I) W(x) dx$$

$$+ \sum_{I=1}^{J} \int_{\mathscr{H}_I} \theta_I^2(x) W(x)^T [P(x) - P(y^I)] W(x) dx .$$

By virtue of (5.3.36) and (5.3.37),

$$(5.3.39) \qquad \sum_{I=1}^{J} \int_{\mathscr{H}_I} \theta_I^2(x) W(x)^T [P(x) - P(y^I)] W(x) dx \geq -(\mu - 2\delta) \|W\|_{L^2}^2 .$$

Recalling Lemma 5.3.1, we construct the function χ which induces S through (5.3.13). For each $I = 1, \cdots, J$, we split $\theta_I W$ into

$$(5.3.40) \qquad \theta_I W = S_I + T_I ,$$

where

$$(5.3.41) \qquad S_I = \mathscr{L}(\theta_I \chi) ,$$

$$(5.3.42) \qquad T_I(x) = \theta_I(x) X(0) + \theta_I(x) V(x) - \left[\sum_{\beta=1}^{m} \partial_\beta \theta_I(x) DG_\beta(\hat{U}) \right] \chi(x) .$$

Clearly, S_I is square integrable, has compact support in $\mathscr{H}_I$, and

$$(5.3.43) \qquad \int_{\mathscr{H}_I} S_I(x) dx = 0 .$$

Furthermore, by Lemma 5.3.1,

$$(5.3.44) \qquad \sum_{\alpha=1}^{m} M_\alpha \partial_\alpha S_I = 0 .$$

Consequently, S_I may be expanded in Fourier series over $\mathscr{H}_I$,

$$(5.3.45) \qquad S_I(x) = \sum_{\xi} \exp\left\{ \frac{i\pi}{p^I} [\xi \cdot (x - y^I)] \right\} Z(\xi) , \qquad x \in \mathscr{H}_I ,$$

with $Z(0) = 0$ and

$$(5.3.46) \qquad |\xi| N(|\xi|^{-1}\xi) Z(\xi) = \left(\sum_{\alpha=1}^{m} \xi_\alpha M_\alpha \right) Z(\xi) = 0 .$$

Thus $Z(\xi)$ lies in the complexification of $\mathscr{C}$ and so, by Parseval's relation and (5.3.22),

$$(5.3.47) \qquad \int_{\mathscr{K}_I} S_I(x)^T P(y^I) S_I(x) dx = (2p^I)^m \sum_\xi Z(\xi)^* P(y^I) Z(\xi)$$

$$\geq \mu (2p^I)^m \sum_\xi |Z(\xi)|^2 = \mu \int_{\mathscr{K}_I} |S_I(x)|^2 dx \ .$$

Moreover, from (5.3.42), (5.3.28), (5.3.35), (5.3.14), and (5.3.33) we infer

$$(5.3.48) \qquad \int_{\mathscr{K}_I} |T_I(x)|^2 dx \leq c\|W\|_{W^{-1,2}}^2 + c\|Q\|_{W^{-1,2}}^2 \ .$$

We now return to (5.3.38). From (5.3.40), (5.3.47), and (5.3.48) it follows that

$$(5.3.49) \qquad \int_{\mathscr{K}_I} \theta_I^2(x) W(x)^T P(y^I) W(x) dx$$

$$\geq \left(1 - \frac{\delta}{2\mu}\right) \int_{\mathscr{K}_I} S_I(x)^T P(y^I) S_I(x) dx - \frac{2\mu}{\delta} \int_{\mathscr{K}_I} T_I(x)^T P(y^I) T_I(x) dx$$

$$\geq \left(\mu - \frac{\delta}{2}\right) \int_{\mathscr{K}_I} |S_I(x)|^2 dx - c\|W\|_{W^{-1,2}}^2 - c\|Q\|_{W^{-1,2}}^2 \ .$$

Again by (5.3.40) and (5.3.48),

$$(5.3.50)$$
$$\int_{\mathscr{K}_I} |S_I(x)|^2 dx \geq \left(1 - \frac{\delta}{2\mu}\right) \int_{\mathscr{K}_I} \theta_I^2(x) |W(x)|^2 dx - \frac{2\mu}{\delta} \int_{\mathscr{K}_I} |T_I(x)|^2 dx$$

$$\geq \left(1 - \frac{\delta}{2\mu}\right) \int_{\mathscr{K}_I} \theta_I^2(x) |W(x)|^2 dx - c\|W\|_{W^{-1,2}}^2 - c\|Q\|_{W^{-1,2}}^2 \ .$$

Combining (5.3.38), (5.3.39), (5.3.49), (5.3.50) and (5.3.37), we arrive at (5.3.25). This completes the proof.

The following proposition extends Theorem 5.1.1 to the situation where involutions are present and the entropy is convex only in the direction of the involution cone.

Theorem 5.3.1 *Assume the system of conservation laws (5.1.1) is endowed with an involution (5.3.4), with involution cone $\mathscr{C}$, and an entropy η, with $D^2\eta(U)$ positive definite in the direction of $\mathscr{C}$, uniformly on compact subsets of $\mathscr{O}$. Suppose the initial data U_0 are continuously differentiable on $\mathbb{R}^m$, take values in a compact subset of $\mathscr{O}$, are constant, say $\tilde{U}$, outside a bounded subset of $\mathbb{R}^m$, satisfy the involution on $\mathbb{R}^m$, and $\nabla U_0 \in H^\ell$ for some $\ell > m/2$. Then there exists T_∞, $0 < T_\infty \leq \infty$, and a unique continuously differentiable function U on $\mathbb{R}^m \times [0, T_\infty)$, taking values in $\mathscr{O}$, which is a classical solution of the initial-value problem (5.1.1), (5.1.2) on $[0, T_\infty)$. Furthermore,*

$$(5.3.51) \qquad \nabla U(\cdot, t) \in C^0([0, T_\infty); H^\ell) \ .$$

The interval $[0, T_\infty)$ is maximal, in the sense that whenever $T_\infty < \infty$

$$(5.3.52) \qquad \limsup_{t \uparrow T_\infty} \|\nabla U(\cdot, t)\|_{L^\infty} = \infty$$

and/or the range of $U(\cdot, t)$ escapes from every compact subset of $\mathcal{O}$ as $t \uparrow T_\infty$.

Proof. It suffices to retrace the steps of the proof of Theorem 5.1.1. In the definition of the metric space $\mathscr{F}$ the stipulation should be added that its members are constant, $\tilde{U}$, outside a large ball in $\mathbb{R}^m$.

The first snag we hit is that (5.1.15) no longer applies, since $D^2\eta(V)$ may now be positive definite only in the direction of $\mathscr{C}$. To remove this obstacle, we first note that by (5.1.11) and (5.3.6)

$$(5.3.53) \qquad \sum_{\beta=1}^{m} M_\beta \partial_\beta U_r = Q ,$$

where

$$(5.3.54) \qquad Q = - \sum_{\alpha,\beta=1}^{m} M_\beta \int_0^t \partial_\beta [DG_\alpha(V)] \partial_\alpha U_r d\tau$$
$$+ \sum_{\alpha,\beta=1}^{m} M_\beta \partial_\beta \int_0^t \{DG_\alpha(V)\partial^r \partial_\alpha U - \partial^r [DG_\alpha(V)\partial_\alpha U]\}d\tau .$$

Applying Lemma 5.3.2, with $P = D^2\eta(V)$ and $W = U_r$, we obtain

$$(5.3.55) \qquad \int_{\mathbb{R}^m} U_r^T D^2\eta(V) U_r dx \geq \delta \|U_r\|_{L^2}^2 - c\|U_r\|_{W^{-1,2}}^2 - c\|Q\|_{W^{-1,2}}^2 .$$

Integrating (5.1.11) with respect to t yields

$$(5.3.56) \qquad \|U_r(\cdot, t)\|_{W^{-1,2}} \leq c\|\nabla U_0(\cdot)\|_\ell + c\omega \int_0^t \|\nabla U(\cdot, \tau)\|_\ell d\tau .$$

Furthermore, (5.3.54) together with (5.1.6) and (5.1.12) imply

$$(5.3.57) \qquad \|Q(\cdot, t)\|_{W^{-1,2}} \leq c\omega \int_0^t \|\nabla U(\cdot, \tau)\|_\ell d\tau .$$

By employing (5.3.55), (5.3.56), (5.3.57) as a substitute for (5.1.15), we establish, in the place of (5.1.16), the new estimate

$$(5.3.58) \qquad \|\nabla U(\cdot, t)\|_\ell^2 \leq c\|\nabla U_0(\cdot)\|_\ell^2 + c\omega(1 + \omega T) \int_0^t \|\nabla U(\cdot, \tau)\|_\ell^2 d\tau$$

whence we deduce that, when ω is sufficiently large and T is sufficiently small, $\sup_{[0,T]} \|\nabla U(\cdot, t)\|_\ell \leq \omega$, as required for the proof.

A similar procedure is used to compensate for the failure of (5.1.21): We use (5.1.18) and (5.3.6) to get

$$(5.3.59) \qquad \sum_{\beta=1}^{m} M_\beta \partial_\beta (U - \bar{U}) = Q ,$$

where

(5.3.60)
$$Q = -\sum_{\alpha,\beta=1}^{m} M_\beta \int_0^t \partial_\beta[DG_\alpha(V)]\partial_\alpha(U - \overline{U})d\tau$$
$$-\sum_{\alpha,\beta=1}^{m} M_\beta\partial_\beta \int_0^t [DG_\alpha(V) - DG_\alpha(\overline{V})]\partial_\alpha\overline{U}d\tau \,,$$

and then apply Lemma 5.3.2, with $P = D^2\eta(V)$, $W = U - \overline{U}$, to get

(5.3.61)
$$\int_{\mathbb{R}^m} (U - \overline{U})^T D^2\eta(V)(U - \overline{U})dx \geq \delta\|U - \overline{U}\|_{L^2}^2 - c\|U - \overline{U}\|_{W^{-1,2}}^2 - c\|Q\|_{W^{-1,2}}^2 \,.$$

Integrating (5.1.18) with respect to t, we obtain the estimate

(5.3.62)
$$\|(U - \overline{U})(\cdot, t)\|_{W^{-1,2}} \leq c\omega \int_0^t \{\|(U - \overline{U})(\cdot, \tau)\|_{L^2} + \|(V - \overline{V})(\cdot, \tau)\|_{L^2}\}d\tau \,.$$

Moreover, (5.3.60) together with (5.1.6) imply

(5.3.63) $$\|Q(\cdot, t)\|_{W^{-1,2}} \leq c\omega \int_0^t \{\|(U - \overline{U})(\cdot, \tau)\|_{L^2} + \|(V - \overline{V})(\cdot, \tau)\|_{L^2}\}d\tau \,.$$

We employ (5.3.61), (5.3.62) and (5.3.63) as a substitute for (5.1.21). This yields, in the place of (5.1.22), the new estimate

(5.3.64)
$$\|(U - \overline{U})(\cdot, t)\|_{L^2}^2 \leq c\omega(1 + \omega T) \int_0^t \{\|(U - \overline{U})(\cdot, \tau)\|_{L^2}^2 + \|(V - \overline{V})(\cdot, \tau)\|_{L^2}^2\}d\tau \,.$$

From (5.1.9), (5.3.64) and Gronwall's inequality we deduce

(5.3.65) $$\rho(U, \overline{U}) \leq [c\omega T(1 + \omega T)]^{1/2} \exp[c\omega T(1 + \omega T)]\rho(V, \overline{V})$$

which verifies that, for T small, the map that carries $V \in \mathscr{F}$ to the solution $U \in \mathscr{F}$ of (5.1.10), (5.1.2) is a contraction.

Apart from the above modifications, the proofs of Theorems 5.3.1 and 5.1.1 are identical.

To illustrate the use of Theorem 5.3.1, we apply it to the system (3.3.10), with involution (5.3.1). A simple calculation shows that the involution cone $\mathscr{C}$ consists of all vectors in $\mathbb{R}^{m^2+m}$ of the form $(\xi \otimes v, w)$, with arbitrary ξ, v and w in $\mathbb{R}^m$. Thus, the entropy $\eta = \varepsilon(F) + \frac{1}{2}|v|^2$ is convex in the direction of $\mathscr{C}$ provided that $\varepsilon(F)$ satisfies (3.3.7). Consequently, Theorem 5.3.1 establishes local existence of classical solutions for the system of balance laws of isentropic, adiabatic thermoelasticity under the physically natural assumption that the internal energy is a rank-one convex function of the deformation gradient.

Theorem 5.2.1 may also be similarly extended to the situation where the entropy is convex just in the direction of the involution cone:

Theorem 5.3.2 *Assume the system of conservation laws* (5.1.1) *is endowed with an involution* (5.3.4), *with involution cone* $\mathscr{C}$, *and with an entropy* η *with* $D^2\eta(U)$ *positive definite in the direction of* $\mathscr{C}$, *uniformly on compact subsets of* $\mathscr{O}$. *Suppose* $\overline{U}$ *is a classical solution of* (5.1.1) *on a bounded time interval* $[0, T)$, *taking values in a convex, compact subset* $\mathscr{D}$ *of* $\mathscr{O}$, *with initial data* $\overline{U}_0$ *that satisfy the involution. Let* U *be any admissible weak solution of* (5.1.1) *on* $[0, T)$, *with* $t \mapsto U(\cdot, t)$ *continuous in* $L^1_{loc}(\mathbb{R}^m)$, *which takes values in* $\mathscr{D}$, *coincides with* $\overline{U}$ *outside some ball in* $\mathbb{R}^m$, *has local oscillation*

$$(5.3.66) \qquad \limsup_{\varepsilon \downarrow 0} \; \sup_{|y-x|<\varepsilon} \; |U(y,t) - U(x,t)| < \kappa \;, \quad 0 \le t < T \;,$$

and initial data U_0 *satisfying the involution. If* κ *is small, then*

$$(5.3.67) \qquad \int_{\mathbb{R}^m} |U(x,t) - \overline{U}(x,t)|^2 dx \le a \int_{\mathbb{R}^m} |U_0(x) - \overline{U}_0(x)|^2 dx$$

holds for $t \in [0, T)$, *and some constant* a *that depends on* $\mathscr{D}$, *on* T *and on the Lipschitz constant of* $\overline{U}$. *In particular,* $\overline{U}$ *is the unique admissible weak solution of* (5.1.1) *with values in* $\mathscr{D}$, *small local oscillation and initial data* $\overline{U}_0$.

Proof. Retracing the steps of the proof of Theorem 5.2.1, we rederive (5.2.13). Letting $R \uparrow \infty$ and $\varepsilon \downarrow 0$ in (5.2.13), taking into account that $U - \overline{U}$ vanishes outside some ball, $\overline{U}$ is Lipschitz, and $t \mapsto U(\cdot, t)$ is continuous, we arrive again at (5.2.14), with $R = \infty$:

$$(5.3.68) \qquad \int_{\mathbb{R}^m} h(U(x,t), \overline{U}(x,t))dx \le \int_{\mathbb{R}^m} h(U_0(x), \overline{U}_0(x))dx$$

$$- \int_0^t \int_{\mathbb{R}^m} \sum_{\alpha=1}^m \partial_\alpha \overline{U}^T D^2\eta(\overline{U}) Z_\alpha(U, \overline{U}) dx d\tau \;.$$

From (5.2.2),

$$(5.3.69) \qquad h(U, \overline{U}) = (U - \overline{U})^T P(U, \overline{U})(U - \overline{U}) \;,$$

where

$$(5.3.70) \qquad P(U, \overline{U}) = \int_0^1 \int_0^w D^2\eta(\overline{U} + z(U - \overline{U})) dz dw \;.$$

In particular,

$$(5.3.71) \qquad Z^T P(U, \overline{U}) Z \ge \mu |Z|^2 \;, \quad Z \in \mathscr{C} \;,$$

for some $\mu > 0$. Therefore, when κ in (5.3.66) is so small that the local oscillation of $P(U(x,t), \overline{U}(x,t))$ is less than μ, we may apply Lemma 5.3.2, with $W = U - \overline{U}$ and $Q = 0$, to get

$$(5.3.72)$$

$$\int_{\mathbb{R}^m} h(U(x,t), \overline{U}(x,t)) dx \ge \delta \|U(\cdot,t) - \overline{U}(\cdot,t)\|_{L^2}^2 - c\|U(\cdot,t) - \overline{U}(\cdot,t)\|_{W^{-1,2}}^2$$

for some $\delta > 0$. We estimate the second term on the right-hand side of (5.3.72) as follows:

$$(5.3.73) \quad \|U(\cdot, t) - \overline{U}(\cdot, t)\|_{W^{-1,2}} \leq \|U_0(\cdot) - \overline{U}_0(\cdot)\|_{W^{-1,2}}$$
$$+ \int_0^t \|\partial_t \{U(\cdot, \tau) - \overline{U}(\cdot, \tau)\}\|_{W^{-1,2}} d\tau ,$$

(5.3.74)

$$\|\partial_t \{U(\cdot, \tau) - \overline{U}(\cdot, \tau)\}\|_{W^{-1,2}} = \|\sum_{\alpha=1}^m \partial_\alpha \{G_\alpha(U(\cdot, \tau)) - G_\alpha(\overline{U}(\cdot, \tau))\}\|_{W^{-1,2}}$$
$$\leq \sum_{\alpha=1}^m \|G_\alpha(U(\cdot, \tau)) - G_\alpha(\overline{U}(\cdot, \tau))\|_{L^2} \leq c\|U(\cdot, \tau) - \overline{U}(\cdot, \tau)\|_{L^2} .$$

Furthermore, by (5.2.4), $Z_\alpha(U, \overline{U})$ is of quadratic order in $U - \overline{U}$. Therefore, combining (5.3.72), (5.3.73) and (5.3.74), we deduce from (5.3.68):

$$(5.3.75) \quad \|U(\cdot, t) - \overline{U}(\cdot, t)\|_{L^2}^2 \leq c\|U_0(\cdot) - \overline{U}_0(\cdot)\|_{L^2}^2$$
$$+ c\int_0^t \|U(\cdot, \tau) - \overline{U}(\cdot, \tau)\|_{L^2}^2 d\tau + c\left\{\int_0^t \|U(\cdot, \tau) - \overline{U}(\cdot, \tau)\|_{L^2} d\tau\right\}^2$$

whence (5.3.67) follows. This completes the proof.

In the above theorem, the hypothesis that $t \mapsto U(\cdot, t)$ is strongly continuous appears extraneous, so it is natural to inquire what conditions, short of convexity, would have to be imposed on η to render this requirement superfluous.

Definition 5.3.3 An entropy η for the system of conservation laws (5.1.1), endowed with an involution (5.3.4), is called *quasiconvex* if for any $U \in L^\infty(\mathbb{R}^m; \mathscr{O})$, which is $2p$-periodic, satisfies (5.3.4) and has mean

$$(5.3.76) \quad \hat{U} = (2p)^{-m} \int_{\mathscr{H}} U(y) dy ,$$

over the standard hypercube $\mathscr{H}$ in $\mathbb{R}^m$ with edge length $2p$, it is

$$(5.3.77) \quad \eta(\hat{U}) \leq (2p)^{-m} \int_{\mathscr{H}} \eta(U(y)) dy .$$

Roughly, quasiconvexity stipulates that the uniform state minimizes the total entropy, among all states that are compatible with the involution and have the same "mass". This is in the spirit of the fundamental principle of classical thermostatics, that the physical entropy is maximized at the equilibrium state.

The relevance of quasiconvexity is demonstrated by the following proposition, whose proof may be found in the references cited in Section 5.4:

Theorem 5.3.3 *Assume the system of conservation laws* (5.1.1) *is endowed with an entropy* η *and an involution* (5.3.4), *such that the rank of* $N(\nu)$ *is constant,* *for any $v \in \mathscr{S}^{m-1}$, and equal to the dimension of the kernel of $\Lambda(v; U)$. Then $\int_{|x|<R} \eta(U)dx$ is weak* lower semicontinuous on the space of L^∞ vector fields U that satisfy (5.3.4) if and only if η is quasiconvex. Furthermore, any quasiconvex η is necessarily convex in the direction of the involution cone $\mathscr{C}$.*

In particular, when η is quasiconvex the requirement of strong L^1 continuity of $t \mapsto U(\cdot, t)$ may be relaxed into continuity in L^∞ weak*.

Even though Definition 5.3.3 is tailored to equilibrium, quasiconvexity may also be interpreted in the framework of dynamics, as follows. Recalling Lemma 5.3.1, U on the right-hand side of (5.3.77) can be replaced by $\hat{U} + \mathscr{L}\chi$, for some $2p$-periodic Lipschitz function χ. In turn, $\hat{U} + \mathscr{L}\chi$ may be interpreted as an approximation to the value at time ε of the solution of (5.1.1), (5.1.2), with $U_0(x) = \hat{U} + \varepsilon\chi(\varepsilon^{-2}x)$. A more natural extension of the notion of quasiconvexity to the realm of dynamics is provided by the following

Definition 5.3.4 An entropy η for the system of conservation laws (5.1.1), endowed with an involution (5.3.4), is called *dynamically quasiconvex* if for any $U \in L^\infty(\mathbb{R}^{m+1}; \mathscr{O})$, which is $2p$-periodic in the spatial variables, satisfies (5.3.4) and has asymptotic mean

$$(5.3.78) \qquad \hat{U} = (2p)^{-m} \lim_{\varepsilon \downarrow 0} \frac{1}{\varepsilon} \int_0^\varepsilon \int_{\mathscr{H}} U(y, \tau)dyd\tau \ ,$$

at time zero, it is

$$(5.3.79) \qquad \eta(\hat{U}) \le (2p)^{-m} \liminf_{\varepsilon \downarrow 0} \frac{1}{\varepsilon} \int_0^\varepsilon \int_{\mathscr{H}} \eta(U(y, \tau))dyd\tau \ .$$

By mimicking the proof of Theorem 5.3.3, one shows that when η is dynamically quasiconvex no extraneous hypothesis on the continuity of $t \mapsto U(\cdot, t)$ is required, because the weak time regularity induced by Theorem 4.1.1 is sufficient for passing to the $\varepsilon \downarrow 0$ limit in the proof of Theorem 5.3.2.

Clearly, any dynamically quasiconvex entropy is quasiconvex. Moreover, any convex entropy is dynamically quasiconvex, and a fortiori quasiconvex, by virtue of Jensen's inequality. However, whereas convexity may be tested easily, by direct calculation, quasiconvexity, as defined by Definition 5.3.3, is an implicit condition that is hard to verify in practice. In view of Theorem 5.3.3, it is tempting to conjecture that any entropy which is convex in the direction of the involution cone is necessarily quasiconvex. This is indeed the case when the entropy is quadratic: $\eta = U^T A U$. In general, however, quasiconvexity is a stricter condition than mere convexity in the direction of the involution cone.

The above may be illustrated in the framework of our prototypical example, the system of balance laws (3.3.10) of isentropic, adiabatic thermoelasticity, with involution (5.3.1) and entropy $\eta = \varepsilon(F) + \frac{1}{2}|v|^2$. In that case η is quasiconvex when $\varepsilon(F)$ is quasiconvex in the sense of Morrey: For any constant deformation gradient $\hat{F}$ and any Lipschitz function χ from $\mathscr{H}$ to $\mathbb{R}^m$, with compact support in $\mathscr{H}$,

$$(5.3.80) \qquad \varepsilon(\hat{F}) \leq (2p)^{-m} \int_{\mathscr{H}} \varepsilon(\hat{F} + \nabla\chi) dy .$$

In other words, a homogeneous deformation of $\mathscr{H}$ minimizes the total internal energy among all placements of $\mathscr{H}$ with the same boundary values. Any quasi-convex internal energy is rank-one convex (3.3.7), i.e., η is convex in the direction of the involution cone. On the other hand, examples have been constructed of $\varepsilon(F)$ that are rank-one convex but not quasiconvex.

A sufficient condition for $\varepsilon(F)$ to be quasiconvex is that it be *polyconvex*, in the sense of Ball, i.e., in the physically relevant case $m = 3$:

$$(5.3.81) \qquad \varepsilon(F) = \sigma(F, F^*, w) ,$$

where F^* denotes the matrix of cofactors of F, $F^* = (\det F)F^{-1}$, $w = \det F$, and σ is a convex function on $\mathbb{R}^{19}$. When the elastic material is isotropic and the function $\tilde{\varepsilon}$ in (2.5.18) is convex in the principal invariants $(I_1, \cdots, I_m)$, then $\varepsilon(F)$ is polyconvex.

Actually, polyconvex $\varepsilon(F)$ are even dynamically quasiconvex. This follows from Theorem 4.1.1 combined with the observation that, as F and v result from a Lipschitz motion $\chi = \chi(x, t)$ through (2.1.1) and (2.1.2), the L^∞ fields w and F^* satisfy the following, kinematically induced, conservation laws:

$$(5.3.82) \qquad \partial_t w = \sum_{\alpha=1}^{3} \sum_{i=1}^{3} \partial_\alpha (F_{\alpha i}^* v_i) ,$$

$$(5.3.83) \qquad \partial_t F_{\gamma k}^* = \sum_{\alpha,\beta=1}^{3} \sum_{i,j=1}^{3} \partial_\alpha (\epsilon_{\alpha\beta\gamma} \epsilon_{ijk} F_{j\beta} v_i) , \qquad \gamma, k = 1, 2, 3 ,$$

where $\epsilon_{\alpha\beta\gamma}$, ϵ_{ijk} denote the standard permutation symbols. To establish these identities, recall the discussion and notation of Section 2.2. Consider first the (trivial) conservation law (2.2.3), in Eulerian coordinates, with $\Theta^* = 1$, $\Psi^* = v^T$, $P^* = 0$. Its equivalent, Lagrangian form (2.2.1) reduces to (5.3.82), because, by virtue of (2.2.4), $\Theta = \det F = w$, $\Psi = \det F(F^{-1}v)^T = (F^*v)^T$, $P = 0$. To verify (5.3.83), start with the obvious conservation law

$$(5.3.84) \qquad (F^{-1})_t = (\operatorname{grad} x)_t = \operatorname{grad} (x_t) = -\operatorname{grad} (F^{-1}v) ,$$

in Eulerian coordinates, and derive its equivalent Lagrangian form through (2.2.4). The calculation gives

$$(5.3.85) \qquad \partial_t F_{\gamma k}^* = \sum_{\alpha=1}^{3} \sum_{i=1}^{3} \partial_\alpha [(\det F)(F_{\alpha i}^{-1} F_{\gamma k}^{-1} - F_{\gamma i}^{-1} F_{\alpha k}^{-1}) v_i] ,$$

which easily reduces to (5.3.83).

In an alternative approach, one may embed the system (3.3.10), with state vector $U = (F, v)$, into a larger, albeit equivalent, system, consisting of (3.3.10), (5.3.82) and (5.3.83), with new state vector $U = (F, F^*, w, v)$. When the internal energy function is of the form (5.3.81), with σ (locally) uniformly convex, the system in this new realization is endowed with a uniformly convex entropy $\eta = \sigma(F, F^*, w) + \frac{1}{2}|v|^2$, so that local existence of classical solutions may be inferred directly from Theorem 5.1.1, thus circumventing the need for Theorem 5.3.1, while stability of classical solutions within the class of admissible weak solutions follows immediately from Theorem 5.2.1, even without requiring time continuity and small oscillation as in Theorem 5.3.2.

5.4 Notes

A more extensive discussion of the material covered in Section 5.1, including a complete proof of (a slight variation of) Theorem 5.1.1, is contained in Majda [3]. This presentation is in the spirit of the theory of linear symmetric hyperbolic systems developed by Friedrichs. For an alternative, functional analytic, approach to the subject, see Kato [1] and Taylor [1].

The proof of Theorem 5.2.1 combines ideas of DiPerna [5] and Dafermos [9].

Hyperbolic systems of conservation laws with involutions were considered by Boillat [3] and by Dafermos [12]. In particular, Boillat [3] exhibits sufficient conditions that are more general than (5.3.6) and presents examples arising in the theory of general relativity. The analysis in Section 5.3 is intimately related to the theory of compensated compactness as formalized by Murat and Tartar; see Tartar [1,2]. The "involution cone" corresponds to the "characteristic cone", in the terminology of that theory. Theorem 5.3.1 is new; however, typical examples, like the system (3.3.10) of balance laws of isentropic thermoelasticity, have been studied extensively in the literature; see, for example, Hughes, Kato and Marsden [1], or Dafermos and Hrusa [1]. Theorem 5.3.2 is taken from Dafermos [21].

The notion of quasiconvexity introduced by Definition 5.3.3 is a generalization of quasiconvexity in the sense of Morrey [1] due to Dacorogna [1]; for detailed study and a proof of Theorem 5.3.3, see Müller and Fonseca [1]. These notions were originally developed in the framework of the calculus of variations, where lower semicontinuity of functionals is the central issue. In particular, equilibrium (i.e. time-independent) solutions of the system (3.3.10) are minimizers of the internal energy. It is in the context of this problem that Ball [1] introduced the concept of polyconvexity and discussed its connections with quasiconvexity and rank-one convexity. Voluminous literature on the subject has derived from Ball's pioneering paper. The question whether rank-one convexity generally implies quasiconvexity was debated for a long time until finally answered, in the negative, by Šverak [1]. The idea of enlarging the system (3.3.10) is due, independently, to LeFloch and to Qin [1], who derives the kinematical conservation laws (5.3.82), (5.3.83), albeit only for smooth motions. Other state variables, such as $\det F^{-1}$ may be used as well, which also satisfy kinematically induced conservation laws (Wagner [3]).

Chapter VI. The L^1 Theory
of the Scalar Conservation Law

The theory of the scalar balance law has reached a state of virtual completeness. In the framework of classical solutions, the elementary, yet effective, method of characteristics yields a sharper version of Theorem 5.1.1, determining explicitly the life span of solutions with Lipschitz continuous initial data and thereby demonstrating that in general this life span is finite. Thus one has to deal with weak solutions, even when the initial data are very smooth.

In regard to weak solutions, the special feature that sets the scalar balance law apart from systems of more than one equation is the size of its family of entropies. It will be shown that the abundance of entropies induces an effective characterization of admissibe weak solutions as well as very strong L^1-stability and L^∞-monotonicity properties. Armed with such powerful a priori estimates, one can construct admissible weak solutions in a number of ways. As a sample, construction will be effected here by the method of vanishing viscosity, the theory of L^1-contraction semigroups, the layering method, an approach motivated by the kinetic theory, and a relaxation method. It will also be shown that when the initial data are functions of locally bounded variation then so are the solutions. Finally, it will be explained why these methods fail in the case of systems of balance laws.

In order to expose the elegance of the theory, the discussion will be restricted to the homogeneous scalar conservation law, even though the general, inhomogeneous balance law (3.3.1) may be treated by the same methodology, at the expense of rather minor technical complications.

The issue of stability of weak solutions with respect to the topology of L^∞ weak* will be addressed in Chapter XV. The special case of a single space variable, $m = 1$, has a very rich theory of its own, certain aspects of which will be presented in later chapters.

6.1 The Initial-Value Problem:
Perseverance and Demise of Classical Solutions

We consider the initial-value problem for a homogeneous scalar conservation law:

$$(6.1.1) \qquad \partial_t u(x,t) + \sum_{\alpha=1}^m \partial_\alpha g_\alpha(u(x,t)) = 0 , \quad x \in \mathbb{R}^m , \quad t \in [0,T) ,$$

$$(6.1.2) \qquad u(x, 0) = u_0(x) , \quad x \in \mathbb{R}^m .$$

For $\alpha = 1, \cdots, m$, the g_α are given smooth functions on $\mathbb{R}$. We realize the g_α as components of a m-vector g.

A *characteristic* of (6.1.1), associated with a continuously differentiable solution u, is an orbit $\xi : [0, T) \to \mathbb{R}^m$ of the system of ordinary differential equations

$$(6.1.3) \qquad \frac{dx_\alpha}{dt} = g'_\alpha(u(x, t)) , \quad \alpha = 1, \cdots, m .$$

With every characteristic ξ we associate the differential operator

$$(6.1.4) \qquad \frac{d}{dt} = \partial_t + \sum_{\alpha=1}^{m} g'_\alpha(u(\xi(t), t))\partial_\alpha ,$$

which determines the directional derivative along ξ. In particular, since u satisfies (6.1.1), $du/dt = 0$, i.e., u is constant along any characteristic. By virtue of (6.1.3), this implies that the slope of the characteristic is constant. Thus all characteristics are straight lines along which the solution is constant. With the help of this property, one may study classical solutions of (6.1.1), (6.1.2) in minute detail. In particular, for scalar conservation laws Theorem 5.1.1 admits the following refinement:

Theorem 6.1.1 *Assume that u_0, defined on $\mathbb{R}^m$, is bounded and Lipschitz continuous. Let*

$$(6.1.5) \qquad \kappa = \operatorname*{ess\,inf}_{y \in \mathbb{R}^m} \sum_{\beta=1}^{m} g''_\beta(u_0(y))\partial_\beta u_0(y) .$$

Then there exists a classical solution u of (6.1.1), (6.1.2) on the maximal interval $[0, T_\infty)$, where $T_\infty = \infty$ when $\kappa \geq 0$ and $T_\infty = -\kappa^{-1}$ when $\kappa < 0$. Furthermore, if u_0 is C^k so is u.

Proof. Suppose first $\nabla u_0 \in H^\ell$, for ℓ very large, so that, by Theorem 5.1.1, the solution u of (6.1.1), (6.1.2) exists on some maximal interval $[0, T_\infty)$ and is a smooth function. Since u is constant along characteristics, its value at any point (x, t), $x \in \mathbb{R}^m$, $t \in [0, T_\infty)$, satisfies the implicit relation

$$(6.1.6) \qquad u(x, t) = u_0(x - tg'(u(x, t))) .$$

In particular, the range of u coincides with the range of u_0.

Differentiating (6.1.6) yields

$$(6.1.7) \qquad \partial_\alpha u(x, t) = \frac{\partial_\alpha u_0(y)}{1 + t \sum_{\beta=1}^{m} g''_\beta(u_0(y))\partial_\beta u_0(y)} , \quad \alpha = 1, \cdots, m ,$$

where $y = x - tg'(u(x, t))$. Thus, by virtue of Theorem 5.1.1, $T_\infty = \infty$ when $\kappa \geq 0$ and $T_\infty \geq -\kappa^{-1}$ when $\kappa < 0$. On the other hand, if $\kappa < 0$ then derivatives of the solution along characteristics emanating from points $(y, 0)$ with $\sum g''_\beta(u_0(y))\partial_\beta u_0(y) < \kappa + \varepsilon < 0$ will have to blow up no later than $t = -(\kappa + \varepsilon)^{-1}$. Hence $T_\infty = -\kappa^{-1}$.

When u_0 is merely Lipschitz continuous, we approximate it in $L^\infty(\mathbb{R}^m)$, via mollification, by a sequence $\{u_n\}$ of smooth functions with $\nabla u_n \in H^\ell$ and

$$(6.1.8) \qquad \operatorname{ess\,inf} \sum_{\beta=1}^{m} g''_\beta(u_n(y))\partial_\beta u_n(y) \geq \kappa - \frac{1}{n} .$$

Classical solutions of (6.1.1) with initial data u_n are defined on $\mathbb{R}^m \times [0, T_n)$, where $T_n \geq n$ when $\kappa \geq 0$ and $T_n \geq -(\kappa - 1/n)^{-1}$ when $\kappa < 0$, and are Lipschitz equicontinuous on any compact subset of $\mathbb{R}^m \times [0, T_\infty)$. Therefore, we may extract a subsequence which converges, uniformly on compact sets, to some function u on $\mathbb{R}^m \times [0, T_\infty)$. Clearly, u is at least a weak solution of (6.1.1), (6.1.2) and, being locally Lipschitz continuous, is actually a classical solution on $[0, T_\infty)$. The limiting process also implies that u still satisfies (6.1.6) for $x \in \mathbb{R}^m$ and $t \in [0, T_\infty)$. In particular, if u_0 is differentiable at a point $y \in \mathbb{R}^m$ then u is differentiable along the straight line $x = y + tg'(u_0(y))$, the derivatives being given by (6.1.7). Consequently, T_∞ is the life span of the classical solution.

When u_0 is C^k, (6.1.6) together with (6.1.7) and the implicit function theorem imply that u is also C^k on $\mathbb{R}^m \times [0, T_\infty)$. This completes the proof.

From the above considerations becomes clear that the life span of classical solutions is generally finite. It is thus imperative to deal with weak solutions.

6.2 Admissible Weak Solutions and Their Stability Properties

In Section 4.2, we saw that the initial-value problem for a scalar conservation law may admit more than one weak solution, thus raising the need to impose admissibility conditions. In Section 4.3, we discussed how entropy inequalities may serve that purpose. Recall from Section 3.3 that for the scalar conservation law (6.1.1) any smooth function η may serve as an entropy, with associated entropy flux

$$(6.2.1) \qquad q_\alpha(u) = \int_v^u \eta'(\omega)g'_\alpha(\omega)d\omega , \qquad \alpha = 1, \cdots, m ,$$

and entropy production zero. It will be convenient to relax slightly the regularity condition and allow entropies (and thereby entropy fluxes) that are merely locally Lipschitz continuous. Similarly, the g_α need only be locally Lipschitz continuous functions. It turns out that in order to characterize properly admissible weak solutions, one has to impose the entropy inequality

$$(6.2.2) \qquad \partial_t \eta(u(x, t)) + \sum_{\alpha=1}^{m} \partial_\alpha q_\alpha(u(x, t)) \leq 0$$

for every convex entropy-entropy flux pair:

Definition 6.2.1 A bounded measurable function u on $\mathbb{R}^m \times [0, T)$ is an *admissible weak solution* of (6.1.1), (6.1.2), with u_0 in $L^\infty(\mathbb{R}^m)$, if the inequality

$$(6.2.3) \quad \int_0^T \int_{\mathbb{R}^m} \left[\partial_t \psi \eta(u) + \sum_{\alpha=1}^m \partial_\alpha \psi q_\alpha(u) \right] dx dt + \int_{\mathbb{R}^m} \psi(x, 0) \eta(u_0(x)) dx \geq 0$$

holds for every convex function η, with q_α determined through (6.2.1), and all nonnegative Lipschitz continuous test functions ψ on $\mathbb{R}^m \times [0, T)$, with compact support.

Applying (6.2.3) with $\eta(u) = \pm u$, $q_\alpha(u) = \pm g_\alpha(u)$ shows that (6.2.3) implies (4.1.6), i.e., any admissible weak solution in the sense of Definition 6.2.1 is in particular a weak solution as defined in Section 4.1. Also note that if u is a classical solution of (6.1.1), (6.1.2), then (6.2.3) holds automatically, as an equality, i.e., all classical solutions are admissible. Several motivations for (6.2.3) will be presented in subsequent sections.

To verify (6.2.3) for all convex η, it would suffice to test it just for some family of convex η with the property that the set of linear combinations of its members, with nonnegative coefficients, spans the entire set of convex functions. To formulate examples, consider the following standard notation: For $w \in \mathbb{R}$, w^+ denotes $\max\{w, 0\}$ and $\operatorname{sgn} w$ stands for $-1, 1$ or 0, as w is negative, positive or zero. Notice that any Lipschitz continuous function is the limit of a sequence of piecewise linear convex functions

$$(6.2.4) \qquad c_0 u + \sum_{i=1}^k c_i (u - u_i)^+$$

with $c_i > 0$, $i = 1, \cdots, k$. Consequently, it would suffice to verify (6.2.3) for the entropies $\pm u$, with entropy flux $\pm g_\alpha$, $\alpha = 1, \cdots, m$, together with the family of entropy-entropy flux pairs

(6.2.5)

$$\eta(u; \bar{u}) = (u - \bar{u})^+ \,, \quad q_\alpha(u; \bar{u}) = \operatorname{sgn}(u - \bar{u})^+ [g_\alpha(u) - g_\alpha(\bar{u})] \,, \quad \alpha = 1, \cdots, m \,,$$

where $\bar{u}$ is a parameter taking values in $\mathbb{R}$. Equally well, one may use the celebrated family of entropy-entropy flux pairs of Kruzkov:

(6.2.6)

$$\eta(u; \bar{u}) = |u - \bar{u}| \,, \quad q_\alpha(u; \bar{u}) = \operatorname{sgn}(u - \bar{u})[g_\alpha(u) - g_\alpha(\bar{u})] \,, \quad \alpha = 1, \cdots, m \,.$$

The fundamental existence and uniqueness theorem, which will be demonstrated by several methods in subsequent sections, is

Theorem 6.2.1 *For each $u_0 \in L^\infty(\mathbb{R}^m)$, there exists a unique admissible weak solution u of* (6.1.1), (6.1.2) *on* $[0, \infty)$ *and*

$$(6.2.7) \qquad u(\cdot, t) \in C^0([0, \infty); L^1_{loc}(\mathbb{R}^m)) \,.$$

The following proposition establishes the most important properties of admissible weak solutions of the scalar conservation law, namely, stability in L^1 and monotonicity in L^∞:

Theorem 6.2.2 *Let u and $\overline{u}$ be admissible weak solutions of* (6.1.1) *with respective initial data u_0 and $\overline{u}_0$ taking values in a compact interval* [a,b]. *There is $s > 0$, depending solely on* [a,b], *such that, for any $t \in [0, T)$ and $R > 0$*

$$(6.2.8) \qquad \int_{|x|<R} [u(x,t) - \overline{u}(x,t)]^+ dx \le \int_{|x|<R+st} [u_0(x) - \overline{u}_0(x)]^+ dx ,$$

$$(6.2.9) \qquad \|u(\cdot,t) - \overline{u}(\cdot,t)\|_{L^1(|x|<R)} \le \|u_0(\cdot) - \overline{u}_0(\cdot)\|_{L^1(|x|<R+st)} .$$

Furthermore, if

$$(6.2.10) \qquad u_0(x) \le \overline{u}_0(x) , \qquad a.e. \ on \ \mathbb{R}^m$$

then

$$(6.2.11) \qquad u(x,t) \le \overline{u}(x,t) , \qquad a.e. \ on \ \mathbb{R}^m \times [0,T) .$$

In particular, the (essential) range of both u and $\overline{u}$ is contained in [a,b].

Proof. The salient feature of the scalar conservation law that induces (6.2.8) is that the functions $\eta(u; \overline{u})$, $q_\alpha(u; \overline{u})$, defined through (6.2.5), constitute entropy-entropy flux pairs not only in the variable u, for fixed $\overline{u}$, but also in the variable $\overline{u}$, for fixed u.

Consider any nonnegative Lipschitz continuous function $\phi(x, t, \overline{x}, \overline{t})$, defined on $\mathbb{R}^m \times [0, T) \times \mathbb{R}^m \times [0, T)$ and having compact support. Fix $(\overline{x}, \overline{t})$ in $\mathbb{R}^m \times [0, T)$ and write (6.2.3) for the entropy-entropy flux pair $\eta(u; \overline{u}(\overline{x}, \overline{t}))$, $q_\alpha(u; \overline{u}(\overline{x}, \overline{t}))$, and the test function $\psi(x, t) = \phi(x, t, \overline{x}, \overline{t})$:

$$(6.2.12) \qquad \int_0^T \int_{\mathbb{R}^m} \Big\{ \partial_t \phi(x, t, \overline{x}, \overline{t}) \eta(u(x,t); \overline{u}(\overline{x}, \overline{t}))$$
$$+ \sum_{\alpha=1}^m \partial_{x_\alpha} \phi(x, t, \overline{x}, \overline{t}) q_\alpha(u(x,t); \overline{u}(\overline{x}, \overline{t})) \Big\} dxdt$$
$$+ \int_{\mathbb{R}^m} \phi(x, 0, \overline{x}, \overline{t}) \eta(u_0(x); \overline{u}(\overline{x}, \overline{t})) dx \ge 0 .$$

Interchanging the roles of u and $\overline{u}$, we similarly obtain, for any fixed (x, t) in $\mathbb{R}^m \times [0, T)$:

$$(6.2.13) \qquad \int_0^T \int_{\mathbb{R}^m} \Big\{ \partial_{\overline{t}} \phi(x, t, \overline{x}, \overline{t}) \eta(u(x,t); \overline{u}(\overline{x}, \overline{t}))$$
$$+ \sum_{\alpha=1}^m \partial_{\overline{x}_\alpha} \phi(x, t, \overline{x}, \overline{t}) q_\alpha(u(x,t); \overline{u}(\overline{x}, \overline{t})) \Big\} d\overline{x}d\overline{t}$$
$$+ \int_{\mathbb{R}^m} \phi(x, t, \overline{x}, 0) \eta(u(x,t); \overline{u}_0(\overline{x})) d\overline{x} \ge 0 .$$

Integrating over $\mathbb{R}^m \times [0, T)$ (6.2.12), with respect to $(\bar{x}, \bar{t})$, and (6.2.13), with respect to (x, t), and then adding the resulting inequalities yields

$$(6.2.14) \quad \int_0^T \int_{\mathbb{R}^m} \int_0^T \int_{\mathbb{R}^m} \Big\{ (\partial_t + \partial_{\bar{t}}) \phi(x, t, \bar{x}, \bar{t}) \eta(u(x, t); \bar{u}(\bar{x}, \bar{t}))$$

$$+ \sum_{\alpha=1}^m (\partial_{x_\alpha} + \partial_{\bar{x}_\alpha}) \phi(x, t, \bar{x}, \bar{t}) q_\alpha(u(x, t); \bar{u}(\bar{x}, \bar{t})) \Big\} dx dt d\bar{x} d\bar{t}$$

$$+ \int_0^T \int_{\mathbb{R}^m} \int_{\mathbb{R}^m} \phi(x, 0, \bar{x}, \bar{t}) \eta(u_0(x); \bar{u}(\bar{x}, \bar{t})) dx d\bar{x} d\bar{t}$$

$$+ \int_0^T \int_{\mathbb{R}^m} \int_{\mathbb{R}^m} \phi(x, t, \bar{x}, 0) \eta(u(x, t); \bar{u}_0(\bar{x})) dx d\bar{x} d\bar{t} \geq 0 .$$

We fix a smooth nonnegative function ρ on $\mathbb{R}$ with compact support and total mass one:

$$(6.2.15) \qquad\qquad \int_{-\infty}^{\infty} \rho(\xi) d\xi = 1 .$$

Consider any nonnegative Lipschitz continuous test function ψ on $\mathbb{R}^m \times [0, T)$, with compact support. For positive small ε, write (6.2.14) with

$$(6.2.16) \quad \phi(x, t, \bar{x}, \bar{t}) = \varepsilon^{-(m+1)} \psi\left(\frac{x + \bar{x}}{2}, \frac{t + \bar{t}}{2}\right) \rho\left(\frac{t - \bar{t}}{2\varepsilon}\right) \prod_{\alpha=1}^m \rho\left(\frac{x_\alpha - \bar{x}_\alpha}{2\varepsilon}\right)$$

and then let $\varepsilon \downarrow 0$. Noting that

$$(6.2.17) \quad (\partial_t + \partial_{\bar{t}}) \phi(x, t, \bar{x}, \bar{t})$$

$$= \varepsilon^{-(m+1)} \partial_t \psi\left(\frac{x + \bar{x}}{2}, \frac{t + \bar{t}}{2}\right) \rho\left(\frac{t - \bar{t}}{2\varepsilon}\right) \prod_{\alpha=1}^m \rho\left(\frac{x_\alpha - \bar{x}_\alpha}{2\varepsilon}\right) ,$$

$$(6.2.18) \quad \left(\partial_{x_\alpha} + \partial_{\bar{x}_\alpha}\right) \phi(x, t, \bar{x}, \bar{t})$$

$$= \varepsilon^{-(m+1)} \partial_\alpha \psi\left(\frac{x + \bar{x}}{2}, \frac{t + \bar{t}}{2}\right) \rho\left(\frac{t - \bar{t}}{2\varepsilon}\right) \prod_{\alpha=1}^m \rho\left(\frac{x_\alpha - \bar{x}_\alpha}{2\varepsilon}\right) ,$$

$$(6.2.19) \qquad |\eta(u(x, t); \bar{u}_0(\bar{x})) - \eta(u_0(x); \bar{u}_0(\bar{x}))| \leq |u(x, t) - u_0(x)| ,$$

$$(6.2.20) \qquad |\eta(u_0(x); \bar{u}(\bar{x}, \bar{t})) - \eta(u_0(x); \bar{u}_0(\bar{x}))| \leq |\bar{u}(\bar{x}, \bar{t}) - \bar{u}_0(\bar{x})| ,$$

recalling Theorem 4.3.1, and by standard convergence theorems, we conclude

$$(6.2.21)$$

$$\int_0^T \int_{\mathbb{R}^m} \Big\{ \partial_t \psi(x, t) \eta(u(x, t); \bar{u}(x, t)) + \sum_{\alpha=1}^m \partial_\alpha \psi(x, t) q_\alpha(u(x, t); \bar{u}(x, t)) \Big\} dx dt$$

$$+ \int_{\mathbb{R}^m} \psi(x, 0) \eta(u_0(x); \bar{u}_0(x)) dx \geq 0 .$$

From (6.2.5) it is clear that there is $s > 0$ such that

$$(6.2.22) \qquad \left[\sum_{\alpha=1}^{m} |q_\alpha(u; \overline{u})|^2 \right]^{1/2} \leq s\eta(u; \overline{u})$$

for all u and $\overline{u}$ in the range of the solutions.

Fix $R > 0$, $t \in [0, T)$, and $\varepsilon > 0$ small; write (6.2.21) for the test function $\psi(x, \tau) = \chi(x, \tau)\theta(\tau)$, with χ and θ defined by (5.2.12) and (5.2.11), to get

$$(6.2.23) \quad \frac{1}{\varepsilon} \int_t^{t+\varepsilon} \int_{|x|<R} [u(x, \tau) - \overline{u}(x, \tau)]^+ dx d\tau \leq \int_{|x|<R+st} [u_0(x) - \overline{u}_0(x)]^+ dx$$

$$- \frac{1}{\varepsilon} \int_0^t \int_{R+s(t-\tau)<x<R+s(t-\tau)+\varepsilon} \left\{ s\eta(u; \overline{u}) + \sum_{\alpha=1}^{m} \frac{x_\alpha}{|x|} q_\alpha(u; \overline{u}) \right\} dx d\tau + O(\varepsilon) .$$

On account of (6.2.22), the second integral on the right-hand side of (6.2.23) is nonnegative. Thus, letting $\varepsilon \downarrow 0$, recalling Theorem 4.3.1, and using that $(u - \overline{u})^+$ is a convex function of $u - \overline{u}$, we arrive at (6.2.8).

Interchanging the roles of u and $\overline{u}$ in (6.2.8) we deduce a similar inequality which added to (6.2.8) yields (6.2.9).

Clearly, (6.2.10) implies (6.2.11), by virtue of (6.2.8). In particular, applying this monotonicity property, first for $\overline{u}_0(x) \equiv b$ and then for $u_0(x) \equiv a$, we deduce $u(x, t) \leq b$ and $\overline{u}(x, t) \geq a$ a.e. Interchanging the roles of u and $\overline{u}$, we conclude that the essential range of both solutions is contained in [a,b]. Thus s in (6.2.22) depends solely on [a,b]. This completes the proof.

From (6.2.9) we draw immediately the following conclusion on uniqueness and finite dependence:

Corollary 6.2.1 *There is at most one admissible weak solution of* (6.1.1), (6.1.2).

Corollary 6.2.2 *The value of the admissible weak solution at any point $(\overline{x}, \overline{t})$ depends solely on the restriction of the initial data to the ball $\{x \in \mathbb{R}^m : |x - \overline{x}| < s\overline{t}\}$.*

Another important consequence of (6.2.9), recorded in the following proposition, is that any admissible weak solution of (6.1.1) with initial data of locally bounded variation is itself a function of locally bounded variation. At this point the reader may wish to review the material in Sections 1.7 and 1.8.

Theorem 6.2.3 *Let u be an admissible weak solution of (6.1.1) with initial data $u_0 \in BV_{\text{loc}}(\mathbb{R}^m)$ taking values in a compact interval [a,b]. Then $u \in BV_{\text{loc}}(\mathbb{R}^m \times (0, T))$. For any $t \in (0, T)$, the restriction $u(\cdot, t)$ of u to t is in $BV_{\text{loc}}(\mathbb{R}^m)$ and*

$$(6.2.24) \qquad TV_{\{|x|<R\}}u(\cdot, t) \leq TV_{\{|x|<R+st\}}u_0(\cdot) ,$$

for every $R > 0$, where s depends solely on [a,b].

Proof. Let $\{E_\alpha, \alpha = 1, \cdots, m\}$ denote the standard orthonormal basis of $\mathbb{R}^m$. Note that, for $\alpha = 1, \cdots, m$, the function $\bar{u}$, defined by $\bar{u}(x,t) = u(x+hE_\alpha, t), h > 0$, is an admissible weak solution of (6.1.1) with initial data $\bar{u}_0, \bar{u}_0(x) = u_0(x+hE_\alpha)$. Therefore, by virtue of (6.2.9), for any $t \in (0,T)$,

$$(6.2.25) \quad \int_{|x|<R} |u(x+hE_\alpha, t) - u(x,t)|dx \leq \int_{|x|<R+st} |u_0(x+hE_\alpha) - u_0(x)|dx .$$

Since $u_0 \in BV_{\text{loc}}(\mathbb{R}^m)$, Theorem 1.7.1 and (1.7.2) yield that $u(\cdot, t) \in BV_{\text{loc}}(\mathbb{R}^m)$ and (6.2.24) holds.

Thus $\partial_\alpha u(\cdot, t)$ is a Radon measure which is bounded on any ball of radius R in $\mathbb{R}^m$, uniformly on compact time intervals. Since u is bounded, it follows from Theorem 1.7.4 that $\partial_\alpha g_\alpha(u(\cdot, t))$ has the same property. In particular, for $\alpha = 1, \cdots, m$, the distributions $\partial_\alpha u$ and $\partial_\alpha g_\alpha(u)$ are locally finite measures on $\mathbb{R}^m \times (0,T)$. Because (6.1.1) is satisfied in the sense of distributions, $\partial_t u$ will also be a measure on $\mathbb{R}^m \times (0,T)$. Consequently, $u \in BV_{\text{loc}}(\mathbb{R}^m \times (0,T))$. This completes the proof.

The trivial, constant, solutions of (6.1.1) are stable, not only in L^1 but also in any L^p. Since u may be renormalized, it suffices to establish L^p-stability for the zero solution.

Theorem 6.2.4 *Let u be an admissible weak solution of (6.1.1), (6.1.2), with initial data taking values in a compact interval [a,b]. There is $s > 0$, depending solely on [a,b], such that, for any $1 \leq p \leq \infty, t \in [0,T)$, and $R > 0$,*

$$(6.2.26) \quad \|u(\cdot,t)\|_{L^p(|x|<R)} \leq \|u_0(\cdot)\|_{L^p(|x|<R+st)} .$$

Proof. For $1 \leq p < \infty$, consider the convex entropy $\eta(u) = |u|^p$, with entropy flux $(q_1, \cdots, q_m)$ determined through (6.2.1). Note that there is $s > 0$, independent of p, such that

$$(6.2.27) \quad \left[\sum_{\alpha=1}^m |q_\alpha(u)|^2\right]^{1/2} \leq s\eta(u), \quad u \in [a,b] .$$

Fix $R > 0$, $t \in [0,T)$, and $\varepsilon > 0$ small; write (6.2.3) for the above entropy-entropy flux pair and the test function $\psi(x,\tau) = \chi(x,\tau)\theta(\tau)$, with χ and θ defined by (5.2.12) and (5.2.11). This yields

$$(6.2.28) \quad \frac{1}{\varepsilon}\int_t^{t+\varepsilon}\int_{|x|<R} |u(x,\tau)|^p dxd\tau \leq \int_{|x|<R+st} |u_0(x)|^p dx$$

$$-\frac{1}{\varepsilon}\int_0^t \int_{R+s(t-\tau)<|x|<R+s(t-\tau)+\varepsilon} \left\{s\eta(u) + \sum_{\alpha=1}^m \frac{x_\alpha}{|x|}q_\alpha(u)\right\}dxd\tau + O(\varepsilon) .$$

We know that the range of u is contained in [a,b] and so, by (6.2.27), the second integral on the right-hand side of (6.2.28) is nonnegative. Thus, letting $\varepsilon \downarrow 0$ and using that $|u|^p$ is convex, we arrive at (6.2.26). This completes the proof.

The following sections will present various methods of constructing admissible weak solutions of (6.1.1), (6.1.2), inducing alternative proofs of Theorem 6.1.1.

6.3 The Method of Vanishing Viscosity

The aim here is to construct admissible weak solutions of the scalar hyperbolic conservation law (6.1.1) as the $\mu \downarrow 0$ limit of solutions of the family of parabolic equations

$$(6.3.1) \quad \partial_t u(x,t) + \sum_{\alpha=1}^{m} \partial_\alpha g_\alpha(u(x,t)) = \mu \Delta u(x,t) , \quad x \in \mathbb{R}^m , \quad t \in [0, \infty) ,$$

where Δ stands for Laplace's operator with respect to the space variables, $\Delta = \sum_{\alpha=1}^{m} \partial_\alpha^2$, and μ is a positive parameter.

The motivation for this approach has already been presented in Section 4.4. Note that (6.3.1) is not necessarily related to any specific physical model and so the term $\mu \Delta u$ should be regarded as "artificial viscosity".

Because (6.3.1) is parabolic, the initial value problem (6.3.1), (6.1.2) always has a unique solution, which is smooth for $t > 0$ (assuming the g_α are regular) even when the initial data u_0 are merely in L^∞. For example, if the derivatives g'_α are Hölder continuous, then the solution u of (6.3.1), (6.1.2) is continuously differentiable with respect to t and twice continuously differentiable with respect to the space variables, on $\mathbb{R}^m \times (0, \infty)$.

Espousing the premise that "interesting" solutions of (6.1.1), (6.1.2) are $\mu \downarrow 0$ limits of solutions of (6.3.1), (6.1.2), provides the first justification of the notion of admissible weak solution postulated by Definition 6.2.1:

Theorem 6.3.1 *Let* u_μ *denote the solution of* (6.3.1), (6.1.2). *Assume that for some sequence* $\{\mu_k\}$, *with* $\mu_k \downarrow 0$ *as* $k \to \infty$, $\{u_{\mu_k}\}$ *converges to some function* u, *boundedly almost everywhere on* $\mathbb{R}^m \times [0, \infty)$. *Then* u *is an admissible weak solution of* (6.1.1), (6.1.2) *on* $\mathbb{R}^m \times [0, \infty)$.

Proof. Consider any smooth convex entropy function η, with associated entropy flux $(q_1, \cdots, q_m)$ determined through (6.2.1). Multiply (6.3.1) by $\eta'(u_\mu(x,t))$ and use (6.2.1) to get

$$(6.3.2) \qquad \partial_t \eta(u_\mu) + \sum_{\alpha=1}^{m} \partial_\alpha q_\alpha(u_\mu) = \mu \Delta \eta(u_\mu) - \mu \eta''(u_\mu) \sum_{\alpha=1}^{m} |\partial_\alpha u_\mu|^2 .$$

Multiply (6.3.2) by any smooth nonnegative test function ψ, with compact support in $\mathbb{R}^m \times [0, \infty)$, integrate over $\mathbb{R}^m \times [0, \infty)$, and integrate by parts. Taking into account that the last term in (6.3.2) is nonnegative yields the inequality

$$(6.3.3) \quad \int_0^\infty \int_{\mathbb{R}^m} \left[\partial_t \psi \eta(u_\mu) + \sum_{\alpha=1}^m \partial_\alpha \psi q_\alpha(u_\mu) \right] dx dt + \int_{\mathbb{R}^m} \psi(x,0) \eta(u_0(x)) dx$$

$$\geq -\mu \int_0^\infty \int_{\mathbb{R}^m} \Delta \psi \eta(u_\mu) dx dt .$$

Setting $\mu = \mu_k$ in (6.3.3) and letting $k \to \infty$, we conclude that the limit u of $\{u_{\mu_k}\}$ satisfies (6.2.3) for all smooth convex entropy functions η and all smooth nonnegative test functions ψ. By completion we infer that (6.2.3) holds even when η and ψ are merely Lipschitz continuous. This completes the proof.

That (6.1.1) and (6.3.1) are perfectly matched becomes clear by comparing Theorem 6.2.2 with

Theorem 6.3.2 *Let u and $\bar{u}$ be solutions of (6.3.1) with respective initial data u_0 and $\bar{u}_0$ that are in $L^1(\mathbb{R}^m)$ and take values in a compact interval [a,b]. Then, for any $t > 0$,*

$$(6.3.4) \quad \int_{\mathbb{R}^m} [u(x,t) - \bar{u}(x,t)]^+ dx \leq \int_{\mathbb{R}^m} [u_0(x) - \bar{u}_0(x)]^+ dx ,$$

$$(6.3.5) \quad \|u(\cdot,t) - \bar{u}(\cdot,t)\|_{L^1(\mathbb{R}^m)} \leq \|u_0(\cdot) - \bar{u}_0(\cdot)\|_{L^1(\mathbb{R}^m)} .$$

Furthermore, if

$$(6.3.6) \quad u_0(x) \leq \bar{u}_0(x) , \quad a.e. \ on \ \mathbb{R}^m ,$$

then

$$(6.3.7) \quad u(x,t) \leq \bar{u}(x,t) , \quad on \ \mathbb{R}^m \times (0,\infty) .$$

In particular, the range of both u and $\bar{u}$ is contained in [a,b].

Proof. From standard theory of parabolic equations follows that when $u_0(\cdot), \bar{u}_0(\cdot)$ are in $L^1(\mathbb{R}^m) \cap L^\infty(\mathbb{R}^m)$, then $u(\cdot,t), \bar{u}(\cdot,t)$ and their spatial derivatives of any order are also in $L^1(\mathbb{R}^m) \cap L^\infty(\mathbb{R}^m)$, with norms uniformly bounded with respect to t on compact subsets of $(0,\infty)$.

For $\varepsilon > 0$, we define the function η_ε on $\mathbb{R}$ by

$$(6.3.8) \quad \eta_\varepsilon(w) = \begin{cases} 0 & \text{for } -\infty < w \leq 0 \\ \dfrac{w^2}{4\varepsilon} & \text{for } 0 < w \leq 2\varepsilon \\ w - \varepsilon & \text{for } 2\varepsilon < w < \infty . \end{cases}$$

Using that both u and $\bar{u}$ satisfy (6.3.1), one easily verifies the equation

(6.3.9)
$$\partial_t \eta_\varepsilon(u - \overline{u}) + \sum_{\alpha=1}^{m} \partial_\alpha \{\eta_\varepsilon'(u - \overline{u})[g_\alpha(u) - g_\alpha(\overline{u})]\}$$

$$- \sum_{\alpha=1}^{m} \eta_\varepsilon''(u - \overline{u})[g_\alpha(u) - g_\alpha(\overline{u})]\partial_\alpha(u - \overline{u})$$

$$= \mu \Delta \eta_\varepsilon(u - \overline{u}) - \mu \eta_\varepsilon''(u - \overline{u}) \sum_{\alpha=1}^{m} [\partial_\alpha(u - \overline{u})]^2 .$$

Fix $0 < s < t < \infty$ and integrate (6.3.9) over $\mathbb{R}^m \times (s, t)$. Considering that the last term on the right-hand side of (6.3.9) is nonnegative, we thus obtain the inequality

(6.3.10)
$$\int_{\mathbb{R}^m} \eta_\varepsilon(u(x, t) - \overline{u}(x, t))dx - \int_{\mathbb{R}^m} \eta_\varepsilon(u(x, s) - \overline{u}(x, s))dx$$
$$\leq \sum_{\alpha=1}^{m} \int_s^t \int_{\mathbb{R}^m} \eta_\varepsilon''(u - \overline{u})[g_\alpha(u) - g_\alpha(\overline{u})]\partial_\alpha(u - \overline{u})dxd\tau .$$

Notice that $\eta_\varepsilon''(u - \overline{u})[g_\alpha(u) - g_\alpha(\overline{u})]$ is bounded, uniformly for $\varepsilon > 0$. Also, as $\varepsilon \downarrow 0$, $\eta_\varepsilon(u(x, t) - \overline{u}(x, t))$ converges pointwise to $[u(x, t) - \overline{u}(x, t)]^+$ while $\eta_\varepsilon'(u(x, t) - \overline{u}(x, t))[g_\alpha(u(x, t)) - g_\alpha(\overline{u}(x, t))]$ converges pointwise to zero. Therefore, (6.3.10) and the Lebesgue dominated convergence theorem imply

(6.3.11)
$$\int_{\mathbb{R}^m} [u(x, t) - \overline{u}(x, t)]^+dx - \int_{\mathbb{R}^m} [u(x, s) - \overline{u}(x, s)]^+dx \leq 0 ,$$

whence we deduce (6.3.4), by letting $s \downarrow 0$.

Interchanging the roles of u and $\overline{u}$ in (6.3.4) we derive a similar inequality which added to (6.3.4) yields (6.3.5).

Clearly, (6.3.6) implies (6.3.7), by virtue of (6.3.4). In particular, applying this monotonicity property, first for $\overline{u}_0(x) \equiv b$ and then for $u_0(x) \equiv a$, we deduce $u(x, t) \leq b$ and $\overline{u}(x, t) \geq a$. Interchanging the roles of u and $\overline{u}$, we conclude that the range of both solutions is contained in [a,b]. This completes the proof.

Estimate (6.3.5) may be employed to estimate the modulus of continuity in the mean of solutions of (6.3.1) with initial data in $L^\infty(\mathbb{R}^m) \cap L^1(\mathbb{R}^m)$.

Lemma 6.3.1 *Let u be the solution of* (6.3.1), (6.1.2), *where u_0 is in $L^1(\mathbb{R}^m)$ and takes values in a compact interval [a,b]. In particular,*

(6.3.12)
$$\int_{\mathbb{R}^m} |u_0(x + y) - u_0(x)|dx \leq \omega(|y|) , \quad y \in \mathbb{R}^m ,$$

for some nondecreasing function ω on $[0, \infty)$, with $\omega(r) \downarrow 0$ as $r \downarrow 0$. There is a constant c, depending solely on [a,b], such that, for any $t > 0$,

(6.3.13)
$$\int_{\mathbb{R}^m} |u(x + y, t) - u(x, t)|dx \leq \omega(|y|) , \quad y \in \mathbb{R}^m ,$$

(6.3.14)
$$\int_{\mathbb{R}^m} |u(x, t + h) - u(x, t)|dx \leq c(h^{2/3} + \mu h^{1/3})\|u_0\|_{L^1(\mathbb{R}^m)} + 4\omega(h^{1/3}) , \quad h > 0 .$$

Proof. Fix $t > 0$. For any $y \in \mathbb{R}^m$, the function $\bar{u}(x, t) = u(x + y, t)$ is the solution of (6.3.1) with initial data $\bar{u}_0(x) = u_0(x + y)$. Applying (6.3.5) yields

$$(6.3.15) \qquad \int_{\mathbb{R}^m} |u(x + y, t) - u(x, t)| dx \leq \int_{\mathbb{R}^m} |u_0(x + y) - u_0(x)| dx$$

whence (6.3.13) follows.

We now fix $h > 0$. For $\alpha = 1, \cdots, m$, we normalize g_α by subtracting $g_\alpha(0)$ so henceforth we may assume, without loss of generality, that $g_\alpha(0) = 0$. We multiply (6.3.1) by a bounded smooth function ϕ, defined on $\mathbb{R}^m$, and integrate the resulting equation over $\mathbb{R}^m \times (t, t + h)$. Integration by parts yields

$$(6.3.16) \qquad \int_{\mathbb{R}^m} \phi(x)[u(x, t + h) - u(x, t)] dx$$

$$= \int_t^{t+h} \int_{\mathbb{R}^m} \left\{ \sum_{\alpha=1}^m \partial_\alpha \phi(x) g_\alpha(u(x, \tau)) + \mu \Delta \phi(x) u(x, \tau) \right\} dx d\tau .$$

Let us set

$$(6.3.17) \qquad v(x) = u(x, t + h) - u(x, t) .$$

To establish (6.3.14), we would wish to insert $\phi(x) = \operatorname{sgn} v(x)$ in (6.3.16). However, since the function sgn is discontinuous, we have to mollify it first, with the help of a smooth, nonnegative function ρ on $\mathbb{R}$, with support contained in $[-m^{1/2}, m^{1/2}]$ and total mass one, (6.2.15):

$$(6.3.18) \qquad \phi(x) = \int_{\mathbb{R}^m} h^{-m/3} \prod_{\alpha=1}^m \rho \left(\frac{x_\alpha - z_\alpha}{h^{1/3}} \right) \operatorname{sgn} v(z) dz .$$

Notice that $|\partial_\alpha \phi| \leq c_1 h^{-1/3}$ and $|\Delta \phi| \leq c_2 h^{-2/3}$. Moreover, by virtue of (6.3.5), with $\bar{u} \equiv 0$, $\|u(\cdot, \tau)\|_{L^1(\mathbb{R}^m)} \leq \|u_0(\cdot)\|_{L^1(\mathbb{R}^m)}$. Therefore, (6.3.16) implies

$$(6.3.19) \qquad \int_{\mathbb{R}^m} \phi(x) v(x) dx \leq c(h^{2/3} + \mu h^{1/3}) \|u_0\|_{L^1(\mathbb{R}^m)} ,$$

where c depends solely on [a,b]. On the other hand, observing that

(6.3.20)
$$|v(x)| - v(x) \operatorname{sgn} v(z) = |v(x)| - |v(z)| + [v(z) - v(x)] \operatorname{sgn} v(z) \leq 2|v(x) - v(z)| ,$$

we obtain from (6.3.18):

$$(6.3.21) \quad |v(x)| - \phi(x) v(x) = \int_{\mathbb{R}^m} h^{-m/2} \prod_{\alpha=1}^m \rho(\frac{x_\alpha - z_\alpha}{h^{1/3}})[|v(x)| - v(x) \operatorname{sgn} v(z)] dz$$

$$\leq 2 \int_{|\xi| < 1} \prod_{\alpha=1}^m \rho(\xi_\alpha) |v(x) - v(x - h^{1/3}\xi)| d\xi .$$

Combining (6.3.17), (6.3.21), (6.3.19), and (6.3.13), we arrive at (6.3.14). This completes the proof.

We have now laid the groundwork for presenting a

Proof of Theorem 6.2.1 Assume first that $u_0 \in L^\infty(\mathbb{R}^m) \cap L^1(\mathbb{R}^m)$. Let u_μ denote the solution of (6.3.1), (6.1.2). By Theorem 6.3.2 and Lemma 6.3.1, the family $\{u_\mu\}$ is uniformly bounded and equicontinuous in the mean on any compact subset of $\mathbb{R}^m \times (0, \infty)$. Consequently, any sequence $\{\mu_k\}$, with $\mu_k \downarrow 0$ as $k \to \infty$, will contain a subsequence, denoted again by $\{\mu_k\}$, such that $\{u_{\mu_k}\}$ converges in L^1_{loc}, as well as boundedly almost everywhere on $\mathbb{R}^m \times [0, \infty)$, to some function u. On account of Theorem 6.3.1, u is an admissible weak solution of (6.1.1), (6.1.2). Since there may exist at most one such solution (cf. Corollary 6.2.1), we conclude that the whole family $\{u_\mu\}$ converges to u, as $\mu \downarrow 0$. Furthermore, by virtue of Lemma 6.3.1, for $h > 0$,

$$(6.3.22) \qquad \int_{\mathbb{R}^m} |u(x, t + h) - u(x, t)| dx \le ch^{2/3} \|u_0\|_{L^1(\mathbb{R}^m)} + 4\omega(h^{1/3}) ,$$

so $u(\cdot, t) \in C^0([0, \infty); L^1(\mathbb{R}^m))$.

Suppose now $u_0 \in L^\infty(\mathbb{R}^m)$. For $R > 0$, let χ_R denote the function on $\mathbb{R}^m$, with $\chi_R(x) = 1$ if $|x| \le R$ and $\chi_R(x) = 0$ if $|x| > R$. Let u^R denote the admissible weak solution of (6.1.1), with initial data $\chi_R u_0 \in L^\infty(\mathbb{R}^m) \cap L^1(\mathbb{R}^m)$. As $R \to \infty$, $\chi_R u_0 \to u_0$ in $L^1_{\text{loc}}(\mathbb{R}^m)$. Therefore, by account of estimate (6.2.9), the family $\{u^R\}$ will converge in L^1_{loc} to some function u. Clearly, u is an admissible weak solution of (6.1.1), (6.1.2). By Corollary 6.2.1, this solution is unique. Moreover, by Corollary 6.2.2, on any compact subset of $\mathbb{R}^m \times [0, \infty)$, $u \equiv u^R$ if R is sufficiently large. Since $u^R(\cdot, t) \in C^0([0, \infty); L^1(\mathbb{R}^m))$, it follows that $u(\cdot, t) \in C^0([0, \infty); L^1_{\text{loc}}(\mathbb{R}^m))$. This completes the proof.

6.4 Solutions as Trajectories of a Contraction Semigroup

For $t \in [0, \infty)$, consider the map $S(t)$ that carries $u_0 \in L^\infty(\mathbb{R}^m) \cap L^1(\mathbb{R}^m)$ to the admissible weak solution u of (6.1.1), (6.1.2) restricted to t, i.e., $S(t)u_0(\cdot) = u(\cdot, t)$. By virtue of the properties of admissible weak solutions demonstrated in the previous two sections, $S(t)$ is well-defined as a map from $L^\infty(\mathbb{R}^m) \cap L^1(\mathbb{R}^m)$ to $L^\infty(\mathbb{R}^m) \cap L^1(\mathbb{R}^m)$ and

$$(6.4.1) \qquad\qquad S(0) = I \quad \text{(the identity)} ,$$

$$(6.4.2) \qquad S(t + \tau) = S(t)S(\tau) , \quad \text{for any } t \text{ and } \tau \text{ in } [0, \infty) ,$$

$$(6.4.3) \qquad\qquad S(\cdot)u_0 \in C^0([0, \infty); L^1(\mathbb{R}^m)) ,$$

$$(6.4.4) \quad \|S(t)u_0 - S(t)\bar{u}_0\|_{L^1(\mathbb{R}^m)} \le \|u_0 - \bar{u}_0\|_{L^1(\mathbb{R}^m)} , \quad \text{for any } t \text{ in } [0, \infty) .$$

Consequently, $S(\cdot)$ is a L^1-contraction semigroup on $L^\infty(\mathbb{R}^m) \cap L^1(\mathbb{R}^m)$.

Naturally, the question arises whether one may construct $S(\cdot)$ *ab initio*, through the theory of nonlinear contraction semigroups in Banach space. This would provide a direct, independent proof of existence of admissible weak solutions of (6.1.1), (6.1.2) as well as an alternative derivation of their properties.

To construct the semigroup, we should realize (6.1.1) as an abstract differential equation

$$(6.4.5) \qquad \frac{du}{dt} + A(u) \ni 0,$$

for a suitably defined nonlinear transformation A, with domain $\mathscr{D}(A)$ and range $\mathscr{R}(A)$ in $L^1(\mathbb{R}^m)$. This operator may, in general, be multivalued, i.e., for $u \in \mathscr{D}(A)$, $A(u)$ will be a nonempty subset of $L^1(\mathbb{R}^m)$ that may contain more than one point.

For u smooth, one should expect $A(u) = \sum \partial_\alpha g_\alpha(u)$. However, the task of extending $\mathscr{D}(A)$ to u that are not smooth is by no means straightforward, because the construction should reflect somehow the admissibility condition encoded in Definition 6.2.1. First we perform a preliminary extension. For convenience, we normalize the g_α so that $g_\alpha(0) = 0$, $\alpha = 1, \cdots, m$.

Definition 6.4.1 The (possibly multivalued) transformation $\hat{A}$, with domain $\mathscr{D}(\hat{A}) \subset L^1(\mathbb{R}^m)$, is determined by $u \in \mathscr{D}(\hat{A})$ and $w \in \hat{A}(u)$ if u, w and $g_\alpha(u)$, $\alpha = 1, \cdots, m$, are all in $L^1(\mathbb{R}^m)$ and the inequality

$$(6.4.6) \qquad \int_{\mathbb{R}^m} \left\{ \sum_{\alpha=1}^m \partial_\alpha \psi(x) q_\alpha(u(x)) + \psi(x) \eta'(u(x)) w(x) \right\} dx \geq 0$$

holds for any convex entropy function η, such that η' is bounded on $\mathbb{R}$, with associated entropy flux $(q_1, \cdots, q_m)$ determined through (6.2.1), and for all nonnegative Lipschitz continuous test functions ψ on $\mathbb{R}^m$, with compact support.

Applying (6.4.6) for the entropy-entropy flux pairs $\pm u$, $\pm g_\alpha(u)$, $\alpha = 1, \cdots, m$, verifies that

$$(6.4.7) \qquad \hat{A}(u) = \sum_{\alpha=1}^m \partial_\alpha g_\alpha(u)$$

holds, in the sense of distributions, for any $u \in \mathscr{D}(\hat{A})$. In particular, $\hat{A}$ is single-valued. Furthermore, the identity

$$(6.4.8) \qquad \int_{\mathbb{R}^m} \left\{ \sum_{\alpha=1}^m \partial_\alpha \psi q_\alpha(u) + \psi \eta'(u) \sum_{\alpha=1}^m \partial_\alpha g_\alpha(u) \right\} dx = 0 \, ,$$

which is valid for any $u \in C_0^1(\mathbb{R}^m)$ and every entropy-entropy flux pair, implies that $C_0^1(\mathbb{R}^m) \subset \mathscr{D}(\hat{A})$. In particular, $\mathscr{D}(\hat{A})$ is dense in $L^1(\mathbb{R}^m)$. For $u \in C_0^1(\mathbb{R}^m)$, $\hat{A}(u)$ is given by (6.4.7). Thus $\hat{A}$ is indeed an extension of (6.4.7).

The reader may have already noticed the similarity between (6.4.6) and (6.2.3). Similar to (6.2.3), to verify (6.4.6) it would suffice to test it just for the entropies $\pm u$ and the family (6.2.5) or (6.2.6) of entropy-entropy flux pairs.

Definition 6.4.2 The (possibly multivalued) transformation A, with domain $\mathscr{D}(A)$ $\subset L^1(\mathbb{R}^m)$, is the graph closure of $\hat{A}$, i.e., $u \in \mathscr{D}(A)$ and $w \in A(u)$ if (u, w) is the limit in $L^1(\mathbb{R}^m) \times L^1(\mathbb{R}^m)$ of a sequence $\{(u_k, w_k)\}$ such that $u_k \in \mathscr{D}(\hat{A})$ and $w_k \in \hat{A}(u_k)$.

The following propositions establish properties of A, implying that it is the generator of a contraction semigroup on $L^1(\mathbb{R}^m)$.

Theorem 6.4.1 *The transformation A is accretive, that is if u and $\bar{u}$ are in $\mathscr{D}(A)$, then*

$$(6.4.9) \qquad \|(u + \lambda w) - (\bar{u} + \lambda \bar{w})\|_{L^1(\mathbb{R}^m)} \geq \|u - \bar{u}\|_{L^1(\mathbb{R}^m)} ,$$
$$\lambda > 0 , \quad w \in A(u) , \quad \bar{w} \in A(\bar{u}) .$$

Proof. It is the property of accretiveness that renders the semigroup generated by A contractive. Consequently, the proof of Theorem 6.4.1 bears close resemblance to the demonstration of the L^1-contraction estimate (6.2.9) in Theorem 6.2.2.

In view of Definition 6.4.2, it would suffice to show that the "smaller" transformation $\hat{A}$ is accretive. Accordingly, fix $u, \bar{u}$ in $\mathscr{D}(\hat{A})$ and let $w = \hat{A}(u)$, $\bar{w} = \hat{A}(\bar{u})$. Consider any nonnegative Lipschitz continuous function ϕ on $\mathbb{R}^m \times \mathbb{R}^m$, with compact support. Fix $\bar{x}$ in $\mathbb{R}^m$ and write (6.4.6) for the entropy-entropy flux pair $\eta(u; \bar{u}(\bar{x})), q_\alpha(u; \bar{u}(\bar{x}))$ of the Kruzkov family (6.2.6) and the test function $\psi(x) = \phi(x, \bar{x})$ to obtain

$$(6.4.10) \qquad \int_{\mathbb{R}^m} \text{sgn}\,[u(x) - \bar{u}(\bar{x})]\left\{ \sum_{\alpha=1}^m \partial_{x_\alpha}\phi(x, \bar{x})[g_\alpha(u(x)) - g_\alpha(\bar{u}(\bar{x}))] \right.$$
$$\left. + \phi(x, \bar{x})w(x) \right\} dx \geq 0 .$$

We may interchange the roles of u and $\bar{u}$ and derive the analog of (6.4.10), for any fixed x in $\mathbb{R}^m$:

$$(6.4.11) \qquad \int_{\mathbb{R}^m} \text{sgn}\,[\bar{u}(\bar{x}) - u(x)]\left\{ \sum_{\alpha=1}^m \partial_{\bar{x}_\alpha}\phi(x, \bar{x})[g_\alpha(\bar{u}(\bar{x})) - g_\alpha(u(x))] \right.$$
$$\left. + \phi(x, \bar{x})\bar{w}(\bar{x}) \right\} d\bar{x} \geq 0 .$$

Integrating over $\mathbb{R}^m$ (6.4.10), with respect to $\bar{x}$, and (6.4.11), with respect to x, and then adding the resulting inequalities yields

$$(6.4.12) \quad \int_{\mathbb{R}^m}\int_{\mathbb{R}^m} \text{sgn}\,[u(x) - \bar{u}(\bar{x})]\left\{ \sum_{\alpha=1}^m (\partial_{x_\alpha} + \partial_{\bar{x}_\alpha})\phi(x, \bar{x})[g_\alpha(u(x)) - g_\alpha(\bar{u}(\bar{x}))] \right.$$
$$\left. + \phi(x, \bar{x})[w(x) - \bar{w}(\bar{x})] \right\} dx\, d\bar{x} \geq 0 .$$

Fix a smooth nonnegative function ρ on $\mathbb{R}$ with compact support and total mass one, (6.2.15). Take any nonnegative Lipschitz continuous test function ψ on $\mathbb{R}^m$, with compact support. For positive small ε, write (6.4.12) with

(6.4.13) $$\phi(x, \bar{x}) = \varepsilon^{-m} \psi \left(\frac{x + \bar{x}}{2} \right) \prod_{\alpha=1}^{m} \rho \left(\frac{x_\alpha - \bar{x}_\alpha}{2\varepsilon} \right) ,$$

and let $\varepsilon \downarrow 0$. Noting that

(6.4.14) $$(\partial_{x_\alpha} + \partial_{\bar{x}_\alpha}) \phi(x, \bar{x}) = \varepsilon^{-m} \partial_\alpha \psi \left(\frac{x + \bar{x}}{2} \right) \prod_{\alpha=1}^{m} \rho \left(\frac{x_\alpha - \bar{x}_\alpha}{2\varepsilon} \right) ,$$

(6.4.15)

$$\int_{\mathbb{R}^m} \sigma(x) \left\{ \sum_{\alpha=1}^{m} \partial_\alpha \psi(x) [g_\alpha(u(x)) - g_\alpha(\bar{u}(x))] + \psi(x) [w(x) - \bar{w}(x)] \right\} dx \geq 0 ,$$

where σ is some function such that

(6.4.16) $$\sigma(x) \begin{cases} = 1 & \text{if } u(x) > \bar{u}(x) \\ \in [-1, 1] & \text{if } u(x) = \bar{u}(x) \\ = -1 & \text{if } u(x) < \bar{u}(x) . \end{cases}$$

In particular, choosing ψ with $\psi(x) = 1$ for $|x| < R$, $\psi(x) = 1 + R - |x|$ for $R \leq |x| < R + 1$ and $\psi(x) = 0$ for $R + 1 \leq |x| < \infty$, and letting $R \to \infty$, we obtain

(6.4.17) $$\int_{\mathbb{R}^m} \sigma(x) [w(x) - \bar{w}(x)] dx \geq 0 ,$$

for some function σ as in (6.4.16).

Take now any $\lambda > 0$ and use (6.4.17), (6.4.16) to conclude

(6.4.18)

$$\|(u + \lambda w) - (\bar{u} + \lambda \bar{w})\|_{L^1(\mathbb{R}^m)} \geq \int_{\mathbb{R}^m} \sigma(x) \{u(x) - \bar{u}(x) + \lambda[w(x) - \bar{w}(x)]\} dx$$

$$\geq \int_{\mathbb{R}^m} \sigma(x) [u(x) - \bar{u}(x)] dx = \|u - \bar{u}\|_{L^1(\mathbb{R}^m)} .$$

This completes the proof.

An immediate consequence (actually an alternative, equivalent restatement) of the assertion of Theorem 6.4.1 is

Corollary 6.4.1 *For any $\lambda > 0$, $(I + \lambda A)^{-1}$ is a well-defined, single-valued, L^1-contractive transformation, defined on the range $\mathscr{R}(I + \lambda A)$ of $I + \lambda A$.*

Theorem 6.4.2 *The transformation A is maximal, that is*

(6.4.19) $$\mathscr{R}(I + \lambda A) = L^1(\mathbb{R}^m) , \quad \text{for any } \lambda > 0 .$$

Proof. By virtue of Definition 6.4.2 and Corollary 6.4.1, it will suffice to show that $\mathscr{R}(I + \lambda \hat{A})$ is dense in $L^1(\mathbb{R}^m)$; for instance that it contains $L^1(\mathbb{R}^m) \cap L^\infty(\mathbb{R}^m)$. We thus fix $f \in L^1(\mathbb{R}^m) \cap L^\infty(\mathbb{R}^m)$ and seek solutions $u \in \mathscr{D}(\hat{A})$ of the equation

$$(6.4.20) \qquad\qquad u + \lambda \hat{A}(u) = f \ .$$

Recall that $\hat{A}(u)$ admits the representation (6.4.7), in the sense of distributions. Thus solving (6.4.20) amounts to determining an admissible weak solution of a first order quasilinear partial differential equation, namely the stationary analog of (6.1.1).

Motivated by the method of vanishing viscosity, discussed in Section 6.3, we shall construct solution to (6.4.20) as the $\mu \downarrow 0$ limit of solutions of the family of elliptic equations

$$(6.4.21) \qquad u(x) + \lambda \sum_{\alpha=1}^{m} \partial_\alpha g_\alpha(u(x)) - \mu \Delta u(x) = f(x) \ , \qquad x \in \mathbb{R}^m \ .$$

For any fixed $\mu > 0$, (6.4.21) admits a solution in $H^2(\mathbb{R}^m)$. We have to show that, as $\mu \downarrow 0$, the family of solutions of (6.4.21) converges, boundedly a.e., to some function u which is the solution of (6.4.20). The proof will be partitioned into the following steps.

Lemma 6.4.1 *Let* u *and* $\overline{u}$ *be solutions of* (6.4.21) *with respective right-hand sides* f *and* $\overline{f}$ *that are in* $L^1(\mathbb{R}^m)$ *and take values in a compact interval* [a,b]. *Then*

$$(6.4.22) \qquad \int_{\mathbb{R}^m} [u(x) - \overline{u}(x)]^+ dx \leq \int_{\mathbb{R}^m} [f(x) - \overline{f}(x)]^+ dx \ ,$$

$$(6.4.23) \qquad \|u - \overline{u}\|_{L^1(\mathbb{R}^m)} \leq \|f - \overline{f}\|_{L^1(\mathbb{R}^m)} \ .$$

Furthermore, if

$$(6.4.24) \qquad\qquad f(x) \leq \overline{f}(x) \ , \qquad a.e. \ on \ \mathbb{R}^m$$

then

$$(6.4.25) \qquad\qquad u(x) \leq \overline{u}(x) \ , \qquad on \ \mathbb{R}^m \ .$$

In particular, the range of both u *and* $\overline{u}$ *is contained in* [a,b].

Proof. It is very similar to the proof of Theorem 6.3.2 and so it shall be left to the reader.

Lemma 6.4.2 *Let* u_μ *denote the solution of* (6.4.21), *with right-hand side* f *in* $L^\infty(\mathbb{R}^m) \cap L^1(\mathbb{R}^m)$. *Then, as* $\mu \downarrow 0$, $\{u_\mu\}$ *converges boundedly a.e. to the solution* u *of* (6.4.20).

Proof. For any $y \in \mathbb{R}^m$, the function $\overline{u}_\mu$, defined by $\overline{u}_\mu(x) = u_\mu(x + y)$, is a solution of (6.4.21) with right-hand side $\overline{f}$, $\overline{f}(x) = f(x + y)$. Hence, by (6.4.23),

$$(6.4.26) \qquad \int_{\mathbb{R}^m} |u_\mu(x + y) - u_\mu(x)| dx \leq \int_{\mathbb{R}^m} |f(x + y) - f(x)| dx \ .$$

Thus the family $\{u_\mu\}$ is uniformly bounded and uniformly equicontinuous in L^1. It follows that every sequence $\{\mu_k\}$, with $\mu_k \to 0$ as $k \to \infty$, will contain a subsequence, labeled again as $\{\mu_k\}$, such that

$$(6.4.27) \qquad u_{\mu_k} \to u , \quad \text{boundedly a.e. on } \mathbb{R}^m ,$$

where u is in $L^\infty(\mathbb{R}^m) \cap L^1(\mathbb{R}^m)$.

Consider now any smooth convex entropy function η, with associated entropy flux $(q_1, \cdots, q_m)$, determined by (6.2.1). Then u_μ will satisfy the identity

$$(6.4.28)$$

$$\eta'(u_\mu)u_\mu + \lambda \sum_{\alpha=1}^m \partial_\alpha q_\alpha(u_\mu) - \mu \Delta \eta(u_\mu) + \mu \eta''(u_\mu) \sum_{\alpha=1}^m (\partial_\alpha u_\mu)^2 = \eta'(u_\mu)f .$$

Multiplying (6.4.28) by any nonnegative smooth test function ψ on $\mathbb{R}^m$, with compact support, and integrating over $\mathbb{R}^m$ yields

$$(6.4.29) \quad \int_{\mathbb{R}^m} \left\{ \lambda \sum_{\alpha=1}^m \partial_\alpha \psi q_\alpha(u_\mu) + \psi \eta'(u_\mu)(f - u_\mu) \right\} dx \geq -\mu \int_{\mathbb{R}^m} \Delta \psi \eta dx .$$

From (6.4.27) and (6.4.29),

$$(6.4.30) \qquad \int_{\mathbb{R}^m} \left\{ \sum_{\alpha=1}^m \partial_\alpha \psi q_\alpha(u) + \psi \eta'(u) \frac{1}{\lambda}(f - u) \right\} dx \geq 0 ,$$

which shows that u is indeed a solution of (6.4.20).

By virtue of Corollary 6.4.1, the solution of (6.4.20) is unique and so the entire family $\{u_\mu\}$ converges to u, as $\mu \downarrow 0$. This completes the proof.

Once accretiveness and maximality have been established, the Crandall-Liggett theory of semigroups in nonreflexive Banach space ensures that A generates a contraction semigroup $S(\cdot)$ on $\mathscr{D}(A) = L^1(\mathbb{R}^m)$. $S(\cdot)u_0$ can be constructed by solving the differential equation (6.4.5) through the implicit difference scheme

$$(6.4.31) \qquad \begin{cases} \dfrac{1}{\varepsilon}[u_\varepsilon(t) - u_\varepsilon(t - \varepsilon)] + A(u_\varepsilon(t)) \ni 0 , & t > 0 , \\ u_\varepsilon(t) = u_0 , & t < 0 . \end{cases}$$

For any $\varepsilon > 0$, a unique solution u_ε of (6.4.31) exists on $[0, \infty)$, by virtue of Theorem 6.4.2 and Corollary 6.4.1. It can be shown, further, that Corollary 6.4.1 provides the necessary stability to ensure that, as $\varepsilon \downarrow 0$, $u_\varepsilon(\cdot)$ converges, uniformly on compact subsets of $[0, \infty)$, to some function that we denote by $S(\cdot)u_0$.

The general properties of $S(\cdot)$ follow from the Crandall-Liggett theory: When $u_0 \in \mathscr{D}(A)$, $S(t)u_0$ stays in $\mathscr{D}(A)$ for all $t \in [0, \infty)$. In general, $S(t)u_0$ may fail to be differentiable, with respect to t, even when $u_0 \in \mathscr{D}(A)$. Thus $S(\cdot)u_0$ should be interpreted as a weak solution of the differential equation (6.4.5).

The special properties of $S(\cdot)$ are consequences of the special properties of A induced by the propositions recorded above (e.g. Lemma 6.4.1). The following theorem, whose proof can be found in the references cited in Section 6.9, summarizes the properties of $S(\cdot)$ and, in particular, provides an alternative proof for the existence of a unique admissible weak solution to (6.1.1), (6.1.2) (Theorem 6.2.1) and its basic properties (Theorems 6.2.2 and 6.2.3).

Theorem 6.4.3 *The transformation* A *generates a contraction semigroup* $S(\cdot)$ *in* $L^1(\mathbb{R}^m)$, *namely, a family of maps* $S(t) : L^1(\mathbb{R}^m) \to L^1(\mathbb{R}^m)$, $t \in [0, \infty)$, *which satisfy the semigroup property* (6.4.1), (6.4.2); *the continuity property* (6.4.3), *for any* $u_0 \in L^1(\mathbb{R}^m)$; *and the contraction property* (6.4.4), *for any* $u_0, \bar{u}_0$ *in* $L^1(\mathbb{R}^m)$. *If*

$$(6.4.32) \qquad u_0 \leq \bar{u}_0 , \quad a.e. \ on \ \mathbb{R}^m ,$$

then

$$(6.4.33) \qquad S(t)u_0 \leq S(t)\bar{u}_0 , \quad a.e. \ on \ \mathbb{R}^m .$$

For $1 \leq p \leq \infty$, *the sets* $L^p(\mathbb{R}^m) \cap L^1(\mathbb{R}^m)$ *are positively invariant under* $S(t)$ *and, for any* $t \in [0, \infty)$,

$$(6.4.34) \qquad \|S(t)u_0\|_{L^p(\mathbb{R}^m)} \leq \|u_0\|_{L^p(\mathbb{R}^m)} , \quad for \ all \ u_0 \in L^p(\mathbb{R}^m) \cap L^1(\mathbb{R}^m) .$$

If $u_0 \in L^\infty(\mathbb{R}^m) \cap L^1(\mathbb{R}^m)$, *then* $S(\cdot)u_0$ *is the admissible weak solution of* (6.1.1), (6.1.2), *in the sense of Definition* 6.2.1.

The reader should note that the approach via semigroups suggests a notion of admissible weak solution to (6.1.1), (6.1.2) for any, even unbounded, u_0 in $L^1(\mathbb{R}^m)$. These are not necessarily distributional solutions of (6.1.1), unless the fluxes g_α exhibit linear growth at infinity.

6.5 The Layering Method

The admissible weak solution of (6.1.1), (6.1.2) will here be determined as the $h \downarrow 0$ limit of a family $\{u_h\}$ of functions constructed by patching together classical solutions of (6.1.1) in a stratified pattern. In addition to providing another method for constructing solutions and thereby an alternative proof of the existence Theorem 6.2.1, this approach also offers a different justification of the admissibility condition, Definition 6.2.1.

The initial data u_0 are in $L^\infty(\mathbb{R}^m)$, taking values in a compact interval $[a, b]$. The construction of approximate solutions will involve mollification of functions on $\mathbb{R}^m$ by forming their convolution with a kernel λ_h constructed as follows. We start out with a nonnegative, smooth function ρ on $\mathbb{R}$, supported in $[-1, 1]$, which is even, $\rho(-\xi) = \rho(\xi)$ for $\xi \in \mathbb{R}$, and has total mass one, (6.2.15). For $h > 0$, we set

$$(6.5.1) \qquad \lambda_h(x) = (ph)^{-m} \prod_{\alpha=1}^{m} \rho\left(\frac{x_\alpha}{ph}\right) ,$$

with

(6.5.2) $$p = \sqrt{m}\, q\gamma \|u_0\|_{L^\infty(\mathbb{R}^m)} ,$$

where q denotes the total variation of the function ρ and γ is the maximum of $|g''(u)|$ over the interval $[a, b]$. We employ λ_h to mollify functions $f \in L^\infty(\mathbb{R}^m)$:

(6.5.3) $$(\lambda_h * f)(x) = \int_{\mathbb{R}^m} \lambda_h(x - y) f(y) dy , \quad x \in \mathbb{R}^m .$$

From (6.5.3) and (6.5.1) follows easily

(6.5.4) $$\inf(\lambda_h * f) \geq \operatorname{ess\,inf} f , \quad \sup(\lambda_h * f) \leq \operatorname{ess\,sup} f ,$$

(6.5.5) $$\|\lambda_h * f\|_{L^1(|x|<R)} \leq \|f\|_{L^1(|x|<R+\sqrt{m}ph)} , \quad \text{for any } R > 0 ,$$

(6.5.6) $$\|\partial_\alpha(\lambda_h * f)\|_{L^\infty(\mathbb{R}^m)} \leq \frac{q}{ph}\|f\|_{L^\infty(\mathbb{R}^m)} , \quad \alpha = 1, \cdots, m .$$

A somewhat subtler estimate, which depends crucially on that λ_h is an even function, and whose proof can be found in the references cited in Section 6.9, is

(6.5.7) $$\left| \int_{\mathbb{R}^m} \chi(x)[(\lambda_h * f)(x) - f(x)]dx \right| \leq ch^2 \|\chi\|_{C^2(\mathbb{R}^m)} \|f\|_{L^\infty(\mathbb{R}^m)} ,$$

for all $\chi \in C_0^\infty(\mathbb{R}^m)$.

The construction of the approximate solutions proceeds as follows. After the parameter $h > 0$ has been fixed, $\mathbb{R}^m \times [0, \infty)$ is partitioned into layers:

(6.5.8) $$\mathbb{R}^m \times [0, \infty) = \bigcup_{\ell=0}^{\infty} \mathbb{R}^m \times [\ell h, \ell h + h) .$$

$u_h(\cdot, 0)$ is determined by

(6.5.9) $$u_h(\cdot, 0) = \lambda_h * u_0(\cdot) .$$

By virtue of (6.5.6) and (6.5.2), $u_h(\cdot, 0)$ is Lipschitz continuous, with Lipschitz constant $\omega = 1/p\gamma$. Hence, by Theorem 6.1.1, (6.1.1) with initial data $u_h(\cdot, 0)$ admits a classical solution u_h on the layer $\mathbb{R}^m \times [0, h)$.

Next we determine $u_h(\cdot, h)$ by mollifying the limit $u_h(\cdot, h-)$ of $u_h(\cdot, t)$ as $t \uparrow h$:

(6.5.10) $$u_h(\cdot, h) = \lambda_h * u_h(\cdot, h-) .$$

We extend u_h to the layer $\mathbb{R}^m \times [h, 2h)$ by solving (6.1.1) with data $u_h(\cdot, h)$ at $t = h$.

Continuing this process, we determine u_h on the general layer $[\ell h, \ell h + h)$ by solving (6.1.1) with data

(6.5.11) $$u_h(\cdot, \ell h) = \lambda_h * u_h(\cdot, \ell h-)$$

at $t = \ell h$. We thus end up with a measurable function u_h on $\mathbb{R}^m \times [0, \infty)$ which takes values in the interval $[a, b]$. Inside each layer $\mathbb{R}^m \times [\ell h, \ell h + h)$, u_h is a classical solution of (6.1.1). However, as one crosses the border $t = \ell h$ between adjacent layers, u_h experiences jump discontinuities, from $u_h(\cdot, \ell h-)$ to $u_h(\cdot, \ell h)$.

Theorem 6.5.1 *As $h \downarrow 0$, the family $\{u_h\}$ constructed above converges boundedly almost everywhere on $\mathbb{R}^m \times [0, \infty)$ to the admissible solution u of* (6.1.1), (6.1.2).

The proof is an immediate consequence of the following two propositions and uniqueness of the admissible solution, Corollary 6.2.1. The fact that the limit of classical solutions yields the admissible weak solution provides another justification of Definition 6.2.1.

Lemma 6.5.1 (*Consistency*). *Assume that for some sequence $\{h_k\}$, with $h_k \to 0$ as $k \to \infty$,*

$$(6.5.12) \qquad u_{h_k}(x, t) \to u(x, t) , \quad a.e. \text{ on } \mathbb{R}^m \times [0, \infty) .$$

Then u is an admissible weak solution of (6.1.1), (6.1.2).

Proof. Consider any convex entropy function η with associated entropy flux $(q_1, \cdots, q_m)$ determined through (6.2.1). In the interior of each layer, u_h is a classical solution of (6.1.1) and so it satisfies the identity

$$(6.5.13) \qquad \partial_t \eta(u_h(x, t)) + \sum_{\alpha=1}^{m} \partial_\alpha q_\alpha(u_h(x, t)) = 0 .$$

Fix any nonnegative smooth test function ψ on $\mathbb{R}^m \times [0, T)$, with compact support. Multiply (6.5.13) by ψ, integrate over each layer, integrate by parts, and then sum the resulting equations over all layers to get

$$(6.5.14) \quad \int_0^\infty \int_{\mathbb{R}^m} \left[\partial_t \psi \eta(u_h) + \sum_{\alpha=1}^{m} \partial_\alpha \psi q_\alpha(u_h) \right] dx dt + \int_{\mathbb{R}^m} \psi(x, 0) \eta(u_h(x, 0)) dx$$

$$= - \sum_{\ell=1}^{\infty} \int_{\mathbb{R}^m} \psi(x, \ell h)[\eta(u_h(x, \ell h)) - \eta(u_h(x, \ell h-))] dx .$$

Combining (6.5.11) with Jensen's inequality and using (6.5.7) yields

$$(6.5.15) \qquad \int_{\mathbb{R}^m} \psi(x, \ell h)[\eta(u_h(x, \ell h)) - \eta(u_h(x, \ell h-))] dx$$

$$\leq \int_{\mathbb{R}^m} \psi(x, \ell h)[\lambda_h * \eta(u_h)(x, \ell h-)) - \eta(u_h(x, \ell h-))] dx \leq C h^2 .$$

The summation on the right-hand side of (6.5.14) contains $O(1/h)$ many nonzero terms. Therefore, passing to the $k \to \infty$ limit along the sequence $\{h_k\}$ in (6.5.14) and using (6.5.12), (6.5.9), and (6.5.15), we conclude that u satisfies (6.2.3). This completes the proof.

Lemma 6.5.2 (*Compactness*). *There is a sequence* $\{h_k\}$, *with* $h_k \to 0$ *as* $k \to \infty$, *and a* L^∞ *function* u *on* $\mathbb{R}^m \times [0, \infty)$ *such that* (6.5.12) *holds.*

Proof. The first step is to establish the weaker assertion that for some sequence $\{h_k\}$, $h_k \to 0$ as $k \to \infty$, and a function u,

$$(6.5.16) \qquad u_{h_k}(\cdot, t) \to u(\cdot, t) , \quad \text{as } k \to \infty , \quad \text{in } L^\infty(\mathbb{R}^m) \text{ weak*} ,$$

for almost all t in $[0, \infty)$. To this end, fix any smooth test function χ on $\mathbb{R}^m$, with compact support, and consider the real-valued function v_h on $[0, \infty)$:

$$(6.5.17) \qquad v_h(t) = \int_{\mathbb{R}^m} \chi(x) u_h(x, t) dx .$$

Notice that v_h is smooth on $[\ell h, \ell h + h)$ and satisfies

$$(6.5.18) \qquad \int_{\ell h}^{\ell h+h} \left| \frac{d}{dt} v_h(t) \right| dt = \int_{\ell h}^{\ell h+h} \left| -\int_{\mathbb{R}^m} \chi(x) \sum_{\alpha=1}^{m} \partial_\alpha g_\alpha(u(x, t)) dx \right| dt$$

$$= \int_{\ell h}^{\ell h+h} \left| \int_{\mathbb{R}^m} \sum_{\alpha=1}^{m} \partial_\alpha \chi(x) g_\alpha(u(x, t)) dx \right| dt \leq Ch .$$

On the other hand, v_h experiences jump discontinuities across the points $t = \ell h$ which can be estimated with the help of (6.5.11) and (6.5.7):

$$(6.5.19) \quad |v_h(\ell h) - v_h(\ell h-)| = \left\| \int_{\mathbb{R}^m} \chi(x) [u_h(x, \ell h) - u_h(x, \ell h-)] dx \right\| \leq Ch^2 .$$

From (6.5.18) and (6.5.19) follows that the total variation of v_h over any compact subinterval of $[0, \infty)$ is bounded, uniformly in h. Therefore, by Helly's theorem (cf. Section 1.7), there is a sequence $\{h_k\}$, $h_k \to 0$ as $k \to \infty$, such that $v_{h_k}(t)$ converges for almost all t in $[0, \infty)$.

By Cantor's diagonal process, we may construct a subsequence of $\{h_k\}$, which will be denoted again by $\{h_k\}$, such that the sequence

$$(6.5.20) \qquad \left\{ \int_{\mathbb{R}^m} \chi(x) u_{h_k}(x, t) dx \right\}$$

converges for almost all t, where χ is any member of any given countable family of test functions. Consequently, the sequence (6.5.20) converges for any χ in $L^1(\mathbb{R}^m)$. Thus, for almost any t in $[0, \infty)$ there is a bounded measurable function on $\mathbb{R}^m$, denoted by $u(\cdot, t)$, such that (6.5.16) holds.

We now strengthen the mode of convergence in (6.5.16). For any $y \in \mathbb{R}^m$, the functions u_h and $\bar{u}_h$, $\bar{u}_h(x, t) = u_h(x + y, t)$, are both solutions of (6.1.1) in every layer. Let us fix $t > 0$ and $R > 0$. Suppose $t \in [\ell h, \ell h + h)$. Applying repeatedly (6.2.9) and (6.5.5) (recalling (6.5.11)), we conclude

(6.5.21)

$$\int_{|x|<R} |u_h(x+y,t) - u_h(x,t)|dx \leq \int_{|x|<R+s(t-\ell h)} |u_h(x+y,\ell h) - u_h(x,\ell h)|dx$$

$$\leq \int_{|x|<R+s(t-\ell h)+\sqrt{m}ph} |u_h(x+y,\ell h-) - u_h(x,\ell h-)|dx$$

$$\leq \cdots \leq \int_{|x|<R+st+\sqrt{m}p(t+h)} |u_0(x+y) - u_0(x)|dx .$$

It follows that the family $\{u_h(\cdot,t)\}$ is equicontinuous in the mean on every compact subset of $\mathbb{R}^m$. Therefore, the convergence in (6.5.16) is upgraded to strongly in $L^1_{\text{loc}}(\mathbb{R}^m)$. Thus, passing to a final subsequence we arrive at (6.5.12). This completes the proof.

6.6 A Kinetic Formulation

Our discussion thus far has been guided by the perspective of Continuum Physics, which, as we have seen, derives systems of field equations by combining balance laws with constitutive relations. An alternative approach is motivated by the kinetic theory of matter. In the classical kinetic theory of gases, the state at the point x and time t is described by the density function $f(v,x,t)$ of the molecular velocity v. The evolution of f is governed by the *Boltzmann equation*, which monitors the changes in the distribution of molecular velocities due to transport and collisions. The connection between the kinetic and the continuum (or phenomenological) approach is established by identifying intensive quantities, like density, velocity, pressure, temperature, heat flux etc., with appropriate moments of the density f and showing that these fields satisfy the balance laws of Continuum Physics presented here. Thus, in principle, one could establish existence and other properties of solutions of (at least certain) hyperbolic systems of balance laws via the corresponding kinetic formulation. This worthwhile research program, which is currently in the stage of active development, lies outside the scope of the present book. However, in order to get at least a taste of the flavor of this approach, we shall discuss here a simple, artificial, kinetic model which is related to the scalar conservation law and may be used to establish existence and other properties of solutions of the initial-value problem (6.1.1), (6.1.2).

In the spirit of the kinetic theory, u should be the mean of a "density" function f, which, however, is allowed to take also negative values. We thus introduce a scalar-valued artificial "velocity" v, and write

$$(6.6.1) \qquad u(x,t) = \int_{-\infty}^{\infty} f(v,x,t)dv .$$

The function f will be determined as the $\mu \downarrow 0$ limit of solutions of the transport equation

$$(6.6.2) \qquad \partial_t f(v,x,t) + \sum_{\alpha=1}^{m} g'_\alpha(v)\partial_\alpha f(v,x,t) = \frac{1}{\mu}[\chi_{u(x,t)}(v) - f(v,x,t)] ,$$

where μ is a small positive parameter and we are employing the notation

(6.6.3)
$$\chi_w(v) = \begin{cases} 1 & \text{if } 0 < v \leq w \\ -1 & \text{if } w \leq v < 0 \\ 0 & \text{otherwise} . \end{cases}$$

Readers familiar with the kinetic theory will recognize in (6.6.2) a model of the BGK approximation to the classical Boltzmann equation. Formally at least, the $\mu \downarrow 0$ limits of solutions of (6.6.2) will satisfy

(6.6.4) $f(v, x, t) = \chi_{u(x,t)}(v)$, $v \in \mathbb{R}$, $x \in \mathbb{R}^m$, $t \in [0, \infty)$,

so that f will be uniformly distributed on the interval bordered by 0 and u, with value -1 or $+1$.

Before verifying that the above procedure does indeed yield admissible weak solutions of (6.1.1), let us discuss the properties of solutions of (6.6.2), (6.6.1).

Theorem 6.6.1 *Assume $u_0 \in L^\infty(\mathbb{R}^m) \cap L^1(\mathbb{R}^m)$. For any $\mu > 0$, there exist bounded measurable functions (f, u), with*

$$f(\cdot, \cdot, t) \in C^0([0, \infty); L^1(\mathbb{R} \times \mathbb{R}^m)), \quad u(\cdot, t) \in C^0([0, \infty); L^1(\mathbb{R}^m)) ,$$

which provide the unique solution of (6.6.2), (6.6.1) under the initial condition

(6.6.5) $f(v, x, 0) = \chi_{u_0(x)}(v)$, $v \in \mathbb{R}$, $x \in \mathbb{R}^m$.

Moreover,

(6.6.6) $0 \leq f(v, x, t) \leq 1$ *for $v \geq 0$* , $-1 \leq f(v, x, t) \leq 0$ *for $v \leq 0$* .

If $\bar{u}_0 \in L^\infty(\mathbb{R}^m) \cap L^1(\mathbb{R}^m)$ are other initial data inducing the solution $(\bar{f}, \bar{u})$, then, for any $t > 0$,

(6.6.7) $\| f(\cdot, \cdot, t) - \bar{f}(\cdot, \cdot, t) \|_{L^1(\mathbb{R} \times \mathbb{R}^m)} \leq \| f(\cdot, \cdot, 0) - \bar{f}(\cdot, \cdot, 0) \|_{L^1(\mathbb{R} \times \mathbb{R}^m)}$,

(6.6.8) $\| u(\cdot, t) - \bar{u}(\cdot, t) \|_{L^1(\mathbb{R}^m)} \leq \| u_0(\cdot) - \bar{u}_0(\cdot) \|_{L^1(\mathbb{R}^m)}$.

Furthermore, if

(6.6.9) $u_0(x) \leq \bar{u}_0(x)$, $x \in \mathbb{R}^m$,

then

(6.6.10) $f(v, x, t) \leq \bar{f}(v, x, t)$, $v \in \mathbb{R}$, $x \in \mathbb{R}^m$, $t \in [0, \infty)$,

(6.6.11) $u(x, t) \leq \bar{u}(x, t)$, $x \in \mathbb{R}^m$, $t \in [0, \infty)$.

Proof. Let us realize the g_α, $\alpha = 1, \cdots, m$, as components of a m-vector g. Taking existence of solutions to (6.6.2), (6.6.1), (6.6.5) for granted, we integrate (6.6.2) along characteristics $dx/dt = g'(v)$, $dv/dt = 0$, to get

$$(6.6.12) \quad f(v, x, t) = e^{-\frac{t}{\mu}} f(x - tg'(v), v, 0) + \frac{1}{\mu} \int_0^t e^{-\frac{t-\tau}{\mu}} \chi_{u(x-(t-\tau)g'(v),\tau)}(v) d\tau \ .$$

By virtue of (6.6.5) and the definition (6.6.3) of χ_w, (6.6.12) immediately yields (6.6.6).

If $(\overline{f}, \overline{u})$ is another solution generated by initial data $\overline{u}_0$, we have

$$(6.6.13) \quad f(v, x, t) - \overline{f}(v, x, t) = e^{-\frac{t}{\mu}}[f(x, tg'(v), v, 0) - \overline{f}(x - tg'(v), v, 0)]$$

$$+ \frac{1}{\mu} \int_0^t e^{-\frac{t-\tau}{\mu}} [\chi_{u(x-(t-\tau)g'(v),\tau)}(v) - \chi_{\overline{u}(x-(t-\tau)g'(v),\tau)}(v)] d\tau$$

whence

$$(6.6.14) \qquad\qquad \|f(\cdot, \cdot, t) - \overline{f}(\cdot, \cdot, t)\|_{L^1(\mathbb{R}\times\mathbb{R}^m)}$$

$$\leq e^{-\frac{t}{\mu}} \|f(\cdot, \cdot, 0) - \overline{f}(\cdot, \cdot, 0)\|_{L^1(\mathbb{R}\times\mathbb{R}^m)}$$

$$+ \frac{1}{\mu} \int_0^t e^{-\frac{t-\tau}{\mu}} \|\chi_{u(\xi(\tau),\tau)}(\cdot) - \chi_{\overline{u}(\xi(\tau),\tau)}(\cdot)\|_{L^1(\mathbb{R}\times\mathbb{R}^m)} d\tau$$

$$\leq e^{-\frac{t}{\mu}} \|f(\cdot, \cdot, 0) - \overline{f}(\cdot, \cdot, 0)\|_{L^1(\mathbb{R}\times\mathbb{R}^m)}$$

$$+ (1 - e^{-\frac{t}{\mu}}) \max_{0\leq\tau\leq t} \|f(\cdot, \cdot, \tau) - \overline{f}(\cdot, \cdot, \tau)\|_{L^1(\mathbb{R}\times\mathbb{R}^m)}.$$

Clearly, (6.6.14) implies (6.6.7) and this in turn yields (6.6.8). In particular, there is at most one solution to (6.6.2), (6.6.1), (6.6.5). Furthermore, the estimate (6.6.14) implies that Picard iteration applied to the family of integral equations (6.6.12) converges and generates the solution of (6.6.2), (6.6.1), (6.6.5).

By its definition, χ_w is an increasing function of w. Therefore, it follows easily from (6.6.13), (6.6.1) that (6.6.9) implies (6.6.10) and (6.6.11). This completes the proof.

We now turn to the limiting behavior of solutions as $\mu \downarrow 0$.

Theorem 6.6.2 *For $\mu > 0$, let (f_μ, u_μ) denote the solution of (6.6.2), (6.6.1), (6.6.5) with $u_0 \in L^\infty(\mathbb{R}^m) \cap L^1(\mathbb{R}^m)$. As $\mu \downarrow 0$, the family (f_μ, u_μ) converges in L^1_{loc} to bounded measurable functions (f, u) such that f satisfies the transport equation*

$$(6.6.15) \qquad \partial_t f(v, x, t) + \sum_{\alpha=1}^m g'_\alpha(v) \partial_\alpha f(v, x, t) = \frac{\partial v}{\partial v},$$

for some nonnegative measure v on $\mathbb{R} \times \mathbb{R}^m \times [0, \infty)$; (6.6.1) and (6.6.4) hold; and u is the admissible weak solution of (6.1.1), (6.1.2).

Proof. The first step is to show that the family (f_μ, u_μ) is equicontinuous in the mean. That this is the case in the v and x directions, is an immediate consequence of the contraction property (6.6.7), (6.6.8): For any $w \in \mathbb{R}$ and $y \in \mathbb{R}^m$, the functions $(\overline{f}_\mu, \overline{u}_\mu)$, $\overline{f}_\mu(v, x, t) = f_\mu(v + w, x + y, t)$, $\overline{u}_\mu(x, t) = u_\mu(x + y, t)$, provide a solution of (6.6.2), (6.6.1) with initial data $\overline{f}_\mu(v, x, 0) = \chi_{u_0(x+y)}(v+w)$, and so

(6.6.16)
$$\int_{\mathbb{R}^m} \int_{\mathbb{R}} |f_\mu(v + w, x + y, t) - f_\mu(v, x, t)| dv dx$$
$$\leq \int_{\mathbb{R}^m} \int_{\mathbb{R}} |\chi_{u_0(x+y)}(v + w) - \chi_{u_0(x)}(v)| dv dx ,$$

(6.6.17)
$$\int_{\mathbb{R}^m} |u_\mu(x + y, t) - u_\mu(x, t)| dx \leq \int_{\mathbb{R}^m} |u_0(x + y) - u_0(x)| dx .$$

Equicontinuity in the t-direction is easily verified with the help of the transport equation; the details are omitted.

Next we consider the function

(6.6.18)
$$\omega_\mu(v, x, t) = \int_{-\infty}^{v} [\chi_{u_\mu(x,t)}(w) - f_\mu(w, x, t)] dw .$$

Let us fix (x, t), assuming for definiteness $u_\mu(x, t) > 0$ (the other cases being similarly treated). Clearly, $\omega_\mu(-\infty, x, t) = 0$. By virtue of (6.6.3) and (6.6.6), $\omega_\mu(\cdot, x, t)$ is nondecreasing on the interval $(-\infty, u_\mu(x, t))$ and nonincreasing on the interval $(u_\mu(x, t), \infty)$. Finally, by account of (6.6.1), $\omega_\mu(\infty, x, t) = 0$. Consequently, we may write

(6.6.19)
$$\frac{1}{\mu}[\chi_{u_\mu} - f_\mu] = \frac{\partial v_\mu}{\partial v} ,$$

where v_μ is a nonnegative measure on $\mathbb{R} \times \mathbb{R}^m \times [0, \infty)$, which is bounded uniformly in $\mu > 0$.

It follows that from any sequence $\{\mu_k\}$, $\mu_k \to 0$ as $k \to \infty$, we may extract a subsequence, denoted again by $\{\mu_k\}$, so that (f_{μ_k}, u_{μ_k}) converges in L^1_{loc} to functions (f, u) and v_{μ_k} converges weakly in the space of measures to a bounded nonnegative measure v. Clearly, (6.6.1), (6.6.4) and (6.6.15) hold.

It remains to show that u is the admissible weak solution of (6.1.1), (6.1.2). Towards that end, fix any convex entropy η, with associated entropy flux $(q_1, \cdots, q_m)$ determined through (6.2.1), and take any nonnegative Lipschitz continuous test function ψ on $\mathbb{R}^m \times [0, \infty)$, with compact support. Without loss of generality (since u is bounded) assume $\eta(0) = 0$ and η' is bounded on $(-\infty, \infty)$. Fix some C^∞ function h on $(-\infty, \infty)$ with support contained in $[-2, 2]$ and equal to 1 on $[-1, 1]$. For $k > \|u_0\|_{L^\infty}$, set $h_k(v) = h(v/k)$. Multiply (6.6.15) by $h_k(v)\eta'(v)\psi(x, t)$, integrate over $\mathbb{R} \times \mathbb{R}^m \times [0, \infty)$ and integrate by parts. Noticing that, for any continuous function p on $(-\infty, \infty)$ and every $w \in (-\infty, \infty)$

(6.6.20)
$$\int_{-\infty}^{\infty} \chi_w(v) p(v) dv = \int_{0}^{w} p(\xi) d\xi ,$$

we deduce, by virtue of (6.6.4), (6.6.5) and (6.2.1),

(6.6.21) $\displaystyle\int_0^\infty \int_{\mathbb{R}^m} \left[\partial_t \psi \eta(u) + \sum_{\alpha=1}^m \partial_\alpha \psi q_\alpha(u) \right] dx dt + \int_{\mathbb{R}^m} \psi(x, 0)\eta(u_0(x)) dx$

$\displaystyle = \int_0^\infty \int_{\mathbb{R}^m} \int_{\mathbb{R}} \psi[\eta''(v)h_k(v) + \eta'(v)h'_k(v)] d\mu(v) dx dt .$

On the right-hand side of (6.6.21), $\psi \eta'' h_k \geq 0$ while $h'_k \to 0$, as $k \to \infty$. Hence, letting $k \to \infty$ we arrive at (6.2.3) thus verifying that u is an admissible weak solution of (6.1.1), (6.1.2). In particular, since u is unique so is f. But then the entire family (f_μ, u_μ) must be convergent, as $\mu \downarrow 0$. This completes the proof.

In setting up the transport equation (6.6.2), the role of the stiff term $\mu^{-1}(\chi_u - f)$ is to enforce, in the limit, (6.6.4) while at the same time accounting for entropy dissipation by generating the term $\partial v / \partial v$ on the right-hand side of (6.6.15). Any other "collision" mechanism with the same features may be used to construct solutions of (6.1.1), (6.1.2). In fact the following proposition is established in the references cited in Section 6.9:

Theorem 6.6.3 *A bounded measurable function u on $\mathbb{R}^m \times [0, \infty)$, with $u(\cdot, t) \in C^0([0, \infty); L^1(\mathbb{R}^m))$, is the admissible weak solution of (6.1.1), (6.1.2) if and only if the function f defined through (6.6.4) satisfies the transport equation (6.6.15), for some nonnegative measure v, together with the initial condition (6.6.5).*

Up to this point we have been facing nonlinearity as an agent that provokes the development of discontinuities in solutions starting out from smooth initial data. It turns out, however, that nonlinearity may also play the opposite role, of smoothing out solutions with rough initial data. In the course of the book, we shall encounter various manifestations of such behavior. The kinetic formulation discussed above provides valuable insight in the compactifying and smoothing effects of nonlinearity in scalar conservation laws. Whenever the g'_α are constant, i.e., (6.1.1) is linear, solutions are as smooth as the initial data. This follows from the discussion of Section 6.1 but also from the kinetic formulation: f is transported along characteristics, uniformly in v (cf. (6.6.12)). By contrast, when the g'_α vary with u, mixing occurs which, averaged through (6.6.1), may improve the regularity of u. This is quantified in the following deep theorem, whose (hard and technical) proof is given in the references cited in Section 6.8.

Theorem 6.6.4 *Assume there is $r \in (0, 1]$ and $C \geq 0$ such that*

(6.6.22) $\displaystyle \text{meas} \left\{ v : |v| \leq \|u_0\|_{L^\infty}, \left| \tau + \sum_{\alpha=1}^m \xi_\alpha g'_\alpha(v) \right| \leq \delta \right\} \leq C\delta^r$

for all $\delta \in (0, 1)$, $\tau \in \mathbb{R}$, $\xi \in \mathbb{R}^m$ with $\tau^2 + |\xi|^2 = 1$. Then the admissible weak solution u of (6.1.1), (6.1.2) satisfies

(6.6.23) $u(\cdot, t) \in C^0((0, \infty); W_{\text{loc}}^{s,1}(\mathbb{R}^m))$

for any $s \in (0, \frac{r}{r+2})$.

It is condition (6.6.22) that encodes the aspect of nonlinearity responsible for the regularizing effect. For example, (6.6.22) fails, for any r, when the g_α are linear, but it is satisfied, with $r = 1$, when the g_α are uniformly convex functions, $g''_\alpha(u) > 0$.

6.7 Relaxation

Another interesting method for constructing admissible weak solutions of (6.1.1) is through *relaxation*. The point of departure is a semilinear system of $m + 1$ equations,

(6.7.1)

$$\begin{cases} \partial_t v(x, t) + \sum_{\alpha=1}^{m} c_\alpha \partial_\alpha v(x, t) = \dfrac{1}{\mu} \sum_{\alpha=1}^{m} [f_\alpha(v(x, t)) - z_\alpha(x, t)] \\ \partial_t z_\alpha(x, t) - c_\alpha \partial_\alpha z_\alpha(x, t) = \dfrac{1}{\mu} [f_\alpha(v(x, t)) - z_\alpha(x, t)] , \quad \alpha = 1, \cdots, m , \end{cases}$$

in the $m + 1$ unknowns $(v, z_1, \cdots, z_m)$, where μ is a small positive parameter while, for $\alpha = 1, \cdots, m$, the c_α are given constants and the f_α are specified smooth functions such that

(6.7.2) $f'_\alpha(v) < 0 , \quad -\infty < v < \infty , \quad \alpha = 1, \cdots, m ,$

(6.7.3) $f_\alpha(0) = 0 , \quad f_\alpha(v) \to \pm\infty \quad \text{as } v \longrightarrow \mp\infty , \quad \alpha = 1, \cdots, m .$

Notice that solutions of (6.7.1) satisfy the conservation law

(6.7.4) $\partial_t [v(x, t) - \sum_{\alpha=1}^{m} z_\alpha(x, t)] + \sum_{\alpha=1}^{m} c_\alpha \partial_\alpha [v(x, t) + z_\alpha(x, t)] = 0 .$

Due to the form of the right-hand side of (6.7.1), one should expect that, as $\mu \downarrow 0$, the variables z_α "relax" to their equilibrium states $f_\alpha(v)$, in which case (6.7.4) reduces to a scalar conservation law (6.1.1) with[1]

(6.7.5) $u = v - \sum_{\alpha=1}^{m} f_\alpha(v) , \quad g_\alpha(u) = c_\alpha [v + f_\alpha(v)] , \quad \alpha = 1, \cdots, m .$

The above considerations suggest a program for constructing solutions of (6.1.1) as asymptotic limits of solutions of (6.7.1). The reader may have already noticed the similarity of this approach with the kinetic formulation presented in Section 6.6. In fact the analogy is not merely formal: It is possible (references in Section 6.9) to interpret (6.7.1) as a system governing, at a mesoscopic scale,

---

[1] By virtue of (6.7.2), the transformation (6.7.5)$_1$ may be inverted to express v as a smooth, increasing function of u, and it is in that sense that g_α, defined by (6.7.5)$_2$, should be realized as function of u.

the evolution of an ensemble of interacting particles. In that connection, the small parameter μ plays the role of *mean free path*.

In realizing our program, the first step is to examine the Cauchy problem for the system (6.7.1), under assigned initial conditions

$$(6.7.6) \quad v(x, 0) = v_0(x) , \quad z_\alpha(x, 0) = z_{\alpha 0}(x) , \quad \alpha = 1, \cdots, m , \quad x \in \mathbb{R}^m .$$

Since (6.7.1) is semilinear hyperbolic, the classical theory guarantees that whenever the initial data $(v_0, z_{10}, \cdots, z_{m0})$ are in $C_0^1(\mathbb{R}^m)$ there exists a unique classical solution $(v, z_1, \cdots, z_m)$ of (6.7.1), (6.7.6), defined on a maximal time interval $[0, T)$, with $0 < T \leq \infty$. For any $t \in [0, T)$, the functions $(v(\cdot, t), z_1(\cdot, t), \cdots, z_m(\cdot, t))$ are in $C_0^1(\mathbb{R}^m)$. Furthermore, when $T < \infty$,

$$(6.7.7) \quad \|v(\cdot, t)\|_{L^\infty(\mathbb{R}^m)} + \sum_{\alpha=1}^m \|z_\alpha(\cdot, t)\|_{L^\infty(\mathbb{R}^m)} \to \infty , \quad \text{as } t \uparrow T .$$

Here we need (possibly weak) solutions, under a broader class of initial data, which exist globally in time. Such solutions do indeed exist because, due to our assumptions (6.7.2), (6.7.3), the effect of the right-hand side in (6.7.1) is dissipative. This is manifested in the following proposition, which should be compared with Theorems 6.3.2 and 6.6.1:

Theorem 6.7.1 *For any initial data $(v_0, z_{10}, \cdots, z_{m0})$ in $L^1(\mathbb{R}^m) \cap L^\infty(\mathbb{R}^m)$, there exists a unique weak solution $(v, z_1, \cdots, z_m)$ of (6.7.1), (6.7.6) on $\mathbb{R}^m \times [0, \infty)$ such that $(v(\cdot, t), z_1(\cdot, t), \cdots, z_m(\cdot, t))$ are in $C^0([0, \infty); L^1(\mathbb{R}^m))$. If*

$$(6.7.8)$$
$$a \leq v_0(x) \leq b , \quad f_\alpha(b) \leq z_{\alpha 0}(x) \leq f_\alpha(\alpha) , \quad \alpha = 1, \cdots, m , \quad x \in \mathbb{R}^m ,$$

then

$$(6.7.9) \quad a \leq v(x, t) \leq b , \quad f_\alpha(b) \leq z_\alpha(x, t) \leq f_\alpha(\alpha) ,$$

$$\alpha = 1, \cdots, m , \quad (x, t) \in \mathbb{R}^m \times [0, \infty) .$$

Furthermore, if $(\bar{v}, \bar{z}_1, \cdots, \bar{z}_m)$ is another such solution, with initial data $(\bar{v}_0, \bar{z}_{10}, \cdots, \bar{z}_{m0})$ in $L^1(\mathbb{R}^m) \cap L^\infty(\mathbb{R}^m)$, then, for any $t \in [0, \infty)$,

$$(6.7.10) \quad \int_{\mathbb{R}^m} \left\{ [v(x, t) - \bar{v}(x, t)]^+ + \sum_{\alpha=1}^m [\bar{z}_\alpha(x, t) - z_\alpha(x, t)]^+ \right\} dx$$

$$\leq \int_{\mathbb{R}^m} \left\{ [v_0(x) - \bar{v}_0(x)]^+ + \sum_{\alpha=1}^m [\bar{z}_{\alpha 0}(x) - z_{\alpha 0}(x)]^+ \right\} dx ,$$

$$(6.7.11) \quad \|v(\cdot, t) - \bar{v}(\cdot, t)\|_{L^1(\mathbb{R}^m)} + \sum_{\alpha=1}^m \|z_\alpha(\cdot, t) - \bar{z}_\alpha(\cdot, t)\|_{L^1(\mathbb{R}^m)}$$

$$\leq \|v_0(\cdot) - \bar{v}_0(\cdot)\|_{L^1(\mathbb{R}^m)} + \sum_{\alpha=1}^m \|z_{\alpha 0}(\cdot) - \bar{z}_{\alpha 0}(\cdot)\|_{L^1(\mathbb{R}^m)} .$$

In particular, if

(6.7.12) $v_0(x) \leq \bar{v}_0(x)$, $z_{\alpha 0}(x) \geq \bar{z}_{\alpha 0}(x)$, $\alpha = 1, \cdots, m$, $x \in \mathbb{R}^m$,

then

(6.7.13)
$v(x, t) \leq \bar{v}(x, t)$, $z_\alpha(x, t) \geq \bar{z}_\alpha(x, t)$, $\alpha = 1, \cdots, m$, $(x, t) \in \mathbb{R}^m \times [0, \infty)$.

Proof. The first objective is to establish (6.7.10) under the assumption that both solutions $(v, z_1, \cdots, z_m)$ and $(\bar{v}, \bar{z}_1, \cdots, \bar{z}_m)$ are classical, with initial data $(v_0, z_{10}, \cdots, z_{m0})$ and $(\bar{v}_0, \bar{z}_{10}, \cdots, \bar{z}_{m0})$ in $C_0^1(\mathbb{R}^m)$. For $\varepsilon > 0$, we recall the function η_ε defined through (6.3.8) and note that

(6.7.14) $\partial_t \left[\eta_\varepsilon(v - \bar{v}) + \sum_{\alpha=1}^m \eta_\varepsilon(\bar{z}_\alpha - z_\alpha) \right] + \sum_{\alpha=1}^m c_\alpha \partial_\alpha [\eta_\varepsilon(v - \bar{v}) - \eta_\varepsilon(\bar{z}_\alpha - z_\alpha)]$

$$= \frac{1}{\mu} \sum_{\alpha=1}^m [\eta'_\varepsilon(v - \bar{v}) - \eta'_\varepsilon(\bar{z}_\alpha - z_\alpha)][f_\alpha(v) - f_\alpha(\bar{v}) + \bar{z}_\alpha - z_\alpha]$$

follows readily from (6.7.1). For fixed values of $v, \bar{v}, z_\alpha, \bar{z}_\alpha$, of any sign, the right-hand side of (6.7.14) has a nonpositive limit as $\varepsilon \downarrow 0$. Therefore, integrating (6.7.14) over $\mathbb{R}^m \times (0, t)$ and letting $\varepsilon \downarrow 0$ we arrive at (6.7.10).

When (6.7.12) holds, (6.7.10) immediately implies (6.7.13). Notice that, for any constants a and b, $(a, f_1(a), \cdots, f_m(a))$ and $(b, f_1(b), \cdots, f_m(b))$ are particular solutions of (6.7.1) and hence (6.7.8) implies (6.7.9). In particular, blow-up (6.7.7) canot occur for any T and thus the solutions exist on $\mathbb{R}^m \times [0, \infty)$.

To derive (6.7.11), it suffices to rewrite (6.7.10) with the roles of $(v, z_1, \cdots, z_m)$ and $(\bar{v}, \bar{z}_1, \cdots, \bar{z}_m)$ reversed and then add the resulting inequality to the original (6.7.10).

We have now verified all the assertions of the theorem, albeit within the context of classical solutions, with initial data in $C_0^1(\mathbb{R}^m)$. Nevertheless, by virtue of the L^1-contraction estimate (6.7.11), weak solutions, with any initial data in $L^1(\mathbb{R}^m) \cap L^\infty(\mathbb{R}^m)$, satisfying the asserted properties, may readily be constructed as L^1 limits of sequences of classical solutions. This completes the proof.

Our next task is to investigate the limiting behavior of solutions of (6.7.1) as $\mu \downarrow 0$. The mechanism that induces the z_α to relax to their equilibrium values $f_\alpha(v)$ will be captured through an entropy-like inequality. We define the family

(6.7.15) $\phi_\alpha(z_\alpha) = -\int_0^{z_\alpha} f_\alpha^{-1}(w) dw$, $\alpha = 1, \cdots, m$

of nonnegative, convex functions on $(-\infty, \infty)$. Assuming $(v, z_1, \cdots, z_m)$ is a classical solution of (6.7.1), with initial data $(v_0, z_{10}, \cdots, z_{m0})$ in $C_0^1(\mathbb{R}^m)$, we readily verify that

$$(6.7.16) \qquad \partial_t \left[\frac{1}{2} v^2 + \sum_{\alpha=1}^{m} \phi_\alpha(z_\alpha) \right] + \sum_{\alpha=1}^{m} c_\alpha \partial_\alpha \left[\frac{1}{2} v^2 - \phi_\alpha(z_\alpha) \right]$$

$$= \frac{1}{\mu} \sum_{\alpha=1}^{m} [v - f_\alpha^{-1}(z_\alpha)][f_\alpha(v) - z_\alpha].$$

Since $v - f_\alpha^{-1}(z_\alpha) = f_\alpha^{-1}(f_\alpha(v)) - f_\alpha^{-1}(z_\alpha)$, the mean-value theorem implies

$$(6.7.17) \qquad -[v - f_\alpha^{-1}(z_\alpha)][f_\alpha(v) - z_\alpha] \geq \frac{1}{k} [f_\alpha(v) - z_\alpha]^2 ,$$

where k is any upper bound of $-f_\alpha'$ over the range of v. Therefore, upon integrating (6.7.16) over $\mathbb{R}^m \times [0, \infty)$ we deduce the inequality

$$(6.7.18) \qquad \int_0^\infty \int_{\mathbb{R}^m} \sum_{\alpha=1}^{m} [f_\alpha(v) - z_\alpha]^2 dx dt \leq k\mu \int_{\mathbb{R}^m} \left[\frac{1}{2} v_0^2 + \sum_{\alpha=1}^{m} \phi_\alpha(z_{\alpha_0}) \right] dx .$$

As explained in the proof of Theorem 6.7.1, weak solutions of (6.7.1) are constructed as L^1 limits of sequences of classical solutions, and hence the inequality (6.7.18) will hold even for weak solutions with initial data in $L^1(\mathbb{R}^m) \cap L^\infty(\mathbb{R}^m)$.

Theorem 6.7.2 *Let* $(v^\mu, z_1^\mu, \cdots, z_m^\mu)$ *denote the family of solutions of the initial-value problem* (6.7.1), (6.7.6), *with parameter* $\mu > 0$, *and initial data* $(v_0, f_1(v_0),$ $\cdots, f_m(v_0))$, *where* v_0 *is in* $L^1(\mathbb{R}^m) \cap L^\infty(\mathbb{R}^m)$. *Then there is a bounded measurable function* v *on* $\mathbb{R}^m \times [0, \infty)$ *such that, as* $\mu \downarrow 0$,

$$(6.7.19) \quad v^\mu(x, t) \longrightarrow v(x, t) , \quad z_\alpha^\mu(x, t) \longrightarrow f_\alpha(v(x, t)) , \quad \alpha = 1, \cdots, m ,$$

almost everywhere on $\mathbb{R}^m \times [0, \infty)$. *The function*

$$(6.7.20) \qquad u(x, t) = v(x, t) - \sum_{\alpha=1}^{m} f_\alpha(v(x, t))$$

is the admissible weak solution of the conservation law (6.1.1), *with flux functions* g_α *defined through* (6.7.5), *and initial data*

$$(6.7.21) \qquad u_0(x) = v_0(x) - \sum_{\alpha=1}^{m} f_\alpha(v_0(x)) , \quad x \in \mathbb{R}^m .$$

Proof. Let us set, for $(x, t) \in \mathbb{R}^m \times [0, \infty)$,

$$(6.7.22) \qquad u^\mu(x, t) = v^\mu(x, t) - \sum_{\alpha=1}^{m} z_\alpha^\mu(x, t) ,$$

$$(6.7.23) \qquad g_\alpha^\mu(x, t) = c_\alpha[v^\mu(x, t) + z_\alpha^\mu(x, t)] .$$

By virtue of (6.7.4),

$$(6.7.24) \qquad \partial_t u^\mu(x, t) + \sum_{\alpha=1}^{m} \partial_\alpha g_\alpha^\mu(x, t) = 0 .$$

First we show that there is a bounded measurable function u on $\mathbb{R}^m \times [0, \infty)$ and some sequence $\{\mu_n\}$, $\mu_n \downarrow 0$ as $n \to \infty$, such that

(6.7.25) $$u^{\mu_n}(\cdot, t) \longrightarrow u(\cdot, t) , \quad n \to \infty ,$$

in $L^\infty(\mathbb{R}^m)$ weak*, for all $t \in [0, \infty)$. To that end, let us fix any test function $\chi \in C_0^\infty(\mathbb{R}^m)$ and define the family of functions

(6.7.26) $$w^\mu(t) = \int_{\mathbb{R}^m} \chi(x) u^\mu(x, t) dx , \quad t \in [0, \infty) ,$$

which, by account of (6.7.24), are continuously differentiable with derivative

(6.7.27) $$\frac{d}{dt} w^\mu(t) = \sum_{\alpha=1}^m \int_{\mathbb{R}^m} \partial_\alpha \chi(x) g_\alpha^\mu(x, t)$$

bounded, uniformly in $\mu > 0$. It then follows from Arzela's theorem that there is a sequence $\{\mu_n\}$, with $\mu_n \downarrow 0$ as $n \to \infty$, such that $\{w^{\mu_n}\}$ converges for all $t \in [0, \infty)$. By Cantor's diagonal process we may construct a subsequence of $\{\mu_n\}$, denoted again by $\{\mu_n\}$, such that the sequence

(6.7.28) $$\left\{ \int_{\mathbb{R}^m} \chi(x) u^{\mu_n}(x, t) dx \right\}$$

is convergent for all $t \in [0, \infty)$ and every member χ of any given countable family of test functions. Consequently, (6.7.28) is convergent for any $\chi \in L^1(\mathbb{R}^m)$. Thus, for each $t \in [0, \infty)$ there is a bounded measurable function on $\mathbb{R}^m$, denoted by $u(\cdot, t)$, such that (6.7.25) holds in $L^\infty(\mathbb{R}^m)$ weak*.

Next we note that, by the L^1 contraction estimate (6.7.11), for any fixed $t \in [0, \infty)$, the family of functions $(v^\mu(\cdot, t), z_1^\mu(\cdot, t), \cdots, z_m^\mu(\cdot, t))$ is equicontinuous in the mean. Hence, the convergence in (6.7.25) is upgraded to strongly in $L^1(\mathbb{R}^m)$. In particular,

(6.7.29) $$u^{\mu_n}(x, t) \longrightarrow u(x, t) , \quad n \to \infty ,$$

almost everywhere on $\mathbb{R}^m \times [0, \infty)$.

We now apply (6.7.18) for our solutions $(v^{\mu_n}, z_1^{\mu_n}, \cdots, z_m^{\mu_n})$ and, passing if necessary to a subsequence, denoted again by $\{\mu_n\}$, we obtain

(6.7.30) $$f_\alpha(v^{\mu_n}(x, t)) - z_\alpha^{\mu_n}(x, t) \to 0 , \quad n \to \infty , \quad \alpha = 1, \cdots, m ,$$

almost everywhere on $\mathbb{R}^m \times [0, \infty)$.

Combining (6.7.22), (6.7.29) and (6.7.30), we deduce

(6.7.31) $$v^{\mu_n}(x, t) - \sum_{\alpha=1}^m f_\alpha(v^{\mu_n}(x, t)) \to u(x, t) , \quad n \to \infty ,$$

almost everywhere on $\mathbb{R}^m \times [0, \infty)$. Because of the monotonicity assumption (6.7.2), (6.7.31) implies that the sequence $\{v^{\mu_n}\}$ itself must be convergent, say

$$(6.7.32) \qquad v^{\mu_n}(x,t) \longrightarrow v(x,t) , \quad n \to \infty ,$$

almost everywhere on $\mathbb{R}^m \times [0, \infty)$, where v is a function related to u through (6.7.20). Furthermore, (6.7.30) and (6.7.32) together imply

$$(6.7.33) \qquad z_\alpha^{\mu_n}(x,t) \to f_\alpha(v(x,t)) , \quad n \to \infty , \quad \alpha = 1, \cdots, m ,$$

almost everywhere on $\mathbb{R}^m \times [0, \infty)$.

By virtue of (6.7.22), (6.7.23), (6.7.24), (6.7.32) and (6.7.33), u is a weak solution of (6.1.1), with fluxes g_α defined through (6.7.5). We proceed to show that this solution is admissible. We fix any constant $\bar{v}$ and write (6.7.14) for the two solutions $(v^{\mu_n}, z_1^{\mu_n}, \cdots, z_m^{\mu_n})$ and $(\bar{v}, f_1(\bar{v}), \cdots, f_m(\bar{v}))$. We apply this (distributional) equation to any nonnegative Lipschitz continuous test function ψ, with compact support on $\mathbb{R}^m \times [0, \infty)$ and let $\varepsilon \downarrow 0$. Recalling that the $\varepsilon \downarrow 0$ limit of the right-hand side of (6.7.14) is nonpositive, this calculation gives

$$(6.7.34) \qquad \int_0^\infty \int_{\mathbb{R}^m} \partial_t \psi \left[(v^{\mu_n} - \bar{v})^+ + \sum_{\alpha=1}^m (f_\alpha(\bar{v}) - z_\alpha^{\mu_n})^+ \right] dx dt$$
$$+ \int_0^\infty \int_{\mathbb{R}^m} \sum_{\alpha=1}^m c_\alpha \partial_\alpha \psi [(v^{\mu_n} - \bar{v})^+ - (f_\alpha(\bar{v}) - z_\alpha^{\mu_n})^+] dx dt$$
$$+ \int_{\mathbb{R}^m} \psi(x,0) \left[(v_0 - \bar{v})^+ + \sum_{\alpha=1}^m (f_\alpha(\bar{v}) - f_\alpha(v_0))^+ \right] dx \geq 0 .$$

Letting $n \to \infty$ and using (6.7.32) and (6.7.33), (6.7.34) yields

$$(6.7.35) \qquad \int_0^\infty \int_{\mathbb{R}^m} \partial_t \psi \left[(v - \bar{v})^+ + \sum_{\alpha=1}^m (f_\alpha(\bar{v}) - f_\alpha(v))^+ \right] dx dt$$
$$+ \int_0^\infty \int_{\mathbb{R}^m} \sum_{\alpha=1}^m c_\alpha \partial_\alpha \psi [(v - \bar{v})^+ - (f_\alpha(\bar{v}) - f_\alpha(v))^+] dx dt$$
$$+ \int_{\mathbb{R}^m} \psi(x,0) \left[(v_0 - \bar{v})^+ + \sum_{\alpha=1}^m (f_\alpha(\bar{v}) - f_\alpha(v_0))^+ \right] dx \geq 0 .$$

By account of (6.7.2), $v - \bar{v}$ and $f_\alpha(\bar{v}) - f_\alpha(v)$ have the same sign. Furthermore, if we set $\bar{u} = \bar{v} - \sum f_\alpha(\bar{v})$, then $v - \bar{v}$ and $u - \bar{u}$ have also the same sign. Therefore, upon using (6.7.20), (6.7.21), and (6.7.5), we may rewrite (6.7.35) as

$$(6.7.36) \qquad \int_0^\infty \int_{\mathbb{R}^m} \left[\partial_t \psi \eta(u; \bar{u}) + \sum_{\alpha=1}^m \partial_\alpha \psi q_\alpha(u; \bar{u}) \right] dx dt$$
$$+ \int_{\mathbb{R}^m} \psi(x,0) \eta(u_0; \bar{u}) dx \geq 0 ,$$

where $(\eta(u; \bar{u}), q(u; \bar{u}))$ is the entropy-entropy flux pair defined by (6.2.5). As noted in Section 6.2, the set of entropy-entropy flux pairs (6.2.5), with $\bar{u}$ arbitrary, is "complete" and hence (6.7.36) implies that (6.2.3) will hold for any entropy-entropy flux pair (η, q) with η convex. This verifies that u is the admissible weak solution of (6.1.1), with initial data u_0 given by (6.7.21). Since u is unique, the convergence in (6.7.29), (6.7.32) and (6.7.33) applies not only along the particular sequence $\{\mu_n\}$ but also along the whole family $\{\mu\}$, as $\mu \downarrow 0$. This completes the proof.

Theorem 6.7.2 demonstrates how, starting out from a given system (6.7.1), one may construct, by relaxation, admissible solutions of a particular scalar conservation law induced by (6.7.1). Of course, we are interested in the reverse process, namely to determine the appropriate system (6.7.1) whose relaxed form is a given scalar conservation law (6.1.1). This may be accomplished when, given the fluxes $g_\alpha(u)$, it is possible to select constants c_α in such a way that the transformations (6.7.5) determine implicitly functions $f_\alpha(v)$ that satisfy the assumptions (6.7.2) and (6.7.3). Let us normalize the given fluxes by $g_\alpha(0) = 0$, $\alpha = 1, \cdots, m$. Since our solutions will be a priori bounded, let us assume, without loss of generality, that the $g'_\alpha(u)$ are uniformly bounded on $(-\infty, \infty)$. From (6.7.5),

$$(6.7.37) \qquad (m + 1)v = u + \sum_{\alpha=1}^{m} \frac{1}{c_\alpha} g_\alpha(u) \; .$$

Therefore, the first constraint is to fix the $|c_\alpha|$ so large that

$$(6.7.38) \qquad (m+1)\frac{dv}{du} = 1 + \sum_{\alpha=1}^{m} \frac{1}{c_\alpha} g'_\alpha(u) \geq \frac{1}{2}$$

in order to secure that the map $v \mapsto u$ will possess a smooth inverse. Next we note

$$(6.7.39) \quad f'_\alpha(v) = -1 + \frac{1}{c_\alpha} g'_\alpha(u)\frac{du}{dv} = -1 + \frac{m+1}{c_\alpha}\left[1 + \sum_{\beta=1}^{m} \frac{1}{c_\beta} g'_\beta(u)\right]^{-1} g'_\alpha(u) \; ,$$

so that, by selecting the $|c_\alpha|$ sufficiently large, we can satisfy the assumptions (6.7.2) and (6.7.3). Restrictions on c_α which maintain that the convective characteristic speeds c_α should be high relative to the characteristic speeds g'_α of the relaxed conservation law are called *subcharacteristic conditions*.

6.8 The L^1 Theory for Systems of Balance Laws

The successful treatment of the scalar conservation law, based on L^1 and L^∞ estimates, that we witnessed in the preceeding sections, naturally raises the expectation that a similar approach may also be effective for systems of conservation laws. Unfortunately, this does not seem to be the case. In order to gain some insight of the difficulty, let us consider the initial-value problem (5.1.1), (5.1.2) for a homogeneous symmetric system of conservation laws in canonical form, on some neighborhood $\mathscr{O}$ of the origin. In analogy to Definition 6.2.1, for the scalar case, we shall call admissible those weak solutions of (5.1.1), (5.1.2) that satisfy the inequality (4.3.4) for any convex entropy η with associated entropy flux $(q_1, \ldots, q_m)$. The first test should be to investigate whether admissible solutions satisfy the relatively modest stability estimate

$$(6.8.1) \qquad \|U(\cdot, t)\|_{L^p(|x|<R)} \le c_p \|U_0(\cdot)\|_{L^p(|x|<R+st)} ,$$

namely, the analog of (6.2.26).

Since (5.1.1) is endowed with the uniformly convex entropy $\eta(U) = |U|^2$, (6.8.1) is satisfied for $p = 2$, by virtue of Theorem 5.2.1. The question is whether (6.8.1) may also hold for any $p \ne 2$, the cases $p = 1$ and $p = \infty$ being of particular interest.

When the system (5.1.1) is linear, it is known that the following three statements are equivalent: (a) (6.8.1) is satisfied for some $p \ne 2$; (b) (6.8.1) holds for all $1 \le p \le \infty$; and (c) the Jacobians of the G_α commute:

$$(6.8.2) \qquad DG_\alpha DG_\beta = DG_\beta DG_\alpha , \quad \alpha, \beta = 1, \cdots, m .$$

It has also been shown that if (6.8.1) is satisfied by all solutions of a quasilinear system (5.1.1) then it must also hold for solutions of the system resulting from linearization of (5.1.1) about any constant state. It thus follows that (6.8.2) is necessary for (6.8.1) in the quasilinear case as well. Finally, it has been proved that, in systems of two conservation laws, $n = 2$, condition (6.8.2) is also sufficient for (6.8.1) to hold, for any $1 \le p \le 2$, and, under additional assumptions on the system, even for $p = \infty$.

The above discussion suggests that only systems in which the commutativity relation (6.8.2) holds offer any hope for treatment in the framework of L^1. This special class includes the scalar case, $n = 1$, already considered here, and the case of a single space dimension, $m = 1$, which will be discussed at length in subsequent chapters; but beyond that it contains very few systems of (even modest) physical interest. An example is the system with fluxes

$$(6.8.3) \qquad G_\alpha(U) = g_\alpha(|U|^2)U , \quad \alpha = 1, \cdots, m ,$$

which governs the flow of a fluid in an anisotropic porous medium. It would be interesting to know whether the initial-value problem for this system is well-posed in L^1.

6.9 Notes

More extensive discussion on the breakdown of classical solutions of scalar conservation laws can be found in Majda [3]. Theorem 6.1.1 is due to Conway [1]. For a systematic study of the geometric features of shock formation and propagation, see Izumiya and Kossioris [1].

There is voluminous literature on weak solutions of the scalar conservation law. The investigation was initiated in the 1950's, in the framework of the single space dimension, stimulated by the seminal paper of Hopf [1], already cited in Section

4.6. References to this early work will be provided, as they become relevant, in Section 11.9.

The first existence proof in several space dimensions is due to Conway and Smoller [1], who recognized the relevance of the space BV and constructed solutions with bounded variation through the Lax-Friedrichs difference scheme. The definitive treatment in the space BV was later given by Volpert [1], who was apparently the first to realize the L^1 contraction property. Building on Volpert's work, Kruzkov [1] proposed the characterization of admissible weak solutions recorded in Section 6.2, derived the L^1 contraction estimate and established the convergence of the method of vanishing viscosity along the lines of our discussion in Section 6.3. More delicate treatment is needed when the flux is merely continuous in u; see Bénilan and Kruzkov [1]. On the other hand, the analysis extends routinely to inhomogeneous systems of balance laws (3.3.1), though solutions may blow up in finite time when the production grows superlinearly with u; see Natalini, Sinestrari and Tesei [1]. In particular, the inhomogeneous conservation law of "transport type," with flux $g_\alpha(u, x) = f(u)v_\alpha(x)$, has interesting structure, especially when div $v = 0$; see Caginalp [1] and Otto [2]. The existence of solutions to the initial-boundary value problem has also been established by the method of vanishing viscosity; see Bardos, Leroux and Nédélec [1] for the BV space, Otto [1] for the L^∞ space, as well as the book of Málek, Nečas and Rokyta [1].

The theory of nonlinear contraction semigroups in general, not necessarily reflexive, Banach space is due to Crandall and Liggett [1]. The application to the scalar conservation law presented in Section 6.4 is taken from Crandall [1].

The construction of solutions by the layering method, discussed in Section 6.5, was suggested by Roždestvenskii [1] and was carried out by Kuznetsov [1] and Douglis [1].

The kinetic formulation described in Section 6.6 is due to Perthame and Tadmor [1] and Lions, Perthame and Tadmor [2]. For related results, see Brenier [1], James, Peng and Perthame [1], Natalini [2], Perthame [1] and Perthame and Pulvirenti [1]. The mechanism that induces the regularizing effect stated in Theorem 6.6.4 plays a prominent role in the theory of nonlinear transport equations in general, including the classical Boltzmann equation (cf. DiPerna and Lions [1]). Cheverry [4] discusses regularity, in detail, by a different approach.

Relaxation phenomena are widely studied in Continuum Physics. The program of constructing solutions to hyperbolic conservation laws via relaxation schemes is undergoing active development. The presentation in Section 6.7 follows Katsoulakis and Tzavaras [1]. See also Natalini [2], and Jin and Xin [1]. Further discussion of relaxation algorithms is found in Chapter XV.

There are several other methods for constructing solutions, most notably by fractional stepping, spectral viscosity approximation, or through various difference schemes that may also be employed for efficient computation. See, for example, Bouchut and Perthame [1], Chen, Du and Tadmor [1], Coquel and LeFloch [1] and Crandall and Majda [1]. For references on the numerics the reader should consult LeVeque [1], Kröner [1] and Godlewski and Raviart [1,2].

In addition to L^1 and BV, other function spaces are relevant to the theory. DeVore and Lucier [1] show that solutions of (6.1.1) reside in Besov spaces.

The large time behavior of solutions of (6.1.1), (6.1.2) is discussed in Conway [1], Engquist and E [1], Bauman and Phillips [1], and Feireisl and Petzeltová [1]. Chen and Frid [1,3,4,5,7] set a framework for investigating, in general systems of conservation laws, decay of solutions induced by scale invariance and compactness. In particular, this theory establishes the long time behavior of solutions of (6.1.1), (6.1.2) when u_0 is either periodic or of the form $u_0(x) = v(|x|^{-1}x) + w(x)$, with $w \in L^1(\mathbb{R}^m)$.

The proof that (6.8.2) is necessary and sufficient for (6.8.1) to hold, in symmetric linear systems, is due to Brenner [1]. Rauch [1] demonstrated that (6.8.2) is necessary for (6.8.1) in the quasilinear case as well. Dafermos [19] proved that (6.8.2) is also sufficient for (6.8.1), at least when $n = 2$ and $1 \le p \le 2$.

Chapter VII. Hyperbolic Systems of Balance Laws in One-Space Dimension

The remainder of the book will be devoted to the study of systems of balance laws in one-space dimension. This narrowing of focus is principally dictated by necessity: At the present time the theory of multidimensional systems is terra incognita. Eventually, research should turn to that vastly unexplored area, which is replete with fascinating problems. In any event, the reader should bear in mind that certain multidimensional phenomena, with special symmetry, like wave focussing, may be studied in the context of the one-space dimensional theory.

This chapter introduces many of the concepts that serve as foundation of the theory of hyperbolic systems of balance laws in one space dimension: Strict hyperbolicity; Riemann invariants and their relation to entropy; simple waves; genuine nonlinearity and its role in the breakdown of classical solutions.

7.1 Balance Laws in One-Space Dimension

When $m = 1$, the general system of balance laws (3.1.1) reduces to

$$(7.1.1) \qquad \partial_t H(U(x,t), x, t) + \partial_x F(U(x,t), x, t) = \Pi(U(x,t), x, t) \ .$$

Systems (7.1.1) naturally arise in the study of gas flow in ducts, vibration of elastic bars or strings, etc., in which the medium itself is modeled as one-dimensional. The simplest examples are homogeneous systems of conservation laws, beginning with the scalar conservation law

$$(7.1.2) \qquad \partial_t u + \partial_x f(u) = 0 \ .$$

A very important example is the one-space dimensional version of the system of balance laws (3.3.4) of adiabatic thermoelasticity, in Lagrangian coordinates. The deformation gradient is now scalar-valued and will be denoted by u. By virtue of (2.3.3), $u = 1/\rho$. Thus when the medium is fluid it is natural to view u as *specific volume*. On the other hand, when the medium is solid, that is an elastic bar undergoing longitudinal oscillation, u measures the *strain*. The scalar-valued stress will be denoted by σ. Note that the physical range of u is $(0, \infty)$, with σ becoming unbounded as $u \downarrow 0$. In this notation and under the assumption of zero body force and heat source, (3.3.4) takes the form

$$(7.1.3) \qquad \begin{cases} \partial_t u - \partial_x v = 0 \\[2mm] \partial_t v - \partial_x \sigma(u, s) = 0 \\[2mm] \partial_t \left[\varepsilon(u, s) + \frac{1}{2} v^2 \right] - \partial_x [v \sigma(u, s)] = 0 \; , \end{cases}$$

with

$$(7.1.4) \qquad \sigma(u, s) = \varepsilon_u(u, s) \; , \qquad \theta(u, s) = \varepsilon_s(u, s) \; .$$

The conditions (3.3.6), (3.3.7) for hyperbolicity here reduce to

$$(7.1.5) \qquad \varepsilon_s(u, s) > 0 \; , \qquad \varepsilon_{uu}(u, s) > 0 \; ,$$

i.e., the absolute temperature θ is positive and the internal energy ε is convex in u. Equivalently, σ is increasing in u, $\sigma_u(u, s) > 0$. When the medium is fluid, it is customary to use pressure $p = -\sigma$ in the place of stress σ.

In the isentropic case, (7.1.3) reduces to

$$(7.1.6) \qquad \begin{cases} \partial_t u - \partial_x v = 0 \\[2mm] \partial_t v - \partial_x \sigma(u) = 0 \; , \end{cases}$$

which is hyperbolic when $\sigma'(u) > 0$. Again, in the context of gas dynamics, the stress σ in (7.1.6) is replaced by $-p$. This results in the so called "p-system". As with (7.1.3), when (7.1.6) is interpreted as governing the longitudinal oscillation of elastic bars, the natural range of u is $(0, \infty)$, with σ becoming unbounded as $u \downarrow 0$. However, system (7.1.6) governs equally well the shearing motion of an elastic layer, where v, σ and u stand for velocity, shear stress and shearing, in the direction of the motion. In that context, u is no longer constrained by $u > 0$ but may take any value in $(-\infty, \infty)$. Accordingly, in our use of (7.1.6) as a mathematical model we shall be assuming that σ is defined as a smooth monotone increasing function on $(-\infty, \infty)$.

The system of conservation laws of one-dimensional isentropic flow of a thermoelastic fluid, in Eulerian coordinates, namely the $m = 1$ version of (3.3.17), with zero body force reads

$$(7.1.7) \qquad \begin{cases} \partial_t \rho + \partial_x(\rho v) = 0 \\[2mm] \partial_t(\rho v) + \partial_x[\rho v^2 + p(\rho)] = 0 \; , \end{cases}$$

which is hyperbolic when $p'(\rho) > 0$. In particular, when the fluid is a polytropic gas (3.3.20), then (7.1.7) reduces to

$$(7.1.8) \qquad \begin{cases} \partial_t \rho + \partial_x(\rho v) = 0 \\[2mm] \partial_t(\rho v) + \partial_x[\rho v^2 + \kappa \rho^\gamma] = 0 \; . \end{cases}$$

Another instructive example is provided by the system that governs the oscillation of a flexible, extensible elastic string. The reference configuration of the string lies along the x-axis, and is assumed to be a natural state of density one.

The motion $\chi = \chi(x, t)$ is monitored through the velocity $V = \partial_t \chi$ and the stretching $W = \partial_x \chi$, which take values in $\mathbb{R}^3$ or in $\mathbb{R}^2$, depending on whether the string is free to move in 3-dimensional space or is constrained to undergo planar oscillations. The tension τ of the string is assumed to depend solely on the length of W, which measures the *stretch*, $v = |W|$. Since the string cannot sustain any compression, the natural range of v is $[1, \infty)$ and τ is assumed to satisfy $\tau(v) > 0, [\tau(v)/v]' > 0$, for $v > 1$. The compatibility relation between V and W together with balance of momentum, in Lagrangian coordinates, yield the hyperbolic system

(7.1.9)
$$\begin{cases} \partial_t W - \partial_x V = 0 \\ \partial_t V - \partial_x \left[\dfrac{\tau(v)}{v} W \right] = 0 . \end{cases}$$

Systems with interesting features govern the propagation of planar electromagnetic waves through special isotropic dielectrics in which the electromagnetic energy depends on the magnetic induction B and the electric displacement D solely through the scalar $r = (B \cdot B + D \cdot D)^{\frac{1}{2}}$; i.e., in the notation of Section 3.3 (f), $\eta(B, D) = \psi(r)$, with $\psi'(0) = 0, \psi''(0) > 0$, and $\psi'(r) > 0, \psi''(r) > 0$ for $r > 0$. Waves propagating in the direction of the 3-axis are represented by solutions of Maxwell's equations (3.3.22), with $J = 0$, in which the fields B, D, E and H depend solely on the single spatial variable $x = x_3$ and on time t. In particular, (3.3.22) imply $B_3 = 0$ and $D_3 = 0$ so that B and D should be regarded as vectors in $\mathbb{R}^2$ satisfying the hyperbolic system

(7.1.10)
$$\begin{cases} \partial_t B - \partial_x \left[\dfrac{\psi'(r)}{r} A D \right] = 0 \\ \partial_t D + \partial_x \left[\dfrac{\psi'(r)}{r} A B \right] = 0 , \end{cases}$$

where A denotes the alternating 2×2 matrix, with entries $A_{11} = A_{22} = 0$, $A_{12} = -A_{21} = 1$.

Returning to the general balance law (7.1.1), it should be noted that H and/or F may depend explicitly on x, to account for inhomogeneity of the medium. For example, isentropic gas flow through a duct of (slowly) varying cross section $a(x)$ is governed by the system

(7.1.11)
$$\begin{cases} \partial_t [a(x)\rho] + \partial_x [a(x)\rho v] = 0 \\ \partial_t [a(x)\rho v] + \partial_x [a(x)\rho v^2 + a(x)p(\rho)] = a'(x)p(\rho) \end{cases}$$

which reduces to (7.1.7) in the homogeneous case $a = $ constant. On the other hand, explicit dependence of H or F on t, indicating "ageing" of the medium, is fairly rare. By contrast, dependence of Π on t is not uncommon, because external forcing is generally time-dependent.

One-space dimensional systems (7.1.1) also derive from multi-space dimensional systems (3.1.1), in the presence of symmetry (planar, cylindrical, radial, etc.)

that reduces spatial dependence to a single parameter. In that case, even when the parent, multidimensional system is homogeneous, the resulting one-dimensional system may be inhomogeneous, to reflect multidimensional geometric effects. For example, the one-space dimensional system governing radial, isentropic gas flow, that results from the homogeneous Euler equations (3.3.17) is inhomogeneous:

$$(7.1.12) \qquad \begin{cases} \partial_t \rho + \partial_r (\rho v) + \dfrac{2\rho v}{r} = 0 \\[2ex] \partial_t (\rho v) + \partial_r [\rho v^2 + p(\rho)] + \dfrac{2\rho v^2}{r} = 0 \ . \end{cases}$$

In particular, certain multidimensional phenomena, like wave focusing, may be investigated in the framework of one-space dimension.

7.2 Hyperbolicity and Strict Hyperbolicity

As in earlier chapters, to avoid inessential technical complications, the theory shall be developed in the context of homogeneous systems of conservation laws in canonical form:

$$(7.2.1) \qquad\qquad \partial_t U(x, t) + \partial_x F(U(x, t)) = 0 \ .$$

F is a smooth map from an open convex subset $\mathcal{O}$ of $\mathbb{R}^n$ to $\mathbb{R}^n$.

Often in the applications, systems (7.2.1) govern planar front solutions $U = U(v \cdot x, t)$, in the spatial direction $v \in \mathcal{S}^{m-1}$, of multispace-dimensional systems of conservation laws (4.1.1). In that connection,

$$(7.2.2) \qquad\qquad F(U) = \sum_{\alpha=1}^{m} v_\alpha G_\alpha(U) \ , \qquad U \in \mathcal{O} \ .$$

Referring to the examples introduced in Section 7.1, in order to cast the system (7.1.3) of thermoelasticity to canonical form, we have to switch from (u, v, s) to new state variables (u, v, e), where $e = \varepsilon + \frac{1}{2} v^2$ is the total energy. Similarly, the system (7.1.7) of isentropic gas flow is written in canonical form in terms of the state variables (ρ, m), where $m = \rho v$ is the momentum.

By Definition 3.1.1, the system (7.2.1) is hyperbolic if for every $U \in \mathcal{O}$ the $n \times n$ Jacobian matrix $DF(U)$ has real eigenvalues $\lambda_1(U) \leq \cdots \leq \lambda_n(U)$ and n lineary independent eigenvectors $R_1(U), \cdots, R_n(U)$. For future use, we also introduce left (row) eigenvectors $L_1(U), \cdots, L_n(U)$ of $DF(U)$, normalized by

$$(7.2.3) \qquad\qquad L_i(U) R_j(U) = \begin{cases} 0 \ \text{ if } \ i \neq j \\[1ex] 1 \ \text{ if } \ i = j \ . \end{cases}$$

Notice that the multispace-dimensional system (4.1.1) is hyperbolic if and only if all one-space dimensional systems (7.2.1) resulting from it through (7.2.2), for arbitrary $\nu \in \mathscr{S}^{m-1}$, are hyperbolic. Thus hyperbolicity is essentially a one-space dimensional notion.

For the system (7.1.6) of one-dimensional isentropic elasticity, in Lagrangian coordinates, which will serve throughout as a demo for illustrating the general concepts, we have

$$(7.2.4) \qquad \lambda_1 = -\sigma'(u)^{1/2} , \quad \lambda_2 = \sigma'(u)^{1/2} ,$$

$$(7.2.5) \qquad R_1 = \frac{1}{2} \begin{pmatrix} -\sigma'(u)^{-1/2} \\ -1 \end{pmatrix} , \quad R_2 = \frac{1}{2} \begin{pmatrix} -\sigma'(u)^{-1/2} \\ 1 \end{pmatrix} ,$$

$$(7.2.6) \qquad L_1 = (-\sigma'(u)^{1/2}, -1) , \quad L_2 = (-\sigma'(u)^{1/2}, 1) .$$

The eigenvalue λ_i of DF, $i = 1, \cdots, n$, is called the *i-characteristic speed* of the system (7.2.1). The term derives from the following

Definition 7.2.1 An *i-characteristic*, $i = 1, \cdots, n$, of the system (7.2.1), associated with a classical solution U, is a C^1 function $x = x(t)$, with graph contained in the domain of U, which is an integral curve of the ordinary differential equation

$$(7.2.7) \qquad \frac{dx}{dt} = \lambda_i(U(x,t)) .$$

The standard existence-uniqueness theory for ordinary differential equations (7.2.7) implies that through any point $(\bar{x}, \bar{t})$ in the domain of a classical solution of (7.2.1) passes precisely one characteristic of each characteristic family.

Characteristics are carriers of waves of various types. For example, Eq. (1.6.1), for the general system (1.4.3) of balance laws, specialized to (7.2.1), implies that weak fronts propagate along characteristics. As a result, the presence of multiple eigenvalues of DF may induce severe complexity in the behavior of solutions, due to resonance. It is thus natural to single out systems that are free from such complication:

Definition 7.2.2 The system (7.2.1) is *strictly hyperbolic* if for any $U \in \mathcal{O}$ the Jacobian $DF(U)$ has real, distinct eigenvalues

$$(7.2.8) \qquad \lambda_1(U) < \cdots < \lambda_n(U) .$$

By virtue of (7.2.4), the system (7.1.6) of isentropic elasticity in Lagrangian coordinates is strictly hyperbolic. The same is true for the system (7.1.3) of adiabatic thermoelasticity, for which the characteristic speeds are

$$(7.2.9) \qquad \lambda_1 = -\sigma_u(u,s)^{1/2} , \quad \lambda_2 = 0 , \quad \lambda_3 = \sigma_u(u,s)^{1/2} .$$

The system (7.1.8) for the polytropic gas has characteristic speeds

$$(7.2.10) \qquad \lambda_1 = v - (\kappa\gamma)^{1/2}\rho^{\frac{\gamma-1}{2}} , \quad \lambda_2 = v + (\kappa\gamma)^{1/2}\rho^{\frac{\gamma-1}{2}} ,$$

and so it is strictly hyperbolic on the part of the state space with $\rho > 0$. However (strict) hyperbolicity fails at the vacuum state, $\rho = 0$.

In view of the above examples, the reader may form the impression that strict hyperbolicity is the norm in systems arising in Continuum Physics. However, this is not the case. It has been shown that in one-space dimensional systems (7.2.1), of size $n = \pm 2, \pm 3, \pm 4$ (mod 8), which result from parent three-space dimensional systems (4.1.1) through (7.2.2), strict hyperbolicity necessarily fails, at least in some spatial direction $\nu \in \mathscr{S}^2$. In fact, failure of strict hyperbolicity has been documented in many important systems, including the system governing planar fronts in isentropic isotropic thermoelasticity, the system modelling flow in porous media, and the system of equations of magnetohydrodynamics.

In systems of size $n = 2$, strict hyperbolicity typically fails at isolated *umbilic points*, at which DF reduces to a multiple of the identity matrix. Even the presence of a single umbilic point is sufficient to create havoc in the behavior of solutions. This will be demonstrated in following chapters by means of the simple demo system

$$(7.2.11) \qquad \begin{cases} \partial_t u + \partial_x[(u^2 + v^2)u] = 0 \\ \partial_t v + \partial_x[(u^2 + v^2)v] = 0 \ , \end{cases}$$

which is a caricature of the system (7.1.9). The characteristic speeds of (7.2.11) are

$$(7.2.12) \qquad \lambda_1 = u^2 + v^2 \ , \quad \lambda_2 = 3(u^2 + v^2) \ ,$$

with corresponding eigenvectors

$$(7.2.13) \qquad R_1 = \begin{pmatrix} v \\ -u \end{pmatrix} \ , \quad R_2 = \begin{pmatrix} u \\ v \end{pmatrix} \ ,$$

so this system is strictly hyperbolic, except at the origin $(0,0)$ which is an umbilic point.

We close this section with the derivation of a useful identity. We apply D to both sides of the equation $DFR_j = \lambda_j R_j$ and then multiply, from the left, by R_k^T; also we apply D to $DFR_k = \lambda_k R_k$ and then multiply, from the left, by R_j^T. Upon subtracting the resulting two equations, we deduce

$$(7.2.14) \qquad \begin{aligned} (D\lambda_j R_k)R_j &- (D\lambda_k R_j)R_k \\ &= DF[R_j, R_k] - \lambda_j DR_j R_k + \lambda_k DR_k R_j \ , \quad j,k = 1, \cdots, n \ , \end{aligned}$$

where $[R_j, R_k]$ denotes the Lie bracket:

$$(7.2.15) \qquad [R_j, R_k] = DR_j R_k - DR_k R_j \ .$$

In particular, at a point $U \in \mathscr{O}$ where strict hyperbolicity fails, say $\lambda_j(U) = \lambda_k(U)$, (7.2.14) yields

$$(7.2.16) \qquad (D\lambda_j R_k)R_j - (D\lambda_k R_j)R_k = (DF - \lambda_j I)[R_j, R_k] \ .$$

Upon multiplying (7.2.16), from the left, by $L_j(U)$ and by $L_k(U)$, we conclude from (7.2.3):

$$(7.2.17) \qquad D\lambda_j(U)R_k(U) = D\lambda_k(U)R_j(U) = 0 .$$

7.3 Riemann Invariants

Consider a hyperbolic system (7.2.1) of conservation laws on $\mathcal{O} \subset \mathbb{R}^n$. A very important concept is introduced by the following

Definition 7.3.1 An *i-Riemann invariant* of (7.2.1) is a smooth scalar-valued function w on $\mathcal{O}$ such that

$$(7.3.1) \qquad Dw(U)R_i(U) = 0 , \quad U \in \mathcal{O} .$$

For example, recalling (7.2.5), we readily verify that the functions

$$(7.3.2) \qquad w = -\int^u \sigma'(\omega)^{\frac{1}{2}}d\omega + v , \quad z = -\int^u \sigma'(\omega)^{\frac{1}{2}}d\omega - v$$

are, respectively, 1- and 2-Riemann invariants of the system (7.1.6). Similarly, it can be shown that

$$(7.3.3) \qquad w = v + \frac{2(\kappa\gamma)^{1/2}}{\gamma - 1}\rho^{\frac{\gamma-1}{2}} , \quad z = v - \frac{2(\kappa\gamma)^{1/2}}{\gamma - 1}\rho^{\frac{\gamma-1}{2}}$$

are 1- and 2-Riemann invariants of the system (7.1.8) of isentropic flow of a polytropic gas.

By solving the first order linear differential equation (7.3.1) for w, one may construct in the vicinity of any point $U \in \mathcal{O}$ $n-1$ i-Riemann invariants whose gradients are linearly independent and span the orthogonal complement of R_i. For example, the reader may verify as an exercise that the three pairs of functions

$$(7.3.4) \qquad \begin{cases} s, -\int^u \sigma_u(\omega, s)^{\frac{1}{2}}d\omega + v \\ v, \sigma(u, s) \\ s, -\int^u \sigma_u(\omega, s)^{\frac{1}{2}}d\omega - v \end{cases}$$

are, respectively, 1-, 2-, and 3-Riemann invariants of the system (7.1.3) of adiabatic thermoelasticity.

Riemann invariants are particularly useful in systems with the following special structure:

Definition 7.3.2 The system (7.2.1) is endowed with a *coordinate system of Riemann invariants* if there exist n scalar-valued functions $(w_1, \cdots, w_n)$ on $\mathcal{O}$ such that, for any $i, j = 1, \cdots, n$, with $i \neq j$, w_j is an i-Riemann invariant of (7.2.1).

An immediate consequence of Definitions 7.3.1 and 7.3.2 is

Theorem 7.3.1 *The functions* $(w_1, \cdots, w_n)$ *form a coordinate system of Riemann invariants for* (7.2.1) *if and only if*

$$(7.3.5) \qquad Dw_i(U)R_j(U) \begin{cases} = 0 & \text{if } i \neq j \\ \neq 0 & \text{if } i = j \end{cases}$$

i.e., if and only if, for $i = 1, \cdots, n$, $Dw_i(U)$ *is a left eigenvector of the matrix* $DF(U)$, *associated with the characteristic speed* $\lambda_i(U)$.

Assuming (7.2.1) is endowed with a coordinate system $(w_1, \cdots, w_n)$ of Riemann invariants and multiplying, from the left, by $Dw_i, i = 1, \cdots, n$, we reduce this system to diagonal form:

$$(7.3.6) \qquad \partial_t w_i + \lambda_i \partial_x w_i = 0 , \quad i = 1, \cdots, n ,$$

which is equivalent to the original form (7.2.1), albeit only in the context of classical solutions. The left-hand side of (7.3.6) is just the derivative of w_i in the i-characteristic direction. Therefore,

Theorem 7.3.2 *Assume* $(w_1, \cdots, w_n)$ *form a coordinate system of Riemann invariants for* (7.2.1). *For* $i = 1, \cdots, n$, w_i *stays constant along every* i-*characteristic associated with any classical solution* U *of* (7.2.1).

Clearly, any hyperbolic system of two conservation laws is endowed with a coordinate system of Riemann invariants. By contrast, in systems of size $n \geq 3$ coordinate systems of Riemann invariants will exist only in the exceptional case where the formally overdetermined system (7.3.5), with $n(n-1)$ equations for the n unknown $(w_1, \cdots, w_n)$, has a solution. Notice that the geometric interpretation of (7.3.5) is that, for every fixed $i = 1, \cdots, n$, the level surface of w_i at any U is spanned by the $n - 1$ vectors $R_1(U), \cdots, R_{i-1}(U), R_{i+1}(U), \cdots, R_n(U)$. By the Frobenius theorem, this may happen if and only if, for $i \neq j \neq k \neq i$, the Lie bracket $[R_j, R_k]$ (cf. (7.2.15)) lies in the span of $\{R_1, \cdots, R_{i-1}, R_{i+1}, \cdots, R_n\}$. Consequently, the system (7.2.1) is endowed with a coordinate system of Riemann invariants if and only if

$$(7.3.7) \qquad [R_j, R_k] = \alpha_j R_j - \alpha_k R_k , \quad j, k = 1, \cdots, n ,$$

where $\alpha_1, \cdots, \alpha_n$ are scalar fields.

When a coordinate system $(w_1, \cdots, w_n)$ of Riemann invariants exists for (7.2.1), it is convenient to normalize the eigenvectors $R_1, \cdots, R_n$ so that

$$(7.3.8) \qquad Dw_i(U)R_j(U) = \begin{cases} 0 & \text{if } i \neq j \\ 1 & \text{if } i = j . \end{cases}$$

In that case we note the identity

$$
\begin{aligned}
Dw_i\, DR_j\, R_k &= D(Dw_i\, R_j) R_k - R_j^T D^2 w_i R_k \\
&= -R_j^T D^2 w_i R_k \ , \quad i,j,k = 1, \cdots, n \ ,
\end{aligned}
$$
(7.3.9)

which implies, in particular, $Dw_i[R_j, R_k] = 0$, $i = 1, \cdots, n$, i.e.,

(7.3.10)
$$
[R_j, R_k] = 0 \ , \quad j,k = 1, \cdots, n \ .
$$

Recalling the identity (7.2.14) and using (7.2.15), (7.3.10), we deduce that whenever $\lambda_j(U) \neq \lambda_k(U)$, $DR_j(U) R_k(U)$ lies in the span of $\{R_j(U), R_k(U)\}$. This, together with (7.3.8) and (7.3.9) yields

(7.3.11)
$$
R_j^T D^2 w_i R_k = -Dw_i\, DR_j\, R_k = 0 \ , \quad i \neq j \neq k \neq i \ .
$$

When (7.2.1) possesses a coordinate system $(w_1, \cdots, w_n)$ of Riemann invariants, the map that carries U to $W = (w_1, \cdots, w_n)^T$ is locally a diffeomorphism. It is often convenient to regard W rather than U as the state vector. To avoid proliferation of symbols, we shall be using, when there is no danger of confusion, the same symbol to denote fields as functions of either U or W. By virtue of (7.3.8), $\partial U / \partial w_i = R_i$ and so the chain rule yields, for the typical function ϕ,

(7.3.12)
$$
\frac{\partial \phi}{\partial w_i} = D\phi R_i \ , \quad i = 1, \cdots, n \ .
$$

For example, (7.3.10) reduces to $\partial R_j / \partial w_k = \partial R_k / \partial w_j = \partial^2 U / \partial w_j \partial w_k$.

We proceed to derive certain identities that will help us later to establish other remarkable properties of systems endowed with a coordinate system of Riemann invariants. Upon combining (7.2.14), (7.2.15), (7.3.10) and (7.3.12), we deduce

(7.3.13)
$$
-\frac{\partial R_j}{\partial w_k} = g_{jk} R_j + g_{kj} R_k \ , \quad j,k = 1, \cdots, n \ ; \quad j \neq k \ ,
$$

where we have set

(7.3.14)
$$
g_{jk} = \frac{1}{\lambda_j - \lambda_k} \frac{\partial \lambda_j}{\partial w_k} \ , \quad j,k = 1, \cdots, n \ ; \quad j \neq k \ .
$$

Notice that g_{jk} may be defined even when $\lambda_j = \lambda_k$, because at such points $\partial \lambda_j / \partial w_k = 0$, by virtue of (7.2.17) and (7.3.12). From (7.3.13),

(7.3.15)
$$
\begin{aligned}
-\frac{\partial^2 R_j}{\partial w_i \partial w_k} &= \frac{\partial g_{jk}}{\partial w_i} R_j - g_{jk}(g_{ji} R_j + g_{ij} R_i) \\
&\quad + \frac{\partial g_{kj}}{\partial w_i} R_k - g_{kj}(g_{ki} R_k + g_{ik} R_i) \ .
\end{aligned}
$$

Since R_i, R_j, R_k are linearly independent for $i \neq j \neq k \neq i$, and the right-hand side of (7.3.15) has to be symmetric in (i,k), we deduce

(7.3.16)
$$
\frac{\partial g_{jk}}{\partial w_i} = \frac{\partial g_{ji}}{\partial w_k} \ , \quad i \neq j \neq k \neq i \ ,
$$

$$(7.3.17) \qquad \frac{\partial g_{ij}}{\partial w_k} + g_{ij}g_{jk} - g_{ij}g_{ik} + g_{ik}g_{kj} = 0 , \qquad i \neq j \neq k \neq i .$$

Of the hyperbolic systems of conservation laws of size $n \geq 3$ that arise in the applications, few possess coordinate systems of Riemann invariants. A noteworthy example is the system of *electrophoresis*:

$$(7.3.18) \qquad \partial_t U_i + \partial_x \frac{c_i U_i}{\displaystyle\sum_{j=1}^{n} U_j} = 0 , \qquad i = 1, \cdots, n ,$$

where $c_1 < c_2 < \cdots < c_n$ are positive constants. This system governs the process used to separate n ionized chemical compounds in solution by applying an electric field. In that context, U_i denotes the concentration and c_i measures the electrophoretic mobility of the i-th species. In particular, $U_i \geq 0$. As an exercise, the reader may verify that the characteristic speeds of (7.3.18) are given by

$$(7.3.19) \qquad \lambda_i = \mu_i \sum_{j=1}^{n} U_j , \qquad i = 1, \cdots, n ,$$

where $\mu_n = 0$ while, for $i = 1, \cdots, n-1$, the value of μ_i at U is the solution of the equation

$$(7.3.20) \qquad \sum_{j=1}^{n} \frac{c_j U_j}{c_j - \mu} = \sum_{j=1}^{n} U_j$$

lying in the interval (c_i, c_{i+1}). Moreover, (7.3.18) is endowed with a coordinate system $(w_1, \cdots, w_n)$ of Riemann invariants where

$$(7.3.21) \qquad w_n = \sum_{j=1}^{n} \frac{1}{c_j} U_j$$

while, for $i = 1, \cdots, n-1$, the value of w_i at U is the solution of the equation

$$(7.3.22) \qquad \sum_{j=1}^{n} \frac{U_j}{c_j - w} = 0$$

that lies in the interval (c_i, c_{i+1}). In following sections we shall see that the system (7.3.18) has very special structure and a host of interesting properties.

Another interesting system endowed with coordinate systems of Riemann invariants is (7.1.10), which, as we recall, governs the propagation of planar electromagnetic waves through special isotropic dielectrics. This is seen by passing from (B_1, B_2, D_1, D_2) to the new state vector (p, q, a, b) defind through

$$(7.3.23) \qquad \begin{cases} \sqrt{2}p \exp(ia) = B_2 + D_1 - i(B_1 - D_2) \\ \sqrt{2}q \exp(ib) = -B_2 + D_1 + i(B_1 + D_2) . \end{cases}$$

In particular, $p^2 + q^2 = r^2$. A simple calculation shows that, at least in the context of classical solutions, (7.1.10) reduces to

(7.3.24)
$$\begin{cases} \partial_t p + \partial_x \left[\dfrac{\psi'(r)}{r} p \right] = 0 \\[2ex] \partial_t q - \partial_x \left[\dfrac{\psi'(r)}{r} q \right] = 0 \,, \end{cases}$$

(7.3.25)
$$\begin{cases} \partial_t a + \dfrac{\psi'(r)}{r} \partial_x a = 0 \\[2ex] \partial_t b - \dfrac{\psi'(r)}{r} \partial_x b = 0 \,. \end{cases}$$

Notice that (7.3.24) constitutes a closed system of two conservation laws, from which p, q, and thereby r, may be determined. Subsequently (7.3.25) may be resolved, as two independent nonhomogeneous scalar conservation laws, to determine a and b. In particular, a and b together with any pair of Riemann invariants of (7.3.24) will constitute a coordinate system of Riemann invariants for (7.1.10).

7.4 Entropy-Entropy Flux Pairs

Entropies play a central role in the theory of hyperbolic systems of conservation laws in one-space dimension. Adapting the discussion of Section 3.2 to the present setting, we infer that functions η and q on $\mathcal{O}$ constitute an entropy-entropy flux pair for the system (7.2.1) if

(7.4.1)
$$Dq(U) = D\eta(U)DF(U) \,, \quad U \in \mathcal{O} \,.$$

Furthermore, the integrability condition (3.2.4) here reduces to

(7.4.2)
$$D^2\eta(U)DF(U) = DF(U)^T D^2\eta(U) \,, \quad U \in \mathcal{O} \,.$$

Upon multiplying (7.4.2) from the left by $R_j(U)^T$ and from the right by $R_k(U)$, $j \neq k$, we deduce that (7.4.2) is equivalent to

(7.4.3)
$$R_j(U)^T D^2\eta(U)R_k(U) = 0 \,, \quad j, k = 1, \cdots, n \,; \quad j \neq k \,,$$

with the understanding that (7.4.3) holds automatically when $\lambda_j(U) \neq \lambda_k(U)$ but may require renormalization of eigenvectors R_i associated with multiple characteristic speeds. Note that the requirement that some entropy η is convex may now be conveniently expressed as

(7.4.4)
$$R_j(U)^T D^2\eta(U)R_j(U) > 0 \,, \quad j = 1, \cdots, n \,.$$

When the system (7.2.1) is symmetric,

(7.4.5)
$$DF(U)^T = DF(U) \,, \quad U \in \mathcal{O} \,,$$

it admits two interesting entropy-entropy flux pairs:

$$(7.4.6) \qquad \eta = \frac{1}{2} U^T U \,, \quad q = U^T F(U) - h(U) \,,$$

$$(7.4.7) \qquad \eta = h(U) \,, \quad q = \frac{1}{2} F(U)^T F(U) \,,$$

where h is defined by the condition

$$(7.4.8) \qquad Dh(U) = F(U)^T \,.$$

As explained in Chapter III, our model systems (7.1.3), (7.1.6), (7.1.8) are endowed with entropy-entropy flux pairs, respectively,

$$(7.4.9) \qquad \eta = -s \,, \quad q = 0 \,,$$

$$(7.4.10) \qquad \eta = \frac{1}{2} v^2 + \Sigma(u) \,, \quad q = -v\sigma(u) \,, \quad \Sigma(u) = \int^u \sigma(\omega) d\omega \,,$$

$$(7.4.11) \qquad \eta = \frac{1}{2} \rho v^2 + \frac{\kappa}{\gamma - 1} \rho^\gamma \,, \quad q = \frac{1}{2} \rho v^3 + \frac{\kappa \gamma}{\gamma - 1} \rho^\gamma v \,,$$

induced by the Second Law of thermodynamics. When expressed as functions of the canonical state variables, that is (u, v, e) for (7.4.9), (u, v) for (7.4.10), and (ρ, m) for (7.4.11), the above entropies are convex.

In developing the theory of systems (7.2.1), it will be very useful to be able to construct entropies with given specifications. These must be solutions of (7.4.2), which is a linear, second order system of $\frac{1}{2}n(n-1)$ partial differential equations in a single unknown η. Thus, when $n = 2$, (7.4.2) reduces to a single linear hyperbolic equation that may be solved to produce an abundance of entropies. By contrast, for $n \geq 3$, (7.4.2) is formally overdetermined. Notwithstanding the presence of special solutions like (7.4.6) and (7.4.7), one should not be expecting an abundance of entropies, unless (7.2.1) is special. It is remarkable that the overdeterminacy of (7.4.2) is nullified when (7.2.1) is endowed with a coordinate system $(w_1, \cdots, w_n)$ of Riemann invariants. In that case it is convenient to seek η and q as functions of the state vector $W = (w_1, \cdots, w_n)^T$. Upon multiplying (7.4.1), from the right, by $R_j(U)$ and using (7.3.12), we deduce that (7.4.1) is now equivalent to

$$(7.4.12) \qquad \frac{\partial q}{\partial w_j} = \lambda_j \frac{\partial \eta}{\partial w_j} \,, \quad j = 1, \cdots, n \,.$$

The integrability condition associated with (7.4.12) takes the form

$$(7.4.13) \qquad \frac{\partial^2 \eta}{\partial w_j \partial w_k} + g_{jk} \frac{\partial \eta}{\partial w_j} + g_{kj} \frac{\partial \eta}{\partial w_k} = 0 \,, \quad j, k = 1, \cdots, n \,; \quad j \neq k \,,$$

where g_{jk}, g_{kj} are the functions defined through (7.3.14). An alternative, useful expression for g_{jk} obtains if one derives (7.4.13) directly from (7.4.3). Indeed, for $j, k = 1, \cdots, n$,

$$R_j^T D^2 \eta R_k = D(D\eta R_j)R_k - D\eta DR_j R_k$$

(7.4.14)
$$= D(D\eta R_j)R_k - \sum_{i=1}^{n} \frac{\partial \eta}{\partial w_i} Dw_i DR_j R_k \ .$$

Combining (7.4.3), (7.3.12), (7.3.10) and (7.3.9), we arrive at an equation of the form (7.4.13) with

(7.4.15)
$$g_{jk} = R_j^T D^2 w_j R_k \ , \quad j, k = 1, \cdots, n \ ; \quad j \neq k \ .$$

The reader may verify directly, as an exercise, with the help of (7.2.14), (7.3.8), (7.3.11), (7.3.10), (7.3.9) and (7.3.12) that (7.3.14) and (7.4.15) are equivalent.

Applying (7.4.14) with $k = j$, using (7.3.12), (7.3.9) and recalling (7.4.4), we deduce that, in terms of Riemann invariants, the convexity condition on η is expressed by the set of inequalities

(7.4.16)
$$\frac{\partial^2 \eta}{\partial w_j^2} + \sum_{i=1}^{n} a_{ij} \frac{\partial \eta}{\partial w_i} \geq 0 \ , \quad j = 1, \cdots, n \ ,$$

where

(7.4.17)
$$a_{ij} = R_j^T D^2 w_i R_j \ , \quad i, j = 1, \cdots, n \ .$$

The system (7.4.13) contains $\frac{1}{2}n(n-1)$ equations in the single unknown η and thus looks overdetermined when $n \geq 3$. It turns out, however, that this set of equations is automatically compatible. To see this, differentiate (7.4.13) with respect to w_i, $i \neq j \neq k \neq i$, to get

(7.4.18)
$$\frac{\partial^3 \eta}{\partial w_i \partial w_j \partial w_k} = \frac{\partial g_{jk}}{\partial w_i} \frac{\partial \eta}{\partial w_j} - g_{jk}\left(g_{ji} \frac{\partial \eta}{\partial w_j} + g_{ij} \frac{\partial \eta}{\partial w_i} \right)$$
$$+ \frac{\partial g_{kj}}{\partial w_i} \frac{\partial \eta}{\partial w_k} - g_{kj}\left(g_{ki} \frac{\partial \eta}{\partial w_i} + g_{ik} \frac{\partial \eta}{\partial w_i} \right) \ .$$

The system (7.4.13) will be integrable if and only if, for $i \neq j \neq k \neq i$, the right-hand side of (7.4.18) is symmetric in (i, j, k). But this is always the case, by account of the identities (7.3.16) and (7.3.17). Consequently, in a neighborhood of any given point $\overline{W} = (\overline{w}_1, \cdots, \overline{w}_n)^T$ in state space, there exists a unique entropy η with arbitrarily prescribed values $\{\eta(w_1, \overline{w}_2, \cdots, \overline{w}_n), \eta(\overline{w}_1, w_2, \cdots, \overline{w}_n), \cdots, \eta(\overline{w}_1, \cdots, \overline{w}_{n-1}, w_n)\}$ along straight lines parallel to the coordinate axes. When $n = 2$, this amounts to solving a classical Goursat problem.

We have thus shown that systems endowed with coordinate sets of Riemann invariants are also endowed with an abundance of entropies. For this reason, such systems are called *rich*. In particular, the system (7.3.18) of electrophoresis and the system (7.1.10) of electromagnetic waves are rich. The reader will find how to construct the family of its entropies in the references cited in Section 7.9.

7.5 Genuine Nonlinearity and Linear Degeneracy

The feature distinguishing the behavior of linear and nonlinear hyperbolic systems of conservation laws is that in the former, characteristic speeds being constant, all waves of the same family propagate with fixed speed; while in the latter wave speeds vary with wave-amplitude. As we proceed with our study, we will encounter various manifestations of nonlinearity and in every case we shall notice that its effects will be particularly pronounced when the characteristic speeds λ_i vary in the direction of the corresponding eigenvectors R_i. This motivates the following

Definition 7.5.1 For the hyperbolic system (7.2.1) of conservation laws on $\mathcal{O}$, $U \in \mathcal{O}$ is called a *point of genuine nonlinearity of the i-characteristic family* if

$$(7.5.1) \qquad\qquad D\lambda_i(U)R_i(U) \neq 0 ,$$

or a point of *linear degeneracy of the i-characteristic family* if

$$(7.5.2) \qquad\qquad D\lambda_i(U)R_i(U) = 0 .$$

When (7.5.1) holds for all $U \in \mathcal{O}$, i is a *genuinely nonlinear characteristic family* while if (7.5.2) is satisfid for all $U \in \mathcal{O}$, then i is a *linearly degenerate characteristic family*. When every characteristic family is genuinely nonlinear, (7.2.1) is a *genuinely nonlinear system*.

It is clear that the i-characteristic family is linearly degenerate if and only if the i-characteristic speed λ_i is constant along the integral curves of the vector field R_i.

The scalar conservation law (7.1.2), with characteristics speed $\lambda = f'(u)$, is genuinely nonlinear when f has no inflection points: $f''(u) \neq 0$. In particular, the Burgers equation (4.2.1) is genuinely nonlinear.

Using (7.2.4) and (7.2.5), one readily checks that the system (7.1.6) is genuinely nonlinear when $\sigma''(u) \neq 0$. As an exercise, the reader may verify that the system (7.1.7) is genuinely nonlinear if $2p'(\rho) + \rho p''(\rho) > 0$ so, in particular, the system (7.1.8) for the polytropic gas is genuinely nonlinear.

By account of (7.2.9), the 2-characteristic family of the system (7.1.3) of thermoelasticity is linearly degenerate. It turns out that the other two characteristic families are genuinely nonlinear, provided $\sigma_{uu}(u, s) \neq 0$.

The system (7.1.9) of planar oscillations of elastic strings possesses four characteristic families. Two of them, associated with transverse oscillations with characteristic speeds $\pm\sqrt{T(v)/v}$ are linearly degenerate. The other two, associated with longitudinal oscillations with characteristic speeds $\pm\sqrt{T'(v)}$ are genuinely nonlinear when $T''(v) \neq 0$ and linearly degenerate when T is a linear function of v. Similarly, the 1-characteristic family of the system (7.2.11) is linearly degenerate, while the 2-characteristic family is genuinely nonlinear, except at the origin.

Finally, in the system (7.3.18) of electrophoresis the n-characteristic family is linearly degenerate while the rest are genuinely nonlinear.

Quite often, linear degeneracy results from the loss of strict hyperbolicity. Indeed, an immediate consequence of (7.2.17) is

Theorem 7.5.1 *In the hyperbolic system (7.2.1) of conservation laws, assume that the j- and k-characteristic speeds coincide: $\lambda_j(U) = \lambda_k(U)$, $U \in \mathcal{O}$. Then both the j- and the k-characteristic families are linearly degenerate.*

When the system (7.2.1) is endowed with a coordinate system $(w_1, \cdots, w_n)$ of Riemann invariants and we use $W = (w_1, \cdots, w_n)^T$ as our state vector, the conditions of genuine nonlinearity and linear degeneracy assume an elegant and suggestive form. Indeed, upon using (7.3.12), we deduce that (7.5.1) and (7.5.2) are respectively equivalent to

$$(7.5.3) \qquad \frac{\partial \lambda_i}{\partial w_i} \neq 0$$

and

$$(7.5.4) \qquad \frac{\partial \lambda_i}{\partial w_i} = 0 \ .$$

7.6 Simple Waves

In the context of classical solutions, the scalar conservation law (7.1.2), with characteristic speed $\lambda = f'(u)$, takes the form

$$(7.6.1) \qquad \partial_t u(x, t) + \lambda(u(x, t))\partial_x u(x, t) = 0 \ .$$

As noted already in Section 6.1, by virtue of (7.6.1) u stays constant along characteristics and this, in turn, implies that each characteristic propagates with constant speed, i.e., it is a straight line. It turns out that general hyperbolic systems (7.2.1) of conservation laws admit special solutions with the same features:

Definition 7.6.1 A classical, C^1 solution U of the hyperbolic system (7.2.1) of conservation laws is called an *i-simple wave* if U stays constant along any i-characteristic associated with it.

Thus a C^1 function U, defined on an open subset of $\mathbb{R}^2$ and taking values in $\mathcal{O}$, is an i-simple wave if it satisfies (7.2.1) together with

$$(7.6.2) \qquad \partial_t U(x, t) + \lambda_i(U(x, t))\partial_x U(x, t) = 0 \ .$$

In particular, in an i-simple wave each i-characteristic propagates with constant speed and so it is a straight line.

If U is an i-simple wave, combining (7.2.1) with (7.6.2) we deduce

(7.6.3)
$$\begin{cases} \partial_x U(x,t) = a(x,t) R_i(U(x,t)) \\ \partial_t U(x,t) = -a(x,t) \lambda_i(U(x,t)) R_i(U(x,t)) \,, \end{cases}$$

where a is a scalar field. Conversely, any C^1 function U that satisfies (7.6.3) is necessarily an i-simple wave.

It is possible to give still another characterization of simple waves, in terms of Riemann invariants:

Theorem 7.6.1 *A classical, C^1 solution U of (7.2.1) is an i-simple wave if and only if every i-Riemann invariant is constant on each connected component of the domain of U.*

Proof. For any i-Riemann invariant w, $\partial_x w = Dw \partial_x U$ and $\partial_t w = Dw \partial_t U$. If U is an i-simple wave, $\partial_x w$ and $\partial_t w$ vanish identically, by virtue of (7.6.3) and (7.3.1), so that w is constant on any connected component of the domain of U.

Conversely, recalling that the gradients of i-Riemann invariants span the orthogonal complement of R_i, we infer that when $\partial_x w = Dw \partial_x U$ vanishes identically for all i-Riemann invariants w, $\partial_x U$ must satisfy (7.6.3)$_1$. Substituting (7.6.3)$_1$ into (7.2.1) we conclude that (7.6.3)$_2$ holds as well, i.e. U is an i-simple wave. This completes the proof.

Any constant function $U = \overline{U}$ qualifies, according to Definition 7.6.1, to be viewed as an i-simple wave, for every $i = 1, \cdots, n$. It is expedient, however, to refer to such trivial solutions as *constant states* and reserve the term simple wave for solutions that are not constant on any open subset of their domain. The following proposition, which demonstrates that simple waves are the natural neighbors of constant states, is stated informally, in physical rather than mathematical terminology. The precise meaning of assumptions and conclusions may be extracted from the proof.

Theorem 7.6.2 *Any weak front moving into a constant state propagates with constant characteristic speed of some family i. Furthermore, the wake of this front is necessarily an i-simple wave.*

Proof. The setting is as follows: The system (7.2.1) is assumed strictly hyperbolic. U is a classical, Lipschitz solution which is C^1 on its domain, except along the graph of a C^1 curve $x = \chi(t)$. U is constant, $\overline{U}$, at any point of its domain lying on one side, say to the right, of the graph of χ. By contrast, $\partial_x U$ and $\partial_t U$ attain nonzero limits from the left along the graph of χ. Thus, according to the terminology of Section 1.6, χ is a weak front propagating with speed $\dot{\chi} = d\chi/dt$. In particular, (1.6.1) here reduces to

(7.6.4)
$$(DF(\overline{U}) - \dot{\chi} I)[\partial U / \partial N] = 0 \,,$$

which shows that $\dot{\chi}$ is constant and equal to $\lambda_i(\overline{U})$ for some i.

Next we show that to the left of, and suffcently close to, the graph of χ the solution U is an i-simple wave. By virtue of Theorem 7.6.1, it suffices to prove that $n-1$ independent i-Riemann invariants, which will be denoted by $w_1, \cdots, w_{i-1}, w_{i+1}, \cdots, w_n$, are constant.

For U near $\overline{U}$, the $n-1$ vectors $\{Dw_1(U), \cdots, Dw_{i-1}(U), Dw_{i+1}(U), \cdots, Dw_n(U)\}$ span the orthogonal complement of $R_i(U)$ and so do the vectors $\{L_1(U), \cdots, L_{i-1}(U), L_{i+1}(U), \cdots, L_n(U)\}$. Consequently, there is a nonsingular $(n-1) \times (n-1)$ matrix $B(U)$ such that

$$(7.6.5) \qquad L_j(U) = \sum_{k \neq i} B_{jk}(U) Dw_k(U) , \quad j = 1, \cdots, i-1, i+1, \cdots, n .$$

Multiplying (7.2.1), from the left, by $L_j(U)$ yields

$$(7.6.6) \qquad L_j(U)\partial_t U + \lambda_j(U)L_j(U)\partial_x U = 0 , \quad j = 1, \cdots, n .$$

Combining (7.6.5) with (7.6.6), we conclude

$$(7.6.7) \quad \sum_{k \neq i} B_{jk}\partial_t w_k + \sum_{k \neq i} \lambda_j B_{jk}\partial_x w_k = 0 , \quad j = 1, \cdots, i-1, i+1, \cdots, n .$$

We regard (7.6.7) as a first order linear inhomogeneous system of $n-1$ equations in the $n-1$ unknowns $w_1, \cdots, w_{i-1}, w_{i+1}, \cdots, w_n$. In that sense, (7.6.7) is strictly hyperbolic, with characteristic speeds $\lambda_1, \cdots, \lambda_{i-1}, \lambda_{i+1}, \cdots, \lambda_n$. Along the graph of χ, the $n-1$ Riemann invariants are constant, $w_1(\overline{U}), \cdots, w_{i-1}(\overline{U}), w_{i+1}(\overline{U}), \cdots, w_n(\overline{U})$. Also the graph of χ is non-characteristic for the system (7.6.7). Consequently, the standard uniqueness theorem for the Cauchy problem for linear hyperbolic systems implies that (7.6.7) may admit only one solution compatible with the Cauchy data, namely the trivial one: $w_1 = w_1(\overline{U}), \cdots, w_{i-1} = w_{i-1}(\overline{U})$, $w_{i+1} = w_{i+1}(\overline{U}), \cdots, w_n = w_n(\overline{U})$. This completes the proof.

At any point (x, t) in the domain of an i-simple wave U of (7.2.1) we let $\xi(x, t)$ denote the slope at (x, t) of the i-characteristic associated with U, i.e.,

$$(7.6.8) \qquad\qquad \xi(x, t) = \lambda_i(U(x, t)) .$$

The derivative of ξ in the direction of the line with slope ξ is zero, that is

$$(7.6.9) \qquad\qquad \partial_t \xi + \xi \partial_x \xi = 0 .$$

Thus ξ satisfies the Burgers equation (4.2.1).

In the vicinity of any point $(\overline{x}, \overline{t})$ in the domain of U, we shall say that the i-simple wave is an i-rarefaction wave if $\partial_x \xi(\overline{x}, \overline{t}) > 0$, i.e., when the i-characteristics diverge, or an i-compression wave if $\partial_x \xi(\overline{x}, \overline{t}) < 0$, i.e., when the i-characteristics converge. This terminology originated in the context of gas dynamics.

Since in an i-simple wave U stays constant along i-characteristics, on a small neighborhood $\mathscr{X}$ of any point $(\overline{x}, \overline{t})$ where $\partial_x \xi(\overline{x}, \overline{t}) \neq 0$ we may use the single variable ξ to label U, i.e., there is a function V_i, defined on some open interval $(\bar{\xi} - \varepsilon, \bar{\xi} + \varepsilon)$, with $\bar{\xi} = \lambda_i(U(\bar{x}, \bar{t}))$, taking values in $\mathcal{O}$ and such that

$$(7.6.10) \qquad U(x, t) = V_i(\xi(x, t)) , \qquad (x, t) \in \mathcal{X} .$$

Furthermore, by account of (7.6.3) and (7.6.8), V_i satisfies

$$(7.6.11) \qquad \dot{V}_i(\xi) = b(\xi) R_i(V(\xi)) , \qquad \xi \in (\bar{\xi} - \varepsilon, \bar{\xi} + \varepsilon) ,$$

$$(7.6.12) \qquad \lambda_i(V_i(\xi)) = \xi , \qquad \xi \in (\bar{\xi} - \varepsilon, \bar{\xi} + \varepsilon) ,$$

where b is a scalar function and an overdot denotes derivative with respect to ξ.

Conversely, if V_i satisfies (7.6.11), (7.6.12) and ξ is any classical C^1 solution of the Burgers equation (7.6.9) taking values in the interval $(\bar{\xi} - \varepsilon, \bar{\xi} + \varepsilon)$, then the composition $U = V_i(\xi(x, t))$ is an i-simple wave. The above considerations motivate the following

Definition 7.6.2 An i-*rarefaction wave curve* in the state space $\mathbb{R}^n$, for the hyperbolic system (7.2.1), is a curve $U = V_i(\cdot)$, where the function V_i satisfies (7.6.11) and (7.6.12).

Rarefaction wave curves will be one of the principal tools for solving the Riemann problem in Chapter IX. The construction of these curves is particularly simple in the neighborhood of points of genuine nonlinearity:

Theorem 7.6.3 *Assume* $\overline{U} \in \mathcal{O}$ *is a point of genuine nonlinearity of the i-characteristic family of the hyperbolic system* (7.2.1) *of conservation laws. Then there exists a unique i-rarefaction wave curve* V_i *through* $\overline{U}$. *If* R_i *is normalized on a neighborhood of* $\overline{U}$ *through*

$$(7.6.13) \qquad D\lambda_i(U) R_i(U) = 1 ,$$

and V_i *is reparametrized by* $\tau = \xi - \bar{\xi}$, *where* $\bar{\xi} = \lambda_i(\overline{U})$, *then* V_i *is the solution of the ordinary differential equation*

$$(7.6.14) \qquad \dot{V}_i = R_i(V)$$

with initial condition $V_i(0) = \overline{U}$. *The more complicated notation* $V_i(\tau; \overline{U})$ *shall be employed when one needs to display the point of origin of this rarefaction wave curve.*

Proof. Any solution V_i of (7.6.14) clearly satisfies (7.6.11) with $b = 1$. At $\xi = \bar{\xi}$, i.e. $\tau = 0$, $\lambda_i(V_i) = \lambda_i(\overline{U}) = \bar{\xi}$. Furthermore, $\dot{\lambda}_i(V_i) = D\lambda_i(V_i)\dot{V}_i = 1$, by virtue of (7.6.14) and (7.6.13). This establishes (7.6.12) and completes the proof.

By contrast, when the i-characteristic family is linearly degenerate, differentiating (7.6.12) with respect to ξ and combining the resulting equation with (7.6.11), yields a contradiction: $0 = 1$. In that case, i-characteristics in any i-simple wave are necessarily parallel straight lines. It is still true, however, that any i-simple wave takes values along some integral curve of the differential equation (7.6.14).

Motivated by Theorem 7.6.1, we may characterize rarefaction wave curves in terms of Riemann invariants:

Theorem 7.6.4 *Every i-Riemann invariant is constant along any i-rarefaction wave curve of the system (7.2.1). Conversely, if $\overline{U}$ is any point of genuine nonlinearity of the i-characteristic family of (7.2.1) and $w_1, \cdots, w_{i-1}, w_{i+1}, \cdots, w_n$ are independent i-Riemann invariants on some neighborhood of $\overline{U}$, then the i-rarefaction curve through $\overline{U}$ is determined implicitly by the system of equations $w_j(U) = w_j(\overline{U})$, $j = 1, \cdots, i-1, i+1, \cdots, n$.*

Proof. Any i-rarefaction curve V_i satisfies (7.6.11). If w is an i-Riemann invariant of (7.2.1), multiplying (7.6.11), from the left, by $Dw(V_i(\xi))$ and using (7.3.1) yields $\dot{w}(V_i(\xi)) = 0$, i.e., w stays constant along V_i.

Assume now $w_1, \cdots, w_{i-1}, w_{i+1}, \cdots, w_n$ are i-Riemann invariants with $Dw_1, \cdots, Dw_{i-1}, Dw_{i+1}, \cdots, Dw_n$ linearly independent. Then the $n-1$ surfaces $w_j(U) = w_j(\overline{U})$, $j = 1, \cdots, i-1, i+1, \cdots, n$, intersect transversely to form a C^1 curve V_i through $\overline{U}$, parametrized by arc-length s, whose tangent V_i' must satisfy, by account of Definition 7.3.1, $V_i'(s) = c(s) R_i(V(s))$, for some nonzero scalar function c. For as long as V_i is a point of genuine nonlinearity of the i-characteristic field, $\lambda_i'(V_i) = D\lambda_i V_i' = c D\lambda_i R_i \neq 0$. We may thus find the proper parametrization $s = s(\xi)$ so that V_i satisfies both (7.6.11) and (7.6.12). This completes the proof.

As an application of Theorem 7.6.4, we infer that the 1- and 2-rarefaction wave curves of the system (7.1.6) through a point $(\overline{u}, \overline{v})$, with $\sigma''(\overline{u}) \neq 0$, are determined, in terms of the Riemann invariants (7.3.2), by the equations

$$(7.6.15) \qquad v = \overline{v} + \int_{\overline{u}}^{u} \sqrt{\sigma'(\omega)}\, d\omega\,, \qquad v = \overline{v} - \int_{\overline{u}}^{u} \sqrt{\sigma'(\omega)}\, d\omega\,.$$

When the system (7.2.1) is endowed with a coordinate system $(w_1, \cdots, w_n)$ of Riemann invariants and we use $W = (w_1, \cdots, w_n)^T$, instead of U, as our state variable, the rarefaction wave curves assume a very simple form. Indeed, by virtue of Theorem 7.6.4, the i-rarefaction wave curve through the point $\overline{W} = (\overline{w}_1, \cdots, \overline{w}_n)^T$ is the straight line $w_j = \overline{w}_j$, $j \neq i$, parallel to the i-axis.

7.7 Breakdown of Classical Solutions

When the system (7.2.1) is equiped with a convex entropy, Theorem 5.1.1 guarantees the existence of a unique, locally defined, classical solution, with initial data U_0 in the Sobolev space H^2. In one-space dimension, however, there is a sharper existence theory which applies to quasilinear hyperbolic systems in general, not necessarily conservation laws, and does not rely on the existence of entropies:

Theorem 7.7.1 *Let U_0 be a C^1 function, defined on $(-\infty, \infty)$ and taking values in a ball of $\mathbb{R}^n$ with closure contained in $\mathcal{O}$. Assume, further, that dU_0/dx is bounded on $(-\infty, \infty)$. Then there exists a unique C^1 function U defined on $(-\infty, \infty) \times [0, T_\infty)$, for some T_∞, $0 < T_\infty \leq \infty$, and taking values in $\mathcal{O}$, which satisfies (7.2.1) on $(-\infty, \infty) \times (0, T_\infty)$ together with the initial condition $U(x, 0) = U_0(x)$ on $(-\infty, \infty)$. Furthermore, the interval $[0, T_\infty)$ is maximal in the sense that if $T_\infty < \infty$, then, as $t \uparrow T_\infty$, $\|\partial_x U(\cdot, t)\|_{L^\infty} \to \infty$ and/or the range of $U(\cdot, t)$ escapes from every compact subset of $\mathcal{O}$.*

The proof of the above theorem, which may be found in the references cited in Section 7.9, relies on pointwise bounds for U and $\partial_x U$ obtained by monitoring the evolution of U and its derivatives along characteristics. Estimates of this nature will be established below but they will not be employed for establishing the existence of classical solutions but rather for demonstrating that classical solutions break down in finite time.

In Section 6.1, we saw that, in any number of space dimensions, classical solutions of the scalar conservation law generally break down in finite time, as a result of collisions of characteristics. This effect is particularly pronounced in one-space dimension, in which characteristics have less room to manoeuver and are thus forced to collide. Breakdown of this nature is not peculiar to scalar conservation laws but occurs in general systems (7.2.1) as well. Indeed, in Section 7.6 we saw that (7.2.1) admits i-simple wave solutions U which obtain by taking the composition (7.6.10) of a (smooth) solution V_i to the ordinary differential equation (7.6.11) with a classical solution ξ to Burgers' equation (7.6.9). When the solution of (7.6.9) breaks down, so does the i-simple wave. Below we shall look into this phenomenon more closely and, in particular, we shall investigate the effect of interactions of simple waves of different characteristic families.

Any classical, C^2 solution U of (7.2.1) on $(-\infty, \infty) \times [0, T)$ may be written as

$$(7.7.1) \qquad \begin{cases} \partial_x U = \displaystyle\sum_{j=1}^n a_j R_j(U) \\ \partial_t U = -\displaystyle\sum_{j=1}^n a_j \lambda_j(U) R_j(U) \end{cases}$$

with

$$(7.7.2) \qquad a_j = L_j(U)\partial_x U , \quad j = 1, \cdots, n .$$

In view of (7.6.3), one may interpret (7.7.1) as a decomposition of U into simple waves, one for each characteristic family, with respective strength $a_1, \cdots, a_n$. Our aim is to study the evolution of a_i along the i-characteristics associated with U. We let

$$(7.7.3) \qquad \frac{d}{dt} = \partial_t + \lambda_i \partial_x$$

denote differentiation in the i-characteristic direction. Combining (7.7.2) with (7.7.1) yields

(7.7.4)
$$\partial_t a_i = L_i \partial_t \partial_x U + \partial_x U^T DL_i^T \partial_t U$$
$$= \partial_x (L_i \partial_t U) - \partial_t U^T DL_i^T \partial_x U + \partial_x U^T DL_i^T \partial_t U$$
$$= \partial_x (L_i \partial_t U) + \sum_{j,k=1}^{n} (\lambda_j - \lambda_k) R_j^T DL_i^T R_k a_j a_k ,$$

(7.7.5)
$$\lambda_i \partial_x a_i = \partial_x (\lambda_i L_i \partial_x U) - (D\lambda_i \partial_x U)(L_i \partial_x U)$$
$$= \partial_x (\lambda_i L_i \partial_x U) - \sum_{j,k=1}^{n} (D\lambda_i R_j) \delta_{ik} a_j a_k ,$$

where δ_{ik} is the Kronecker delta. From (7.2.1), $L_i \partial_t U + \lambda_i L_i \partial_x U = 0$. Also, by virtue of (7.2.3), $R_j^T DL_i^T R_k = -L_i DR_j R_k$. Therefore, combining (7.7.3), (7.7.4), (7.7.5) and symmetrizing we conclude

(7.7.6)
$$\frac{da_i}{dt} = \sum_{j,k=1}^{n} \gamma_{ijk} a_j a_k$$

with

(7.7.7)
$$\gamma_{ijk} = -\frac{1}{2}(\lambda_j - \lambda_k) L_i [R_j, R_k] - (D\lambda_i R_j) \delta_{ik} ,$$

where $[R_j, R_k]$ denotes the Lie bracket (7.2.15). Note, in particular, that

(7.7.8)
$$\gamma_{iii} = -D\lambda_i R_i ,$$

(7.7.9)
$$\gamma_{ijj} = 0 , \quad j \neq i .$$

It is clear that in any argument showing blow-up of a_i through (7.7.6), the coefficient γ_{iii} will play a pivotal role. By virtue of (7.7.8), γ_{iii} never vanishes when the i-characteristic family is genuinely nonlinear, and vanishes identically when the i-characteristic family is linearly degenerate.

To gain some insight, let us consider first the case where U is just an i-simple wave, i.e., $a_i \neq 0$ and $a_j = 0$ for $j \neq i$. In that case, (7.7.6) reduces to

(7.7.10)
$$\frac{da_i}{dt} = \gamma_{iii} a_i^2 .$$

Furthermore, since U is constant along characteristics, γ_{iii} in (7.7.10) is a constant. When $\gamma_{iii} \neq 0$ and a_i has the same sign as γ_{iii}, (7.7.10) induces blow-up of a_i in a finite time.

Another noteworthy special case is when the system (7.2.1) is endowed with a coordinate system $(w_1, \cdots, w_n)$ of Riemann invariants. In that case $L_j = Dw_j$ and so, by (7.7.2),

(7.7.11)
$$a_j = \partial_x w_j .$$

Moreover, in virtue of (7.7.7), (7.3.10) and (7.3.12), (7.7.6) reduces to

$$(7.7.12) \qquad \frac{da_i}{dt} = -\sum_{j=1}^{n} \frac{\partial \lambda_i}{\partial w_j} a_i a_j .$$

We seek an integrating factor for (7.7.12). If ϕ is any smooth scalar function of U, we get from (7.7.1):

$$(7.7.13) \qquad \begin{aligned} \frac{d\phi}{dt} &= D\phi (\partial_t U + \lambda_i \partial_x U) = \sum_{j \neq i} (\lambda_i - \lambda_j)(D\phi R_j) a_j \\ &= \sum_{j \neq i} (\lambda_i - \lambda_j) \frac{\partial \phi}{\partial w_j} a_j . \end{aligned}$$

Combining (7.7.12) with (7.7.13) yields

$$(7.7.14) \qquad \frac{d}{dt}(e^\phi a_i) = -e^\phi \frac{\partial \lambda_i}{\partial w_i} a_i^2 - \sum_{j \neq i} e^\phi \left[\frac{\partial \lambda_i}{\partial w_j} - (\lambda_i - \lambda_j) \frac{\partial \phi}{\partial w_j} \right] a_i a_j .$$

From (7.3.14) and (7.3.16), follows that there exists ϕ which satisfies

$$(7.7.15) \qquad \frac{\partial \phi}{\partial w_j} = \frac{1}{\lambda_i - \lambda_j} \frac{\partial \lambda_i}{\partial w_j}, \qquad j = 1, \cdots, i-1, i+1, \cdots, n .$$

For that ϕ, (7.7.14) reduces to

$$(7.7.16) \qquad \frac{d}{dt}(e^\phi a_i) = -e^{-\phi} \frac{\partial \lambda_i}{\partial w_i} (e^\phi a_i)^2 .$$

When the i-characteristic family is genuinely nonlinear, $\partial \lambda_i / \partial w_i \neq 0$. Whenever $e^{-\phi} \partial \lambda_i / \partial w_i$ is bounded away from zero, uniformly on the range of the solution, (7.7.16) will induce blow-up of a_i, in finite time, along any characteristic emanating from a point $\bar{x}$ of the x-axis where a_i has the opposite sign of $\partial \lambda_i / \partial w_i$. Uniform boundedness of $e^{-\phi} \partial \lambda_i / \partial w_i$ is maintained, because, by Theorem 7.3.2, the range of any classical solution in the state space of Riemann invariants coincides with the range of its initial values. We have thus established

Theorem 7.7.2 *Assume (7.2.1) is endowed with a coordinate system $(w_1, \cdots, w_n)$ of Riemann invariants. Suppose the i-characteristic family is genuinely nonlinear. Then any classical solution U with initial data U_0 taking values in a compact subset of $\mathcal{O}$ and satisfying, at some point $\bar{x} \in (-\infty, \infty)$,*

$$(7.7.17) \qquad \frac{dw_i(U_0(\bar{x}))}{dx} \frac{\partial \lambda_i}{\partial w_i} < 0 ,$$

breaks down in finite time.

We now return to the general situation. When the i-characteristic field is genuinely nonlinear, and thus, by (7.7.8), $\gamma_{iii} \neq 0$, the term $\gamma_{iii} a_i^2$ in (7.7.6) will have a destabilizing effect. Any expectation that this may be neutralized by the remaining terms in (7.7.6), which account for the effect of the other characteristic fields, is not likely to be fulfilled, at least when the initial data have compact support, for the following reasons. Equation (7.7.9) rules out the possibility of selfinteractions of the remaining characteristic fields: All interactions, other than $\gamma_{iii}a_i^2$, involve two distinct characteristic families. Now, when the initial data have compact support, mutual interactions eventually become insignificant, because waves of distinct characteristic families propagate with different speeds and thus eventually separate. Consequently, in the long run the term $\gamma_{iii}a_i^2$ becomes the dominant factor and drives a_i to infinity in finite time. The above heuristic arguments can be formalized and lead to the following

Theorem 7.7.3 *Assume* (7.2.1) *is a genuinely nonlinear strictly hyperbolic system of conservation laws. When the initial data* U_0 *are* C^2, *have compact support, and* $\max |dU_0/dx|$ *is sufficiently small, the classical solution of the initial-value problem breaks down in finite time.*

The long and technical proof of Theorem 7.7.3, together with various extensions, addressing the situation where some (or all) of the characteristic fields are linearly degenerate or weakly linearly degenerate, may be found in the references cited in Section 7.9.

7.8 Weak Solutions

In view of the breakdown of classical solutions, demonstrated in the previous section, to solve the initial-value problem in the large, for nonlinear hyperbolic systems of conservation laws, one has to resort to weak solutions. As explained in Chapter IV, the issue of the admissibility of weak solutions will have to be addressed.

In earlier chapters, we mainly considered weak solutions that are merely bounded measurable functions. Existence in that function class will indeed be established, for certain systems, in Chapter XV through the functional analytic method of compensated compactness. Nevertheless, the function class of choice for hyperbolic systems of conservation laws is BV, which provides the natural framework for envisioning the most important features of weak solutions, namely shocks and their interactions.

The finite domain of dependence property for solutions of hyperbolic systems combined with the fact that our system (7.2.1) is invariant under uniform stretching of coordinates: $x = \bar{x} + ay$, $t = \bar{t} + a\tau$, $a > 0$, suggests that the admissibility of BV weak solutions may be decided locally, through examination of shocks and wave fans. These issues will be discussed thoroughly in the following two chapters.

7.9 Notes

The general mathematical framework of the theory of hyperbolic systems of conservation laws in one-space dimension was set in the seminal paper of Lax [2], which distills the material collected over the years in the context of special systems. The notions of Riemann invariants, genuine nonlinearity, simple waves and rarefaction wave curves, at the level of generality considered here, were introduced in that paper. The books of Smoller [1] and Serre [9] contain expositions of these topics, illustrated by interesting examples.

A systematic, rigorous exposition of the theory of one-dimensional elastic continua (strings, rods, etc.) is found in the book of Antman [1]. See also Antman [2]. The system (7.1.10) for planar electromagnetic waves was studied thoroughly by Serre [4].

The failure of strict hyperbolicity in one-space dimensional systems deriving from three-space dimensional parent systems is discussed by Lax [6]. The system (7.2.11) has been used extensively as a vehicle for demonstrating the features of non-strictly hyperbolic systems of conservation laws, beginning with the work of Keyfitz and Kranzer [1].

Riemann invariants were first considered by Earnshaw [1] and by Riemann [1], in the context of the system (7.1.7) of isentropic gas dynamics. Conditions for existence of coordinate systems of Riemann invariants and its implications on the existence of entropies were investigated by Conlon and Liu [1] and by Sévennec [1]. The calculation of the characteristic speeds and Riemann invariants of the system (7.3.18) of electrophoresis is due to Alekseyevskaya [1] and Fife and Geng [1]. A detailed exposition of the noteworthy properties of this system is contained in Serre [9]. Serre [4], shows that the system (7.1.10) is equivalent to (7.3.24), (7.3.25) even within the realm of weak solutions.

As already mentioned in Section 1.10, the special entropy-entropy flux pair (7.4.6), for symmetric systems, was noted by Godunov [1,2,3] and by Friedrichs and Lax [1]. Over the years, a great number of entropy-entropy flux pairs with special properties have been constructed, mainly for systems of two conservation laws, beginning with the pioneering paper of Lax [4]. We shall see some of that work in later chapters. The characterization of systems of size $n \geq 3$, endowed with an abundance of entropies is due to Tsarev [1], who calls them *semi-Hamiltonian*, and Serre [6], who named them *rich*. A comprehensive exposition of their theory is contained in Serre [9].

Theorem 7.5.1 is due to Boillat [2].

The earliest example of a simple wave, in the context of the system of isothermal gas dynamics, appears in a memoir by Poisson [1]. See also Earnshaw [1]. Theorem 7.6.2 is taken from Lax [2], who attributes the proof to Friedrichs.

Local existence of C^1 solutions to the initial-value problem in one-space dimension was first established by Schauder [1] and Friedrichs [1]. For a comprehensive treatment of the initial as well as the initial-boundary value problem see the monograph by Li Ta-tsien and Yu Wen-ci [1].

The breakdown of classical solutions was first noticed by Challis [1], in the context of the compressible simple wave solution of the system of isothermal gas dynamics derived by Poisson [1]. It is this paper that provided the stimulus for the introduction of weak solutions with shocks, by Stokes [1] (see Sections 1.10 and 4.6). The earliest result on generic breakdown of solutions is due to Lax [3], who proved directly the case $n = 2$ of Theorem 7.7.2. This work was extended in several directions: Klainerman and Majda [1] established breakdown in the case $n = 2$ so long as none of the two characteristic families is linearly degenerate. John [1] derived[1] (7.7.6) and used it to prove Theorem 7.7.3. Liu [9] gives an extension of Theorem 7.7.3 covering the case where some of the characteristic families are linearly degenerate. Li Ta-tsien, Zhou Yi and Kong De-xing [1] consider the case of weakly linearly degenerate characteristic families. A direct proof of Theorem 7.7.2, for any n, is found in Serre [9]. Additional results are presented in the monograph of Alinhac [1]. It is also possible (John [2], Sideris [1]) to establish blow-up in the Sobolev norm of solutions, by use of energy methods, and some of that work extends to systems of conservation laws in several space dimensions. A class of systems, with applications to elastodynamics, for which the breakdown of smooth solutions may be averted is studied in Li Ta-tsien [1]. For global classical solutions to the Euler equations, see Serre [11] and Grassin and Serre [1].

---

[1] John's formula for γ_{ijk} is different from (7.7.7) but, of course, the two expressions are equivalent.

Chapter VIII. Admissible Shocks

Shock fronts were introduced in Section 1.6, for general systems of balance laws, and were placed in the context of BV solutions in Section 1.8. They were encountered again, briefly, in Section 3.1, where the governing Rankine-Hugoniot condition was recorded.

Since shock fronts have codimension one, important aspects of their local behavior may be investigated, without loss of generality, within the framework of systems in one-space dimension. This will be the object of the present chapter. The discussion will begin with an exploration of the geometric features of the Rankine-Hugoniot condition, leading to the introduction of the Hugoniot locus.

The necessity of imposing admissibility conditions on weak solutions was pointed out in Chapter IV. These in turn induce, or at least motivate, admissibility conditions on shocks. Indeed, the prevailing view is that the issue of admissibility of general BV weak solutions should by resolved through a test applied to every point of the shock set. In particular, the shock admissibility conditions associated with the entropy condition of Section 4.3 and the vanishing viscosity approach of Section 4.4 will be introduced and will be compared with each other as well as with other important shock admissibility conditions proposed by Lax and by Liu.

8.1 Strong Shocks, Weak Shocks, and Shocks of Moderate Strength

For the system

$$(8.1.1) \qquad \partial_t U + \partial_x F(U) = 0 ,$$

in one-space dimension, the Rankine-Hugoniot jump condition (3.1.3) reduces to

$$(8.1.2) \qquad F(U_+) - F(U_-) = s(U_+ - U_-) .$$

Actually, (8.1.2) is as general as the multi-space dimensional version (3.1.3), once the direction N of propagation of the shock has been fixed and F has been defined through (7.2.2).

When (8.1.2) holds, we say that *the state U_-, on the left, is joined to the state U_+, on the right, by a shock of speed s*. Note that "left" and "right" may be interchanged in (8.1.2), in consequence of the invariance of (8.1.1) under the transformation $(x, t) \mapsto (-x, -t)$. Nevertheless, later on we shall introduce admissibility conditions inducing irreversibility, as a result of which the roles of U_- and U_+ cannot be reversed.

The jump $U_+ - U_-$ is the *amplitude* and its size $|U_+ - U_-|$ is the *strength* of the shock. Properties established without restriction on the strength, are said to hold even for *strong shocks*. Quite often, however, we shall have to impose limitations on the strength of shocks: $|U_+ - U_-| < \delta$, with δ positive small which depends typically on $DF(U_-)$, $D^2F(U_-)$ and the modulus of continuity of D^2F at U_-. In that case the shock is called *weak*. When the bound δ on the strength depends solely on $DF(U_-)$ and the modulus of continuity of DF at U_-, we say the shock has *moderate strength*.

Notice that (8.1.2) may be written as

(8.1.3) $$[A(U_-, U_+) - sI](U_+ - U_-) = 0 ,$$

where we are using the notation

(8.1.4) $$A(V, U) = \int_0^1 DF(\tau U + (1 - \tau)V)d\tau .$$

For $i = 1, \cdots, n$, we let $\mu_i(V, U)$ denote the eigenvalues and $S_i(V, U)$ the corresponding eigenvectors of $A(V, U)$. In particular, $A(U, U) = DF(U)$ and so $\mu_i(U, U) = \lambda_i(U)$, $S_i(U, U) = R_i(U)$. Note that $A(V, U)$, and thereby also $\mu_i(V, U)$, $S_i(V, U)$ are symmetric in (V, U). Therefore, (finite) Taylor expanding of these functions about the midpoint $\frac{1}{2}(V + U)$ yields

(8.1.5) $$\mu_i(V, U) = \lambda_i(\tfrac{1}{2}(V + U)) + O(|V - U|^2) ,$$

(8.1.6) $$S_i(V, U) = R_i(\tfrac{1}{2}(V + U)) + O(|V - U|^2) .$$

Clearly, (8.1.3) will hold if and only if

(8.1.7) $$s = \mu_i(U_-, U_+) ,$$

(8.1.8) $$U_+ - U_- = \zeta S_i(U_-, U_+) ,$$

for some $i = 1, \cdots, n$ and some $\zeta \in \mathbb{R}$. In particular, the speed s of any shock of moderate strength must be close to some characteristic speed λ_i. Such a shock is then called an *i-shock*.

An interesting implication of (8.1.5), (8.1.7) is the useful identity

(8.1.9) $$s = \frac{1}{2}[\lambda_i(U_-) + \lambda_i(U_+)] + O(|U_- - U_+|^2) .$$

In special systems it is possible to associate even strong shocks with a particular characteristic family. For example, the Rankine-Hugoniot condition

(8.1.10) $$\begin{cases} v_+ - v_- + s(u_+ - u_-) = 0 \\ \sigma(u_+) - \sigma(u_-) + s(v_+ - v_-) = 0 \end{cases}$$

for the system (7.1.6) of isentropic elasticity implies

$$(8.1.11) \qquad s = \pm\sqrt{\frac{\sigma(u_+) - \sigma(u_-)}{u_+ - u_-}} \; .$$

In view of the characteristic speeds (7.2.4) of this system, it is natural to call shocks propagating to the left $(s < 0)$ 1-shocks and shocks propagating to the right $(s > 0)$ 2-shocks.

8.2 The Hugoniot Locus

The set of points U in state space that may be joined to a fixed point $\overline{U}$ by a weak shock is called the *Hugoniot locus* of $\overline{U}$. It has a simple geometric structure, so long as $\overline{U}$ is a point of strict hyperbolicity of the system.

Theorem 8.2.1 *For a given state $\overline{U} \in \mathcal{O}$, assume that the characteristic speed $\lambda_i(\overline{U})$ is a simple eigenvalue of $DF(\overline{U})$. Then there is a C^2 curve $U = W_i(\tau)$ in state space, called the i-shock curve through $\overline{U}$, and a C^2 function $s = s_i(\tau)$, both defined for τ in some neighborhood of 0, with the following property: A state U can be joined to $\overline{U}$ by a weak i-shock of speed s if and only if $U = W_i(\tau)$, $s = s_i(\tau)$, for some τ. Furthermore, $W_i(0) = \overline{U}$ and*

$$(8.2.1) \qquad s_i(0) = \lambda_i(\overline{U}) \; ,$$

$$(8.2.2) \qquad \dot{s}_i(0) = \tfrac{1}{2} D\lambda_i(\overline{U}) R_i(\overline{U}) \; ,$$

$$(8.2.3) \qquad \dot{W}_i(0) = R_i(\overline{U}) \; ,$$

$$(8.2.4) \qquad \ddot{W}_i(0) = DR_i(\overline{U}) R_i(\overline{U}) \; .$$

The more complicated notation $W_i(\tau; \overline{U})$, $s_i(\tau; \overline{U})$ shall be employed when one needs to display the point of origin of this shock curve.

Proof. Recall the notation developed in Section 8.1 and, in particular, Equations (8.1.7), (8.1.8). A state U may be joined to $\overline{U}$ by an i-shock of speed s if and only if

$$(8.2.5) \qquad U = \overline{U} + \tau S_i(\overline{U}, U) \; ,$$

$$(8.2.6) \qquad s = \mu_i(\overline{U}, U) \; .$$

Accordingly, we consider the function

$$(8.2.7) \qquad H(U, \tau) = U - \overline{U} - \tau S_i(\overline{U}, U) \; ,$$

defined on $\mathcal{O} \times \mathbb{R}$, and note that $H(\overline{U}, 0) = 0$, $DH(\overline{U}, 0) = I$. Consequently, by the implicit function theorem, there is a curve $U = W_i(\tau)$ in state space, with

$W_i(0) = \overline{U}$, such that $H(U, \tau) = 0$ for τ near 0 if and only if $U = W_i(\tau)$. We then define

$$(8.2.8) \qquad\qquad s_i(\tau) = \mu_i(\overline{U}, W_i(\tau)) .$$

In particular, $s_i(0) = \mu_i(\overline{U}, \overline{U}) = \lambda_i(\overline{U})$. Furthermore, differentiating (8.2.5) with respect to τ and setting $\tau = 0$, we deduce $\dot{W}_i(0) = S_i(\overline{U}, \overline{U}) = R_i(\overline{U})$. To establish the remaining equations (8.2.2) and (8.2.4), we appeal to (8.1.5) and (8.1.6) to get

$$(8.2.9) \qquad \begin{aligned} s_i(\tau) &= \lambda_i(\tfrac{1}{2}(\overline{U} + W_i(\tau))) + O(\tau^2) \\ &= \lambda_i(\overline{U}) + \tfrac{1}{2}\tau D\lambda_i(\overline{U})R_i(\overline{U}) + O(\tau^2) , \end{aligned}$$

$$(8.2.10) \qquad \begin{aligned} W_i(\tau) &= \overline{U} + \tau R_i(\tfrac{1}{2}(\overline{U} + W_i(\tau))) + O(\tau^3) \\ &= \overline{U} + \tau R_i(\overline{U}) + \tfrac{1}{2}\tau^2 DR_i(\overline{U})R_i(\overline{U}) + O(\tau^3) . \end{aligned}$$

This completes the proof.

In particular, if $\overline{U}$ is a point of strict hyperbolicity of the system (8.1.1), Theorem 8.2.1 implies that the Hugoniot locus of $\overline{U}$ is the union of n shock curves, one for each characteristic family.

The shock curve constructed above is generally confined in the regime of weak shocks, because of the use of the implicit function theorem, which applies only when the strength of the shock, measured by $|\tau|$, is sufficiently small: $|\tau| < \delta$ with δ depending on the Lipschitz constant of S_i, which in turn depends on $D^2 F$. Nevertheless, in special systems one may often use more delicate analytical or topological arguments or explicit calculation to extend shock curves to the range of shocks of moderate strength or even to the range of strong shocks. For example, in the case of the system (7.1.6), combining (8.1.10) with (8.1.11) we deduce that the Hugoniot locus of any point $(\overline{u}, \overline{v})$ in state space consists of two curves

$$(8.2.11) \qquad\qquad v = \overline{v} \pm \sqrt{[\sigma(u) - \sigma(\overline{u})](u - \overline{u})} ,$$

defined on the whole range of u.

The i-shock curves introduced here have common features with the i-rare-faction wave curves defined in Section 7.6. Indeed, recalling Theorems 7.6.3 and 8.2.1, and, in particular, comparing (7.6.14) with (8.2.3), (8.2.4), we deduce

Theorem 8.2.2 *Assume $\overline{U} \in \mathcal{O}$ is a point of genuine nonlinearity of the i-characteristic family of the hyperbolic system (8.1.1) of conservation laws, and $\lambda_i(\overline{U})$ is a simple eigenvalue of $DF(\overline{U})$. Normalize R_i so that (7.6.13) holds on some neighborhood of $\overline{U}$. Then the i-rarefaction wave curve V_i, defined through Theorem 7.6.3, and the i-shock curve W_i, defined through Theorem 8.2.1, have a second order contact at $\overline{U}$.*

Recall that, by Theorem 7.6.4, i-Riemann invariants are constant along i-rarefaction wave curves. At the same time, as shown above, i-shock curves are very close to i-rarefaction wave curves. It is then to be expected that i-Riemann invariants vary very slowly along i-shock curves. Indeed,

Theorem 8.2.3 *The jump of any i-Riemann invariant across a weak i-shock is of third order in the strength of the shock.*

Proof. Assume $\lambda_i(\overline{U})$ is a simple eigenvalue of $DF(\overline{U})$ and consider the i-shock curve W_i through $\overline{U}$. For any i-Riemann invariant w, differentiating along the curve $W_i(\cdot)$,

$$(8.2.12) \qquad \dot{w} = Dw\dot{W}_i ,$$

$$(8.2.13) \qquad \ddot{w} = \dot{W}_i^T D^2 w \dot{W}_i + Dw\ddot{W}_i .$$

By virtue of (8.2.3) and (7.3.1), $\dot{w} = 0$ at $\tau = 0$.

We now apply D to (7.3.1) and then multiply the resulting equation from the right by R_i to deduce the identity

$$(8.2.14) \qquad R_i^T D^2 w R_i + DwDR_i R_i = 0 .$$

Combining (8.2.13), (8.2.3), (8.2.4) and (8.2.14), we conclude that $\ddot{w} = 0$ at $\tau = 0$. This completes the proof.

In the special case where the system (8.1.1) is endowed with a coordinate system $(w_1, \cdots, w_n)$ of Riemann invariants, we may calculate the leading term in the jump of w_j across a weak i-shock, $i \neq j$, as follows. The Rankine-Hugoniot condition reads

$$(8.2.15) \qquad F(W_i(\tau)) - F(\overline{U}) = s_i(\tau)[W_i(\tau) - \overline{U}] .$$

Differentiating with respect to τ yields

$$(8.2.16) \qquad [DF(W_i(\tau)) - s_i(\tau)I]\dot{W}_i(\tau) = \dot{s}_i(\tau)[W_i(\tau) - \overline{U}] .$$

Multiplying (8.2.16), from the left, by $Dw_j(W_i)$ gives

$$(8.2.17) \qquad (\lambda_j - s_i)\dot{w}_j = \dot{s}_i Dw_j[W_i - \overline{U}] .$$

Next we differentiate (8.2.17), with respect to τ, thus obtaining

$$(8.2.18) \quad (\lambda_j - s_i)\ddot{w}_j + (\dot{\lambda}_j - 2\dot{s}_i)\dot{w}_j = \ddot{s}_i Dw_j[W_i - \overline{U}] + \dot{s}_i \dot{W}_i^T D^2 w_j[W_i - \overline{U}] .$$

We differentiate (8.2.18), with respect to τ, and then set $\tau = 0$. We use (8.2.1), (8.2.2), (8.2.3), (7.3.12) and that both $\dot{w}_j$ and $\ddot{w}_j$ vanish at 0, by virtue of Theorem 8.2.3, to conclude

$$(8.2.19) \qquad \dddot{w}_j = \frac{1}{2}\frac{1}{\lambda_j - \lambda_i}\frac{\partial \lambda_i}{\partial w_i} R_i^T D^2 w_j R_i ,$$

where $\dddot{w}_j$ is evaluated at 0 and the right-hand side is evaluated at $\overline{U}$.

Returning to the general case, we next investigate how the shock speed function $s_i(\tau)$ evolves along the i-shock curve. We multiply (8.2.16), from the left, by $L_i(W_i(\tau))$ to get

(8.2.20) $[\lambda_i(W_i(\tau)) - s_i(\tau)]L_i(W_i(\tau))\dot{W}_i(\tau) = \dot{s}_i(\tau)L_i(W_i(\tau))[W_i(\tau) - \overline{U}]$.

For τ sufficiently close to 0, but $\tau \neq 0$,

(8.2.21) $L_i(W_i(\tau))\dot{W}_i(\tau) > 0$, $\tau L_i(W_i(\tau))[W_i(\tau) - \overline{U}] > 0$,

by virtue of (8.2.3). In the applications it turns out that (8.2.21) continue to hold for a broad range of τ, often extending to the regime of strong shocks. In that case, (8.2.20) and (8.2.16) immediately yield the following

Theorem 8.2.4 *Assume* (8.2.21) *hold. Then*

(8.2.22) $\dot{s}_i(\tau) > 0$ *if and only if* $\tau[\lambda_i(W_i(\tau)) - s_i(\tau)] > 0$,

(8.2.23) $\dot{s}_i(\tau) = 0$ *if and only if* $\lambda_i(W_i(\tau)) = s_i(\tau)$.

Moreover, $\dot{s}_i(\tau) = 0$ *implies that* $\dot{W}_i(\tau)$ *is collinear to* $R_i(W_i(\tau))$.

In particular, s_i constant implies that the i-shock curve is an integral curve of the vector field R_i, along which λ_i is constant. Consequently, all points along such a shock curve are states of linear degeneracy of the i-characteristic family. The converse of this statement is also valid:

Theorem 8.2.5 *Assume the* i-*characteristic family of the hyperbolic system* (8.1.1) *of conservation laws is linearly degenerate and* $\lambda_i(\overline{U})$ *is a simple eigenvalue of* $DF(\overline{U})$. *Then the* i-*shock curve* W_i *through* $\overline{U}$ *is the integral curve of* R_i *through* $\overline{U}$. *In fact, under the proper parametrization,* W_i *is the solution of the differential equation*

(8.2.24) $\dot{W}_i = R_i(W_i)$

with initial condition $W_i(0) = \overline{U}$. *Along* W_i, *the characteristic speed* λ_i *and all* i-*Riemann invariants are constant. The shock speed function* s_i *is also constant:*

(8.2.25) $s_i(\tau) = \lambda_i(W_i(\tau)) = \lambda_i(\overline{U})$.

Proof. Let W_i denote the solution of (8.2.24) with initial condition $W_i(0) = \overline{U}$. Then

(8.2.26) $[DF(W_i(\tau)) - \lambda_i(W_i(\tau))I]\dot{W}_i(\tau) = 0$.

Since $D\lambda_i(U)R_i(U) = 0$, $\dot{\lambda}_i = 0$ and so $\lambda_i(W_i(\tau)) = \lambda_i(\overline{U})$. Integrating (8.2.26) from 0 to τ yields

(8.2.27) $F(W_i(\tau)) - F(\overline{U}) = \lambda_i(\overline{U})[W_i(\tau) - \overline{U}]$,

which establishes that W_i is the i-shock curve through $\overline{U}$, with corresponding shock speed function s_i given by (8.2.25). This completes the proof.

It is natural to inquire whether an i-shock curve may be an integral curve of the vector field R_i in the absence of linear degeneracy. It turns out that this may only occur under very special circumstances:

Theorem 8.2.6 *For the hyperbolic system* (8.1.1), *assume* $\overline{U}$ *is a state of genuine nonlinearity for the* i-*characteristic family and* $\lambda_i(\overline{U})$ *is a simple eigenvalue of* $DF(\overline{U})$. *The* i-*shock curve through* $\overline{U}$ *coincides with the integral curve of the field* R_i, *i.e. the* i-*rarefaction wave curve, through* $\overline{U}$ *if and only if the latter is a straight line in state space.*

Proof. If the i-shock curve W_i through $\overline{U}$ coincides with the integral curve of R_i through $\overline{U}$, then $\dot{W}_i(\tau)$ must be collinear to $R_i(W_i(\tau))$. In that case, (8.2.16) imples

$$(8.2.28) \qquad [\lambda_i(W_i(\tau)) - s_i(\tau)]\dot{W}_i(\tau) = \dot{s}_i(\tau)[W_i(\tau) - \overline{U}] \ .$$

For τ near 0, but $\tau \neq 0$, it is $\lambda_i(W_i(\tau)) \neq s_i(\tau)$, by genuine nonlinearity. Therefore, (8.2.28) implies that the graph of W_i is a straight line through $\overline{U}$.

Conversely, assume the integral curve of R_i through $\overline{U}$ is a straight line, which may be parametrized as $U = W_i(\tau)$, where W_i is some smooth function satisfying $W_i(0) = \overline{U}$, as well as (8.2.3) and (8.2.4) (note that $DR_i(\overline{U})R_i(\overline{U})$ is necessarily collinear to $R_i(\overline{U})$). We may then determine a scalar-valued function $s_i(\tau)$ such that

$$
\begin{aligned}
(8.2.29) \qquad F(W_i(\tau)) - F(\overline{U}) &= \int_0^\tau DF(W_i(\zeta))\dot{W}_i(\zeta)d\zeta \\
&= \int_0^\tau \lambda_i(W_i(\zeta))\dot{W}_i(\zeta)d\zeta = s_i(\tau)[W_i(\tau) - \overline{U}] \ .
\end{aligned}
$$

Thus W_i is the i-shock curve through $\overline{U}$. This completes the proof.

Special as it may be, the class of hyperbolic systems of conservation laws with coinciding shock and rarefaction wave curves of each characteristic family includes some noteworthy examples. Consider, for instance, the system (7.3.18) of electrophoresis. Notice that, for $i = 1, \cdots, n$, the level surfaces of the i-Riemann invariant W_i, determined through (7.3.21) or (7.3.22), are hyperplanes. In particular, for $i = 1, \cdots, n$, the integral curves of the vector field R_i are the straight lines produced by the intersection of the level hyperplanes of the $n - 1$ Riemann invariants $w_1, \cdots, w_{i-1}, w_{i+1}, \cdots, w_n$. Consequently, the conditions of Theorem 8.2.6 apply to the system (7.3.18).

In the presence of multiple characteristic speeds, the Hugoniot locus may contain multi-dimensional varieties, in the place of shock curves. In that connection it is instructive to consider the model system (7.2.11), for which the origin is an umbilic point. When a state $(\overline{u}, \overline{v})$ is joined to a state (u, v) by a shock of speed s, the Rankine-Hugoniot condition reads

$$(8.2.30) \quad \begin{cases} (u^2 + v^2)u - (\overline{u}^2 + \overline{v}^2)\overline{u} = s(u - \overline{u}) \\ (u^2 + v^2)v - (\overline{u}^2 + \overline{v}^2)\overline{v} = s(v - \overline{v}) \ . \end{cases}$$

Notice that when $(\overline{u}, \overline{v}) \neq (0, 0)$, the Hugoniot locus of $(\overline{u}, \overline{v})$ consists of the circle $u^2 + v^2 = \overline{u}^2 + \overline{v}^2$, along which the shock speed is constant, $s = \overline{u}^2 + \overline{v}^2$, and the straight line $\overline{v}u = \overline{u}v$, which connects $(\overline{u}, \overline{v})$ with the origin. Thus, the 1-characteristic family provides an example of the application of Theorem 8.2.5 while the 2-characteristic family satisfies the assumptions of Theorem 8.2.6. On the other hand, the Hugoniot locus of the umbilic point $(0, 0)$ is the entire plane, because any point (u, v) can be joined to $(0, 0)$ by a shock of speed $s = u^2 + v^2$.

Not all systems in which strict hyperbolicity fails exhibit the same behavior. For instance, for the system

$$(8.2.31) \quad \begin{cases} \partial_t u + \partial_x[2(u^2 + v^2)u] = 0 \\ \partial_t v + \partial_x[(u^2 + v^2)v] = 0 \ , \end{cases}$$

in which strict hyperbolicity also fails at the origin, the Hugoniot locus of $(0, 0)$ consists of two lines, namely the u-axis and the v-axis.

8.3 The Lax Shock Admissibility Criterion

An i-shock of speed s which joins the state U_-, on the left, to the state U_+, on the right, is said to satisfy the *Lax E-condition* if

$$(8.3.1) \quad \lambda_i(U_-) \geq s \geq \lambda_i(U_+) \ .$$

In particular, when the left or the right part of (8.3.1) is satisfied as an equality, the shock is called a *left* or a *right i-contact discontinuity*; and when both parts of (8.3.1) hold as equalities, the shock is called an *i-contact discontinuity*. For example, by account of Theorem 8.2.5, any weak shock associated with a linearly degenerate characteristic family is necessarily a contact discontinuity. Notice that, with the exception of contact discontinuities, (8.3.1) induces an *irreversibility* condition that fixes the roles of U_- and U_+ as left and right states of the shock.

When the above shock is embedded in an otherwise smooth solution, the meaning of (8.3.1) is that i-characteristics from the left catch up with i-characteristics from the right and they collide at the shock. Thus "information" from the past propagating along i-characteristics is absorbed and lost into admissible shocks. In contrast, shocks that violate (8.3.1) become sources of new "information" which is then carried along i-characteristics into the future. Postulating the Lax E-condition may appear ad hoc at this point, but justification is provided by its implications on stability of weak solutions as well as through its connection with other, physically motivated, shock admissibility criteria. These issues will be discussed at length in following sections.

Let us begin the investigation with the scalar conservation law (7.1.2). The characteristic speed is $\lambda(u) = f'(u)$ and so (8.3.1) takes the form

(8.3.2)
$$f'(u_-) \geq s \geq f'(u_+) \,,$$

where s is the shock speed computed through the Rankine-Hugoniot jump condition:

(8.3.3)
$$s = \frac{f(u_+) - f(u_-)}{u_+ - u_-} \,.$$

The reader will immediately realize the geometric interpretation of (8.3.2) upon noticing that $f'(u_-)$ and $f'(u_+)$ are the slopes of the graph of f at the points $(u_-, f(u_-))$ and $(u_+, f(u_+))$ while s is the slope of the chord that connects $(u_-, f(u_-))$ with $(u_+, f(u_+))$. In particular, when (7.1.2) is genuinely nonlinear, i.e., $f''(u) \neq 0$ for all u, then (8.3.2) reduces to $u_- < u_+$ if $f''(u) < 0$, and $u_- > u_+$ if $f''(u) > 0$.

Next we consider the system (7.1.6) of isentropic elasticity. The characteristic speeds are recorded in (7.2.4) and the shock speeds in (8.1.11), so that (8.3.1) assumes the form

(8.3.4)
$$\sigma'(u_-) \leq \frac{\sigma(u_+) - \sigma(u_-)}{u_+ - u_-} \leq \sigma'(u_+) \text{ or } \sigma'(u_-) \geq \frac{\sigma(u_+) - \sigma(u_-)}{u_+ - u_-} \geq \sigma'(u_+) \,,$$

for 1-shocks or 2-shocks, respectively. The geometric interpretation of (8.3.4) is again clear. When (7.1.6) is genuinely nonlinear, i.e., $\sigma''(u) \neq 0$ for all u, (8.3.4) reduces to $u_- < u_+$ or $u_- > u_+$ if $\sigma''(u) > 0$, and $u_- > u_+$ or $u_- < u_+$ if $\sigma''(u) < 0$. Equivalently, in terms of velocity, by virtue of (8.1.10): $v_- < v_+$ if $\sigma''(u) > 0$ and $v_- > v_+$ if $\sigma''(u) < 0$, for both shock families.

A similar analysis applies to the system (7.1.8) of isentropic flow of a polytropic gas, with characteristic speeds given by (7.2.10), and yields that a 1-shock (or 2-shock) that joins the state (ρ_-, v_-), on the left, to the state (ρ_+, v_+), on the right, satisfies the Lax E-condition if and only if $\rho_- < \rho_+$ (or $\rho_- > \rho_+$). In other words, the passing of an admissible shock front compresses the gas. Because classical gas dynamics has served as the prototype for the development of the general theory, shocks that satisfy the Lax E-condition (8.3.1) as strict inequalities are often called *compressive*.

When λ_i is a simple eigenvalue of DF and we are dealing with i-shocks of (at most) moderate strength, the remaining characteristic speeds are well-separated and do not interfere, i.e., (8.3.1) may be extended into

(8.3.5)
$$\lambda_j(U_\pm) > \lambda_i(U_-) \geq s \geq \lambda_i(U_+) > \lambda_k(U_\pm) \,, \quad j > i > k \,.$$

In many special systems, like those considered above, (8.3.5) may hold even in the realm of strong shocks. On the other hand, in the presence of umbilic points and/or strong shocks, one may encounter the situation in which a shock satisfies the Lax E-condition simultaneously for two distinct characteristic families i and j, say

(8.3.6)
$$\lambda_j(U_-) > \lambda_i(U_-) > s > \lambda_j(U_+) > \lambda_i(U_+) \,.$$

Such shocks are called *overcompressive*. An example is provided by our model system (7.2.11). Recalling the form of the Hugoniot locus, described in Section 8.2, we consider a shock of speed s, joining, on the left, a state (u_-, v_-), lying on the unit circle, to a state $(u_+, v_+) = a(u_-, v_-)$, on the right, where a is some constant. From (7.2.12), $\lambda_1(u_-, v_-) = 1$, $\lambda_2(u_-, v_-) = 3$, $\lambda_1(u_+, v_+) = a^2$, $\lambda_2(u_+, v_+) = 3a^2$. Furthermore, the Rankine-Hugoniot condition (8.2.30) yields $s = a^2 + a + 1$. Therefore, if $a \in (-\frac{1}{2}, 0)$,

$$(8.3.7) \qquad \lambda_2(u_-, v_-) > \lambda_1(u_-, v_-) > s > \lambda_2(u_+, v_+) > \lambda_1(u_+, v_+) ,$$

i.e., the shock is overcompressive.

The occurrence of overcompressive shocks raises serious difficulties in the theory, which, at the time of this writing, have only been partially resolved. To avoid such complications, we shall limit our investigation to the range of shock strength in which the assumptions of Theorem 8.2.2 are satisfied. In particular, this will encompass the case of weak shocks. Thus, with reference to the system (8.1.1), let us consider a state U_-, on the left, which is joined to a state U_+, on the right, by an i-shock of speed s. Assuming $\lambda_i(U_-)$ is a simple eigenvalue of $DF(U_-)$, let W_i denote the i-shock curve through U_- (cf. Theorem 8.2.1), so that $U_- = W_i(0)$ and $U_+ = W_i(\tau)$. Furthermore, $\lambda_i(U_-) = s_i(0)$ and $s = s_i(\tau)$. We show that if $\tau < 0$ and $\dot{s}_i(\cdot) \geq 0$ on $(\tau, 0)$, then the shock satisfies the Lax E-condition. Indeed, $\dot{s}_i(\cdot) \geq 0$ implies $s = s_i(\tau) \leq s_i(0) = \lambda_i(U_-)$, which is the left half of (8.3.1). At the same time, so long as (8.2.22) and (8.2.23) hold at τ, $\dot{s}_i(\cdot) \geq 0$ implies, by virtue of Theorem 8.2.4, that $s = s_i(\tau) \geq \lambda_i(W_i(\tau)) = \lambda_i(U_+)$, namely, the right half of (8.3.1). A similar argument demonstrates that the Lax E-condition also holds when $\tau > 0$ and $\dot{s}_i(\cdot) \leq 0$ on $(0, \tau)$, but it is violated if either $\tau < 0$ and $\dot{s}_i(\cdot) < 0$ on $(\tau, 0)$ or $\tau > 0$ and $\dot{s}_i(\cdot) > 0$ on $(0, \tau)$. The implications of the above statements to the genuine nonlinear case, in which, by virtue of (8.2.2), $\dot{s}_i(\cdot)$ does not change sign across 0, are recorded in the following

Theorem 8.3.1 *Assume U_- is a point of genuine nonlinearity of the i-characteristic family of the system (8.1.1), with $D\lambda_i(U_-)R_i(U_-) > 0$ (or < 0). Suppose $\lambda_i(U_-)$ is a simple eigenvalue of $DF(U_-)$ and let W_i denote the i-shock curve through U_-, with $U_- = W_i(0)$. Then a weak i-shock that joins U_- to a state $U_+ = W_i(\tau)$ satisfies the Lax E-condition if and only if $\tau < 0$ (or $\tau > 0$).*

Thus, in the genuinely nonlinear case, one half of the shock curve is compatible with the Lax E-condition (8.3.1), as strict inequalities, and the other half is incompatible with it. When U_- is a point of linear degeneracy of the i-characteristic field, so that $\dot{s}_i(0) = 0$, the situation is more delicate: If $\ddot{s}_i(0) < 0$, $\dot{s}_i(\tau)$ is positive for $\tau < 0$ and negative for $\tau > 0$, so that weak i-shocks that join U_- to $U_+ = W_i(\tau)$ are admissible, regardless of the sign of τ. On the other hand, if $\ddot{s}_i(0) > 0$, $\dot{s}_i(\tau)$ is negative for $\tau < 0$ and positive for $\tau > 0$, in which case all (sufficiently) weak i-shocks violate the Lax E-condition. As noted above, when the i-characteristic family itself is linearly degenerate, i-shocks are i-contact discontinuities satisfying (8.3.1) as equalities.

Experience indicates that the primary role of the Lax E-condition is to secure the stability of the interaction of the shock, as an entity, with its adjacent "smoother" parts of the solution. This view is corroborated by the following

Theorem 8.3.2 *Assume the system* (8.1.1) *is strictly hyperbolic and the i-characteristic family is genuinely nonlinear. Consider initial data U_0 such that $U_0(x) = U_\ell(x)$ for $x \in (-\infty, 0)$ and $U_0(x) = U_r(x)$ for $x \in (0, \infty)$, where U_ℓ and U_r are smooth functions which are bounded, together with their first derivatives, on $(-\infty, \infty)$. Assume, further, that the state $U_- = U_\ell(0)$, on the left, is joined to the state $U_+ = U_r(0)$, on the right, by a weak i-shock of speed s, which satisfies the Lax E-condition* (8.3.1), *as strict inequalities. Then there are: $T > 0$; a smooth function $x = \chi(t)$ on $[0, T)$; and a function U on $(-\infty, \infty) \times [0, T)$ with the following properties. U is smooth and satisfies* (8.1.1), *in the classical sense, for any (x, t), with $t \in [0, T)$ and $x \neq \chi(t)$. Furthermore, for $t \in [0, T)$ one-sided limits $U(\chi(t)-, t)$ and $U(\chi(t)+, t)$ exist and are joined by a weak i-shock of speed $\dot{\chi}(t)$, which satisfies the Lax E-condition.*

The proof employs techniques similar to those involved in the proof of Theorem 7.7.1 and can be found in the references cited in Section 8.7. One may get a rough idea through the considerably simpler, special case $n = 1$.

We thus consider the scalar conservation law (7.1.2), assuming it is genuinely nonlinear, say $f''(u) > 0$ for $u \in (-\infty, \infty)$. We assign initial data u_0 such that $u_0(x) = u_\ell(x)$ for $x \in (-\infty, 0)$ and $u_0(x) = u_r(x)$ for $x \in (0, \infty)$, where u_ℓ and u_r are bounded and uniformly Lipschitz continuous functions on $(-\infty, \infty)$. Furthermore, $u_- = u_\ell(0)$ and $u_+ = u_r(0)$ satisfy $u_- > u_+$. Let $u_-(x, t)$ and $u_+(x, t)$ be the classical solutions of (7.1.2) with initial data u_ℓ and u_r, respectively, which, by virtue of Theorem 6.1.1, exist on $(-\infty, \infty) \times [0, T)$, for some $T > 0$. On $[0, T)$ we define the function χ as solution of the ordinary differential equation

$$(8.3.8) \qquad \frac{dx}{dt} = \frac{f(u_+(x, t)) - f(u_-(x, t))}{u_+(x, t) - u_-(x, t)}$$

with initial condition $\chi(0) = 0$. Finally, we define the function u on $(-\infty, \infty) \times [0, T)$ by

$$(8.3.9) \qquad u(x, t) = \begin{cases} u_-(x, t), & t \in [0, T), \ x < \chi(t) \\ u_+(x, t), & t \in [0, T), \ x > \chi(t) \ . \end{cases}$$

Clearly, u satisfies (7.1.2), in the classical sense, for any (x, t) with $t \in [0, T)$ and $x \neq \chi(t)$. Furthermore, $u(\chi(t)-, t)$ and $u(\chi(t)+, t)$ are joined by a shock of speed $\dot{\chi}(t)$. Finally, for T sufficiently small, the Lax E-condition $u(\chi(t)-, t) > u(\chi(t)+, t)$ holds by continuity, since it is satisfied at $t = 0$. Notice that it is due to the Lax condition that the solution u solely depends on the initial data, i.e., it is independent of the "extraneous" information carried by $u_\ell(x)$ for $x > 0$ and $u_r(x)$ for $x < 0$.

Another serious issue of concern is the internal stability of shocks. It turns out that the Lax E-condition is effective in that direction as well, so long as the system is genuinely nonlinear and the shocks are weak; however, it is insufficient in more general situations. For that purpose, we have to consider additional, more discriminating shock admissibility criteria, which will be introduced in the following sections.

8.4 The Liu Shock Admissibility Criterion

The Liu shock admissibility test is more discriminating than the Lax E-condition and strives to capture the internal stability of shocks. By its design, it only makes sense in the context of shocks joining states that may be connected by shock curves. Thus, for general systems, its applicability is limited to weak shocks. Nevertheless, in special systems it also applies to shocks of moderate strength or even to strong shocks.

For a given state U_-, assume $\lambda_i(U_-)$ is a simple eigenvalue of $DF(U_-)$ so that the i-shock curve W_i through U_- is well defined, by Theorem 8.2.1, and satisfies $W_i(0) = U_-$. An i-shock that joins U_-, on the left, to a state $U_+ = W_i(\tau)$, on the right, of speed $s = s_i(\tau)$, satisfies the *Liu E-condition* if

$$(8.4.1) \qquad\qquad s \le s_i(\xi) , \qquad \text{for all } \xi \text{ between 0 and } \tau .$$

Similar to the Lax E-condition, the justification of the above admissibility criterion will be established a posteriori, through its connection to other, physically motivated, shock admissibility criteria, as well as by its role in the construction of stable solutions to the Riemann problem, in Chapter IX.

We proceed to discuss the relationship between the Liu E-condition and the Lax E-condition:

Theorem 8.4.1 *Assume, in the notation above, the state $U_- = W_i(0)$, on the left, is joined to the state $U_+ = W_i(\tau)$, on the right, by an i-shock of speed s, satisfying the Liu E-condition (8.4.1). Suppose (8.2.21) hold at τ. Then the shock also satisfies the Lax E-condition.*

Proof. By (8.4.1), $s \le s_i(0) = \lambda_i(U_-)$ which is the left half of (8.3.1). Furthermore, since $s = s_i(\tau)$, (8.4.1) implies $\tau \dot{s}_i(\tau) \le 0$. It then follows from Theorem 8.2.4 that $s = s_i(\tau) \ge \lambda_i(W_i(\tau)) = \lambda_i(U_+)$, namely the right half of (8.3.1). This completes the proof.

We have thus shown that the Liu E-condition implies the Lax E-condition. Indeed, when the system is genuinely nonlinear, these two criteria coincide, at least in the realm of weak shocks:

Theorem 8.4.2 *Assume the i-characteristic family is genuinely nonlinear and* λ_i *is a simple characteristic speed. Then weak i-shocks satisfy the Liu E-condition if and only if they satisfy the Lax E-condition.*

Proof. The Liu E-condition implies the Lax E-condition by Theorem 8.4.1. To show the converse, assume the state U_-, on the left, is joined to the state U_+, on the right, by a weak i-shock of speed s, which satisfies the Lax E-condition (8.3.1). Suppose, for definiteness, $D\lambda_i(U_-)R_i(U_-) > 0$ (the case of the opposite sign is similarly treated). By virtue of Theorem 8.3.1, $\tau < 0$. Since the shock is weak, by Theorem 8.2.1, $\dot{s}_i(\xi) > 0$ on the interval $(\tau, 0)$. Then $s = s_i(\tau) < s_i(\xi)$ for $\xi \in (\tau, 0)$, i.e., the Liu E-condition holds. This completes the proof.

When the system is not genuinely nonlinear and/or the shocks are not weak, the Liu E-condition is stricter than the Lax E-condition. This will be demonstrated by means of the following examples.

Let us first consider the scalar conservation law (7.1.2). The shock curve is the u-axis and we may use u as the parameter ξ. The shock speed is given by (8.3.3). It is then clear that a shock joining the states u_- and u_+ will satisfy the Liu E-condition (8.4.1) if and only if

$$(8.4.2) \qquad \frac{f(u_0) - f(u_-)}{u_0 - u_-} \geq \frac{f(u_+) - f(u_-)}{u_+ - u_-} \geq \frac{f(u_+) - f(u_0)}{u_+ - u_0}$$

holds for every u_0 between u_- and u_+. This is the celebrated *Oleinik E-condition*. It is easily memorized as a geometric statement: When $u_- < u_+$ (or $u_- > u_+$) the shock that joins u_-, on the left, to u_+, on the right, is admissible if the arc of the graph of f with endpoints $(u_-, f(u_-))$ and $(u_+, f(u_+))$ lies above (or below) the chord that connects the points $(u_-, f(u_-))$ and $(u_+, f(u_+))$. Letting u_0 converge to u_- and to u_+, we deduce that (8.4.2) implies (8.3.2). The converse, of course, is generally false, unless f is convex or concave. We have thus demonstrated that in the scalar conservation law the Liu E-condition is stricter than the Lax E-condition when f contains inflection points. In the genuinely nonlinear case, the Liu and Lax E-conditions are equivalent.

We now turn to the system (7.1.6) of isentropic elasticity. The shock curves are determined by (8.2.11) so we may use u as parameter instead of ξ. The shock speed is given by (8.1.11). Therefore, a shock joining the states (u_-, v_-) and (u_+, v_+) will satisfy the Liu E-condition (8.4.1) if and only if

$$(8.4.3) \qquad \frac{\sigma(u_0) - \sigma(u_-)}{u_0 - u_-} \begin{smallmatrix}\leq\\>\end{smallmatrix} \frac{\sigma(u_+) - \sigma(u_-)}{u_+ - u_-} \begin{smallmatrix}\leq\\>\end{smallmatrix} \frac{\sigma(u_+) - \sigma(u_0)}{u_+ - u_0}$$

holds for all u_0 between u_- and u_+, where "$\leq$" applies for 1-shocks and "$\geq$" applies for 2-shocks. This is called the *Wendroff E-condition*. In geometric terms, it may be stated as follows: When $s(u_+ - u_-) < 0$ (or > 0) the shock that joins (u_-, v_-), on the left, to (u_+, v_+), on the right, is admissible if the arc of the graph of σ with endpoints $(u_-, \sigma(u_-))$ and $(u_+, \sigma(u_+))$ lies below (or above) the chord that connects the points $(u_-, \sigma(u_-))$ and $(u_+, \sigma(u_+))$. Clearly, there is close analogy with the Oleinik E-condition. Letting u_0 in (8.4.3) converge to u_- and to u_+, we deduce that the Wendroff E-condition implies the Lax E-condition (8.3.4). The converse is true when σ is convex or concave, but false otherwise. Thus, for the system (7.1.6) the Liu E-condition is stricter than the Lax E-condition when σ contains inflection points. In the genuinely nonlinear case, the Liu and Lax E-conditions are equivalent.

As we shall see, the Oleinik E-condition and the Wendroff E-condition follow naturally from other admissibility criteria. To a great extent these special E-conditions provided the motivation for postulating the general Liu E-condition.

8.5 The Entropy Shock Admissibility Criterion

The idea of employing entropy inequalities to weed out spurious weak solutions of general hyperbolic systems of conservation laws was introduced in Section 4.3 and was used repeatedly in Chapters IV, V, and VI. It was observed that in the context of BV weak solutions the entropy condition reduces to the set of inequalities (4.3.9), to be tested pointwise at every point of the shock set. For the system (8.1.1), in one-space dimension, (4.3.9) assumes the form

$$(8.5.1) \qquad -s[\eta(U_+) - \eta(U_-)] + q(U_+) - q(U_-) \leq 0 ,$$

where (η, q) is an entropy-entropy flux pair satisfying (7.4.1), $Dq = D\eta DF$. The quantity on the left-hand side of (8.5.1) will be called henceforth the *entropy production across the shock*.

The fact that the entropy condition reduces to a pointwise test on shocks has played a dominant role in shaping the prevailing view that admissibility need only be tested at the level of shocks, i.e., that a general BV weak solution will be admissible if and only if each one of its shocks is admissible.

In setting up an entropy admissibility condition (8.5.1), the first task is to designate the appropriate entropy-entropy flux pair (η, q). Whenever (8.1.1) arises in connection to physics, the physically appropriate entropy should always be designated. In particular, the pairs (7.4.9), (7.4.10) and (7.4.11) must be designated for the systems (7.1.3), (7.1.6) and (7.1.8), respectively[1].

In the absence of guidelines from physics or when the entropy-entropy flux pair supplied by physics is inadequate to rule out all spurious shocks, additional entropy-entropy flux pairs must be designated (whenever available), motivated by other admissibility criteria, like viscosity. In that connection, we should bear in mind that, as demonstrated in earlier chapters, convexity of the entropy function is a desirable feature.

---

[1] In applying (8.5.1) to the system (7.1.3), with entropy-entropy flux pair (7.4.9), one should not confuse s in (7.4.9), namely the physical entropy, with s in (8.5.1), the shock speed. Since $q = 0$, (8.5.1) here states that "after a shock passes, the physical entropy must increase." The reader is warned that this statement is occasionally misinterpreted as a general physical principle and is applied even when it is no longer relevant.

Let us begin the investigation with the scalar conservation law (7.1.2). The shock speed s is given by (8.3.3). In accordance with the discussion in Chapter VI, admissible shocks must satisfy (8.5.1) for all convex functions η. However, as explained in Section 6.2, (8.5.1) need only be tested for the family (6.2.5) of entropy-entropy flux pairs, namely

$$(8.5.2) \qquad \eta(u; \bar{u}) = (u - \bar{u})^+ , \quad q(u; \bar{u}) = \text{sgn}\,(u - \bar{u})^+[f(u) - f(\bar{u})] .$$

It is immediately seen that (8.5.1) will be satisfied for every (η, q) in the family (8.5.2) if and only if (8.4.2) holds for all u_0 between u_- and u_+. We have thus rederived the Oleinik E-condition encountered in Section 8.4. This implies that, for the scalar conservation law, the entropy admissibility condition, applied for all convex entropies, is equivalent to the Liu E-condition.

It is generally impossible to recover the Oleinik E-condition from the entropy condition (8.5.1) for a single entropy-entropy flux pair. Take for example

$$(8.5.3) \qquad \eta(u) = \tfrac{1}{2}u^2 , \quad q(u) = \int_0^u \omega f'(\omega)d\omega .$$

By virtue of (8.3.3) and after a short calculation, (8.5.1) takes the form

$$(8.5.4) \qquad \tfrac{1}{2}[f(u_+) + f(u_-)](u_+ - u_-) - \int_{u_-}^{u_+} f(\omega)d\omega \leq 0 .$$

Notice that the entropy production across the shock is here measured by the signed area of the domain bordered by the arc of the graph of f with endpoints $(u_-, f(u_-))$, $(u_+, f(u_+))$, and the chord that connects $(u_-, f(u_-))$, $(u_+, f(u_+))$. Clearly, the Oleinik E-condition (8.4.2) implies (8.5.4) but the converse is generally false. Moreover, neither (8.5.4) generally implies the Lax E-condition (8.3.2) nor the other way around. However, when f is convex or concave, (8.5.4), (8.4.2) and (8.3.2) are all equivalent.

Next we turn to the system (7.1.6) of isentropic elasticity. We employ the entropy-entropy flux pair (η, q) given by (7.4.10). An interesting, rather lengthy, calculation, which involves the Rankine-Hugoniot condition (8.1.10), shows that (8.5.1) here reduces to

$$(8.5.5) \qquad s\left\{ \tfrac{1}{2}[\sigma(u_+) + \sigma(u_-)](u_+ - u_-) - \int_{u_-}^{u_+} \sigma(\omega)d\omega \right\} \leq 0 .$$

The quantity in braces on the left-hand side of (8.5.5) measures the signed area of the domain bordered by the arc of the graph of σ with endpoints $(u_-, \sigma(u_-))$, $(u_+, \sigma(u_+))$ and the chord that connects $(u_-, \sigma(u_-))$, $(u_+, \sigma(u_+))$. Hence, the Wendroff E-condition (8.4.3) implies (8.5.5) but the converse is generally false. Neither (8.5.5) necessarily implies the Lax E-condition (8.3.4) nor the other way around. However, when σ is convex or concave, (8.5.5), (8.4.3) and (8.3.4) are all equivalent. Of course, the system (7.1.6) is endowed with a rich collection of entropies so one may employ additional entropy-entropy flux pairs to recover the

Wendroff E-condition from the entropy condition, but this shall not be attempted here.

We now consider the entropy shock admissibility condition (8.5.1) for a general system (8.1.1), under the assumption that U_- and U_+ are connected through a shock curve. In particular, this will encompass the case of weak shocks. We thus assume $\lambda_i(U_-)$ is a simple characteristic speed, we consider the i-shock curve W_i through U_-, with $U_- = W_i(0)$, and we let $U_+ = W_i(\tau)$, $s = s_i(\tau)$, for some τ. The entropy production along the i-shock curve is given by

$$(8.5.6) \qquad E(\cdot) = -s_i(\cdot)[\eta(W_i(\cdot)) - \eta(U_-)] + q(W_i(\cdot)) - q(U_-) .$$

Differentiating (8.5.6) and using (7.4.1) yields

$$(8.5.7) \qquad \dot{E} = -\dot{s}_i[\eta(W_i) - \eta(U_-)] - s_i D\eta(W_i)\dot{W}_i + D\eta(W_i)DF(W_i)\dot{W}_i .$$

Combining (8.5.7) with (8.2.16) (for $\overline{U} = U_-$), we deduce

$$(8.5.8) \qquad \dot{E} = -\dot{s}_i\{\eta(W_i) - \eta(U_-) - D\eta(W_i)[W_i - U_-]\} .$$

Notice that the right-hand side of (8.5.8) is of quadratic order in the strength of the shock. Therefore, the entropy production $E(\tau)$ across the shock, namely the integral of $\dot{E}(\cdot)$ from 0 to τ, is of cubic order in τ. We have thus established the following

Theorem 8.5.1 *The entropy production across a weak shock is of third order in the strength of the shock.*

When U_- is a point of linear degeneracy of the i-characteristic family, $\dot{s}_i(0) = 0$ and so the entropy production across the shock will be of (at most) fourth order in the strength of the shock. In particular, when the i-characteristic family is linearly degenerate, $\dot{s}_i$ vanishes identically, by Theorem 8.2.5, and so

Theorem 8.5.2 *When the i-characteristic family is linearly degenerate, the entropy production across any i-shock (i-contact discontinuity) is zero.*

Turning now to the issue of admissibility of the shock, we observe that when η is a convex function the expression in braces on the right-hand side of (8.5.8) is nonpositive. Thus $\dot{E}$ and $\dot{s}_i$ have the same sign. Consequently, the entropy admissibility condition $E(\tau) \leq 0$ will hold if $\tau < 0$ and $\dot{s}_i(\cdot) \geq 0$ on $(\tau, 0)$, or if $\tau > 0$ and $\dot{s}_i(\cdot) \leq 0$ on $(0, \tau)$; while it will be violated when either $\tau < 0$ and $\dot{s}_i(\cdot) < 0$ on $(\tau, 0)$ or $\tau > 0$ and $\dot{s}_i(\cdot) > 0$ on $(0, \tau)$. Recalling our discussion in Section 8.3, we conclude that the entropy admissibility condition and the Lax E-condition are equivalent in the range of τ, on either side of 0, where $\dot{s}_i(\tau)$ does not change sign. In particular, this will be the case when the characteristic family is genuinely nonlinear and the shocks are weak:

Theorem 8.5.3 *When the i-characteristic family is genuinely nonlinear and λ_i is a simple characteristic speed, the entropy admissibility condition and the Lax E-condition for weak i-shocks are equivalent.*

In order to escape from the realm of genuine nonlinearity and weak shocks, let us consider the condition

(8.5.9) $\xi \dot{W}_i^T(\xi) D^2 \eta(W_i(\xi))[W_i(\xi) - U_-] \geq 0 .$

Recalling (7.4.3), (7.4.4) and Theorem 8.2.1, we conclude that when the entropy η is convex (8.5.9) will always hold for weak i-shocks; it will also be satisfied for shocks of moderate strength when i-shock curves extend into that regime; and may even hold for strong shocks, so long as $\dot{W}_i$ and $W_i - U_-$ keep pointing roughly in the direction of R_i.

Theorem 8.5.4 *Assume that the i-shock curve $W_i(\cdot)$ through $U_- = W_i(0)$, and corresponding shock speed function $s_i(\cdot)$, are defined on an interval (α, β) containing 0, and satisfy (8.5.9) for $\xi \in (\alpha, \beta)$, where η is a convex entropy of the system. Then any i-shock joining U_-, on the left, to $U_+ = W_i(\tau)$, on the right, with speed $s = s_i(\tau)$, which satisfies the Liu E-condition (8.4.1) also satisfies the entropy admissibility condition (8.5.1).*

Proof. We set

(8.5.10) $Q(\xi) = \eta(W_i(\xi)) - \eta(U_-) - D\eta(W_i(\xi))[W_i(\xi) - U_-] .$

By virtue of (8.5.9),

(8.5.11) $\xi \dot{Q}(\xi) \leq 0 .$

Integrating (8.5.8) from 0 to τ, integrating by parts and using (8.5.10), (8.5.11) and (8.4.1) we obtain

$$E(\tau) = -\int_0^\tau \dot{s}_i(\xi) Q(\xi) d\xi = -s_i(\tau) Q(\tau) + \int_0^\tau s_i(\xi) \dot{Q}(\xi) d\xi$$

(8.5.12)

$$\leq -s Q(\tau) + s \int_0^\tau \dot{Q}(\xi) d\xi = 0 ,$$

which shows that the shock satisfies (8.5.1). This completes the proof.

8.6 Viscous Shock Profiles

The idea of using the vanishing viscosity approach for identifying admissible weak solutions of hyperbolic systems of conservation laws was introduced in Section 4.4. In the present setting of one-space dimension, for the system (8.1.1), Equation (4.4.1) reduces to

(8.6.1) $\partial_t U(x,t) + \partial_x F(U(x,t)) = \mu \partial_x [B(U(x,t)) \partial_x U(x,t)]$.

As already explained in Section 4.4, the selection of the $n \times n$ matrix-valued function B may be suggested by the physical context of the system or it may just be an artifact of the analysis. Consider for example the dissipative systems

(8.6.2) $\partial_t u + \partial_x f(u) = \mu \partial_x^2 u$,

(8.6.3) $\begin{cases} \partial_t u - \partial_x v = 0 \\ \partial_t v - \partial_x \sigma(u) = \mu \partial_x \left(\dfrac{1}{u} \partial_x v \right) , \end{cases}$

(8.6.4) $\begin{cases} \partial_t u + \partial_x [(u^2 + v^2)u] = \mu \partial_x^2 u \\ \partial_t v + \partial_x [(u^2 + v^2)v] = \mu \partial_x^2 v , \end{cases}$

associated with the hyperbolic systems (7.1.2), (7.1.6), and (7.2.11). In so far as (7.1.6) is interpreted as the system of isentropic gas dynamics, the selection of viscosity in (8.6.3) is dictated by physics[2]. On the other hand, in (8.6.2) and (8.6.4) the viscosity is artificial.

In contrast to the entropy criterion, it is not at all clear that admissibility of weak solutions by means of the vanishing viscosity criterion is decided solely at the level of the shock set. Taking, however, that premise for granted, it will suffice to test admissibility in the context of solutions in the simple form

(8.6.5) $U(x,t) = \begin{cases} U_-, & x < st \\ U_+, & x > st , \end{cases}$

namely a shock of constant speed s joining the constant state U_-, on the left, to the constant state U_+, on the right. Presumably, functions (8.6.5) may be approximated, as $\mu \downarrow 0$, by a family of solutions U_μ of (8.6.1) in the form of *traveling waves*, namely functions of the single variable $x - st$. Taking advantage of the scaling in (8.6.1), we seek a family of solutions in the form

(8.6.6) $U_\mu(x,t) = V \left(\dfrac{x - st}{\mu} \right)$.

Substituting in (8.6.1), we deduce that V should satisfy the ordinary differential equation

(8.6.7) $[B(V(\tau))\dot{V}(\tau)]\dot{} = \dot{F}(V(\tau)) - s\dot{V}(\tau)$,

where the overdot denotes differentiation with respect to $\tau = \mu^{-1}(x - st)$. We are interested in solutions in which $\dot{V}$ vanishes at $V = U_-$ and so, upon integrating (8.6.7) once with respect to τ:

---

[2] Compare with (4.4.2). The variable viscosity coefficient μ/u is adopted so that in the spatial setting, where measurements are usually performed, viscosity will be constant μ. Of course this will make sense only when $u > 0$.

(8.6.8) $$B(V)\dot{V} = F(V) - F(U_-) - s[V - U_-] \, .$$

Notice that the right-hand side of (8.6.8) vanishes on the set of V that may be joined to U_- by a shock of speed s. This set contains, in particular, the state U_+.

We say that U_-, on the left, is connected to U_+, on the right, by a *viscous shock profile* if there is a smooth arc joining U_- to U_+ which is an invariant set for the differential equation (8.6.8) and, in addition, at any point where there is motion, the flow is directed from U_- to U_+.

The shock that joins U_-, on the left, to U_+, on the right, is said to satisfy the *viscous shock admissibility criterion* if U_- can be connected to U_+ by a viscous shock profile.

Determining viscous shock profiles is important not only because they shed light on the issue of admissibility but also because they provide information (at least when the matrix B is physically motivated) on the nature of the sharp transition modelled by the shock, the so called *structure of the shock*. Indeed, the stretching of coordinates involved in (8.6.6), as $\mu \downarrow 0$, allows us, so to say, to observe the shock under the microscope.

Any contact discontinuity associated with a linearly degenerate characteristic family satisfies the viscous shock admissibility criterion. Indeed, in that case, by virtue of Theorem 8.2.5, the shock curve itself serves as the viscous shock profile and all of its points are equilibria of the differential equation (8.6.8). The opposite extreme arises when U_- and U_+ are the only equilibrium points on the viscous shock profile, in which case U_- is the α-limit set and U_+ is the ω-limit set of an orbit of the differential equation (8.6.8). In the general situation, the viscous shock profile may contain a (finite or infinite) number of equilibrium points with any two consecutive ones connected by orbits of (8.6.8).

Let us illustrate the above through the scalar conservation law (7.1.2) and the corresponding dissipative equation (8.6.2). System (8.6.8) now reduces to the single equation

(8.6.9) $$\dot{u} = f(u) - f(u_-) - s(u - u_-) \, .$$

It is clear that u_- will be connected to u_+ by a viscous shock profile if and only if the right-hand side of (8.6.9) does not change sign between u_- and u_+ and indeed it is nonnegative when $u_- < u_+$ and nonpositive when $u_- > u_+$. Recalling (8.3.3), we conclude that in the scalar conservation law (7.1.2) a shock satisfies the viscous shock admissibility criterion if and only if the Oleinik E-condition (8.4.2) holds. When (8.4.2) holds as a strict inequality for any u_0 (strictly) between u_- and u_+, then u_- is connected to u_+ with a single orbit. By contrast, when (8.4.2) becomes equality for a set of intermediate u_0, we need more than one orbit and perhaps even a number of contact discontinuities in order to build the viscous shock profile. In that case one may prefer to visualize the shock as a composite of several shocks and/or contact discontinuities, all travelling with the same speed.

Next we turn to the system (7.1.6) and the corresponding dissipative system (8.6.3). In that case (8.6.8) reads

$$(8.6.10) \quad \begin{cases} 0 = -v + v_- - s(u - u_-) \\ \dfrac{1}{u}\dot{v} = -\sigma(u) + \sigma(u_-) - s(v - v_-) \ . \end{cases}$$

The reason we end up here with a combination of algebraic and differential equations rather than just differential equations is that B is a singular matrix. In any event, upon eliminating v between the two equations in (8.6.10), we deduce

$$(8.6.11) \quad s\dfrac{\dot{u}}{u} = \sigma(u) - \sigma(u_-) - s^2(u - u_-) \ .$$

Since $u > 0$, (u_-, v_-) will be connected to (u_+, v_+) by a viscous shock profile if and only if the right-hand side of (8.6.11) does not change sign between u_- and u_+ and is in fact nonnegative when $s(u_+ - u_-) > 0$ and nonpositive when $s(u_+ - u_-) < 0$. In view of (8.1.11), we conclude that in the system (7.1.6) of isentropic elasticity a shock satisfies the viscous shock admissibility criterion if and only if the Wendroff E-condition (8.4.3) holds.

It was the Oleinik E-condition and the Wendroff E-condition, originally derived through the above argument, that motivated the general Liu E-condition. We now proceed to show that the viscous shock admissibility criterion is generally equivalent to the Liu E-condition, at least in the range of shocks of moderate strength. For simplicity, only the special case $B = I$ will be discussed here; the case of more general B is treated in the references cited in Section 8.7.

Theorem 8.6.1 *Assume λ_i is a simple eigenvalue of DF. Then an i-shock of moderate strength satisfies the viscous shock admissibility criterion, with $B = I$, if and only if it satisfies the Liu E-condition.*

Proof. Assume the state U_-, on the left, is joined to the state U_+, on the right, by an i-shock of moderate strength and speed s. In order to apply the viscous shock admissibility test, the first task is to construct a curve in state space which connects U_+ with U_- and is invariant under the flow generated by (8.6.8), for $B = I$. To that end, we embed (8.6.8) into a larger system, by introducing a new (scalar) variable r:

$$(8.6.12) \quad \begin{cases} \dot{V} = F(V) - F(U_-) - r[V - U_-] \\ \dot{r} = 0 \ . \end{cases}$$

Notice that the Jacobian matrix of the right-hand side of (8.6.12), evaluated at the equilibrium point $V = U_-$, $r = \lambda_i(U_-)$ is

$$(8.6.13) \quad J = \begin{pmatrix} DF(U_-) - \lambda_i(U_-)I & 0 \\ 0 & 0 \end{pmatrix} ,$$

with eigenvalues $\lambda_j(U_-) - \lambda_i(U_-)$, $j = 1, \cdots, n$, and 0; the corresponding eigenvectors being

$$(8.6.14) \qquad \begin{pmatrix} R_j(U_-) \\ 0 \end{pmatrix} , \quad j = 1, \cdots, n , \quad \text{and} \quad \begin{pmatrix} 0 \\ 1 \end{pmatrix} .$$

We see that J has two zero eigenvalues, associated with a two-dimensional eigenspace, while the remaining eigenvalues are nonzero real numbers. The center manifold theorem then implies that any trajectory of (8.6.12) which is confined in a small neighborhood of the point $(U_-, \lambda_i(U_-))$ must lie on a two-dimensional manifold $\mathscr{M}$, which is invariant under the flow generated by (8.6.12), and may be parametrized by

$$(8.6.15) \qquad V = \Phi(\zeta, r) = U_- + \zeta R_i(U_-) + S(\zeta, r) , \quad r = r ,$$

with

$$(8.6.16) \qquad S(0, \lambda_i(U_-)) = 0 , \quad S_\zeta(0, \lambda_i(U_-)) = 0 , \quad S_r(0, \lambda_i(U_-)) = 0 .$$

In particular, the equilibrium point (U_+, s) of (8.6.12) must lie on $\mathscr{M}$, say $U_+ = \Phi(\rho, s)$, for some ρ near zero. Thus U_- and U_+ are connected by the curve $V = \Phi(\zeta, s)$, for ζ between 0 and ρ, and this curve is invariant under the flow generated by (8.6.8), for $B = I$.

Next we note that the flow induced by $(8.6.12)_1$ along the invariant curve $V = \Phi(\cdot, r)$ is governed by a function $\zeta = \zeta(\cdot)$ which satisfies the scalar ordinary differential equation

$$(8.6.17) \qquad \dot{\zeta} = g(\zeta, r) ,$$

with g defined through

$$(8.6.18) \qquad g(\zeta, r)\Phi_\zeta(\zeta, r) = F(\Phi(\zeta, r)) - F(U_-) - r[\Phi(\zeta, r) - U_-] .$$

In particular, recalling (8.6.15) and (8.6.16),

$$(8.6.19) \qquad g(0, r) = 0 , \quad g_\zeta(0, r) = \lambda_i(U_-) - r .$$

Clearly, the viscous shock admissibility criterion will be satisfied if and only if $\rho g(\zeta, s) \geq 0$ for all ζ between 0 and ρ.

Suppose now the shock satisfies the Liu E-condition. Thus, if W_i denotes the i-shock curve through U_- and s_i is the corresponding shock speed function, so that $U_- = W_i(0)$, $U_+ = W_i(\tau)$, $s = s_i(\tau)$, we must have $s_i(\xi) \geq s$ for ξ between 0 and ρ. For definiteness, let us assume $U_+ - U_-$ points in the general direction of $R_i(U_-)$, in which case both ρ and τ are positive.

We fix $r < s$, with $s - r$ very small, consider the curve $\Phi(\cdot, r)$ and identify $\kappa > 0$ such that $[\Phi(\kappa, r) - U_+]^T R_i(U_-) = 0$. We show that $g(\zeta, r) > 0$, and $0 < \zeta < \kappa$. Indeed, if $g(\zeta, r) = 0$ for some ζ, $0 < \zeta < \kappa$, then, by virtue of (8.6.18), the state $\Phi(\zeta, r)$ may be joined to the state U_- by a shock of speed r. Thus, $\Phi(\zeta, r)$ lies on the shock curve W_i, say $\Phi(\zeta, r) = W_i(\xi)$, for some ξ. By the construction of κ, since $0 < \zeta < \kappa$, it is necessarily $0 < \xi < \tau$. However, in that case $r = s_i(\xi) \geq s$, which is a contradiction to our assumption $r < s$.

This establishes that $g(\zeta, r)$ does not change sign on $(0, \kappa)$. At the same time, by account of (8.6.19), $g_\zeta(0, r) = s_i(0) - r \geq s - r > 0$, which shows that $g(\zeta, r) > 0$, $0 < \zeta < \kappa$. Finally, we let $r \uparrow s$, in which case $\kappa \to \rho$. Hence $g(\zeta, s) \geq 0$ for $\zeta \in (0, \rho)$.

By a similar argument one shows the converse, namely that $\rho g(\zeta, s) \geq 0$, for ζ between 0 and ρ, implies $s_i(\xi) \geq s$, for ξ between 0 and τ. This completes the proof.

Combining Theorems 8.4.1, 8.4.2 and 8.6.1, we conclude that the viscous shock admissibility criterion generally implies the Lax E-condition but the converse is generally false, unless the system is genuinely nonlinear and the shocks are weak.

Our next task is to compare the viscous shock admissibility criterion with the entropy shock admissibility criterion. We thus assume that the system (8.1.1) is equipped with an entropy-entropy flux pair (η, q). The natural compatibility condition between the entropy and the viscosity matrix B was already discussed in Section 4.4. We write (a weaker form of) the condition (4.4.4) in the present, one-dimensional setting:

$$(8.6.20) \qquad H^T D^2 \eta(U) B(U) H \geq 0 , \qquad H \in \mathbb{R}^n , \qquad U \in \mathcal{O} .$$

As already noted in Section 4.4, when $B = I$, (8.6.20) will hold if and only if η is convex.

Theorem 8.6.2 *When* (8.6.20) *holds, any shock that satisfies the viscous shock admissibility criterion also satisfies the entropy shock admissibility criterion.*

Proof. Consider a shock of speed s which joins the state U_-, on the left, to the state U_+, on the right, and satisfies the viscous shock admissibility condition.

Assume first U_- is connected to U_+ with a single orbit of (8.6.8), i.e., there is a function V which satisfies (8.6.8), and thereby also (8.6.7), on $(-\infty, \infty)$, together with the conditions $V(\tau) \to U_\pm$, as $\tau \to \pm\infty$. We multiply (8.6.7), from the left, by $D\eta(V(\tau))$ and use (7.4.1) to get

$$(8.6.21) \qquad [D\eta(V) B(V) \dot{V}]^\cdot - \dot{V}^T D^2 \eta(V) B(V) \dot{V} = \dot{q}(V) - s\dot{\eta}(V) .$$

Integrating (8.6.21) over $(-\infty, \infty)$ and using (8.6.20) we arrive at (8.5.1). We have thus proved that the shock satisfies the entropy condition.

In the general case where the viscous shock profile contains intermediate equilibrium points, we realize the shock as a composite of a (finite or infinite) number of simple shocks of the above type and/or contact discontinuities, all propagating with the same speed s. As shown above, the entropy production across each simple shock is nonpositive. On the other hand, by Theorem 8.5.2, the entropy production across any contact discontinuity will be zero. Therefore, combining the partial entropy productions we conclude that the total entropy production (8.5.1) is nonpositive. This completes the proof.

The converse of Theorem 8.6.2 is generally false. Consider for example the system (7.1.6) of isentropic elasticity, with corresponding dissipative system (8.6.3) and entropy-entropy flux pair (7.4.10), which satisfy the compatibility condition (8.6.20). As shown in Section 8.5, the entropy shock admissibility criterion is tested through the inequality (8.5.5), which follows from, but does not generally imply, the Wendroff E-condition (8.4.3).

One may plausibly argue that mere existence of a viscous shock profile should not constitute grounds for admissibility of the shock unless the profile itself is stable under perturbations of the states $U_\pm$ and perhaps even under perturbations of the flux function F. For simplicity, let us focus attention to the case $B = I$ and let us consider weak shocks, of speed s, joining U_-, on the left, to U_+, on the right, with shock profile consisting of a single connecting orbit of (8.6.8). Clearly, the shock profile must lie on the intersection of the unstable manifold $\mathscr{U}$ of (8.6.8) at U_- and the stable manifold $\mathscr{S}$ of (8.6.8) at U_+. For stability of the profile, $\mathscr{U}$ and $\mathscr{S}$ must intersect transversely. In particular, we would need $\dim \mathscr{U} + \dim \mathscr{S} \geq n + 1$. Now the Jacobian of the right-hand side of (8.6.8) at V is the matrix $DF(V) - sI$, with eigenvalues $\lambda_1(V) - s, \cdots, \lambda_n(V) - s$, and corresponding eigenvectors $R_1(V), \cdots, R_n(V)$. Therefore, $\mathscr{U}$ is equidimensional, and tangential at U_-, to the subspace spanned by $R_j(U_-)$ for all $j = 1, \cdots, n$ with $\lambda_j(U_-) > s$; and $\mathscr{S}$ is equidimensional, and tangential at U_+, to the subspace spanned by $R_k(U_+)$ for all $k = 1, \cdots, n$ with $\lambda_k(U_+) < s$. In a strictly hyperbolic system with weak shocks, we have $\lambda_1(U_\pm) < \lambda_2(U_\pm) < \cdots < \lambda_n(U_\pm)$ and so the stability condition $\dim \mathscr{U} + \dim \mathscr{S} = n + 1$ can be met if and only if

$$(8.6.22) \qquad \lambda_n(U_-) > \cdots > \lambda_i(U_-) > s > \lambda_i(U_+) > \cdots > \lambda_1(U_+)$$

holds for some i. This provides additional support to the thesis that the Lax E-condition is principally a guarantee that the interaction of the shock with its adjacent states is stable. The reader may find in the references cited in Section 8.7 how the above ideas extend to the case of more general dissipative viscosity matrices B.

One may argue, further, that viscous shock profiles employed to test the admissibility of shocks must derive from traveling wave solutions of the system (8.6.1) that are asymptotically stable. This issue has been investigated thoroughly in recent years and a complete theory has emerged, warranting the writing of a monograph on the subject. A detailed presentation would lie beyond the scope of the present book so only the highlights shall be reported here. For details and proofs the reader may consult the references cited in Section 8.7.

For simplicity, we limit our discussion to viscosity matrix $B = I$ and rescale (8.6.1) so that $\mu = 1$. We consider a weak i-shock, joining the states U_-, on the left, and U_+, on the right, which admits a viscous shock profile V. A change of variable $x \mapsto x + st$ renders the shock stationary. The viscous shock profile V is called *asymptotically stable* if the solution $U(x, t)$ of (8.6.1) with initial values $U(x, 0) = V(x) + U_0(x)$, where U_0 is a "small" perturbation decaying at $\pm\infty$, satisfies

$$(8.6.23) \qquad U(x, t) \to V(x + h), \quad \text{as } t \to \infty,$$

for some appropriate phase shift $h \in \mathbb{R}$.

Motivated by the observation that the total mass of solutions of (8.6.1) is conserved, it seems natural to require that the convergence in (8.6.23) be in $L^1(-\infty, \infty)$. In particular, this would imply that $V(x + h)$ carries the excess mass introduced by the perturbation:

$$(8.6.24) \qquad \int_{-\infty}^{\infty} U_0(x)dx = \int_{-\infty}^{\infty} [V(x + h) - V(x)]dx = h[U_+ - U_-] .$$

In the scalar case, $n = 1$, any viscous shock profile is asymptotically stable in $L^1(-\infty, \infty)$, under arbitrary perturbations $U_0 \in L^1(-\infty, \infty)$, with h determined through (8.6.24).

For systems, $n \geq 2$, the single scalar parameter h is generally inadequate to balance the vectorial equation (8.6.24), in which case (8.6.23) cannot hold in $L^1(-\infty, \infty)$, as no h-translate of V alone may carry the excess mass. Insightful analysis of the asymptotics of (8.6.1) suggests that, for large t, the solution U should develop a viscous shock profile accompanied by a family of so called diffusion waves, which share the burden of carrying the mass:

$$(8.6.25) \qquad U(x, t) \sim V(x + h) + W(x, t) + \sum_{j \neq i} \theta_j(x, t)R_j .$$

The j-term in the summation on the right-hand side of (8.6.25) represents a *decoupled j-diffusion wave*, with amplitude collinear to R_j, evaluated at U_-, for $j = 1, \cdots, i - 1$, or at U_+, for $j = i + 1, \cdots, n$. The scalar function θ_j is a self-similar solution,

$$(8.6.26) \qquad \theta_j(x, t) = \frac{1}{\sqrt{t}}\phi_j\left(\frac{x - \lambda_j t}{\sqrt{t}}\right) ,$$

of the nonlinear diffusion equation

$$(8.6.27) \qquad \partial_t\theta_j + \partial_x\left[\lambda_j\theta_j + \frac{1}{2}(D\lambda_j R_j)\theta_j^2\right] = \partial_x^2\theta_j .$$

In (8.6.26) and (8.6.27), λ_j, $D\lambda_j$ and R_j are again evaluated at U_-, for $j = 1, \cdots, i - 1$, or at U_+, for $j = i + 1, \cdots, n$. Thus the j-diffusion wave has a bell-shaped profile which propagates at characteristic speed λ_j, its peak decays like $O(t^{-\frac{1}{2}})$, while its mass stays constant, say $m_j R_j$. The remaining term W on the right-hand side of (8.6.25) represents the *coupled diffusion wave*, which satisfies a complicated linear diffusion equation, not to be recorded here, decays at the same rate as the uncoupled diffusion waves, but carries no mass. Therefore, mass conservation as $t \to \infty$ yields, in lieu of (8.6.24), the equation

$$(8.6.28) \qquad \int_{-\infty}^{\infty} U_0(x)dx = \sum_{j < i} m_j R_j(U_-) + h[U_+ - U_-] + \sum_{j > i} m_j R_j(U_+) ,$$

which dictates how the excess mass is distributed among the viscous shock profile and the decoupled diffusion waves. Since $U_+ - U_-$ and $R_i(U_\pm)$ are nearly collinear, (8.6.28) determines explicitly and uniquely the phase shift h of the viscous shock profile as well as the masses m_j of the j-diffusion waves.

On the other hand, it has been established that the viscous shock profile V is asymptotically stable (8.6.23) in $L^\infty(-\infty, \infty)$, for the h determined through (8.6.28), under any perturbation $U_0 \in H^1(-\infty, \infty)$ of V with

$$(8.6.29) \qquad \int_{-\infty}^\infty |U_0(x)| dx + \int_{-\infty}^\infty (1 + x^2) |U_0(x)|^2 dx \ll 1 \,,$$

provided only that the eigenvalue λ_i is simple and the shock satisfies the strict form of the Lax E-condition. It should be noted that the assertion holds even when the i-characteristic family fails to be genuinely nonlinear.

The orderly structure depicted above disintegrates when dealing with strong shocks and/or systems that are not strictly hyperbolic. In order to get a glimpse of the geometric complexity that may arise in such cases, let us discuss the construction of viscous shock profiles for 2-shocks of the simple system (7.2.11), with dissipative form (8.6.4). The properties of shocks were already discussed in Section 8.3. Taking advantage of symmetry under rotations and scaling properties of the system, we may fix, without loss of generality, the left state (u_-, v_-) at the point $(1, 0)$. The right state (u_+, v_+) will be located at a point $(a, 0)$, with $a \in (-\frac{1}{2}, 0)$. In that case, as shown in Section 8.3, the shock speed is $s = a^2 + a + 1$ and the shock is overcompressive (8.3.7). Notice that the state $(b, 0)$, where $b = -1 - a$, is also joined to $(1, 0)$ by a 2-shock of the same speed s, which satisfies the Lax E-condition, is not overcompressive, but does not satisfy the Liu E-condition.

The system (8.6.8) associated with (8.6.4) reads:

$$(8.6.30) \qquad \begin{cases} \dot{u} = -s(u - 1) + u(u^2 + v^2) - 1 \\ \dot{v} = -sv + v(u^2 + v^2) \,; \end{cases}$$

or, equivalently, in polar coordinates (ρ, θ), $u = \rho \cos\theta$, $v = \rho \sin\theta$:

$$(8.6.31) \qquad \begin{cases} \dot{\rho} = \rho(\rho^2 - s) + (s - 1) \cos\theta \\ \rho\dot{\theta} = -(s - 1) \sin\theta \,. \end{cases}$$

Notice that (8.6.30) possesses three equilibrium points: (a) $(1, 0)$ which is an unstable node; (b) $(a, 0)$ which is a stable node; and (c) $(b, 0)$ which is a saddle. The phase portrait, that may be easily determined through elementary analysis of (8.6.30) and (8.6.31), is depicted in Fig. 8.6.1.

Even though the shock joining $(1, 0)$ to $(b, 0)$ violates the Liu E-condition, these states are connected by two viscous shock profiles, symmetric with respect to the u-axis. By contrast, the states $(1, 0)$ and $(a, 0)$ are connected by infinitely many viscous shock profiles. To test the asymptotic stability of any one of these viscous shock profiles, say $(\bar{u}(\tau), \bar{v}(\tau))$, in the light of our discussion above, we introduce a small perturbation $(u_0(x), v_0(x))$ and inquire whether the solution $(u, v)(x, t)$ of (8.6.4) with initial values

$$(8.6.32) \qquad (u, v)(x, 0) = \left(\bar{u}\left(\frac{x}{\mu}\right) + u_0(x), \bar{v}\left(\frac{x}{\mu}\right) + v_0(x)\right)$$

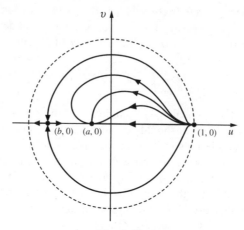

Fig. 8.6.1.

satisfies

$$(8.6.33) \qquad (u, v)(x, t) \to \left(\hat{u} \left(\frac{x - st}{\mu} \right), \hat{v} \left(\frac{x - st}{\mu} \right) \right), \qquad \text{as } t \to \infty,$$

where $(\hat{u}(\tau), \hat{v}(\tau))$ is a (generally different) viscous shock profile. Because no diffusion waves are possible here, the convergence in (8.6.33) must be in $L^1(-\infty, \infty)$. In particular, the v-component of the excess mass conservation yields

$$(8.6.34)$$

$$\int_{-\infty}^{\infty} v_0(x) dx = \int_{-\infty}^{\infty} \left[\hat{v} \left(\frac{x - st}{\mu} \right) - \overline{v} \left(\frac{x - st}{\mu} \right) \right] dx = \mu \int_{-\infty}^{\infty} \left[\hat{v}(\tau) - \overline{v}(\tau) \right] d\tau .$$

It may be shown that the integral on the right-hand side of (8.6.34) is uniformly bounded, independently of the choice of $\overline{v}$ and $\hat{v}$. Consequently, when v_0 is fixed so that $\int v_0 dx \neq 0$, (8.6.34) cannot hold when μ is sufficiently small. Thus, in so far as shock admissibility hinges on stability of the connecting shock profiles, the overcompressive shocks of the system (7.2.11) should be termed inadmissible.

8.7 Notes

The study of shock waves originated in the context of gas dynamics. The book by Courant and Friedrichs [1], already cited in Section 3.4, presented a coherent, mathematical exposition of material from the physical and engineering literature, accumulated over the past 150 years, paving the way for the development of a general theory by Lax [2].

For as long as gas dynamics remained the prototypical example, the focus of the research effort was set on strictly hyperbolic, genuinely nonlinear systems. The intricacy of shock patterns in nonstrictly hyperbolic systems was not recognized until recently, and this subject is currently undergoing active development.

Expositions of most of the topics covered in this chapter are also contained in the books of Smoller [1] and Serre [9].

The notion of Hugoniot locus, in gas dynamics, is traced back to the work of Riemann [1] and Hugoniot [1]; but the definition of shock curves in the general setting is due to Lax [2], who first established the properties stated in Theorems 8.2.1, 8.2.2 and 8.2.3. The elegant proof of Theorem 8.2.1 is here taken from Serre [9]. The significance of systems with coinciding shock and rarefaction wave curves was first recognized by Temple [1], who conducted a thorough study of their noteworthy properties. A detailed discussion is also contained in Serre [9].

For gas dynamics, the statement that admissible shocks should be subsonic relative to their left state and supersonic relative to their right state is found in the pioneering paper of Riemann [1]. This principle was postulated as a general shock admissibility criterion, namely the Lax E-condition, by Lax [2], who also proved Theorem 8.3.1. A proof of Theorem 8.3.2 is given in Li and Yu [1]. See also Hsiao and Chang [1]. A multi-space dimensional version of Theorem 8.3.2, in a Sobolev space setting, was established by Majda [2] by methods akin to those used in the proof of Theorem 5.1.1. Extensive literature has derived from this seminal work, see e.g. Godin [1]. The important issue of linearized stability of multi-space dimensional shock fronts is addressed in Majda [1]. A detailed, systematic presentation, based on the Lopatinski condition, is contained in Serre [9]. See also Corli and Sablé-Tougeron [1]. A different connection between the Lax E-condition and stability is established in Smoller, Temple and Xin [1].

Shock admissibility in the absense of genuine nonlinearity was first discussed by Bethe [1] and Weyl [1], for the system of gas dynamics. The Liu E-condition and related Theorems 8.2.4, 8.4.1, 8.4.2 and 8.5.4 are due to Liu [2]. The motivation was provided by the Oleinik E-condition, derived in Oleinik [4], and the Wendroff E-condition, established in Wendroff [1]. This admissibility criterion seems to have been anticipated in the 1960's by Chang and Hsiao [1,2] (see also Hsiao and Zhang [1]) but their work was not published until much later.

The entropy shock admissibility condition has been part of the basic theory of Continuum Thermomechanics since the turn of the century. The form (8.5.1), for general systems (8.1.1), was postulated by Lax [4], who established Theorems 8.5.1, 8.5.3, and 8.6.2. The proofs of Theorems 8.5.1, 8.5.2, 8.5.3 and 8.5.4 here, based on Equation (8.5.8), are taken from Dafermos [10].

The notion of viscous shock profile was introduced to gas dynamics by Rankine [1] and by Rayleigh [3]. For the physical background, see e.g. Zeldovich and Raizer [1]. A seminal reference is Gilbarg [1]. The general form (8.6.8), for systems (8.1.1), was postulated by Gelfand [1]. Theorem 8.6.1 is due to Majda and Pego [1]. An earlier paper by Foy [1] had established the result in the special case where the system is genuinely nonlinear and the shocks are weak. Also Mock [1] has proved a similar result under the assumption that the system is genuinely nonlinear and it is endowed with a uniformly convex entropy. The issue of characterizing appropriate viscosity matrices B has been discussed by several authors, including Conley and Smoller [1,2], Majda and Pego [1], Pego [1] and Serre [9]. For a detailed study of viscous shock profiles in isentropic (or isothermal) elastodynamics, under physically appropriate assumptions, see Antman and Malek-Madani [1]. The case of general, nonisentropic gas dynamics, with nonconvex equation of state, was investigated by Pego [2], who established that strong shocks satisfying the Liu E-condition do not necessarily admit viscous shock profiles when heat conductivity dominates viscosity. For viscous shock profiles in quadratic systems of conservation laws that are not strictly hyperbolic, see Čanić and Plohr [1].

The literature on asymptotic stability of viscous shock profiles is so vast that it would be impossible to provide here a comprehensive list of references. Of the seminal papers in that area, it will suffice to cite Ilin and Oleinik [1], on the scalar case; Goodman [1], on systems for perturbations with zero excess mass; Liu [12], which introduces the decoupled diffusion waves; and Szepessy and Xin [1], which adds the coupled diffusion waves. For further developments, including the case of contact discontinuities, overcompressive shocks and boundary shock layers, see Gardner and Zumbrun [1], Goodman, Szepessy and Zumbrun [1], Hubert and Serre [1], Kawashima and Matsumura [1], Liu [17], Liu and Nishihara [1], Liu and Xin [2,3], Liu and Yu [1], Liu and Zeng [1,2], Liu and Zumbrun [1], Xin [1,3], Zeng [1,2,3] and Zumbrun and Howard [1]. Multidimensional shocks are considered in Serre and Zumbrun [1]. Profiles obtained via relaxation are discussed in Luo and Serre [1]. For a systematic exposition, the reader may consult the book of Serre [9]. In reference to the specific results reported here, the definitive treatment of the scalar case is in Freistühler and Serre [1] (see also Serre [11]); the stability of (not necessarily genuinely nonlinear) weak shocks, satisfying the strict form of the Lax E-condition, is established in Fries [1,2]; and the study of the stability of overcompressive shocks for the system (7.2.11) is taken from Liu [16].

The class of hyperbolic systems of conservation laws with rotational invariance has interesting mathematical structure as well as applications to elasticity and magnetohydrodynamics. Various aspects of the existence and stability of shock waves in that class are discussed in Brio and Hunter [1], Freistühler [1,3], Freistühler and Liu [1] and Freistühler and Szmolyan [1].

The question of admissibility of jump discontinuities, representing phase boundaries, also arises in systems of conservation laws of mixed type which model phase transitions. Entropy, viscosity and viscosity-capillarity admissibility criteria have been tried in that context, in combination with a new criterion based on "kinetic relations" motivated by considerations at the microscale. See Abeyaratne and Knowles [1,2], Benzoni-Gavage [2], Hagan and Slemrod [1], R.D. James [1], Keyfitz [2], LeFloch [3], Rosakis [1], Slemrod [1,2] and Truskinovsky [1,2].

There is no unique, natural way of defining weak solutions with jump discontinuities for systems that are not in divergence form. However, jump relations extending the Rankine Hugoniot conditions to such systems may be motivated either by physical applications or by purely mathematical considerations. A framework of such theories was introduced by LeFloch [2] and Dal Maso, LeFloch and Murat [1]. For developments and applications of these ideas, see Amadori, Baiti, LeFloch and Piccoli [1], Hayes and LeFloch [1,2] and LeFloch and Tzavaras [1]. For a survey, see LeFloch [4].

Chapter IX. Admissible Wave Fans and the Riemann Problem

The property of systems of conservation laws to be invariant under uniform stretching of the space-time coordinates induces the existence of self-similar solutions, which stay constant along straight-line rays emanating from some focal point in space-time. Such solutions depict a collection of waves converging to the focal point and interacting there to produce a jump discontinuity which is in turn resolved into an outgoing wave fan.

This chapter investigates the celebrated Riemann problem, whose object is the resolution of jump discontinuities into wave fans. A solution will be constructed by the classical method of piecing together elementary centered solutions encountered in earlier chapters, namely, constant states, shocks joining constant states, and centered rarefaction waves bordered by constant states or contact discontinuities. A vanishing viscosity method will also be considered, which employs time-dependent viscosity so that the resulting dissipative system is invariant under stretching of coordinates, just like the original hyperbolic system.

The issue of admissibility of wave fans will be raised and the role of entropy and viscosity will be discussed. It will be shown that under certain conditions the resolution of a jump discontinuity into an admissible wave fan minimizes the total entropy production.

Finally, the interaction of two wave fans will be considered and the resultant wave fan will be determined.

9.1 Self-similar Solutions and the Riemann Problem

The system of conservation laws

$$(9.1.1) \qquad \partial_t U + \partial_x F(U) = 0$$

is invariant under uniform stretching of coordinates: $(x, t) \mapsto (\alpha x, \alpha t)$; hence it admits *self-similar solutions*, defined on the space-time plane and constant along straight-line rays emanating from the origin. Since (9.1.1) is also invariant under translations of coordinates: $(x, t) \mapsto (x + \overline{x}, t + \overline{t})$, the focal point of self-similar solutions may be translated from the origin to any fixed point $(\overline{x}, \overline{t})$ in space-time.

If U is a (generally weak) self-similar solution of (9.1.1), focused at the origin, its restriction to $t > 0$ admits the representation

(9.1.2) $$U(x,t) = V\left(\frac{x}{t}\right), \quad -\infty < x < \infty, \quad 0 < t < \infty,$$

where V is a bounded measurable function on $(-\infty, \infty)$, which satisfies the ordinary differential equation

(9.1.3) $$[F(V(\xi)) - \xi V(\xi)]^{\cdot} + V(\xi) = 0,$$

in the sense of distributions. Indeed, if U is given by (9.1.2) and ϕ is any C^∞ test function with compact support on $(-\infty, \infty) \times (0, \infty)$, then, after a short calculation,

(9.1.4) $$\int_0^\infty \int_{-\infty}^\infty [\partial_t \phi(x,t) U(x,t) + \partial_x \phi(x,t) F(U(x,t))] dx dt$$
$$= \int_{-\infty}^\infty \{\dot\psi(\xi)[F(V(\xi)) - \xi V(\xi)] - \psi(\xi) V(\xi)\} d\xi,$$

where

(9.1.5) $$\psi(\xi) = \int_0^\infty \phi(\xi t, t) dt, \quad -\infty < \xi < \infty.$$

The restriction of U to $t < 0$ similarly admits a representation like (9.1.2), for a (generally different) function V, which also satisfies (9.1.3).

From (9.1.3) we infer that $F(V) - \xi V$ is Lipschitz continuous on $(-\infty, \infty)$ and (9.1.3) holds, in the classical sense, at any Lebesgue point ξ of V.

Henceforth, we shall consider self-similar solutions U of class BV_{loc}. In that case, the function V, above, has bounded variation on $(-\infty, \infty)$. We assume V is normalized, as explained in Section 1.7, so that one-sided limits $V(\xi\pm)$ exist for every $\xi \in (-\infty, \infty)$ and $V(\xi) = V(\xi-) = V(\xi+)$ except possibly on a countable set of ξ.

By account of Theorem 1.7.4, (9.1.3) may be written as

(9.1.6) $$[\widetilde{DF}(V) - \xi I]\dot{V} = 0,$$

in the sense of measures, with

(9.1.7) $$\widetilde{DF}(V)(\xi) = \int_0^1 DF(\tau V(\xi-) + (1-\tau)V(\xi+)) d\tau.$$

Furthermore, as a function of bounded variation V is differentiable almost everywhere on $(-\infty, \infty)$ and (9.1.6) will be satisfied at any point ξ of continuity of V where $\dot{V}(\xi)$ exists.

In view of the above, $(-\infty, \infty)$ is decomposed into the union of three pairwise disjoint sets $\mathscr{C}$, $\mathscr{S}$ and $\mathscr{W}$ as follows:

$\mathscr{C}$ is the maximal open subset of $(-\infty, \infty)$ on which the measure $\dot{V}$ vanishes, i.e., the complement of the support of $\dot{V}$. It is the (at most) countable union of disjoint open intervals, on each of which V is constant.

$\mathscr{S}$ is the (at most) countable set of points of jump discontinuity of V. The Rankine-Hugoniot jump condition

(9.1.8) $$F(V(\xi+)) - F(V(\xi-)) = \xi[V(\xi+) - V(\xi-)]$$

holds at any $\xi \in \mathscr{S}$. This may be inferred from the (Lipschitz) continuity of $F(V) - \xi V$, noted above, or it may be deduced by comparing (9.1.6), (9.1.7) with (8.1.3), (8.1.4).

$\mathscr{W}$ is the (possibly empty) set of points of continuity of V that lie in the support of the measure $\dot{V}$. When $\xi \in \mathscr{W}$, then

$$(9.1.9) \qquad \lambda_i(V(\xi)) = \xi \,,$$

for some $i \in \{1, \cdots, n\}$. Indeed, if ξ is the limit of a sequence $\{\xi_m\}$ in $\mathscr{S}$, then $V(\xi_m+) - V(\xi_m-) \to 0$, as $m \to \infty$, and (9.1.9) follows from the Rankine-Hugoniot condition (9.1.8). On the other hand, if ξ is in the interior of the set of points of continuity of V, and (9.1.9) fails for $i = 1, \cdots, n$, then $\lambda_i(V(\zeta)) \neq \zeta$ for $\zeta \in (\xi - \varepsilon, \xi + \varepsilon)$ and $i = 1, \cdots, n$, in which case, by virtue of (9.1.6), the measure $\dot{V}$ would vanish on $(\xi - \varepsilon, \xi + \varepsilon)$, contrary to our hypothesis that $\xi \in spt \dot{V}$. If ξ is a point of differentiability of V, (9.1.6) implies

$$(9.1.10) \qquad \dot{V}(\xi) = b(\xi) R_i(V(\xi)) \,,$$

where the scalar $b(\xi)$ is determined by combining (9.1.9) with (9.1.10):

$$(9.1.11) \qquad [D\lambda_i(V(\xi))R_i(V(\xi))]b(\xi) = 1 \,.$$

In particular, $V(\xi)$ is a point of genuine nonlinearity of the i-characteristic family

We have thus shown that self-similar solutions are composites of constant states, shocks, and centered simple waves. The simple waves will be centered rarefaction waves, when V is an outgoing wave fan, or centered compression waves, when V depicts a focusing collection of waves. The two configurations are differentiated by time irreversibility, induced by admissibility conditions on weak solutions. More stringent conditions are imposed on outgoing wave fans, so these are generally simpler.

Of central importance will be to understand how a jump discontinuity at the origin, introduced by the initial data, is resolved into an outgoing wave fan. This is the object of the

Riemann Problem. Determine a self-similar (generally weak) solution U of (9.1.1) on $(-\infty, \infty) \times (0, \infty)$, with initial condition

$$(9.1.12) \qquad U(x, 0) = \begin{cases} U_\ell, & \text{for } x < 0 \\ U_r, & \text{for } x > 0 \,, \end{cases}$$

where U_ℓ and U_r are given states in $\mathcal{O}$.

Following our discussion, above, we shall seek a solution of the Riemann problem in the form (9.1.2), where V satisfies the ordinary differential equation (9.1.3), on $(-\infty, \infty)$, together with boundary conditions

$$(9.1.13) \qquad V(-\infty) = U_\ell \,, \quad V(\infty) = U_r \,.$$

The specter of nonuniqueness raises again the issue of admissibility, which will be the subject of discussion in the following sections.

9.2 Wave Fan Admissibility Criteria

Various aspects of admissibility have already been discussed, for general weak solutions, in Chapter IV, and for single shocks, in Chapter VIII. We have thus encountered a number of admissibility criteria and we have seen that they are strongly interrelated but not quite equivalent. As we shall see later, the most discriminating among these criteria, namely viscous shock profiles and the Liu E-condition, are sufficently powerful to weed out all spurious solutions, so long as we are confined to strictly hyperbolic systems and shocks of moderate strength. However, once we move to systems that are not strictly hyperbolic and/or to solutions with strong shocks the situation becomes murky. The question of admissibility is still open.

Any rational new admissibility criterion should adhere to certain basic principles, the fruits of the long experience with the subject. They include:

(a) Localization. The test of admissibility of a solution should apply individually to each point $(\bar{x}, \bar{t})$ in the domain and only involve the restriction of the solution to an arbitrarily small neighborhood of $(\bar{x}, \bar{t})$, say the circle $\{(x, t) : |x - \bar{x}|^2 + |t - \bar{t}|^2 < r^2\}$ where r is fixed but arbitrarily small. This is compatible with the general principle that solutions of hyperbolic systems should have the local dependence property.

(b) Evolutionarity. The test of admissibility should be forward-looking into the future, without regard for the past. Thus, admissibility of a solution at the point $(\bar{x}, \bar{t})$ should depend solely on its restriction to the semicircle $\{(x, t) : |x - x|^2 + |t - \bar{t}|^2 < r^2, t \geq \bar{t}\}$. This is in line with the principle of time irreversibility, which pervades the admissibility criteria we have encountered thus far, like entropy, viscosity, etc.

(c) Invariance Under Translations. A solution U will be admissible at $(\bar{x}, \bar{t})$ if and only if the translated solution $\overline{U}$, $\overline{U}(x, t) = U(x + \bar{x}, t + \bar{t})$, is admissible at the origin $(0, 0)$.

(d) Invariance Under Dilations. A solution U will be admissible at $(0, 0)$ if and only if, for each $\alpha > 0$, the dilated solution $\overline{U}_\alpha$, $\overline{U}_\alpha(x, t) = U(\alpha x, \alpha t)$, is admissible at $(0, 0)$.

Let us focus attention to weak solutions U with the property that, for each fixed point $(\bar{x}, \bar{t})$ in the domain, the limit

$$(9.2.1) \qquad \overline{U}(x, t) = \lim_{\alpha \downarrow 0} U(\bar{x} + \alpha x, \bar{t} + \alpha t)$$

exists for almost all $(x, t) \in (-\infty, \infty) \times [0, \infty)$. Notice that in that case $\overline{U}$ is necessarily a self-similar solution of (9.1.1). In the spirit of the principles listed above, one may use the admissibility of $\overline{U}$ at the origin as a test for the admissibility of U at the point $(\bar{x}, \bar{t})$. Since $\overline{U}$ depicts a fan of waves radiating from the origin, such tests constitute *wave fan admissibility criteria*.

Passing to the limit in (9.2.1), amounts to observing, so to say, the solution U under a microscope focused at the point $(\bar{x}, \bar{t})$. The limit certainly exists if U is piecewise smooth: When the stretching is performed about a point $(\bar{x}, \bar{t})$ of continuity, the resulting $\bar{U}$ will be the constant state $U(\bar{x}, \bar{t})$. When the stretching is performed about a point $(\bar{x}, \bar{t})$ lying on a shock, the resulting $\bar{U}$ will consist of a single shock joining the constant states $U(\bar{x}-, \bar{t})$ and $U(\bar{x}+, \bar{t})$. More complex wave fans $\bar{U}$ will emerge when the stretching is effected about a point $(\bar{x}, \bar{t})$ of wave interactions. The issue whether the limit (9.2.1) exists for general BV solutions will be addressed in Chapter XI, for genuinely nonlinear scalar conservation laws, in Chapter XII, for genuinely nonlinear systems of two conservation laws, and in Chapter XIV, for general genuinely nonlinear systems of conservation laws.

As we saw in Section 9.1, the wave fan $\bar{U}$ is generally a composite of constant states, shocks, and centered rarefaction waves. The simplest wave fan admissibility criterion postulates that the fan is admissible if each one of its shocks, individually, satisfies the shock admissibility conditions discussed in Chapter VIII. As we shall see in the following section, this turns out to be adequate in many cases. Other fan admissibility criteria, which regard the wave fan as an entity rather than as a collection of individual waves, include the entropy rate condition and the viscous fan profile test. These will be discussed later.

9.3 Solution of the Riemann Problem with Admissible Shocks

The aim here is to construct a solution of the Riemann problem by piecing together constant states, centered rarefaction waves, and shocks that satisfy the Liu E-condition. We limit our investigation to the case where wave speeds of different characteristic families are strictly separated. This will cover waves of small amplitude in general strictly hyperbolic systems as well as waves of any amplitude in special systems like (7.1.6) in which all 1-waves travel to the left and all 2-waves travel to the right.

Let us then consider an outgoing wave fan (9.1.2), of bounded variation. Following the discussion in Section 9.1, $(-\infty, \infty)$ is decomposed into the union of the shock set $\mathscr{S}$, the rarefaction wave set $\mathscr{W}$ and the constant state set $\mathscr{C}$. Since the wave speeds of distinct characteristic families are strictly separated, $\mathscr{S} = \bigcup_{i=1}^{n} \mathscr{S}_i$ and $\mathscr{W} = \bigcup_{i=1}^{n} \mathscr{W}_i$, where $\mathscr{S}_i$ is the (at most countable) set of points of jump discontinuity of V that are i-shocks and $\mathscr{W}_i$ is the (possibly empty) set of points of continuity of V in the support of the measure $\dot{V}$ that satisfy (9.1.9). The set $\mathscr{S}_i \bigcup \mathscr{W}_i$ is closed and contains points in the range of wave speeds of the i-characteristic family.

We now assume that the shocks satisfy the Lax E-condition, i.e., for all $\xi \in \mathscr{S}_i$,

$$(9.3.1) \qquad \lambda_i(V(\xi-)) \geq \xi \geq \lambda_i(V(\xi+)) .$$

Then $\mathscr{S}_i \bigcup \mathscr{W}_i$ is necessarily a closed interval $[\alpha_i, \beta_i]$. Indeed, suppose $\mathscr{S}_i \bigcup \mathscr{W}_i$ is disconnected. Then there is an open interval $(\xi_1, \xi_2) \subset \mathscr{C}$ with endpoints ξ_1 and

ξ_2 contained in $\mathscr{S} \bigcup \mathscr{W}_i$. In particular, $V(\xi_1+) = V(\xi_2-)$. On the other hand, by virtue of (9.1.9) and (9.3.1), $\xi_1 \geq \lambda_i(V(\xi_1+))$, $\xi_2 \leq \lambda_i(V(\xi_2-))$, which is a contradiction to $\xi_1 < \xi_2$. Notice further that any $\xi \in \mathscr{S}$ with $\xi > \alpha_i$ (or $\xi < \beta_i$) is the limit of an increasing (or decreasing) sequence of points of $\mathscr{W}_i$ and so $\lambda_i(V(\xi-)) = \xi$ (or $\lambda_i(V(\xi+)) = \xi$). We have thus established the following

Theorem 9.3.1 *Assume the wave speeds of distinct characteristic families are strictly separated. Any self-similar solution (9.1.2) of the Riemann Problem (9.1.1), (9.1.12) comprises $n + 1$ constant states $U_\ell = U_0, U_1, \cdots, U_{n-1}, U_n = U_r$. For $i = 1, \cdots, n$, U_{i-1} is joined to U_i by an i-wave fan, namely a composite of centered i-rarefaction waves and/or i-shocks with the property that i-shocks bordered from the left (and/or the right) by i-rarefaction waves are left (and/or right) i-contact discontinuities (Fig. 9.3.1).*

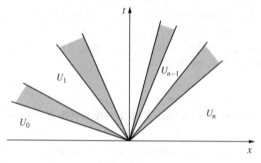

Fig. 9.3.1.

It will be shown below that the locus of states that may be joined on the right (or left) of a fixed state $\overline{U} \in \mathscr{O}$ by an admissible i-wave fan, composed of i-rarefaction waves and admissible i-shocks, as prescribed by Theorem 9.3.1, is a C^2 curve $\Phi_i(\tau; \overline{U})$ (or $\Psi_i(\tau; \overline{U})$), called the *forward* (or *backward*) *i-wave fan curve through* $\overline{U}$, which may be parametrized so that

$$(9.3.2) \qquad \Phi_i(0; \overline{U}) = \overline{U}, \quad \dot{\Phi}_i(0; \overline{U}) = R_i(\overline{U}),$$

$$(9.3.3) \qquad \Psi_i(0; \overline{U}) = \overline{U}, \quad \dot{\Psi}_i(0; \overline{U}) = R_i(\overline{U}).$$

Taking, for the time being, the existence of wave fan curves with the above properties for granted, we note that to solve the Riemann problem we have to determine an n-tuple $\varepsilon = (\varepsilon_1, \cdots, \varepsilon_n)$, realized as a vector in $\mathbb{R}^n$, such that, starting out from $U_0 = U_\ell$ and computing successively $U_i = \Phi_i(\varepsilon_i; U_{i-1})$, $i = 1, \cdots, n$, we end up with $U_n = U_r$. Accordingly, we define the function

$$(9.3.4) \qquad \Omega(\varepsilon; \overline{U}) = \Phi_n(\varepsilon_n; \Phi_{n-1}(\varepsilon_{n-1}; \cdots \Phi_1(\varepsilon_1; \overline{U}) \cdots)).$$

Clearly,

(9.3.5) $$\Omega(0; \overline{U}) = \overline{U} \ .$$

Furthermore, since $\Omega(0, \cdots, 0, \varepsilon_i, 0, \cdots, 0; \overline{U}) = \Phi_i(\varepsilon_i; \overline{U})$,

(9.3.6) $$\frac{\partial \Omega}{\partial \varepsilon_i}(0; \overline{U}) = R_i(\overline{U}) \ , \quad 1 \le i \le n \ ,$$

(9.3.7) $$\frac{\partial^2 \Omega}{\partial \varepsilon_i^2}(0; \overline{U}) = DR_i(\overline{U})R_i(\overline{U}) \ , \quad 1 \le i \le n \ .$$

Finally, for $j < k$, $\Omega(0, \cdots, 0, \varepsilon_j, 0, \cdots, 0, \varepsilon_k, 0, \cdots, 0; \overline{U}) = \Phi_k(\varepsilon_k; \Phi_j(\varepsilon_j; \overline{U}))$
and hence

(9.3.8) $$\frac{\partial^2 \Omega}{\partial \varepsilon_j \partial \varepsilon_k}(0; \overline{U}) = DR_k(\overline{U})R_j(\overline{U}) \ , \quad 1 \le j < k \le n \ .$$

In particular, from (9.3.5), (9.3.6) and the implicit function theorem follows that
when U_r is sufficiently close to U_ℓ there exists a unique ε near 0 with $\Omega(\varepsilon; U_\ell) =
U_r$. This generates a solution to the Riemann problem, which is unique within the
class of self-similar solutions with weak waves. The wave fan joining U_ℓ with U_r
is conveniently described through its left state U_ℓ and the n-tuple $\varepsilon = (\varepsilon_1, \cdots, \varepsilon_n)$.
The value of ε_i determines the i-wave amplitude. In particular, $|\varepsilon_i|$ measures the
i-wave strength while $sgn\varepsilon_i$ provides information on the nature of the i-wave
(shock, rarefaction wave, etc.). When F is C^4, Ω is in $W^{3,\infty}$ and, by virtue of
(9.3.6), (9.3.7) and (9.3.8):

(9.3.9)
$$U_r = U_\ell + \sum_{i=1}^{n} \varepsilon_i R_i(U_\ell) + \frac{1}{2} \sum_{i=1}^{n} \varepsilon_i^2 DR_i(U_\ell)R_i(U_\ell)$$
$$+ \sum_{j=1}^{n} \sum_{k=j+1}^{n} \varepsilon_j \varepsilon_k DR_k(U_\ell)R_j(U_\ell) + O(|\varepsilon|^3) \ .$$

Clearly, we may also synthesize the solution of the Riemann problem in
the reverse order, starting out from $U_n = U_r$ and computing successively
$U_{i-1} = \Psi_i(\varepsilon_i; U_i)$, $i = n, \cdots, 1$, until we reach $U_0 = U_\ell$. Under certain circumstances, a mixed strategy may be advantageous. For example, the most efficient
procedure for solving the Riemann problem for a system of two conservation
laws, $n = 2$, is to draw the forward 1-wave curve $\Phi_1(\varepsilon_1; U_\ell)$ through the left state
U_ℓ and the backward 2-wave curve $\Psi_2(\varepsilon_2; U_r)$ through the right state U_r. The
intersection of these two curves will determine the intermediate constant state:
$U_m = \Phi_1(\varepsilon_1; U_\ell) = \Psi_2(\varepsilon_2; U_r)$.

Our next project is to construct the wave fan curves. We begin our investigation
with systems in which fans are particularly simple. When the i-characteristic family
is linearly degenerate, no centered i-rarefaction waves exist and hence, by Theorem
8.2.5, any i-wave fan is necessarily an i-contact discontinuity. In that case the
forward and backward i-wave fan curves coincide with the shock curve W_i in
Theorem 8.2.5, i.e., $\Phi_i(\tau; \overline{U}) = \Psi_i(\tau; \overline{U}) = W_i(\tau; \overline{U})$.

When the i-characteristic family is genuinely nonlinear, i-contact discontinuities are ruled out by Theorem 8.2.1, and so any i-wave fan of small amplitude must be either a single centered i-rarefaction wave or a single compressive i-shock. Let us normalize the field R_i so that (7.6.13) holds, $D\lambda_i R_i = 1$. The states that may be joined to $\overline{U}$ by a weak i-shock lie on the i-shock curve $W_i(\tau; \overline{U})$ described by Theorem 8.2.1. On account of Theorem 8.3.1, the shock that joins $\overline{U}$, on the left, with $W_i(\tau; \overline{U})$, on the right, is compressive if and only if $\tau < 0$. On the other hand, by Theorem 7.6.3, the state $\overline{U}$ may be joined on the right (or left) by centered i-rarefaction waves to states $V_i(\tau; \overline{U})$ for $\tau > 0$ (or $\tau < 0$). It then follows that we may construct the forward i-wave fan curve by $\Phi_i(\tau; \overline{U}) = W_i(\tau; \overline{U})$, for $\tau < 0$, and $\Phi_i(\tau; \overline{U}) = V_i(\tau; \overline{U})$, for $\tau > 0$. Similarly, the backward i-wave fan curve is defined by $\Psi_i(\tau; \overline{U}) = V_i(\tau; \overline{U})$, for $\tau < 0$, and $\Psi_i(\tau; \overline{U}) = W_i(\tau; \overline{U})$, for $\tau > 0$. These curves are C^2, by account of Theorem 8.2.2, and satisfy (9.3.2), (9.3.3), by Theorem 8.2.1.

In view of the above discussion, we have now established the existence of solution to the Riemann problem for systems with characteristic families that are either genuinely nonlinear or linearly degenerate:

Theorem 9.3.2 *Assume the system* (9.1.1) *is strictly hyperbolic and each characteristic family is either genuinely nonlinear or linearly degenerate. For* $|U_r - U_\ell|$ *sufficiently small, there exists a unique self-similar solution* (9.1.2) *of the Riemann problem* (9.1.1), (9.1.12), *with small total variation. This solution comprises* $n+1$ *constant states* $U_\ell = U_0, U_1, \cdots, U_{n-1}, U_n = U_r$. *When the i-characteristic family is linearly degenerate,* U_i *is joined to* U_{i-1} *by an i-contact discontinuity, while when the i-characteristic family is genuinely nonlinear,* U_i *is joined to* U_{i-1} *by either a centered i-rarefaction wave or a compressive i-shock.*

In particular, Theorem 9.3.2 establishes the existence of solutions, with small total variation, to the Riemann problem for the system (7.1.7) of isentropic gas dynamics, when $2p'(\rho) + \rho p''(\rho) > 0$, so that both characteristic families are genuinely nonlinear; also for the system (7.1.3) of adiabatic thermoelasticity, under the assumption $\sigma_{uu}(u, s) \neq 0$, in which case the 1- and the 3-characteristic families are genuinely nonlinear while the 2-characteristic family is linearly degenerate. As noted earlier, shock and rarefaction wave curves for the above systems exist even in the range of strong shocks and thus one may attempt to construct solutions of the Riemann problem even when U_ℓ and U_r are far apart. The range of U_ℓ and U_r for which the construction is possible depends on the asymptotic behavior of shock and rarefaction wave curves as the state variables ρ and u approach the boundary points of their physical range, namely zero and infinity. An exhaustive discussion of these issues is contained in the literature cited in Section 9.7 so it will not be necessary to reproduce the analysis here. However, in order to illustrate the above ideas by means of a simple example, let us consider the system (7.1.6), assuming that $\sigma(u)$ is defined on $(-\infty, \infty)$ and $0 < a \leq \sigma'(u) \leq b < \infty$, $\sigma''(u) < 0$. It is convenient to reparametrize the wave curves, employing u as the new parameter. In that case, the forward 1-wave curve Φ_1 and the backward 2-wave curve Ψ_2 through the typical point $(\bar{u}, \bar{v})$ of the state space may be represented as $v = \phi(u; \bar{u}, \bar{v})$ and $v = \psi(u; \bar{u}, \bar{v})$, respectively. Recalling the form of the Hugoniot locus (8.2.11) and rarefaction wave curves (7.6.15) for this system, we deduce that

$$(9.3.10) \qquad \phi(u; \bar{u}, \bar{v}) = \begin{cases} \bar{v} - \sqrt{[\sigma(u) - \sigma(\bar{u})](u - \bar{u})}\,, & u \leq \bar{u} \\[2mm] \bar{v} + \displaystyle\int_{\bar{u}}^{u} \sqrt{\sigma'(\omega)}\,d\omega\,, & u > \bar{u} \end{cases}$$

$$(9.3.11) \qquad \psi(u; \bar{u}, \bar{v}) = \begin{cases} \bar{v} + \sqrt{[\sigma(u) - \sigma(\bar{u})](u - \bar{u})}\,, & u \leq \bar{u} \\[2mm] \bar{v} - \displaystyle\int_{\bar{u}}^{u} \sqrt{\sigma'(\omega)}\,d\omega\,, & u > \bar{u}\,. \end{cases}$$

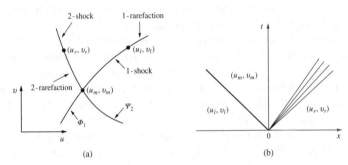

Fig. 9.3.2.

Figure 9.3.2 depicts a solution of the Riemann problem which comprises a compressive 1-shock and a centered 2-rarefaction wave. The intermediate constant state (u_m, v_m) is determined on the u-v plane as the intersection of the forward 1-wave fan curve Φ_1 through (u_ℓ, v_ℓ) with the backward 2-wave fan curve Ψ_2 through (u_r, v_r), namely by solving the equation

$$(9.3.12) \qquad v_m = \phi(u_m; u_\ell, v_\ell) = \psi(u_m; u_r, v_r)\,.$$

For systems of two conservation laws it is often expedient to perform the construction of the intermediate constant state on the plane of Riemann invariants rather than in the original state space. The reason is that, as noted in Section 7.6, in the plane of Riemann invariants rarefaction wave curves become straight lines parallel to the coordinate axes. This facilitates considerably the task of locating the intersection of wave curves of different characteristic families. Figure 9.3.3 depicts the configuration of the wave curves of Fig. 9.3.2 in the plane w-z of Riemann invariants.

Our next task is to describe admissible wave fans, and construct the corresponding wave fan curves, for systems with characteristic families that are neither genuinely nonlinear nor linearly degenerate. In that case, the Lax E-condition is

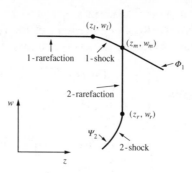

Fig. 9.3.3.

no longer sufficiently selective to single out a unique solution to the Riemann problem so the more stringent Liu E-condition will be imposed on shocks.

We begin with the scalar conservation law (7.1.2), where $f(u)$ may have inflection points. The Liu E-condition is now expressed by the Oleinik E-condition (8.4.2). By Theorem 9.3.1, the solution of the Riemann problem comprises two constant states u_ℓ and u_r joined by a wave fan which is a composite of shocks and/or centered rarefaction waves. There exists precisely one such wave fan with shocks satisfying Oleinik's E-condition, and it is constructed by the following procedure: When $u_\ell < u_r$ (or $u_\ell > u_r$), we let g denote the *convex* (or *concave*) *envelope* of f over the interval $[u_\ell, u_r]$ (or $[u_r, u_\ell]$); namely, g is the maximal (or minimal) element of the partially ordered set $\{h : h$ convex, $h(u) \leq f(u)$, $u_\ell \leq u \leq u_r\}$ (or $\{h : h$ concave, $h(u) \geq f(u)$, $u_r \leq u \leq u_\ell\}$). Thus the graph of g may be visualized as the configuration of a flexible string anchored at the points $(u_\ell, f(u_\ell))$, $(u_r, f(u_r))$ and stretched under (or over) the "obstacle" $\{(u, v) : u_\ell \leq u \leq u_r, v \geq f(u)\}$ (or $\{(u, v) : u_r \leq u \leq u_\ell, v \leq f(u)\}$). The slope $\xi = g'(u)$ is a continuous nondecreasing (or nonincreasing) function whose inverse $u = \omega(\xi)$ generates the wave fan $u = \omega(x/t)$. In particular, the flat parts of $g'(u)$ give rise to the shocks while the intervals over which $g'(u)$ is strictly monotone generate the rarefaction waves. Figure 9.3.4 depicts an example in which the resulting wave fan consists of a centered rarefaction wave bordered by one-sided contact discontinuities.

To prepare the ground for the investigation of systems, we construct wave fans, and corresponding wave fan curves, for the simple system (7.1.6), where $\sigma(u)$ may have inflection points. The Liu E-condition here reduces to the Wendroff E-condition (8.4.3). Similar to the genuinely nonlinear case, we shall employ u as parameter and determine the forward 1-wave fan curve Φ_1 and the backward 2-wave fan curve Ψ_2, through the state $(\bar{u}, \bar{v})$, in the form $v = \phi(u; \bar{u}, \bar{v})$ and $v = \psi(u; \bar{u}, \bar{v})$, respectively. Recalling the equations (8.2.11) for the Hugoniot locus, the equations (7.6.15) for the rarefaction wave curves, and (8.4.3), we easily verify that

$$(9.3.13) \qquad \phi(u; \bar{u}, \bar{v}) = \bar{v} + \int_{\bar{u}}^{u} \sqrt{g'(\omega; u, \bar{u})}\, d\omega ,$$

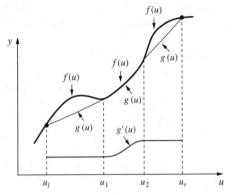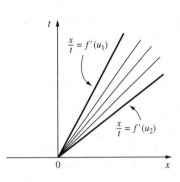

Fig. 9.3.4.

$$(9.3.14) \qquad \psi(u; \overline{u}, \overline{v}) = \overline{v} - \int_{\overline{u}}^{u} \sqrt{g'(\omega; u, \overline{u})} d\omega \ ,$$

where $g'(\omega; u, \overline{u})$ is the derivative, with respect to ω, of the monotone increasing, continuously differentiable function $g(\omega; u, \overline{u})$ which is constructed by the following procedure: For fixed $u \leq \overline{u}$ (or $u \geq \overline{u}$), $g(\cdot, u, \overline{u})$ is the convex (or concave) envelope of $\sigma(\cdot)$ over the interval $[u, \overline{u}]$ (or $[\overline{u}, u]$). Indeed, as in the case of the scalar conservation law discussed above, the states $(\overline{u}, \overline{v})$ and (u, v), $v = \phi(u; \overline{u}, \overline{v})$, are joined by a 1-wave fan $(\omega(x/t), \upsilon(x/t))$, where $\omega(\xi)$ is the inverse of the function $\xi = \sqrt{g'(\omega; u, \overline{u})}$ and

$$(9.3.15) \qquad \upsilon(\xi) = \overline{v} + \int_{\overline{u}}^{\omega(\xi)} \sqrt{g'(\omega; u, \overline{u})} d\omega \ .$$

Again, the flat parts of g' give rise to shocks while the intervals over which g' is strictly monotone generate the rarefaction waves. In the genuinely nonlinear case, $\sigma''(u) < 0$, (9.3.13) and (9.3.14) reduce to (9.3.10) and (9.3.11). Once ϕ and ψ have been determined, the Riemann problem is readily solved, as in the genuinely nonlinear case, by locating the intermediate constant state (u_m, v_m) through the equation (9.3.12).

After this preparation, we continue with a somewhat sketchy and informal description of the construction of wave fan curves for general systems. To avoid aggravating complications induced by various degeneracies, we limit the investigation to i-characteristic families that are *piecewise genuinely nonlinear* in the sense that if U is any state of linear degeneracy, $D\lambda_i(U)R_i(U) = 0$, then $D(D\lambda_i(U)R_i(U))R_i(U) \neq 0$. This implies, in particular, that the set of states of linear degeneracy of the i-characteristic family is locally a smooth manifold of codimension one, which is transversal to the vector field R_i. The scalar conservation law (7.1.2) and the system (7.1.6) of isentropic elasticity will satisfy this assumption when the functions $f(u)$ and $\sigma(u)$ have isolated, nondegenerate inflection points, i.e., $f'''(u)$ and $\sigma'''(u)$ are nonzero at any point u where $f''(u)$

and $\sigma''(u)$ vanish. Even after these simplifications, the construction is very complicated. The ideas may become more transparent if the reader employs the model system (7.1.6) to illustrate each step.

Assuming the i-characteristic family is piecewise genuinely nonlinear, we consider the forward i-wave fan curve $\Phi_i(\tau; \overline{U})$ through a point $\overline{U}$ of genuine nonlinearity, say $D\lambda_i(\overline{U})R_i(\overline{U}) = 1$. Then Φ_i starts out as in the genuinely nonlinear case, namely, for τ positive small it coincides with the i-rarefaction wave curve $V_i(\tau; \overline{U})$ through $\overline{U}$, while for τ negative, near zero, it coincides with the i-shock curve $W_i(\tau; \overline{U})$ through $\overline{U}$. In particular, (9.3.2) holds. We shall follow Φ_i along the positive τ-direction; the description for $\tau < 0$ is quite analogous.

For $\tau > 0$, $\Phi_i(\tau; \overline{U})$ will stay with the i-rarefaction wave curve $V_i(\tau; \overline{U})$ for as long as the latter sojourns in the region of genuine nonlinearity: $D\lambda_i(V_i)R_i(V_i) > 0$. Suppose now $V_i(\tau; \overline{U})$ first encounters the set of states of linear degeneracy of the i-characteristic family at the state $\tilde{U} = V_i(\tilde{\tau}; \overline{U}) : D\lambda_i(\tilde{U})R_i(\tilde{U}) = 0$. The set of states of linear degeneracy in the vicinity of $\tilde{U}$ forms a manifold $\mathcal{M}$ of codimension 1, transversal to the vector field R_i; see Fig. 9.3.5a,b.

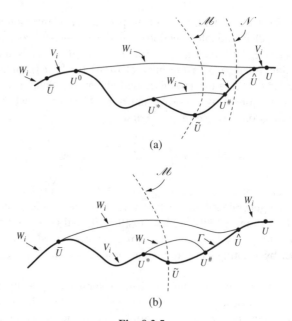

(a)

(b)

Fig. 9.3.5.

The extension of Φ_i beyond $\tilde{U}$ is constructed as follows: For $\tau^* < \tilde{\tau}$, with $\tilde{\tau} - \tau^*$ small, we draw the i-shock curve $W_i(\zeta; U^*)$ through the state $U^* = V_i(\tau^*, \overline{U})$. By account of (8.2.1), $s_i(0; U^*) = \lambda_i(U^*)$ and since $D\lambda_i(U^*)R_i(U^*) > 0$, (8.2.2) implies that for ζ negative, near 0, $\dot{s}_i(\zeta; U^*) > 0$ and $s_i(\zeta; U^*) < \lambda_i(W_i(\zeta; U^*))$. However, after crossing $\mathcal{M}$, $W_i(\zeta; U^*)$ enters the region where $D\lambda_i(U)R_i(U) < 0$ and thus $\lambda_i(W_i(\zeta; U^*))$ will become decreasing. Eventually, ζ^* will be reached where $s_i(\zeta^*; U^*) = \lambda_i(W_i(\zeta^*; U^*))$. For $\zeta < \zeta^*$, by virtue of Theorem 8.2.4, $s_i(\zeta, U^*) > \lambda_i(W_i(\zeta; U^*))$ and $\dot{s}_i(\zeta, U^*) < 0$. Finally, a value $\zeta^\sharp$ will be attained with $s_i(\zeta^\sharp; U^*) = \lambda_i(U^*)$. Then the state $U^\sharp = W_i(\zeta^\sharp; U^*)$, on the right, is joined to U^*, on the left, by a left i-contact discontinuity with speed $\lambda_i(U^*)$. This shock satisfies the Liu E-condition, since $s_i(\zeta; U^*) > \lambda_i(U^*)$ for $\zeta < \zeta^\sharp$. In particular, $\lambda_i(U^*) = s_i(\zeta^\sharp; U^*) > \lambda_i(U^\sharp)$. Consequently, $\overline{U}$, on the left, is joined to $U^\sharp$, on the right, by an admissible i-wave fan, comprising the i-rarefaction wave that joins U^* to $\overline{U}$ and the admissible left i-contact discontinuity that joins $U^\sharp$ to U^*. It can be shown that as U^* moves along the curve $V_i(\tau; \overline{U})$ from $\tilde{U}$ towards $\overline{U}$, the corresponding $U^\sharp$ traces a smooth curve, say Γ. If $U^* = \tilde{U}$, then $U^\sharp = \tilde{U}$ so Γ starts out from $\tilde{U}$. Also Γ at $\tilde{U}$ is tangential to $R_i(\tilde{U})$. We adjoin Γ to $V_i(\tau; \overline{U})$ and consider it as the continuation of $\Phi_i(\tau; \overline{U})$ beyond $\tilde{U}$, with the proper parametrization.

$\Phi_i(\tau; \overline{U})$ will stay with Γ up until a state $\hat{U}$ is reached at which one of the following two events first occurs:

One possibility is depicted in Fig. 9.3.5(a): Γ crosses another manifold $\mathcal{N}$ of states of linear degeneracy of the i-characteristic family, entering the region $D\lambda_i(U)R_i(U) > 0$, and eventually U^* backs up to a position U^0 so that the corresponding $U^\sharp$, denoted by $\hat{U}$, satisfies $\lambda_i(\hat{U}) = \lambda_i(U^0)$. In that case, $\Phi_i(\tau; \overline{U})$ is extended beyond $\hat{U}$ as the i-rarefaction curve $V_i(\zeta; \hat{U})$ through $\hat{U}$, properly reparametrized. Any state U on that curve is joined, on the right, to $\overline{U}$ by a wave fan comprising an i-rarefaction wave that joins U^0 to $\overline{U}$, an i-contact discontinuity that joins $\hat{U}$ to U^0 and a second i-rarefaction wave that joins U to $\hat{U}$.

The other possibility is depicted in Fig. 9.3.5(b): U^* backs up all the way to $\overline{U}$ and the corresponding $U^\sharp$, denoted by $\hat{U}$, satisfies $\lambda_i(\hat{U}) < \lambda_i(\overline{U})$. In that case $\hat{U}$ lies on the i-shock curve through $\overline{U}$, say $\hat{U} = W_i(\hat{\tau}; \overline{U})$. Since $s_i(\hat{\tau}; \overline{U}) = \lambda_i(\overline{U}) > \lambda_i(\hat{U})$, Theorem 8.2.4 implies $\dot{s}_i(\hat{\tau}; \overline{U}) < 0$. Then $\Phi_i(\tau; \overline{U})$ is extended beyond $\hat{U}$ along the i-shock curve $W_i(\tau; \overline{U})$. Any state U on this arc of the curve is joined, on the right, to $\overline{U}$ by a single shock that satisfies the Liu E-condition.

By continuing this process we complete the construction of $\Phi_i(\tau; \overline{U})$ within the range of weak waves, and for certain systems within the range of waves of moderate strength or even for strong waves. Furthermore, careful review of the construction verifies that the graph of Φ_i contains all states in a small neighborhood of $\overline{U}$ that may be joined to $\overline{U}$ by an admissible i-wave fan.

Once wave fan curves satisfying (9.3.2) are in place, we may employ the construction of the solution to the Riemann problem, described above, thus arriving at the following generalization of Theorem 9.3.2:

Theorem 9.3.3 *Assume the system* (9.1.1) *is strictly hyperbolic and each characteristic family is either piecewise genuinely nonlinear or linearly degenerate. For* $|U_r - U_\ell|$ *sufficiently small, there exists a unique self-similar solution* (9.1.2) *of the Riemann problem* (9.1.1), (9.1.12), *with small total variation. This solution comprises* $n + 1$ *constant states* $U_\ell = U_0, U_1, \cdots, U_{n-1}, U_n = U_r$. *When the i-characteristic family is linearly degenerate,* U_i *is joined to* U_{i-1} *by an i-contact discontinuity. When the i-characteristic family is piecewise genuinely nonlinear,*

U_i is joined to U_{i-1} by an admissible i-wave fan, composed of a finite number of i-rarefaction waves and i-shocks that satisfy the Liu E-condition.

The orderly picture painted by the above theorem breaks down when one leaves the realm of strictly hyperbolic systems and waves of small amplitude. The Liu E-condition in no longer sufficiently discriminating to single out a unique solution. To illustrate this, let us consider the Rieman problem for the model system (7.2.11), with data $(u_\ell, v_\ell) = (1, 0)$ and $(u_r, v_r) = (a, 0)$, where $a \in (-\frac{1}{2}, 0)$. One solution comprises the two constant states $(1, 0)$ and $(a, 0)$ joined by an overcompressive shock, of speed $s = 1 + a + a^2$, which satisfies the Liu E-condition. There is, however, another solution comprising three constant states, $(1, 0), (-1, 0)$ and $(a, 0)$, where $(-1, 0)$ is joined to $(1, 0)$ by a 1-contact discontinuity of speed 1 and $(a, 0)$ is joined to $(-1, 0)$ by a 2-shock of speed $s = 1 - a + a^2$. Both shocks satisfy the Liu E-condition. Following the discussion at the end of Section 8.6, one may be inclined to disqualify overcompressive shocks for this system, in which case the second solution of the Riemann problem emerges as the admissible one. This of course hinges on the assumption that (8.6.4) is the proper dissipative form of (7.2.11).

9.4 The Entropy Rate Admissibility Criterion

Here a wave fan admissibility condition will be introduced, which is a logical extension of the entropy shock admissibility criterion discussed in Section 8.5. We thus assume that our system (9.1.1) is endowed with a particular entropy-entropy flux pair (η, q), which is designated to satisfy an entropy inequality, expressing the Second Law of thermodynamics or motivated by alternative physical or mathematical considerations.

The entropy production of a shock of speed s that joins the state U_+, on the right, to the state U_-, on the left, is given by the left-hand side of (8.5.1). Consequently, the total entropy production contributed by all the shocks contained in a wave fan $U = V(x/t)$, with $V(\xi)$ a function of bounded variation on $(-\infty, \infty)$, is given by

(9.4.1) $$P_V = \sum_\xi \{q(V(\xi+)) - q(V(\xi-)) - \xi[\eta(V(\xi+)) - \eta(V(\xi-))]\},$$

where the summation runs over the at most countable set of points ξ of jump discontinuity of V.

The *entropy rate admissibility criterion* stipulates that a wave fan $U = V(x/t)$ is admissible if $P_V \leq P_{\tilde{V}}$, for any other wave fan $U = \tilde{V}(x/t)$ with the same end-states: $\tilde{V}(\pm\infty) = V(\pm\infty)$. The term derives from the observation that the rate of change of the total entropy associated with a wave fan $U = V(x/t)$ is given by

(9.4.2) $$\frac{d^+}{dt} \int_{-1}^{1} \eta(U(x, t))dx|_{t=0} = P_V + q(V(-\infty)) - q(V(+\infty)).$$

To see this, note first that

(9.4.3)
$$\frac{d}{dt}\int_{-1}^{1}\eta(U(x,t)) = \frac{d}{dt}\int_{-1/t}^{1/t} t\eta(V(\xi))d\xi$$

$$= \int_{-1/t}^{1/t}\eta(V(\xi))d\xi - \frac{1}{t}\eta(V(t^{-1})) - \frac{1}{t}\eta(V(-t^{-1})) .$$

Now, in the sense of measures,

(9.4.4) $\eta(V(\xi)) = [\xi\eta(V(\xi)) - q(V(\xi))]^{\boldsymbol{\cdot}} + \dot{q}(V(\xi)) - \xi\dot{\eta}(V(\xi)) .$

The generalized chain rule, Theorem 1.7.4, yields

(9.4.5) $\dot{q}(V) - \xi\dot{\eta}(V) = [\widetilde{Dq(V)} - \xi\widetilde{D\eta(V)}]\dot{V} .$

From (9.4.5), (7.4.1), and (9.1.6) follows that the measure $\dot{q}(V) - \xi\dot{\eta}(V)$ is concentrated in the set of points of jump discontinuity of V. Therefore, combining (9.4.3) with (9.4.4) and letting $t \downarrow 0$ we arrive at (9.4.2).

In its connections to Continuum Physics, the entropy rate admissibility criterion is a more stringent form of the Second Law of thermodynamics: Not only should the physical entropy increase but in fact it should be increasing at the maximum rate allowed by the balance laws of mass, momentum and energy. The kinetic theory seems to provide some support to that thesis, at least for waves of small amplitude (references in Section 9.7). However, the efficacy of the entropy rate admissibility criterion may only be established by examining its implications in the context of familiar systems, and by comparing it with other, established, admissibility conditions.

We begin our investigation by testing the entropy rate criterion on the scalar conservation law:

Theorem 9.4.1 *For the scalar conservation law* (7.1.2), *with designated entropy-entropy flux pair* (8.5.3), *a wave fan satisfies the entropy rate admissibility criterion if and only if every shock satisfies the Oleinik E-condition.*

Proof. Let us fix some wave fan $u = \omega(\xi)$, with bounded variation on $(-\infty, \infty)$. As ξ runs from $-\infty$ to $+\infty$, $y = \omega(\xi)$ traces, on the graph of $y = f(u)$, a (finite or infinite) number of "arcs", separated by gaps induced by the shocks: When ξ is a point of jump discontinuity of ω, $f(\omega)$ jumps from $f(\omega(\xi-))$ to $f(\omega(\xi+))$. We produce a continuous curve by filling these gaps with the chord that connects $(\omega(\xi-), f(\omega(\xi-))$ with $(\omega(\xi+), f(\omega(\xi+)))$. This may be effected by the following procedure: We let $v(\xi)$ denote the variation of ω over the interval $(-\infty, \xi)$. Note that v is a left-continuous nondecreasing function. We now construct the curve $y = \gamma_\omega(\tau)$, $\tau \in [0, v(\infty)]$, as follows: If $\tau = v(\xi)$, for some $\xi \in (-\infty, \infty)$, then $\gamma_\omega(\tau) = f(\omega(\xi))$. On the other hand, if $v(\xi-) < \tau < v(\xi+)$, for some $\xi \in (-\infty, \infty)$, then

$$(9.4.6) \qquad \gamma_\omega(\tau) = \frac{v(\xi+) - \tau}{v(\xi+) - v(\xi-)} f(\omega(\xi-)) + \frac{\tau - v(\xi-)}{v(\xi+) - v(\xi-)} f(\omega(\xi+)) .$$

Notice that γ_ω is a (possibly self-intersecting) curve with endpoints $(\omega(\pm\infty), f(\omega(\pm\infty)))$ having the property that, as τ runs from 0 to $v(\infty)$, the u-slope $d^+\gamma_\omega/du = (d^+\gamma_\omega/d\tau)(du/d\tau)^{-1}$ is nondecreasing.

We recall, from Section 8.5, that the entropy production of a shock that joins u_-, on the left, to u_+, on the right, is given by the left-hand side of (8.5.4), which measures the signed area of the domain bordered by the arc of the graph of f with endpoints $(u_-, f(u_-))$, $(u_+, f(u_+))$, and the chord that connects $(u_-, f(u_-))$, $(u_+, f(u_+))$. It follows that the entropy production P_ω of the wave fan ω is measured by the signed area of the domain bordered by the arc of the graph of f with endpoints $(\omega(-\infty), f(\omega(-\infty)))$, $(\omega(+\infty), f(\omega(+\infty)))$ and the graph of the curve γ_ω. Consequently, the difference $P_\omega - P_{\tilde\omega}$ in the entropy production of two wave fans $u = \omega(x/t)$ and $u = \tilde\omega(x/t)$ with the same endstates, $\tilde\omega(\pm\infty) = \omega(\pm\infty)$, is measured by the signed area of the domain bordered by the corresponding curves γ_ω and $\gamma_{\tilde\omega}$. We conclude that a wave fan $u = \omega(x/t)$ with given end-states u_ℓ and u_r, such that $u_\ell < u_r$ (or $u_\ell > u_r$), minimizes the total entropy production if and only if the curve γ_ω is the convex (or concave) envelope of f over the interval $[u_\ell, u_r]$ (or $[u_r, u_\ell]$). As we saw already in Section 9.3, this is the unique wave fan whose shocks satisfy the Oleinik E-condition. The proof is complete. $\square$

It is interesting that just one entropy suffices to rule out all spurious solutions. The situation is similar with the system (7.1.6) of isentropic elasticity:

Theorem 9.4.2 *For the system* (7.1.6), *with designated entropy-entropy flux pair* (7.4.10), *a wave fan satisfies the entropy rate admissibility criterion if and only if every shock satisfies the Wendroff E-condition.*

The proof of the above theorem, which can be found in the references cited in Section 9.7, is based on the observation that the entropy production of a shock is given by the left-hand side of (8.5.5) and thus, similar to the scalar case, may be interpreted as an area.

We now turn to general strictly hyperbolic systems but limit our investigation to shocks with small amplitude:

Theorem 9.4.3 *For any strictly hyperbolic system* (9.1.1) *of conservation laws, with designated entropy-entropy flux pair* (η, q), *where η is (locally) uniformly convex, a wave fan with waves of moderate strength may satisfy the entropy rate admissibility criterion only if every shock satisfies the Liu E-condition.*

Proof. The assertion is established by contradiction: Assuming some shock in a wave fan $U = V(x/t)$ violates the Liu E-condition, one constructs another wave fan $U = \tilde V(x/t)$, with the same end-states, $\tilde V(\pm\infty) = V(\pm\infty)$, but lower entropy production, $P_{\bar{V}} < P_V$. Here it will suffice to illustrate the idea in the special case where V comprises one i-shock joining constant states U_-, on the left, and U_+, on the right. The general proof is found in the literature cited in Section 9.7.

Let $W_i(\cdot)$ denote the i-shock curve through U_- and let $s_i(\cdot)$ be the corresponding shock speed function, with properties listed in Theorem 8.2.1. In particular, $U_- = W_i(0)$, $U_+ = W_i(\tau)$ and the speed of the shock is $s = s_i(\tau)$. For definiteness, assume $\tau > 0$. When the shock violates the Liu E-condition, there are ξ in $(0, \tau)$ with $s_i(\xi) < s$. The case where $s_i(\xi) < s$ for all $\xi \in (0, \tau)$ is simpler; so let us consider the more interesting situation where there is $\xi_0 \in (0, \tau)$ such that $s_i(\xi) > s$ for $\xi \in (0, \xi_0)$, $s_i(\xi_0) = s$, and $\dot{s}_i(\xi_0) < 0$. We identify the state $U_m = W_i(\xi_0)$. Since U_- may be joined to both U_m and U_+ by shocks of speed s, it follows that U_m and U_+ can also be joined by a shock of speed s. Consequently, one may visualize the shock that joins U_- and U_+ as a composite of two shocks, one that joins U_- and U_m and one that joins U_m and U_+, both propagating with the same speed s. The plan of the proof is to perform a perturbation that splits the original shock into two shocks, one with speed slightly less than s, the other with speed slightly more than s, and to show that the resulting wave fan has lower entropy production.

To that end, we construct $n+2$ families of constant states $U_0(\varepsilon) = U_-, U_1(\varepsilon)$, $\cdots, U_{i-1}(\varepsilon), U_*(\varepsilon), U_i(\varepsilon), \cdots, U_n(\varepsilon)$, depending smoothly on the parameter ε that takes values in a small neighborhood $(-a, a)$ of 0, having the following properties: $U_j(0) = U_-$, for $j = 0, \cdots, i - 1$; $U_*(0) = U_m$; $U_j(0) = U_+$, for $j = i, \cdots, n$. For $j = 1, \cdots, i-1, i+1, \cdots, n$, $U_{j-1}(\varepsilon)$ is joined to $U_j(\varepsilon)$ by a (not necessarily admissible) j-shock of speed $\sigma_j(\varepsilon)$; $U_{i-1}(\varepsilon)$ is joined to $U_*(\varepsilon)$ by an i-shock of speed $s_-(\varepsilon)$; and $U_*(\varepsilon)$ is joined to $U_i(\varepsilon)$ by an i-shock of speed $s_+(\varepsilon)$. The corresponding Rankine-Hugoniot conditions read

(9.4.7)
$$F(U_j(\varepsilon)) - F(U_{j-1}(\varepsilon)) = \sigma_j(\varepsilon)[U_j(\varepsilon) - U_{j-1}(\varepsilon)], \quad j = 1, \cdots, i-1, i+1, \cdots, n,$$

(9.4.8)
$$F(U_*(\varepsilon)) - F(U_{i-1}(\varepsilon)) = s_-(\varepsilon)[U_*(\varepsilon) - U_{i-1}(\varepsilon)],$$

(9.4.9)
$$F(U_i(\varepsilon)) - F(U_*(\varepsilon)) = s_+(\varepsilon)[U_i(\varepsilon) - U_*(\varepsilon)].$$

In particular, $\sigma_j(0) = \lambda_j(U_-)$, for $j = 1, \cdots, i - 1$; $s_-(0) = s_+(0) = s$; and $\sigma_j(0) = \lambda_j(U_+)$, for $j = i, \cdots, n$. The construction may be effected by the method employed in Section 9.3 for constructing solutions to the Riemann problem, with j-shock curves playing here the role of the wave fan curves Φ_j used there. The implicit function theorem here yields a one-parameter family of states (rather than a single state, as in Section 9.3), due to the additional degree of freedom, namely the intermediate state U_*. We choose the parametrization in such a way that $s'_-(0) = -1$. Here and below the prime denotes differentiation with respect to ε.

Our first task is to show that, for ε positive small, the constant states $U_0(\varepsilon), \cdots, U_{i-1}(\varepsilon), U_*(\varepsilon), U_i(\varepsilon), \cdots, U_n(\varepsilon)$, together with the connecting shocks, may be assembled into a wave fan V_ε. For that purpose it suffices to prove that

$s_-(\varepsilon) < s_+(\varepsilon)$. We differentiate (9.4.7), (9.4.8), (9.4.9) with respect to ε and set $\varepsilon = 0$ to get

$$(9.4.10) \quad [DF(U_-) - \lambda_j(U_-)I][U_j'(0) - U_{j-1}'(0)] = 0 , \quad j = 1, \cdots, i-1 ,$$

$$(9.4.11) \quad [DF(U_+) - \lambda_j(U_+)I][U_j'(0) - U_{j-1}'(0)] = 0 , \quad j = i+1, \cdots, n ,$$

$$(9.4.12) \quad [DF(U_m) - sI]U_*'(0) - [DF(U_-) - sI]U_{i-1}'(0) = s_-'(0)[U_m - U_-] ,$$

$$(9.4.13) \quad [DF(U_+) - sI]U_i'(0) - [DF(U_m) - sI]U_*'(0) = s_+'(0)[U_+ - U_m] .$$

Upon combining (9.4.12) with (9.4.13) we deduce

$$(9.4.14) \quad \begin{aligned} s_-'(0)[U_m - U_-] &+ s_+'(0)[U_+ - U_m] \\ &= [DF(U_+) - sI]U_i'(0) - [DF(U_-) - sI]U_{i-1}'(0) . \end{aligned}$$

Both vectors on the left-hand side of (9.4.14) are almost collinear to $R_i(U_\pm)$. On the other hand, by virtue of (9.4.10), $U_{i-1}'(0)$ lies in the span of $\{R_1(U_-), \cdots, R_{i-1}(U_-)\}$, while by account of (9.4.11), $U_i'(0)$ lies in the span of $\{R_{i+1}(U_+), \cdots, R_n(U_+)\}$. Therefore, the right-hand side of (9.4.14) is almost orthogonal to $L_i(U_\pm)$. Let us set $\ell = |U_m - U_-|$. Recalling that $s_-'(0) = -1$, we deduce that in (9.4.14) both terms on the right-hand side are $o(\ell)$ and thus the two terms on the left-hand side must cancel each other out to leading order. In particular, $s_+'(0) > 0$ so that, for ε positive small, $s_-(\varepsilon) < s < s_+(\varepsilon)$, which establishes the desired separation of shocks.

The total entropy production of the wave fan V_ε is

$$(9.4.15) \quad \begin{aligned} P(\varepsilon) = \sum_{j \neq i} &\{q(U_j(\varepsilon)) - q(U_{j-1}(\varepsilon)) - \sigma_j(\varepsilon)[\eta(U_j(\varepsilon)) - \eta(U_{j-1}(\varepsilon))]\} \\ &+ q(U_*(\varepsilon)) - q(U_{i-1}(\varepsilon)) - s_-(\varepsilon)[\eta(U_*(\varepsilon)) - \eta(U_{i-1}(\varepsilon))] \\ &+ q(U_i(\varepsilon)) - q(U_*(\varepsilon)) - s_+(\varepsilon)[\eta(U_i(\varepsilon)) - \eta(U_*(\varepsilon))] . \end{aligned}$$

To establish that for ε positive small the wave fan V_ε dissipates entropy at a higher rate than V, it suffices to show that $P'(0) < 0$. The derivative, with respect to ε, of the summation term on the right-hand side of (9.4.15), evaluated at $\varepsilon = 0$, reduces to

$$(9.4.16) \quad \begin{aligned} \sum_{j=1}^{i-1} &[Dq(U_-) - \lambda_j(U_-)D\eta(U_-)][U_j'(0) - U_{j-1}'(0)] \\ &+ \sum_{j=i+1}^{n} [Dq(U_+) - \lambda_j(U_+)D\eta(U_+)][U_j'(0) - U_{j-1}'(0)] \end{aligned}$$

which vanishes by virtue of (7.4.1), (9.4.10) and (9.4.11). We evaluate, at $\varepsilon = 0$, the derivative of the remaining terms on the right-hand side of (9.4.15). After a straightforward calculation, making use of (7.4.1), (9.4.12) and (9.4.13), we conclude

(9.4.17)
$$P'(0) = -s'_-(0)[\eta(U_m) - \eta(U_-) - D\eta(U_m)(U_m - U_-)]$$
$$- s'_+(0)[\eta(U_+) - \eta(U_m) - D\eta(U_m)(U_+ - U_m)]$$
$$+ [D\eta(U_m) - D\eta(U_-)][DF(U_-) - sI]U'_{i-1}(0)$$
$$+ [D\eta(U_+) - D\eta(U_m)][DF(U_+) - sI]U'_i(0) .$$

We examine the four terms on the right-hand side of (9.4.17), in the light of the scaling analysis of (9.4.14), discussed earlier in the proof. Since η is convex, $s'_-(0) = -1$, $s'_+(0) > 0$ and $|U_m - U_-| = \ell$, it follows that the first term is majorized by $-\beta\ell^2$ and the second term is majorized by $-\beta\ell|U_+ - U_m|$, for some $\beta > 0$. On the other hand, the third term is $o(\ell)\ell$ and the fourth term is $o(\ell)|U_+ - U_m|$. Consequently, for ℓ sufficiently small, $P'(0) < 0$. This completes the proof.

Beyond the range of shocks of moderate strength, the entropy rate admissibility criterion is no longer generally equivalent to the Liu E-condition. The issue has been discussed in detail (references in Section 9.7) in the context of the system (7.1.3) of adiabatic thermoelasticity for a polytropic gas with internal energy $\varepsilon = e^s u^{1-\gamma}$, which induces, through (7.1.4), pressure $p = -\sigma = (\gamma - 1)e^s u^{-\gamma}$. The designated entropy-entropy flux pair is given by (7.4.9), namely $(-s, 0)$. For this system, with 1- and 3-characteristic families that are genuinely nonlinear and 2-characteristic family that is linearly degenerate, the Lax E-condtion and the Liu E-condition are equivalent.

It has been shown that when $\gamma \geq 5/3$ a wave fan, of arbitrary strength, satisfies the entropy rate admissibility criterion if and only if its shocks satisfy the Lax E-condition. The reader should note that $5/3$ is the value for the adiabatic exponent γ predicted by the kinetic theory in the case of a monatomic ideal gas.

When $\gamma < 5/3$ (polyatomic gases), the situation is different. Consider a wave fan comprising three constant states (u_ℓ, v_ℓ, s_ℓ), (u_m, v_m, s_m) and (u_r, v_r, s_r), where the first two are joined by a stationnary 2-contact discontinuity, while the second and the third are joined by a 3-rarefaction wave. In particular, $v_m = v_\ell$, $p(u_m, s_m) = p(u_\ell, s_\ell)$, $s_m = s_r$, and $z(u_m, v_m, s_m) = z(u_r, v_r, s_r)$, where $z(u, v, s)$ denotes the second 3-Riemann invariant listed in (7.3.4). The total entropy production of this wave fan is of course zero. For u_r/u_ℓ in a certain range, there is a second wave fan with the same end-states, which comprises four constant states (u_ℓ, v_ℓ, s_ℓ), (u_1, v_1, s_1), (u_2, v_2, s_2) and (u_r, v_r, s_r), where the first two are joined by a 1-shock that satisfies the Lax E-condition, the second is joined to the third by a 2-contact discontinuity, while the last two are joined by a 3-shock that violates the Lax E-condition. It turns out that when u_m/u_ℓ is not too large, i.e., the contact discontinuity is not too strong, the total entropy production of the second wave fan is positive and hence the first wave fan has lower entropy rate. By contrast, when u_m/u_ℓ is sufficiently large, the total entropy production of the second wave fan is negative and so the first wave fan no longer satisfies the entropy rate criterion.

Similar issues arise for systems that are not strictly hyperbolic. Let us consider our model system (7.2.11). Recall the two wave fans with the same end-states

$(1, 0)$ and $(a, 0)$, $a \in (-\frac{1}{2}, 0)$, described at the end of Section 9.3: The first one comprises the states $(1, 0)$ and $(a, 0)$, joined by an overcompressive shock of speed $1 + a + a^2$. The second comprises three states, $(1, 0), (-1, 0)$ and $(a, 0)$, where the first two are joined by a 1-contact discontinuity of speed 1, while the second is joined to the third by a 2-shock of speed $1 - a + a^2$. If we designate the entropy-entropy flux pair

$$(9.4.18) \qquad \eta = \frac{1}{2}(u^2 + v^2) , \quad q = \frac{3}{4}(u^2 + v^2)^2 ,$$

the entropy production of the overcompressive shock is $\frac{1}{4}(a^2 - 1)(1 - a)^2$ while the entropy production of the second wave fan is $\frac{1}{4}(a^2 - 1)(1 + a)^2$. Thus the entropy rate criterion favors the overcompressive shock, even though, as we saw in Section 8.6, this is incompatible with the stable shock profile condition. The reader should bear in mind, however, that these conclusions are tied to our selections for artificial viscosity and entropy. Whether (8.6.4) is the proper dissipative form, and (9.4.18) is the natural entropy-entropy flux pair for (7.2.11) may only be decided when this system is considered in the context of some physical model.

9.5 Viscous Wave Fans

The viscous shock admissibility criterion, introduced in Section 8.6, characterizes admissible shocks for the hyperbolic system of conservation laws (9.1.1) as $\mu \downarrow 0$ limits of travelling wave solutions of the associated dissipative system (8.6.1). The aim here is to extend this principle from single shocks to general wave fans. The difficulty is that, in contrast to (9.1.1), the system (8.6.1) is not invariant under uniform stretching of the space-time coordinates and thus it does not possess travelling wave fans as solutions. To remedy this, it has been proposed that in the place of (8.6.1) one should employ a system with time-varying viscosity,

$$(9.5.1) \qquad \partial_t U(x, t) + \partial_x F(U(x, t)) = \mu t \partial_x^2 U(x, t) ,$$

which is invariant under the transformation $(x, t) \mapsto (\alpha x, \alpha t)$. Clearly, $U = V_\mu(x/t)$ is a self-similar solution of (9.5.1) if and only if $V_\mu(\xi)$ satisfies the ordinary differential equation

$$(9.5.2) \qquad \mu \ddot{V}_\mu(\xi) = \dot{F}(V_\mu(\xi)) - \xi \dot{V}_\mu(\xi) .$$

A self-similar solution $U = V(x/t)$ of (9.1.1) is said to satisfy the *viscous wave fan admissibility criterion* if V is the almost everywhere limit, as $\mu \downarrow 0$, of a uniformly bounded family of solutions V_μ of (9.5.2).

In addition to serving as a test of admissibility, the wave fan criterion suggests an alternative approach for constructing solutions to the Riemann problem (9.1.1), (9.1.12). Towards that end, one has to show that for any fixed $\mu > 0$ there exists some solution $V_\mu(\xi)$ of (9.5.2) on $(-\infty, \infty)$, with boundary conditions

(9.5.3) $V_\mu(-\infty) = U_\ell$, $V_\mu(+\infty) = U_r$,

and then prove that the family $\{V_\mu(\xi) : 0 < \mu < 1\}$ has uniformly bounded variation on $(-\infty, \infty)$. In that case, by Helly's theorem (cf. Section 1.7), a convergent
sequence $\{V_{\mu_m}\}$ may be extracted, with $\mu_m \downarrow 0$ as $m \to \infty$, whose limit V induces
the solution $U = V(x/t)$ to the Riemann problem.

The above program has been implemented successfully under a variety of
conditions. One may solve the Riemann problem under quite general data U_ℓ
and U_r albeit for special systems, most notably for pairs of conservation laws.
Alternatively, one may treat general systems but only in the context of weak
waves, requiring that $|U_r - U_\ell|$ be sufficiently small. It is this last situation that
will be discussed here. The analysis is lengthy and technical so only the main
ideas will be outlined. For the details, the reader may consult the references cited
in Section 9.7.

The crucial step is to establish a priori bounds on the total variation of $V_\mu(\xi)$
over $(-\infty, \infty)$, independent of μ. To prepare the ground for systems, let us begin
with the scalar conservation law (7.1.2). Setting $\lambda(u) = f'(u)$ and $\dot{V}_\mu(\xi) = a(\xi)$,
we write (9.5.2) in the form

(9.5.4) $\mu\dot{a} + [\xi - \lambda(V_\mu(\xi))]a = 0$.

The solution of (9.5.4) is $a(\xi) = \tau\phi(\xi)$ with

(9.5.5) $\phi(\xi) = \dfrac{\exp[-\frac{1}{\mu}g(\xi)]}{\int_{-\infty}^\infty \exp[-\frac{1}{\mu}g(\zeta)]d\zeta}$,

(9.5.6) $g(\xi) = \displaystyle\int_s^\xi [\zeta - \lambda(V_\mu(\zeta))]d\zeta$.

The lower limit of integration s is selected so that $g(\xi) \geq 0$ for all ξ in
$(-\infty, \infty)$. The amplitude τ is determined with the help of the boundary conditions $V_\mu(-\infty) = u_\ell$, $V_\mu(\infty) = u_r$, $\tau = u_r - u_\ell$. From (9.5.5) follows that the
L^1 norm of $a(\xi)$ is bounded, uniformly in μ, and so the family $\{V_\mu : 0 < \mu < 1\}$
has uniformly bounded variation on $(-\infty, \infty)$.

Turning now to general strictly hyperbolic systems (9.1.1), we realize $V_\mu(\xi)$
as the composition of wave fans associated with distinct characteristic families,
by writing

(9.5.7) $\dot{V}_\mu(\xi) = \displaystyle\sum_{j=1}^n a_j(\xi)R_j(V_\mu(\xi))$.

We substitute $\dot{V}_\mu$ from (9.5.7) into (9.5.2). Upon multiplying the resulting equation,
from the left, by $L_i(V_\mu(\xi))$, we deduce

(9.5.8) $\mu\dot{a}_i + [\xi - \lambda_i(V_\mu(\xi))]a_i = \mu \displaystyle\sum_{j,k=1}^n \beta_{ijk}(V_\mu(\xi))a_j a_k$,

where

(9.5.9) $$\beta_{ijk}(U) = -L_i(U)DR_j(U)R_k(U) \ .$$

In (9.5.8), the left-hand side coincides with the left-hand side of (9.5.4), for the scalar conservation law, while the right-hand side accounts for the interactions of distinct characteristic families. The reader should notice the analogy between (9.5.8) and (7.7.6). It should also be noted that when our system is endowed with a coordinate system $(w_1, \cdots, w_n)$ of Riemann invariants, $a_i(\xi) = \dot{w}_i(V_\mu(\xi))$. In that case, as shown in Section 7.3, for $j \neq k$, $DR_j R_k$ lies in the span of $\{R_j, R_k\}$ and so (9.5.9) implies $\beta_{ijk} = 0$ when $i \neq j \neq k \neq i$. For special systems, like (7.3.18), with coinciding shock and rarefaction wave curves, $DR_j R_j$ is collinear to R_j and so $\beta_{ijk} = 0$ even when $i \neq j = k$ so that the equations in (9.5.8) decouple. In general, the thrust of the analysis is to demonstrate that in the context of solutions with small oscillation, i.e., a_i small, the effect of interactions, of quadratic order, will be even smaller.

The solution of (9.5.8) may be partitioned into

(9.5.10) $$a_i(\xi) = \tau_i \phi_i(\xi) + \theta_i(\xi) \ ,$$

where

(9.5.11) $$\phi_i(\xi) = \frac{\exp[-\frac{1}{\mu} g_i(\xi)]}{\int_{-\infty}^{\infty} \exp[-\frac{1}{\mu} g_i(\zeta)] d\zeta} \ ,$$

(9.5.12) $$g_i(\xi) = \int_{s_i}^{\xi} [\zeta - \lambda_i(V_\mu(\zeta))] d\zeta \ ,$$

and $\theta_i(\xi)$ satisfy the equations

(9.5.13) $$\mu \dot{\theta}_i + [\xi - \lambda_i(V_\mu(\xi))] \theta_i = \mu \sum_{j,k=1}^{n} \beta_{ijk}(V_\mu(\xi))[\tau_j \phi_j(\xi) + \theta_j][\tau_k \phi_k(\xi) + \theta_k] \ .$$

The differential equations (9.5.13) may be transformed into an equivalent system of integral equations by means of the variation of parameters formula:

(9.5.14)
$$\theta_i(\xi) = \phi_i(\xi) \int_{c_i}^{\xi} \phi_i^{-1}(\zeta) \beta_{ijk}(V_\mu(\zeta))[\tau_j \phi_j(\zeta) + \theta_j(\zeta)][\tau_k \phi_k(\zeta) + \theta_k(\zeta)] d\zeta \ .$$

Careful estimation shows that

(9.5.15) $$|\theta_i(\xi)| \leq c(\tau_1^2 + \cdots + \tau_n^2) \sum_{j=1}^{n} \phi_j(\xi) \ ,$$

which verifies that, in (9.5.10), θ_i is subordinate to $\tau_i \phi_i$, i.e., the characteristic families decouple to leading order.

It can be shown, by means of a contraction argument, that for any fixed $(\tau_1, \cdots, \tau_n)$ in a small neighborhood of the origin, there exists some solution

$V_\mu(\xi)$ of (9.5.2) on $(-\infty, \infty)$, which satisfies (9.5.7), (9.5.10) and (9.5.15). To solve the boundary-value problem (9.5.2), (9.5.3), the $(\tau_1, \cdots, \tau_n)$ have to be selected so that

$$(9.5.16) \qquad \sum_{j=1}^{n} \int_{-\infty}^{\infty} [\tau_j \phi_j(\xi) + \theta_j(\xi)] R_j(V_\mu(\xi)) d\xi = U_r - U_\ell .$$

It has been proved that (9.5.16) admits a unique solution $(\tau_1, \cdots, \tau_n)$, at least when $|U_r - U_\ell|$ is sufficiently small. The result is summarized in the following

Theorem 9.5.1 *Assume the system* (9.1.1) *is strictly hyperbolic on $\mathcal{O}$ and fix any state $U_- \in \mathcal{O}$. There is $\delta > 0$ such that for any $U_r \in \mathcal{O}$ with $|U_r - U_\ell| < \delta$ and every $\mu > 0$ the boundary-value problem* (9.5.2), (9.5.3) *possesses a solution $V_\mu(\xi)$, which admits the representation* (9.5.7), (9.5.10) *with $(\tau_1, \cdots, \tau_n)$ close to the origin and θ_i obeying* (9.5.15). *Furthermore, the family $\{V_\mu(\xi) : 0 < \mu < 1\}$ of solutions has uniformly bounded (and small) total variation on $(-\infty, \infty)$. In particular, one may extract a sequence $\{V_{\mu_m}(\xi)\}$, with $\mu_m \downarrow 0$ as $m \to \infty$, which converges, boundedly almost everywhere, to a function $V(\xi)$ such that the wave fan $U = V(x/t)$ solves the Riemann problem* (9.1.1), (9.1.12).

It should be noted that the above theorem establishes the existence of solutions to the Riemann problem, with waves of small amplitude, for general strictly hyperbolic systems of conservation laws, without any restrictions, like piecewise genuine nonlinearity required in Theorem 9.3.3.

Careful analysis of the process that generates $V(\xi)$ as the limit of the sequence $\{V_{\mu_m}(\xi)\}$ reveals that $V(\xi)$ has the structure described in Theorem 9.3.1. Furthermore, for any point ξ of jump discontinuity of V, $V(\xi-)$, on the left, is connected to $V(\xi+)$, on the right, by a viscous shock profile, and so the viscous shock admissibility criterion is satisfied (with $B = I$), as discussed in Section 8.6.

Following up on the discussion in Section 8.6, one may argue that wave fan solutions of the Riemann problem, with end-states U_ℓ and U_r, should not be termed admissible unless they are captured through the $t \to \infty$ asymptotics of solutions of parabolic systems (8.6.1), under initial data $U_0(x)$ which decay sufficiently fast to U_ℓ and U_r, as $x \to \mp\infty$. In fact, the results reported in Section 8.6 on the asymptotic stability of viscous shock profiles address a special case of the above issue. The complementary special case, the asymptotic stability of rarefaction waves, has also been studied extensively (references in Section 9.7). The task of combining the above two ingredients so as to synthesize the full solution of the Riemann problem, has not yet been accomplished in a definitive manner.

9.6 Interaction of Wave Fans

Up to this point, we have exploited the invariance of systems of conservation laws under uniform rescaling of the space-time coordinates in order to perform stretchings that reveal the local structure of solutions. However, one may also operate at the opposite end of the scale by performing contractions of the space-time coordinates that will provide a view of solutions from a large distance from the origin. It is plausible that initial data $U_0(x)$ which converge sufficiently fast to states U_ℓ and U_r, as $x \to \mp\infty$, generate solutions that look from afar like centered wave fans joining the state U_ℓ, on the left, with the state U_r, on the right. Actually, as we shall see in later chapters, this turns out to be true. Indeed, it seems that the quintessential property of hyperbolic systems of conservation laws in one-space dimension is that the Riemann problem describes the asymptotics of solutions at both ends of the time scale: instantaneous and long-term.

Our present purpose is to discuss a related question, which, as we shall see in Chapter XIII, is of central importance in the construction of solutions by the random choice method. We consider three wave fans: the first, joining a state U_ℓ, on the left, with a state U_m, on the right; the second, joining the state U_m, on the left, with a state U_r, on the right; and the third, joining the state U_ℓ, on the left, with the state U_r, on the right. These may be represented, with the help of the function Ω (cf. (9.3.4)), as explained in Section 9.3, by their left states U_ℓ, U_m and U_ℓ together with the respective n-tuples $\alpha = (\alpha_1, \cdots, \alpha_n)$, $\beta = (\beta_1, \cdots, \beta_n)$ and $\varepsilon = (\varepsilon_1, \cdots, \varepsilon_n)$ of wave amplitudes. Based on the arguments presented above, it is natural to regard the wave fan ε as the result of the interaction of the wave fan α, on the left, with the wave fan β, on the right. From $U_m = \Omega(\alpha; U_\ell)$, $U_r = \Omega(\beta; U_m)$ and $U_r = \Omega(\varepsilon; U_\ell)$ we deduce

$$(9.6.1) \qquad \Omega(\varepsilon; U_\ell) = \Omega(\beta; \Omega(\alpha; U_\ell)) \ ,$$

which determines implicitly the relation

$$(9.6.2) \qquad \varepsilon = E(\alpha; \beta; U_\ell) \ .$$

Our task is to study the properties of E in the vicinity of $(0; 0; U_\ell)$. By virtue of (9.3.5),

$$(9.6.3) \qquad E(\alpha; 0; U_\ell) = \alpha \ , \qquad E(0; \beta; U_\ell) = \beta \ ,$$

whence

$$(9.6.4) \qquad \frac{\partial E_k}{\partial \alpha_i}(0; 0; U_\ell) = \delta_{ik} \ , \qquad \frac{\partial E_k}{\partial \beta_j}(0; 0; U_\ell) = \delta_{jk} \ ,$$

namely, the Kronecker delta.

Starting from the identity

$$(9.6.5)$$
$$E(\alpha; \beta; U_\ell) - E(\alpha; 0; U_\ell) - E(0; \beta; U_\ell) + E(0; 0; U_\ell)$$
$$= \sum_{i,j=1}^{n} \{ E(\alpha_1, \cdots, \alpha_i, 0, \cdots, 0; 0, \cdots, 0, \beta_{j+1}, \cdots, \beta_n; U_\ell)$$
$$- E(\alpha_1, \cdots, \alpha_{i-1}, 0, \cdots, 0; 0, \cdots, 0, \beta_{j+1}, \cdots, \beta_n; U_\ell)$$
$$- E(\alpha_1, \cdots, \alpha_i, 0, \cdots, 0; 0, \cdots, 0, \beta_j, \cdots, \beta_n; U_\ell)$$
$$+ E(\alpha_1, \cdots, \alpha_{i-1}, 0, \cdots, 0; 0, \cdots, 0, \beta_j, \cdots, \beta_n; U_\ell) \} \ ,$$

one immediately deduces

$$
(9.6.6) \quad \begin{aligned}
E(\alpha; \beta; U_\ell) &= \alpha + \beta \\
&+ \sum_{i,j=1}^{n} \alpha_i \beta_j \int_0^1 \int_0^1 \frac{\partial^2 E}{\partial \alpha_i \partial \beta_j} (\alpha_1, \cdots, \alpha_{i-1}, \rho \alpha_i, 0, \cdots, 0; \\
&\qquad\qquad 0, \cdots, 0, \sigma \beta_j, \beta_{j+1}, \cdots, \beta_n; U_\ell) d\rho d\sigma .
\end{aligned}
$$

Let us focus attention to the case where each characteristic family is either genuinely nonlinear (7.6.13) or linearly degenerate (7.5.2). We say the waves α_i and β_j are *approaching* when either (a) $i > j$ or (b) $i = j$, the i-characteristic family is genuinely nonlinear, and at least one of α_i, β_i is negative, that is corresponds to a shock. The *amount of wave interaction* of α and β will be measured by the quantity

$$
(9.6.7) \quad D(\alpha, \beta) = \sum_{\text{app}} |\alpha_i| |\beta_j| ,
$$

where $\sum_{\text{app}}$ denotes summation over all pairs (i, j) with α_i, β_j approaching. The crucial observation is that when the wave fans α and β do not contain any approaching waves, i.e., $D(\alpha, \beta) = 0$, then the wave fan ε is synthesized by "glueing together" the wave fan α, on the left, and the wave fan β, on the right; that is, $\varepsilon = \alpha + \beta$. In particular, whenever α_i and β_j are not approaching, either because $i < j$ or because $i = j$ and both α_i and β_i are positive, i.e., rarefaction waves, then

$$
(9.6.8) \quad \begin{aligned}
&E(\alpha_1, \cdots, \alpha_i, 0, \cdots, 0; 0, \cdots, 0, \beta_j, \cdots, \beta_n; U_\ell) \\
&= (\alpha_1, \cdots, \alpha_i, 0, \cdots, 0) + (0, \cdots, 0, \beta_j, \cdots, \beta_n) ,
\end{aligned}
$$

whence it follows that the corresponding (i, j)-term in the summation on the right-hand side of (9.6.6) vanishes. Thus (9.6.6) reduces to

$$
(9.6.9) \quad \varepsilon = \alpha + \beta + \sum_{\text{app}} \alpha_i \beta_j \frac{\partial^2 E}{\partial \alpha_i \partial \beta_j} (0; 0; U_\ell) + D(\alpha, \beta) O(|\alpha| + |\beta|) .
$$

The salient feature of (9.6.9), which will play a key role in Chapter XIII, is that the effect of wave interaction is induced solely by pairs of approaching waves and vanishes in the absence of such pairs. In order to determine the leading interaction term, of quadratic order, we first differentiate (9.6.1) with respect to β_j and set $\beta = 0$. Upon using (9.3.6), this yields

$$
(9.6.10) \quad \sum_{k=1}^{n} \frac{\partial E_k}{\partial \beta_j} (\alpha; 0; U_\ell) \frac{\partial \Omega}{\partial \varepsilon_k} (E(\alpha; 0; U_\ell); U_\ell) = R_j(\Omega(\alpha; U_\ell)) .
$$

Next we differentiate (9.6.10) with respect to α_i and set $\alpha = 0$. Recall that we are only interested in the case where α_i and β_j are approaching, so in particular $i \geq j$. Therefore, upon using (9.6.3), (9.6.4), (9.3.6), (9.3.7), (9.3.8) and (7.2.15), we conclude

$$(9.6.11) \qquad \sum_{k=1}^{n} \frac{\partial^2 E_k}{\partial \alpha_i \partial \beta_j}(0; 0; U_\ell) R_k(U_\ell) = -[R_i(U_\ell), R_j(U_\ell)] ,$$

whence

$$(9.6.12) \qquad \frac{\partial^2 E_k}{\partial \alpha_i \partial \beta_j}(0; 0; U_\ell) = -L_k(U_\ell)[R_i(U_\ell), R_j(U_\ell)] .$$

In particular, when the system is endowed with a coordinate system of Riemann invariants, under the normalization (7.3.8) the Lie brackets $[R_i, R_j]$ vanish (cf. (7.3.10)) and hence the quadratic term in (9.6.9) drops out.

Upon combining (9.6.9) with (9.6.12) we arrive at

Theorem 9.6.1 *For systems with characteristic families that are either genuinely nonlinear or linearly degenerate, let* $\varepsilon = (\varepsilon_1, \cdots, \varepsilon_n)$ *be the wave fan generated by the interaction of the wave fan* $\alpha = (\alpha_1, \cdots, \alpha_n)$, *on the left, with the wave fan* $\beta = (\beta_1, \cdots, \beta_n)$, *on the right. Then*

$$(9.6.13) \qquad \varepsilon = \alpha + \beta - \sum_{i>j} \alpha_i \beta_j L[R_i, R_j] + D(\alpha, \beta) O(|\alpha| + |\beta|) ,$$

where L *denotes the* $n \times n$ *matrix with* k-*row vector the left eigenvector* L_k, *and* $D(\alpha, \beta)$ *is the amount of wave interaction of* α *and* β. *When the system is endowed with a coordinate system of Riemann invariants, the quadratic term vanishes.*

9.7 Notes

The Riemann problem was originally formulated, and solved, in the context of the system (7.1.7) of isentropic gas dynamics, in the pathbreaking paper of Riemann [1], already cited in earlier chapters. Early research on the Riemann problem for the equations of isentropic or adiabatic gas dynamics, surveyed in Courant and Friedrichs [1], was stimulated by the need to analyze wave interactions and shock tube experiments. The distillation of this work led to the solution, by Lax [2], of the Riemann problem, with weak waves, for general genuinely nonlinear strictly hyperbolic systems of conservation laws (Theorem 9.3.2).

The Riemann problem with large data has been studied extensively, mainly in the context of systems of two conservation laws and the system of adiabatic thermoelasticity (gas dynamics), in Lagrangian or Eulerian coordinates. A new type of nonuniqueness arising in that system is reported by Smith [1]. The reader may find detailed expositions and references in the book of Smoller [1], and the monograph by Chang and Hsiao [3]. This last book contains, in particular, references to early work by Chinese authors that did not circulate in the international scientific community until much later, e.g. Chang and Hsiao [1,2] and Hsiao and Zhang [1].

Riemann problems for special systems that are not genuinely nonlinear were considered by several authors. In particular, the elegant geometric description of the solution for the scalar conservation law was presented by Gelfand [1] and the construction for the system (7.1.6) of isentropic elasticity, presented here, is due to Leibovich [1]. Research in that direction culminated in the paper of Liu [1], which solves the Riemann problem, with waves of moderate strength, for general piecewise genuinely nonlinar strictly hyperbolic systems of conservation laws (Theorem 9.3.3).

For a novel, variational approach to the Riemann problem, see Heibig and Serre [1].

The following aspects of the theory of the Riemann problem are presently under investigation (for an expository survey, see Glimm [2]):

(a) Systems that are not strictly hyperbolic, with applications to elasticity and multi-phase flows. The reader has already got a taste of the difficulties stemming from nonuniqueness. For the classification of such systems and the construction of admissible solutions to the Riemann problem, that may contain compressive as well as overcompressive or undercompressive shocks, see Azevedo, Marchesin, Plohr and Zumbrun [1], Freistühler [2], Isaacson, Marchesin and Plohr [1], Isaacson, Marchesin, Plohr and Temple [1], Isaacson and Temple [1], Schaeffer and Shearer [1], Schecter, Marchesin and Plohr [1,2], M. Shearer [1,2,3], Shearer and Schaeffer [1], Shearer, Schaeffer, Marchesin and Paes-Leme [1], Tang and Ting [1] and Zhu and Ting [1]. Existence is also an issue, requiring extension of the notion of solution to wave fans comprising singular shocks, oscillations or Dirac masses (the so called delta-shocks). For a sample of recent work see Čanić and Peters [1], Peters and Čanić [1], Ercole [1], Keyfitz and Kranzer [2,3], Tan [1] and Tan, Zhang and Zheng [1].

(b) Systems of mixed type, elliptic-hyperbolic, employed to model phase transitions. A prototypical example is the system (7.1.6) with nonmonotone $\sigma(u)$; in particular the classical van der Waals fluid. See Čanić [1], Keyfitz [1], Hsiao and DeMottoni [1], Holden [1] and Frid and Liu [1].

(c) Systems of conservation laws in two-space dimensions. The self-similar solutions are now functions of two variables $(x_1/t, x_2/t)$. Even for the scalar conservation law, the resulting wave patterns are very intricate. See Guckenheimer [2], Wagner [1], Lindquist [1], Zhang and Zheng [1], Tan and Zhang [1], Chen, Li and Tan [1], Čanić and Keyfitz [1,2] and Zhang and Zhang [1]. Surveys of known solutions are given in the monographs by Chang and Hsiao [3], Li, Zhang and Yang [1] and Y. Zheng [1].

The entropy rate admissibility criterion was proposed by Dafermos [3]. For motivation from the kinetic theory, see Ferziger and Kaper [1], §5.5, and Kohler [1]. Theorems 9.4.1 and 9.4.2 are taken from Dafermos [3], while Theorem 9.4.3 is found in Dafermos [15]. In the context of the system of adiabatic gas dynamics, the entropy rate criterion is discussed by Hsiao [1]. The efficacy of the entropy rate criterion has also been tested on systems that change type, modelling phase transitions (Hattori [1,2,3,4], Pence [1]). See also Sever [1].

The study of self-similar solutions of hyperbolic systems of conservation laws as limits of self-similar solutions of dissipative systems with time-dependent artificial viscosity was initiated, independently, by Kalasnikov [1], Tupciev [1,2], and

Dafermos [4]. This approach has been employed to solve the Riemann problem for systems of two conservation laws that may be strictly hyperbolic (Dafermos [4,5], Dafermos and DiPerna [1], Kim [1], Slemrod and Tzavaras [1,2], Tzavaras [1,2]), nonstrictly hyperbolic (Ercole [1], Keyfitz and Kranzer [2,3], Tan [1] and Tan, Zhang and Zheng [1]), or of mixed type (Slemrod [3], Fan [1,2]). See also Slemrod [4], for solutions with spherical symmetry to the system of isentropic gas dynamics. The treatment of general strictly hyperbolic systems of conservation laws outlined in Section 9.5 follows Tzavaras [2]. It is in that paper that the reader may find additional information, including the details of the proofs omitted here.

The construction of a single rarefaction wave via the standard vanishing viscosity approach, with time-independent viscosity, is effected in Lin and Yang [1].

For self-similar, spherically symmetric solutions representing cavitation in elastodynamics and gas dynamics, see Pericak-Spector and Spector [1,2] and Yan [1].

The asymptotic stability of viscous rarefaction waves is discussed in Liu, Matsumura and Nishihara [1], Liu and Xin [1], Szepessy and Zumbrun [1], and Xin [2]. The asymptotic stability of viscous wave fans, containing both shocks and rarefaction waves, is under investigation by Liu and Yu [2].

The study of interactions of wave fans and the original proof of Theorem 9.6.1, for genuinely nonlinear systems, is due to Glimm [1]. The derivation presented here is taken from Yong [1]. For a description of actual wave interactions, see Greenberg [1,2].

Chapter X. Generalized Characteristics

As already noted in Section 7.8, the function space of choice for weak solutions of hyperbolic systems of conservation laws in one-space dimension is BV, since it is within its confines that one may discern shocks and study their propagation and interactions. The notion of characteristic, introduced in Section 7.2 for classical solutions, will here be extended to the framework of BV weak solutions. It will be established that generalized characteristics propagate with either classical characteristic speed or with shock speed. In particular, it will be shown that the extremal backward characteristics, emanating from any point in the domain of an admissible solution, always propagate with classical characteristic speed. The implications of these properties to the theory of weak solutions will be demonstrated in following chapters.

10.1 BV Solutions

We consider the strictly hyperbolic system

$$(10.1.1) \qquad \partial_t U + \partial_x F(U) = 0$$

of conservation laws. Throughout this chapter, U will denote a bounded measurable function on $(-\infty, \infty) \times (0, \infty)$, of class BV_{loc}, which is a weak solution of (10.1.1). Following the general theory of BV functions in Section 1.7, we infer that $(-\infty, \infty) \times (0, \infty) = \mathcal{C} \bigcup \mathcal{J} \bigcup \mathcal{I}$ where $\mathcal{C}$ is the set of points of approximate continuity of U, $\mathcal{J}$ denotes the set of points of approximate jump discontinuity (shock set) of U, and $\mathcal{I}$ stands for the set of irregular points of U. The one-dimensional Hausdorff measure of $\mathcal{I}$ is zero : $\mathcal{H}^1(\mathcal{I}) = 0$. The shock set $\mathcal{J}$ is essentially covered by the (at most) countable union of C^1 arcs. With any $(\bar{x}, \bar{t}) \in \mathcal{J}$ are associated one-sided approximate limits $U_\pm$ and a "tangent" line of slope (shock speed) s which, as shown in Section 1.8, are related by the Rankine-Hugoniot jump condition (8.1.2).

We shall be assuming throughout that the Lax E-condition, introduced in Section 8.3, holds here in a strong sense: each shock is compressive but not overcompressive. That is, if $U_\pm$ are the one-sided limits and s is the corresponding shock speed associated with any point of the shock set, then there is $i \in \{1, \cdots, n\}$ such that

(10.1.2) $\lambda_{i-1}(U_{\pm}) < \lambda_i(U_+) \leq s \leq \lambda_i(U_-) < \lambda_{i+1}(U_{\pm})$.

In (10.1.2), the first inequality is not needed when $i = 1$ and the last inequality is unnecessary when $i = n$. Moreover, since (10.1.1) is strictly hyperbolic, the first and the last inequalities will hold automatically whenever the oscillation of U is sufficiently small.

For convenience, we normalize U as explained in Section 1.7. In particular, at every point $(\bar{x}, \bar{t}) \in \mathscr{C}$, $U(\bar{x}, \bar{t})$ equals the corresponding approximate limit U_0. Recalling that $\mathscr{H}^1(\mathscr{J}) = 0$ and using Theorem 1.7.1, we easily conclude that there is a subset $\mathscr{N}$ of $(0, \infty)$, of measure zero, having the following properties. For any fixed $\bar{t} \notin \mathscr{N}$, the function $U(\cdot, \bar{t})$ has locally bounded variation on $(-\infty, \infty)$, and $(\bar{x}, \bar{t}) \in \mathscr{C}$ if and only if $U(\bar{x}-, \bar{t}) = U(\bar{x}+, \bar{t})$, while $(\bar{x}, \bar{t}) \in \mathscr{J}$ if and only if $U(\bar{x}-, \bar{t}) \neq U(\bar{x}+, \bar{t})$. In the latter case, $U_- = U(\bar{x}-, \bar{t})$ and $U_+ = U(\bar{x}+, \bar{t})$.

The above properties of U follow just from membership in BV. The fact that U is also a solution of (10.1.1) should induce additional structure. Based on experience with special systems, to be discussed in later chapters, it seems plausible to expect the following: U should be (classically) continuous on $\mathscr{C}$ and the one-sided limits $U_{\pm}$ at points of $\mathscr{J}$ should be attained in the classical sense. Moreover, $\mathscr{J}$ should be the (at most) countable set of endpoints of the arcs that comprise $\mathscr{J}$. Uniform stretching of the (x, t) coordinates about any point of $\mathscr{J}$ should yield, in the limit, a wave fan with the properties described in Section 9.1, i.e., $\mathscr{J}$ should consist of shock generation and shock interaction points. To what extent the picture painted above accurately describes the structure of solutions of general hyperbolic systems of conservation laws will be discussed in later chapters.

10.2 Generalized Characteristics

Characteristics associated with classical, Lipschitz continuous, solutions were introduced in Section 7.2, through Definition 7.2.1. They provide one of the principal tools of the classical theory for the study of analytical and geometric properties of solutions. It is thus natural to attempt to extend the notion to the framework of weak solutions.

Here we opt to define characteristics of the i-characteristic family, associated with the weak solution U, exactly as in the classical case, namely as integral curves of the ordinary differential equation (7.2.7), in the sense of Filippov:

Definition 10.2.1 A *generalized i-characteristic* for the system (10.1.1), associated with the (generally weak) solution U, on the time interval $[\sigma, \tau] \subset [0, \infty)$, is a Lipschitz function $\xi : [\sigma, \tau] \to (-\infty, \infty)$ which satisfies the differential inclusion

(10.2.1) $\dot{\xi}(t) \in \Lambda_i(\xi(t), t)$,   a.e. on $[\sigma, \tau]$,

where

(10.2.2) $\Lambda_i(\bar{x}, \bar{t}) := \bigcap_{\varepsilon > 0} \left[\underset{[\bar{x}-\varepsilon, \bar{x}+\varepsilon]}{\operatorname{ess\,inf}} \lambda_i(U(x, \bar{t})), \ \underset{[\bar{x}-\varepsilon, \bar{x}+\varepsilon]}{\operatorname{ess\,sup}} \lambda_i(U(x, \bar{t})) \right]$.

From the general theory of contingent equations like (10.2.1), one immediately infers the following

Theorem 10.2.1 *Through any fixed point* $(\overline{x}, \overline{t}) \in (-\infty, \infty) \times [0, \infty)$ *pass two (not necessarily distinct) generalized i-characteristics, associated with U and defined on $[0, \infty)$, namely the minimal $\xi_-(\cdot)$ and the maximal $\xi_+(\cdot)$, with $\xi_-(t) \leq \xi_+(t)$ for $t \in [0, \infty)$. The funnel-shaped region confined between the graphs of $\xi_-(\cdot)$ and $\xi_+(\cdot)$ comprises the set of points (x, t) that may be connected to $(\overline{x}, \overline{t})$ by a generalized i-characteristic associated with U.*

Other standard properties of solutions of differential inclusions also have useful implications to the theory of generalized characteristics: If $\{\xi_m(\cdot)\}$ is a sequence of generalized i-characteristics, associated with U and defined on $[\sigma, \tau]$, which converges to some Lipschitz function $\xi(\cdot)$, uniformly on $[\sigma, \tau]$, then $\xi(\cdot)$ is necessarily a generalized i-characteristic associated with U. In particular, if $\xi_m(\cdot)$ is the minimal (or maximal) generalized i-characteristic through a point $(x_m, \overline{t})$ and $x_m \uparrow \overline{x}$ (or $x_m \downarrow \overline{x}$), as $m \to \infty$, then $\{\xi_m(\cdot)\}$ converges to the minimal (or maximal) generalized i-characteristic $\xi_-(\cdot)$ (or $\xi_+(\cdot)$) through the point $(\overline{x}, \overline{t})$.

In addition to classical i-characteristics, i-shocks that satisfy the Lax E-condition are obvious examples of generalized i-characteristics. In fact, it turns out that these are the only possibilities. Indeed, even though Definition 10.2.1 would seemingly allow $\dot{\xi}$ to select any value in the interval Λ_i, the fact that U is a solution of (10.1.1) constrains generalized i-characteristics associated with U to propagate either with classical i-characteristic speed or with i-shock speed:

Theorem 10.2.2 *Let $\xi(\cdot)$ be a generalized i-characteristic, associated with U and defined on $[\sigma, \tau]$. The following holds for almost all $t \in [\sigma, \tau]$: When $(\xi(t), t) \in \mathscr{C}$, then $\dot{\xi}(t) = \lambda_i(U_0)$ with $U_0 = U(\xi(t)\pm, t)$. When $(\xi(t), t) \in \mathscr{J}$, then $\dot{\xi}(t) = s$, where s is the speed of the i-shock that joins U_-, on the left, to U_+, on the right, with $U_\pm = U(\xi(t)\pm, t)$. In particular, s satisfies the Rankine-Hugoniot condition (8.1.2) as well as the Lax E-condition (10.1.2).*

Proof. Let us recall the properties of BV solutions recorded in Section 10.1. It is then clear that for almost all $t \in [\sigma, \tau]$ with $(\xi(t), t) \in \mathscr{C}$ the interval $\Lambda_i(\xi(t), t)$ reduces to the single point $[\lambda_i(U(\xi(t)\pm, t))]$ and so $\dot{\xi}(t) = \lambda_i(U(\xi(t)\pm, t))$, by virtue of (10.2.1).

Applying the measure equality (10.1.1) to arbitrary subarcs of the graph of ξ, and using Theorem 1.7.5 (in particular Equation (1.7.17)), yields

$$(10.2.3) \quad F(U(\xi(t)+, t)) - F(U(\xi(t)-, t)) = \dot{\xi}(t)[U(\xi(t)+, t) - U(\xi(t)-, t)] ,$$

almost everywhere on $[\sigma, \tau]$. Consequently, for almost all $t \in [\sigma, \tau]$ with $(\xi(t), t) \in \mathscr{J}$, we have $\dot{\xi}(t) = s$, where s is the speed of a shock that joins the states $U_- = U(\xi(t)-, t)$ and $U_+ = U(\xi(t)+, t)$. By our assumptions on the structure of solutions, there is $j \in \{1, \cdots, n\}$ such that $\lambda_{j-1}(U_\pm) < \lambda_j(U_+) \leq s \leq \lambda_j(U_-) < \lambda_{j+1}(U_\pm)$. On the other hand, (10.2.1) implies that s lies in the interval with endpoints $\lambda_i(U_-)$ and $\lambda_i(U_+)$. Therefore, $j = i$ and (10.1.2) holds. This completes the proof.

The above theorem motivates the following terminology:

Definition 10.2.2 A generalized i-characteristic $\xi(\cdot)$, associated with U and defined on $[\sigma, \tau]$, is called *shock free* if $(\xi(t), t) \in \mathscr{C}$ or, equivalently, $U(\xi(t)-, t) = U(\xi(t)+, t)$, for almost all $t \in [\sigma, \tau]$.

A consequence of the proof of Theorem 10.2.2 is that (10.2.1) is equivalent to

$$(10.2.4) \qquad \dot{\xi}(t) \in [\lambda_i(U(\xi(t)+, t)), \lambda_i(U(\xi(t)-, t))] , \quad \text{a.e. on } [\sigma, \tau] .$$

In what follows, an important role will be played by the special generalized characteristics that manage to propagate at the maximum or minimum allowable speed:

Definition 10.2.3 A generalized i-characteristic $\xi(\cdot)$, associated with U and defined on $[\sigma, \tau]$, is called a *left i-contact* if

$$(10.2.5) \qquad \dot{\xi}(t) = \lambda_i(U(\xi(t)-, t)) , \quad \text{a.e. on } [\sigma, \tau] ,$$

and/or a *right i-contact* if

$$(10.2.6) \qquad \dot{\xi}(t) = \lambda_i(U(\xi(t)+, t)) , \quad \text{a.e. on } [\sigma, \tau] .$$

Clearly, shock free i-characteristics are left and right i-contacts. Note that, since they are generalized i-characteristics, left (or right) i-contacts should also satisfy the assertion of Theorem 10.2.2, namely $\dot{\xi}(t) = s$ for almost all $t \in [\sigma, \tau]$ with $(\xi(t), t) \in \mathscr{J}$. Of course this is impossible in systems that do not admit left (or right) contact discontinuities. In any such system, left (or right) contacts are necessarily shock free. In particular, recalling Theorem 8.2.1, we conclude that when the i-characteristic family for the system (10.1.1) is genuinely nonlinear and the oscillation of U is sufficiently small, then any left or right i-contact is necessarily shock free.

10.3 Extremal Backward Characteristics

With reference to some point $(\bar{x}, \bar{t}) \in (-\infty, \infty) \times [0, \infty)$, a generalized characteristic through $(\bar{x}, \bar{t})$ is dubbed *backward* when defined on $[0, \bar{t}]$, or *forward* when defined on $[\bar{t}, \infty)$. The extremal, minimal and maximal, backward and forward generalized characteristics through $(\bar{x}, \bar{t})$ propagate at extremal speeds and may thus be considered natural candidates for being contacts. This turns out to be true, at least for the backward extremal characteristics, in consequence of the Lax E-condition:

Theorem 10.3.1 *The minimal (or maximal) backward i-characteristic, associated with U, emanating from any point $(\bar{x}, \bar{t})$ of the upper half-plane is a left (or right) i-contact.*

Proof. Let $\xi(\cdot)$ denote the minimal backward i-characteristic emanating from $(\bar{x}, \bar{t})$ and defined on $[0, \bar{t}]$. We fix $\varepsilon > 0$ and determine numbers $\bar{t} = \tau_0 > \tau_1 > \cdots > \tau_k = 0$, for some $k \geq 1$, through the following algorithm: We start out with $\tau_0 = \bar{t}$. Assuming $\tau_m > 0$ has been determined, we let $\xi_m(\cdot)$ denote the minimal backward i-characteristic emanating from the point $(\xi(\tau_m) - \varepsilon, \tau_m)$. If $\xi_m(t) < \xi(t)$ for $0 < t \leq \tau_m$, we set $\tau_{m+1} = 0$, $m + 1 = k$ and terminate. Otherwise, we locate the number $\tau_{m+1} \in (0, \tau_m)$ with the property $\xi_m(t) < \xi(t)$ for $\tau_{m+1} < t \leq \tau_m$ and $\xi_m(\tau_{m+1}) = \xi(\tau_{m+1})$. Clearly, this algorithm will terminate after a finite number of steps. Next we construct a left-continuous, piecewise Lipschitz function $\xi_\varepsilon(\cdot)$ on $[0, \bar{t}]$, with jump discontinuities (when $k \geq 2$) at $\tau_1, \cdots, \tau_{k-1}$, by setting $\xi_\varepsilon(t) = \xi_m(t)$ for $\tau_{m+1} < t \leq \tau_m$, $m = 0, 1, \cdots, k - 1$, and $\xi_\varepsilon(0) = \xi_{k-1}(0)$. Then

$$(10.3.1) \quad \xi_\varepsilon(\bar{t}) - \xi_\varepsilon(0) = (k-1)\varepsilon + \sum_{m=0}^{k-1} \int_{\tau_{m+1}}^{\tau_m} \dot{\xi}_m(t) dt \geq \int_0^{\bar{t}} \lambda_i(U(\xi_\varepsilon(t)+, t)) dt \ .$$

By standard theory of contingent equations like (10.2.1), $\xi_\varepsilon(t) \uparrow \xi(t)$ as $\varepsilon \downarrow 0$, uniformly on $[0, \bar{t}]$. Therefore, letting $\varepsilon \downarrow 0$, (10.3.1) yields

$$(10.3.2) \qquad\qquad \xi(\bar{t}) - \xi(0) \geq \int_0^{\bar{t}} \lambda_i(U(\xi(t)-, t)) dt \ .$$

On the other hand, $\dot{\xi}(t) \leq \lambda_i(U(\xi(t)-, t))$, almost everywhere on $[0, \bar{t}]$, and so $\dot{\xi}(t) = \lambda_i(U(\xi(t)-, t))$ for almost all $t \in [0, \bar{t}]$, i.e., $\xi(\cdot)$ is a left i-contact.

Similarly one shows that the maximal backward i-characteristic emanating from $(\bar{x}, \bar{t})$ is a right i-contact. This completes the proof.

In view of the closing remarks in Section 10.2, Theorem 10.3.1 has the following corollary:

Theorem 10.3.2 *Assume the i-characteristic family for the system* (10.1.1) *is genuinely nonlinear and the oscillation of U is sufficiently small. Then the minimal and the maximal backward i-characteristics, emanating from any point $(\bar{x}, \bar{t})$ of the upper half-plane, are shock free.*

The implications of the above theorem will be seen in following chapters.

For future use, it will be expedient to introduce here a special class of backward characteristics emanating from infinity:

Definition 10.3.1 A *minimal* (or *maximal*) i-*separatrix*, associated with the solution U, is a Lipschitz function $\xi : [0, \bar{t}) \rightarrow (-\infty, \infty)$ such that $\xi(t) = \lim_{m \to \infty} \xi_m(t)$, uniformly on compact time intervals, where $\xi_m(\cdot)$ is the minimal (or maximal) backward i-characteristic emanating from a point (x_m, t_m), with $t_m \to \bar{t}$, as $m \to \infty$. In particular, when $\bar{t} = \infty$, the i-separatrix $\xi(\cdot)$ is called a *minimal* (or *maximal*) i-*divide*.

Note that the graphs of any two minimal (or maximal) i-characteristics may run into each other but they cannot cross. Consequently, the graph of a minimal (or maximal) backward i-characteristic cannot cross the graph of any minimal (or maximal) i-separatrix. Similarly, the graphs of any two minimal (or maximal) i-separatrices cannot cross. In particular, any minimal (or maximal) i-divide divides the upper half-plane into two parts in such a way that no forward i-characteristic may cross from the left to the right (or from the right to the left).

Minimal or maximal i-separatrices are necessarily generalized i-characteristics, which by virtue of Theorem 10.3.1, are left or right i-contacts. In particular, when the i-characteristic family is genuinely nonlinear and the oscillation of U is sufficiently small, Theorem 10.3.2 implies that minimal or maximal i-separatrices are shock free.

One should not expect that all solutions possess i-divides. An important class that always do, are solutions which are periodic in x, $U(x + L, t) = U(x, t)$ for some $L > 0$ and all $(x, t) \in (-\infty, \infty) \times (0, \infty)$. Indeed, in that case, given any sequence $\{t_m\}$, with $t_m \to \infty$ as $m \to \infty$, it is always possible to locate $\{x_m\}$ with the property that the minimal or maximal backward i-characteristic $\xi_m(\cdot)$ emanating from (x_m, t_m) will be intercepted by the x-axis at a point lying inside any fixed interval of length L, say $\xi_m(0) \in [0, L)$, $m = 1, 2, \cdots$. The Arzela theorem then implies that $\{\xi_m(\cdot)\}$ contains convergent subsequences whose limits are necessarily i-divides.

10.4 Notes

The presentation of the theory of generalized characteristics in this chapter follows Dafermos [16]. An exposition of the general theory of differential inclusions is found in the monograph by Filippov [1]. An early paper introducing generalized characteristics (for scalar conservation laws) as solutions of the classical characteristic equations, in the sense of Filippov, is Wu [1]. See also Hörmander [1]. Glimm and Lax [1] employ an alternative definition of generalized characteristics, namely Lipschitz curves propagating either with classical characteristic speed or with shock speed, constructed as limits of a family of "approximate characteristics". In view of Theorem 10.2.2, the two notions are closely related. This will be discussed in Chapter XIII.

The notion of divide was introduced in Dafermos [18].

Generalized characteristics in several space dimensions are considered by Poupaud and Rascle [1], in the context of linear transport equations with discontinuous coefficients.

Chapter XI. Genuinely Nonlinear
Scalar Conservation Laws

Despite its apparent simplicity, the genuinely nonlinear scalar conservation law in one-space dimension possesses a surprisingly rich theory, which deserves attention, not only for its intrinsic interest, but also because it provides valuable insight in the behavior of systems. The discussion here will employ the theory of generalized characteristics developed in Chapter X. From the standpoint of this approach, the special feature of genuinely nonlinear scalar conservation laws is that the extremal backward generalized characteristics are essentially classical characteristics, that is straight lines along which the solution is constant. This property induces such a heavy constrain that one is able to derive very precise information on regularity and large time behavior of solutions.

Solutions are (classically) continuous at points of approximate continuity and locally Lipschitz continuous in the interior of the set of points of continuity. Points of approximate jump discontinuity lie on classical shocks. The remaining, irregular, points are at most countable and are formed by the collision of shocks and/or the focussing of compression waves. Generically, solutions with smooth initial data are piecewise smooth.

Genuine nonlinearity gives rise to a host of dissipative mechanisms which affect the large time behavior of solutions. Entropy dissipation induces $O(t^{-\frac{p}{p+1}})$ decay of solutions with initial data in $L^p(-\infty, \infty)$. When the initial data have compact support, spreading of characteristics generates N-wave profiles. Confinement of characteristics under periodic initial data induces $O(t^{-t})$ decay in the total variation per period and the formation of sawtoothed profiles.

Another important feature of admissible weak solutions of the genuinely nonlinear scalar conservation law is that they are related explicitly to their initial data, through the Lax function. This property, which will be established here by the method of generalized characteristics, may serve alternatively as the starting point for developing the general theory of solutions.

The chapter will close with the derivation of properties extracted by comparing solutions. It will be shown that the lap number of any admissible solution is nonincreasing with time. Moreover, the L^1 distance of any two solutions is generally nonincreasing, but potentially conserved, whereas a properly weighted L^1 distance is strictly decreasing.

11.1 Admissible *BV* Solutions and Generalized Characteristics

We consider the scalar conservation law

$$(11.1.1) \qquad \partial_t u(x,t) + \partial_x f(u(x,t)) = 0 ,$$

which is genuinely nonlinear, $f''(u) > 0$, $-\infty < u < \infty$. Throughout this chapter we shall be dealing with admissible weak solutions u on $(-\infty, \infty) \times [0, \infty)$ whose initial data u_0 are bounded and have locally bounded variation on $(-\infty, \infty)$. By virtue of Theorem 6.2.3, u is in BV_{loc} and for any $t \in [0, \infty)$ the function $u(\cdot, t)$ has locally bounded variation on $(-\infty, \infty)$.

As noted in Section 8.5, the entropy shock admissibility criterion will be satisfied almost everywhere (with respect to one-dimensional Hausdorff measure) on the shock set $\mathscr{J}$ of the solution u, for any entropy-entropy flux pair (η, q) with η convex. This in turn implies that the Lax E-condition will also hold almost everywhere on $\mathscr{J}$. Consequently, we have

$$(11.1.2) \qquad u(x+, t) \le u(x-, t) ,$$

for almost all $t \in (0, \infty)$ and all $x \in (-\infty, \infty)$.

By account of Theorem 10.2.2, a Lipschitz curve $\xi(\cdot)$, defined on the time interval $[\sigma, \tau] \subset [0, \infty)$, will be a generalized characteristic, associated with the solution u, if for almost all $t \in [\sigma, \tau]$

$$(11.1.3)$$
$$\dot{\xi}(t) = \begin{cases} f'(u(\xi(t)\pm, t)) , & \text{when } u(\xi(t)+, t) = u(\xi(t)-, t) , \\[2mm] \dfrac{f(u(\xi(t)+, t)) - f(u(\xi(t)-, t))}{u(\xi(t)+, t) - u(\xi(t)-, t)} , & \text{when } u(\xi(t)+, t) < u(\xi(t)-, t) . \end{cases}$$

The special feature of genuinely nonlinear scalar conservation laws is that generalized characteristics that are shock free are essentially classical characteristics:

Theorem 11.1.1 *Let $\xi(\cdot)$ be a generalized characteristic for* (11.1.1), *associated with the admissible solution u, on the time interval $[\sigma, \tau]$, which is shock free. Then there is a constant $\bar{u}$ such that*

$$(11.1.4) \qquad u(\xi(\tau)+, \tau) \le \bar{u} \le u(\xi(\tau)-, \tau) ,$$

$$(11.1.5) \qquad u(\xi(t)+, t) = \bar{u} = u(\xi(t)-, t) , \quad \sigma < t < \tau ,$$

$$(11.1.6) \qquad u(\xi(\sigma)-, \sigma) \le \bar{u} \le u(\xi(\sigma)+, \sigma) .$$

In particular, the graph of $\xi(\cdot)$ is a straight line with slope $f'(\bar{u})$.

Proof. Fix r and s, $\sigma \le r < s \le \tau$. For $\varepsilon > 0$, we integrate the measure equality (11.1.1) over the set $\{(x, t) : r < t < s, \xi(t) - \varepsilon < x < \xi(t)\}$ and use Green's theorem to get

$$(11.1.7) \quad \begin{aligned} &\int_{\xi(s)-\varepsilon}^{\xi(s)} u(x,s)dx - \int_{\xi(r)-\varepsilon}^{\xi(r)} u(x,r)dx \\ &= \int_r^s \{f(u(\xi(t)-\varepsilon+,t)) - f(u(\xi(t)-,t)) \\ &\qquad\qquad - \dot{\xi}(t)[u(\xi(t)-\varepsilon+,t) - u(\xi(t)-,t)]\}dt\ . \end{aligned}$$

By virtue of Definition 10.2.2, $\dot{\xi}(t) = f'(u(\xi(t)-,t))$, a.e. on $[r,s]$. Since f is convex, this implies that the right-hand side of (11.1.7) is nonnegative. Consequently, multiplying (11.1.7) by $1/\varepsilon$ and letting $\varepsilon \downarrow 0$ yields

$$(11.1.8) \qquad u(\xi(s)-,s) \geq u(\xi(r)-,r)\ , \qquad \sigma \leq r < s \leq \tau\ .$$

Next we apply (11.1.1) to the set $\{(x,t) : r < t < s, \xi(t) < x < \xi(t)+\varepsilon\}$ and repeat the above procedure to deduce

$$(11.1.9) \qquad u(\xi(s)+,s) \leq u(\xi(r)+,r)\ , \qquad \sigma \leq r < s \leq \tau\ .$$

We now fix t_1, t_2, $\sigma < t_1 < t_2 < \tau$, such that $u(\xi(t_1)-,t_1) = u(\xi(t_1)+,t_1)$, $u(\xi(t_2)-,t_2) = u(\xi(t_2)+,t_2)$; then fix any $t \in (t_1,t_2)$. We apply (11.1.8) and (11.1.9) first with $r = t_1$, $s = t_2$, then with $r = t_1$, $s = t$, and finally with $r = t, s = t_2$. This yields (11.1.5). To complete the proof, we apply (11.1.8), (11.1.9) for $s = \tau, r \in (\sigma,\tau)$, to obtain (11.1.4), and for $r = \sigma, s \in (\sigma,\tau)$, to deduce (11.1.6).

Corollary 11.1.1 *Assume $\xi(\cdot)$ and $\zeta(\cdot)$ are distinct generalized characteristics for* (11.1.1), *associated with the admissible weak solution u, on the time interval $[\sigma,\tau]$, which is shock free. Then $\xi(\cdot)$ and $\zeta(\cdot)$ cannot intersect for any $t \in (\sigma,\tau)$.*

The above two propositions have significant implications on extremal backward characteristics:

Theorem 11.1.2 *Let $\xi_-(\cdot)$ and $\xi_+(\cdot)$ denote the minimal and maximal backward characteristics, associated with some admissible solution u, emanating from any point $(\bar{x},\bar{t}) \in (-\infty,\infty) \times (0,\infty)$. Then*

$$(11.1.10) \qquad \begin{cases} u(\xi_-(t)-,t) = u(\bar{x}-,\bar{t}) = u(\xi_-(t)+,t) \\ u(\xi_+(t)-,t) = u(\bar{x}+,\bar{t}) = u(\xi_+(t)+,t) \end{cases} \quad 0 < t < \bar{t}\ ,$$

$$(11.1.11) \qquad \begin{cases} u_0(\xi_-(0)-) \leq u(\bar{x}-,\bar{t}) \leq u_0(\xi_-(0)+)\ , \\ u_0(\xi_+(0)-) \leq u(\bar{x}+,\bar{t}) \leq u_0(\xi_+(0)+)\ . \end{cases}$$

In particular, $u(\bar{x}+,\bar{t}) \leq u(\bar{x}-,\bar{t})$ holds for all $(\bar{x},\bar{t}) \in (-\infty,\infty) \times (0,\infty)$ and $\xi_-(\cdot), \xi_+(\cdot)$ coincide if and only if $u(\bar{x}+,\bar{t}) = u(\bar{x}-,\bar{t})$.

Proof. By virtue of Theorem 10.3.2, both $\xi_-(\cdot)$ and $\xi_+(\cdot)$ are shock free. We may then apply Theorem 11.1.1, with $\sigma = 0$ and $\tau = \bar{t}$. On account of (11.1.4), if

$u(\overline{x}+, \overline{t}) = u(\overline{x}-, \overline{t})$, then $\overline{u} = u(\overline{x}\pm, \overline{t})$ and thus $\xi_-(\cdot), \xi_+(\cdot)$ coincide. In the general case, consider an increasing (or decreasing) sequence $\{x_n\}$, converging to $\overline{x}$, such that $u(x_n+, \overline{t}) = u(x_n-, \overline{t}), n = 1, 2, \cdots$. Let $\xi_n(\cdot)$ denote the unique backward characteristic emanating from $(x_n, \overline{t})$. Then $u(\xi_n(t)\pm, t) = u(x_n\pm, \overline{t})$ for all $t \in (0, \overline{t})$. As noted in Section 10.2, the sequence $\{\xi_n(\cdot)\}$ converges from below (or above) to $\xi_-(\cdot)$ (or $\xi_+(\cdot)$). Consequently, $u(\xi_-(t)-, t) = \lim u(x_n\pm, \overline{t}) = u(\overline{x}-, \overline{t})$ (or $u(\xi_+(t)+, t) = \lim u(x_n\pm, \overline{t}) = u(\overline{x}+, \overline{t})$). The proof is complete.

We now turn to the properties of forward characteristics:

Theorem 11.1.3 *A unique forward generalized characteristic, associated with an admissible solution u, issues from any point $(\overline{x}, \overline{t}) \in (-\infty, \infty) \times (0, \infty)$.*

Proof. Suppose two distinct forward characteristics $\phi(\cdot)$ and $\psi(\cdot)$ issue from $(\overline{x}, \overline{t})$, such that $\phi(s) < \psi(s)$ for some $s > \overline{t}$. Let $\xi(\cdot)$ denote the maximal backward characteristic emanating from $(\phi(s), s)$ and $\zeta(\cdot)$ denote the minimal backward characteristic emanating from $(\psi(s), s)$, both being shock free. For $t \in [\overline{t}, s]$, $\xi(t) \geq \phi(t)$ and $\zeta(t) \leq \psi(t)$; hence $\xi(\cdot)$ and $\zeta(\cdot)$ must intersect at some $t \in [\overline{t}, s)$, in contradiction to Corollary 11.1.1. This completes the proof.

Note that, by contrast, multiple forward characteristics may issue from points lying on the x-axis. In particular, the focus of any centered rarefaction wave must necessarily lie on the x-axis.

The next proposition demonstrates that, once they form, jump discontinuities propagate as shock waves for eternity:

Theorem 11.1.4 *Let $\chi(\cdot)$ denote the unique forward generalized characteristic, associated with the admissible solution u, issuing from a point $(\overline{x}, \overline{t})$ such that $\overline{t} > 0$ and $u(\overline{x}+, \overline{t}) < u(\overline{x}-, \overline{t})$. Then $u(\chi(s)+, s) < u(\chi(s)-, s)$ for all $s \in [\overline{t}, \infty)$.*

Proof. Let $\xi_-(\cdot)$ and $\xi_+(\cdot)$ denote the minimal and maximal backward characteristics emanating from $(\overline{x}, \overline{t})$. Since $u(\overline{x}+, \overline{t}) < u(\overline{x}-, \overline{t})$, $\xi_-(\cdot)$ and $\xi_+(\cdot)$ are distinct: $\xi_-(0) < \xi_+(0)$.

Fix any $s \in [\overline{t}, \infty)$ and consider the minimal and maximal backward characteristics $\zeta_-(\cdot)$ and $\zeta_+(\cdot)$ emanating from $(\chi(s), s)$. For $t \in [0, \overline{t}]$, necessarily $\zeta_-(t) \leq \xi_-(t)$ and $\zeta_+(t) \geq \xi_+(t)$. Thus $\zeta_-(0) < \zeta_+(0)$ so that $\zeta_-(\cdot)$ and $\zeta_+(\cdot)$ are distinct. Consequently, $u(\chi(s)+, s) < u(\chi(s)-, s)$. This completes the proof.

In view of the above, it is possible to identify the points from which shocks originate:

Definition 11.1.1 We call $(\overline{x}, \overline{t}) \in (-\infty, \infty) \times [0, \infty)$ a *shock generation point* if some forward generalized characteristic $\chi(\cdot)$ issuing from $(\overline{x}, \overline{t})$ is a shock, i.e., $u(\chi(t)+, t) < u(\chi(t)-, t)$, for all $t > \overline{t}$, while every backward characteristic emanating from $(\overline{x}, \overline{t})$ is shock free.

When $(\bar{x}, \bar{t})$ is a shock generation point with $\bar{t} > 0$, there are two possibilities: $u(\bar{x}+, \bar{t}) = u(\bar{x}-, \bar{t})$ or $u(\bar{x}+, \bar{t}) < u(\bar{x}-, \bar{t})$. In the former case, the shock starts out at $(\bar{x}, \bar{t})$ with zero strength and develops as it evolves. In the latter case, distinct minimal and maximal backward characteristics $\xi_-(\cdot)$ and $\xi_+(\cdot)$ emanate from $(\bar{x}, \bar{t})$. The sector confined between the graphs of $\xi_-(\cdot)$ and $\xi_+(\cdot)$ must be filled with characteristics, connecting $(\bar{x}, \bar{t})$ with the x-axis, which, by definition, are shock free and hence are straight lines. Thus in that case the shock is generated at the focus of a centered compression wave so it starts out with positive strength.

11.2 The Spreading of Rarefaction Waves

We are already familiar with the destabilizing role of genuine nonlinearity: Compression wave fronts get steeper and eventually break generating shocks. It turns out, however, that at the same time genuine nonlinearity also exerts a regularizing influence by inducing the spreading of rarefaction wave fronts. It is remarkable that this effect is purely geometric and is totally unrelated to the regularity of the initial data:

Theorem 11.2.1 *For any admissible solution u,*

(11.2.1)
$$\frac{f'(u(y\pm, t)) - f'(u(x\pm, t))}{y - x} \leq \frac{1}{t}, \quad -\infty < x < y < \infty, \quad 0 < t < \infty.$$

Proof. Fix x, y and t with $x < y$ and $t > 0$. Let $\xi(\cdot)$ and $\zeta(\cdot)$ denote the maximal or minimal backward characteristics emanating from (x, t) and (y, t), respectively. By virtue of Theorem 11.1.2, $\xi(0) = x - tf'(u(x\pm, t))$, $\zeta(0) = y - tf'(u(y\pm, t))$. Furthermore, $\xi(0) \leq \zeta(0)$, on account of Corollary 11.1.1. This immediately implies (11.2.1). The proof is complete.

Notice that (11.2.1) establishes a one-sided Lipschitz condition for $f'(u(\cdot, t))$, with Lipschitz constant independent of the initial data. By the general theory of scalar conservation laws, presented in Chapter VI, admissible solutions of (11.1.1) with initial data in $L^\infty(-\infty, \infty)$ may be realized as a.e. limits of sequences of solutions with initial data of locally bounded variation on $(-\infty, \infty)$. Consequently, (11.2.1) should hold even for admissible solutions with initial data that are merely in $L^\infty(-\infty, \infty)$. Clearly, (11.2.1) implies that, for fixed $t > 0$, $f(u(\cdot, t))$, and thereby also $u(\cdot, t)$, have bounded variation over any bounded interval of $(-\infty, \infty)$. We have thus shown that, due to genuine nonlinearity, solutions are generally smoother than their initial data:

Theorem 11.2.2 *Admissible solutions of (11.1.1), with initial data in $L^\infty(-\infty, \infty)$, are in BV_{loc} on $(-\infty, \infty) \times (0, \infty)$ and satisfy the one-sided Lipschitz condition (11.2.1).*

11.3 Regularity of Solutions

The properties of generalized characteristics established in the previous section lead to a precise description of the structure and regularity of admissible weak solutions.

Theorem 11.3.1 *Let $\chi(\cdot)$ be the unique forward generalized characteristic and $\xi_-(\cdot), \xi_+(\cdot)$ the extremal backward characteristics, associated with an admissible solution u, emanating from any point $(\overline{x}, \overline{t}) \in (-\infty, \infty) \times (0, \infty)$. Then $(\overline{x}, \overline{t})$ is a point of continuity of the function $u(x-, t)$ relative to the set $\{(x, t) : 0 \leq t \leq \overline{t}, x \leq \xi_-(t)$ or $\overline{t} < t < \infty, x \leq \chi(t)\}$ and also a point of continuity of the function $u(x+, t)$ relative to the set $\{(x, t) : 0 \leq t \leq \overline{t}, x \geq \xi_+(t)$ or $\overline{t} < t < \infty, x \geq \chi(t)\}$. Furthermore, $\chi(\cdot)$ is differentiable from the right at $\overline{t}$ and*

$$(11.3.1) \quad \frac{d^+}{dt}\chi(\overline{t}) = \begin{cases} f'(u(\overline{x}\pm, \overline{t})) , & \text{if } u(\overline{x}+, \overline{t}) = u(\overline{x}-, \overline{t}) , \\[2mm] \dfrac{f(u(\overline{x}+, \overline{t})) - f(u(\overline{x}-, \overline{t}))}{u(\overline{x}+, \overline{t}) - u(\overline{x}-, \overline{t})} , & \text{if } u(\overline{x}+, \overline{t}) < u(\overline{x}-, \overline{t}) . \end{cases}$$

Proof. Take any sequence $\{(x_n, t_n)\}$ in the set $\{(x, t) : 0 \leq t < \overline{t}, x \leq \xi_-(t)$ or $\overline{t} < t < \infty, x \leq \chi(t)\}$, which converges to $(\overline{x}, \overline{t})$ as $n \to \infty$. Let $\xi_n(\cdot)$ denote the minimal backward characteristic emanating from (x_n, t_n). Clearly, $\xi_n(t) \leq \xi_-(t)$ for $t \leq \overline{t}$. Thus, as $n \to \infty$, $\{\xi_n(\cdot)\}$ converges from below to $\xi_-(\cdot)$. Consequently, $\{u(x_n-, t_n)\}$ converges to $u(\overline{x}-, \overline{t})$.

Similarly, one shows that for any sequence $\{(x_n, t_n)\}$ in the set $\{(x, t) : 0 \leq t < \overline{t}, x \geq \xi_+(t)$ or $\overline{t} < t < \infty, x \geq \chi(t)\}$, converging to $(\overline{x}, \overline{t})$, the sequence $\{u(x_n+, t_n)\}$ converges to $u(\overline{x}+, \overline{t})$.

For $\varepsilon > 0$,

$$(11.3.2) \quad \frac{1}{\varepsilon}[\chi(\overline{t} + \varepsilon) - \chi(\overline{t})] = \frac{1}{\varepsilon} \int_{\overline{t}}^{\overline{t}+\varepsilon} \dot{\chi}(t)dt ,$$

where $\dot{\chi}(t)$ is determined through (11.1.3), with $\xi \equiv \chi$. As shown above, $\dot{\chi}(t)$ is continuous from the right at $\overline{t}$ and so, letting $\varepsilon \downarrow 0$ in (11.3.2), we arrive at (11.3.1). This completes the proof.

The above theorem has the following corollary:

Theorem 11.3.2 *Let u be an admissible solution and assume $u(\overline{x}+, \overline{t}) = u(\overline{x}-, \overline{t})$, for some $(\overline{x}, \overline{t}) \in (-\infty, \infty) \times (0, \infty)$. Then $(\overline{x}, \overline{t})$ is a point of continuity of u. A unique generalized characteristic $\chi(\cdot)$, associated with u, defined on $[0, \infty)$, passes through $(\overline{x}, \overline{t})$. Furthermore, $\chi(\cdot)$ is differentiable at $\overline{t}$ and $\dot{\chi}(\overline{t}) = f'(u(\overline{x}\pm, \overline{t}))$.*

Next we focus attention on points of discontinuity.

Theorem 11.3.3 *Let u be an admissible solution and assume $u(\overline{x}+,\overline{t}) < u(\overline{x}-,\overline{t})$, for some $(\overline{x},\overline{t}) \in (-\infty,\infty) \times (0,\infty)$. When the extremal backward characteristics $\xi_-(\cdot), \xi_+(\cdot)$ are the only backward generalized characteristics emanating from $(\overline{x},\overline{t})$ that are shock free, then $(\overline{x},\overline{t})$ is a point of jump discontinuity of u in the following sense: There is a generalized characteristic $\chi(\cdot)$, associated with u, defined on $[0,\infty)$ and passing through $(\overline{x},\overline{t})$, such that $(\overline{x},\overline{t})$ is a point of continuity of the function $u(x-,t)$ relative to the set $\{(x,t) : 0 < t < \infty, x \le \chi(t)\}$ as well as a point of continuity of the function $u(x+,t)$ relative to the set $\{(x,t) : 0 < t < \infty, x \ge \chi(t)\}$. Furthermore, $\chi(\cdot)$ is differentiable at $\overline{t}$ and*

$$(11.3.3) \qquad \dot{\chi}(\overline{t}) = \frac{f(u(\overline{x}+,\overline{t})) - f(u(\overline{x}-,\overline{t}))}{u(\overline{x}+,\overline{t}) - u(\overline{x}-,\overline{t})} \ .$$

Proof. Fix any point on the x-axis, in the interval $(\xi_-(0), \xi_+(0))$, and connect it to $(\overline{x},\overline{t})$ by a characteristic $\chi(\cdot)$. Continue $\chi(\cdot)$ to $[\overline{t},\infty)$ as the unique forward characteristic issuing from $(\overline{x},\overline{t})$.

Take any sequence $\{(x_n,t_n)\}$ in the set $\{(x,t) : 0 < t < \infty, x \le \chi(t)\}$, which converges to $(\overline{x},\overline{t})$, as $n \to \infty$. Let $\xi_n(\cdot)$ denote the minimal backward characteristic emanating from (x_n,t_n). As $n \to \infty$, $\{\xi_n(\cdot)\}$, or a subsequence thereof, will converge to some backward characteristic emanating from $(\overline{x},\overline{t})$, which is a straight line and shock free. Since $\xi_n(t) \le \chi(t)$, this implies that $\{\xi_n(\cdot)\}$ must necessarily converge to $\xi_-(\cdot)$. Consequently, $\{u(x_n-,t_n)\}$ converges to $u(\overline{x}-,\overline{t})$, as $n \to \infty$.

Similarly, one shows that for any sequence $\{(x_n,t_n)\}$ in the set $\{(x,t) : 0 < t < \infty, x \ge \chi(t)\}$, converging to $(\overline{x},\overline{t})$, the sequence $\{u(x_n+,t_n)\}$ converges to $u(\overline{x}+,\overline{t})$.

To verify (11.3.3), we start out again from (11.3.2), where now ε may be positive or negative. As shown above, $\overline{t}$ is a point of continuity of $\dot{\chi}(t)$ and so, letting $\varepsilon \to 0$, we arrive at (11.3.3). This completes the proof.

Theorem 11.3.4 *The set of irregular points of any admissible solution u is (at most) countable. $(\overline{x},\overline{t}) \in (-\infty,\infty) \times (0,\infty)$ is an irregular point if and only if $u(\overline{x}+,\overline{t}) < u(\overline{x}-,\overline{t})$ and, in addition to the extremal backward characteristics $\xi_-(\cdot), \xi_+(\cdot)$, there is at least another, distinct, backward characteristic $\xi(\cdot)$, associated with u, emanating from $(\overline{x},\overline{t})$, which is shock free. Irregular points are generated by the collision of shocks and/or by the focussing of centered compression waves.*

Proof. Necessity follows from Theorems 11.3.2 and 11.3.3. To show sufficiency, consider the subset $\mathscr{X}$ of the interval $[\xi_-(0), \xi_+(0)]$ with the property that, for $x \in \mathscr{X}$, the straight line segment connecting the points $(x,0)$ and $(\overline{x},\overline{t})$ is a characteristic associated with u, which is shock free.

When $\mathscr{X} \equiv [\xi_-(0), \xi_+(0)]$, $(\overline{x},\overline{t})$ is the focus of a centered compression wave and the assertion of the theorem is clearly valid. In general, however, $\mathscr{X}$ will be a closed proper subset of $[\xi_-(0), \xi_+(0)]$, containing at least the three points $\xi_-(0), \xi(0)$ and $\xi_+(0)$. The complement of $\mathscr{X}$ relative to $[\xi_-(0), \xi_+(0)]$ will then be the (at most) countable union of disjoint open intervals. Let (α_-, α_+) be one of these intervals, contained say in $(\xi_-(0), \xi(0))$. The straight-line segments connecting the points $(\alpha_-, 0)$ and $(\alpha_+, 0)$ with $(\bar{x}, \bar{t})$ will be shock free characteristics $\zeta_-(\cdot)$ and $\zeta_+(\cdot)$ along which u is constant, say u_- and u_+. Necessarily, $u(\bar{x}-, \bar{t}) \geq u_- > u_+ > u(\bar{x}+, \bar{t})$. Consider a characteristic $\chi(\cdot)$ connecting a point of (α_-, α_+) with $(\bar{x}, \bar{t})$. Then $\zeta_-(t) < \chi(t) < \zeta_+(t), 0 \leq t < \bar{t}$. Take any sequence $\{(x_n, t_n)\}$ in the set $\{(x, t) : 0 \leq t < \bar{t}, \zeta_-(t) \leq x \leq \chi(t)\}$, converging to $(\bar{x}, \bar{t})$, as $n \to \infty$. If $\xi_n(\cdot)$ denotes the minimal backward characteristic emanating from (x_n, t_n), the sequence $\{\xi_n(\cdot)\}$ will necessarily converge to $\zeta_-(\cdot)$. In particular, this implies $u(x_n-, t_n) \longrightarrow u_-$, as $n \to \infty$. Similarly one shows that if $\{(x_n, t_n)\}$ is any sequence in the set $\{(x, t) : 0 \leq t < \bar{t}, \chi(t) \leq x \leq \zeta_+(t)\}$ converging to $(\bar{x}, \bar{t})$, then $u(x_n+, t_n) \longrightarrow u_+$, as $n \to \infty$. Thus, near $\bar{t}$ $\chi(\cdot)$ is a shock, which is differentiable from the left at $\bar{t}$ with

$$(11.3.4) \qquad \frac{d^-}{dt} \chi(\bar{t}) = \frac{f(u_+) - f(u_-)}{u_+ - u_-} .$$

Since $f'(u_-) > \frac{d^-}{dt} \chi(\bar{t}) > f'(u_+)$, we conclude that $(\bar{x}, \bar{t})$ is an irregular point of u.

We have thus shown that $(\bar{x}, \bar{t})$ is a point of collision of shocks, one for each open interval of the complement of $\mathscr{X}$, and centered compression waves, when the measure of $\mathscr{X}$ is positive.

For fixed positive ε, we consider irregular points $(\bar{x}, \bar{t})$, as above, with the additional property $\xi_+(0) - \xi(0) > \varepsilon, \xi(0) - \xi_-(0) > \varepsilon$. It is easy to see that one may fit an at most finite set of such points in any bounded subset of the upper half-plane. This in turn implies that the set of irregular points of any admissible solution is (at most) countable. The proof is complete.

The next proposition provides another indication of the regularizing effect of genuine nonlinearity:

Theorem 11.3.5 *Assume the set $\mathscr{C}$ of points of continuity of an admissible solution u has nonempty interior $\mathscr{C}^0$. Then u is locally Lipschitz on $\mathscr{C}^0$.*

Proof. Fix any point $(\bar{x}, \bar{t}) \in \mathscr{C}^0$ and assume that the circle $\mathscr{B}_r$ of radius r, centered at $(\bar{x}, \bar{t})$, is contained in $\mathscr{C}^0$. Consider any point (x, t) at a distance $\rho < r$ from $(\bar{x}, \bar{t})$. The (unique) characteristics, associated with u, passing through $(\bar{x}, \bar{t})$ and (x, t) are straight lines with slopes $f'(u(\bar{x}, \bar{t}))$ and $f'(u(x, t))$, respectively, which cannot intersect inside the circle $\mathscr{B}_r$. Elementary trigonometric estimations then imply that $|f'(u(x, t)) - f'(u(\bar{x}, \bar{t}))|$ cannot exceed $c\rho/r$, where c is an upper bound of $1 + f'(u)^2$ over $\mathscr{B}_r$. Hence, if $a > 0$ is a lower bound of $f''(u)$ over $\mathscr{B}_r$, $|u(x, t) - u(\bar{x}, \bar{t})| \leq \frac{c}{ar} \rho$. This completes the proof.

The reader should be aware that admissible solutions have been constructed whose set of points of continuity has empty interior.

We now investigate the regularity of admissible solutions with smooth initial data. In what follows, it will be assumed that f is C^{k+1} and u is the admissible solution with C^k initial data u_0, for some $k \in \{1, 2, \cdots, \infty\}$.

For $(x, t) \in (-\infty, \infty) \times (0, \infty)$, we let $y_-(x, t)$ and $y_+(x, t)$ denote the interceptors on the x-axis of the minimal and maximal backward characteristics, associated with u, emanating from the point (x, t). In particular,

$$(11.3.5) \qquad x = y_-(x, t) + tf'(u_0(y_-(x, t))) = y_+(x, t) + tf'(u_0(y_+(x, t))) \, ,$$

$$(11.3.6) \qquad u(x-, t) = u_0(y_-(x, t)) \, , \qquad u(x+, t) = u_0(y_+(x, t)) \, .$$

For fixed $t > 0$, both $y_-(\cdot, t)$ and $y_+(\cdot, t)$ are monotone nondecreasing and the first one is continuous from the left while the second is continuous from the right. Consequently,

$$(11.3.7) \qquad 1 + t\frac{d}{dy} f'(u_0(y)) \geq 0 \, , \qquad y = y_\pm(x, t) \, ,$$

holds for all $(x, t) \in (-\infty, \infty) \times (0, \infty)$.

Any point $(\overline{x}, \overline{t}) \in (-\infty, \infty) \times (0, \infty)$ of continuity of u is necessarily also a point of continuity of $y_\pm(x, t)$ and $y_-(\overline{x}, \overline{t}) = y_+(\overline{x}, \overline{t})$. Therefore, by virtue of (11.3.5), (11.3.6) and the implicit function theorem we deduce

Theorem 11.3.6 *If $(\overline{x}, \overline{t}) \in (-\infty, \infty) \times (0, \infty)$ is a point of continuity of u and*

$$(11.3.8) \qquad 1 + \overline{t}\frac{d}{dy} f'(u_0(y)) > 0 \, , \qquad y = y_\pm(\overline{x}, \overline{t}) \, ,$$

then u is C^k on a neighborhood of $(\overline{x}, \overline{t})$.

Referring to Theorem 11.3.3, if $(\overline{x}, \overline{t})$ is a point of jump discontinuity of u, then $(\overline{x}, \overline{t})$ is a point of continuity of $y_-(x, t)$ and $y_+(x, t)$ relative to the sets $\{(x, t) : 0 < t < \infty, \ x \leq \chi(t)\}$ and $\{(x, t) : 0 < t < \infty, \ x \geq \chi(t)\}$, respectively. Consequently, the implicit function theorem together with (11.3.5) and (11.3.6) yield

Theorem 11.3.7 *If (11.3.8) holds at a point $(\overline{x}, \overline{t}) \in (-\infty, \infty) \times (0, \infty)$ of jump discontinuity of u, then, in a neighborhood of $(\overline{x}, \overline{t})$, the shock $\chi(\cdot)$ passing through $(\overline{x}, \overline{t})$ is C^{k+1} and u is C^k on either side of the graph of $\chi(\cdot)$.*

Next we consider shock generation points, introduced by Definition 11.1.1.

Theorem 11.3.8 *If $(\overline{x}, \overline{t}) \in (-\infty, \infty) \times (0, \infty)$ is a shock generation point, then*

$$(11.3.9) \qquad 1 + \overline{t}\frac{d}{dy} f'(u_0(y)) = 0 \, , \qquad y_-(\overline{x}, \overline{t}) \leq y \leq y_+(\overline{x}, \overline{t}) \, .$$

Furthermore, when $k \geq 2$,

$$(11.3.10) \qquad \frac{d^2}{dy^2} f'(u_0(y)) = 0 , \quad y_-(\overline{x}, \overline{t}) \le y \le y_+(\overline{x}, \overline{t}) .$$

Proof. Recall that there are two types of shock generation points: points of continuity, in which case $y_-(\overline{x}, \overline{t}) = y_+(\overline{x}, \overline{t})$, and focusses of centered compression waves, with $y_-(\overline{x}, \overline{t}) < y_+(\overline{x}, \overline{t})$. When $(\overline{x}, \overline{t})$ is a point of continuity, (11.3.9) is a consequence of (11.3.7) and Theorem 11.3.6. When $(\overline{x}, \overline{t})$ is the focus of a compression wave, $\overline{x} = y + \overline{t} f'(u_0(y))$ for any $y \in [y_-(\overline{x}, \overline{t}), y_+(\overline{x}, \overline{t})]$ and this implies (11.3.9).

When $y_-(\overline{x}, \overline{t}) < y_+(\overline{x}, \overline{t})$, differentiation of (11.3.9) with respect to y yields (11.3.10). To establish (11.3.10) for the case $(\overline{x}, \overline{t})$ is a point of continuity, we take any sequence $\{x_n\}$ which converges from below (or above) to $\overline{x}$. Then $\{y_-(x_n, \overline{t})\}$ will converge from below (or above) to $y_\pm(\overline{x}, \overline{t})$. By virtue of (11.3.7), $1 + \overline{t} \frac{d}{dy} f'(u_0(y)) \ge 0$ for $y = y_-(x_n, \overline{t})$ and this together with (11.3.9) imply that $y_\pm(\overline{x}, \overline{t})$ is a critical point of $\frac{d}{dy} f'(u_0(y))$. The proof is complete.

For $k \ge 3$, the set of functions u_0 in C^k with the property that $\frac{d}{dy} f'(u_0(y))$ has infinitely many critical points in a bounded interval is of the first category. Therefore, generically, initial data $u_0 \in C^k$, with $k \ge 3$, induce solutions with a locally finite set of shock generation points and thereby with a locally finite set of shocks. In other words, generically, solutions with initial data in $C^k, k \ge 3$, are piecewise C^k smooth functions, and do not contain any centered compression waves. In particular, solutions with analytic initial data are always piecewise smooth.

11.4 Divides, Invariants and the Lax Formula

The theory of generalized characteristics will be used here to establish interesting and fundamental properties of admissible solutions of (11.1.1). The starting point will be a simple but, as we shall see, very useful identity.

Let us consider two admissible solutions u and u^*, with corresponding initial data u_0 and u_0^*, and trace one of the extremal backward characteristics $\xi(\cdot)$, associated with u, and one of the extremal backward characteristics $\xi^*(\cdot)$, associated with u^*, that emanate from any fixed point $(x, t) \in (-\infty, \infty) \times (0, \infty)$. Thus, $\xi(\cdot)$ and $\xi^*(\cdot)$ will be straight lines, and along $\xi(\cdot)$ u will be constant, equal to $u(x-, t)$ or $u(x+, t)$, while along $\xi^*(\cdot)$ u^* will be constant, equal to $u^*(x-, t)$ or $u^*(x+, t)$. In particular, $\dot{\xi}(\tau) = f'(u(x\pm, t))$ and $\dot{\xi}^*(\tau) = f'(u^*(x\pm, t))$, $0 < \tau < t$.

We write (11.1.1), first for u then for u^*, we subtract the resulting two equations, we integrate over the triangle with vertices (x, t), $(\xi(0), 0)$, $(\xi^*(0), 0)$, and apply Green's theorem thus arriving at the identity

$$\int_0^t \{f(u(x\pm, t)) - f(u^*(\xi(\tau)-, \tau))$$

$$- f'(u(x\pm, t))[u(x\pm, t) - u^*(\xi(\tau)-, \tau)]\}d\tau$$

(11.4.1)
$$+ \int_0^t \{f(u^*(x\pm, t)) - f(u(\xi^*(\tau)-, \tau))$$

$$- f'(u^*(x\pm, t))[u^*(x\pm, t) - u(\xi^*(\tau)-, \tau)]\}d\tau$$

$$= \int_{\xi^*(0)}^{\xi(0)} [u_0(y) - u_0^*(y)]dy .$$

The usefulness of (11.4.1) lies in that, due to the convexity of f, both integrals on the left-hand side are nonpositive.

As a first application of (11.4.1), we use it to locate divides associated with an admissible solution u. The notion of divide was introduced by Definition 10.3.1. In the context of the genuinely nonlinear scalar conservation law, following the discussion in Section 10.3, divides are shock free and hence, by virtue of Theorem 11.1.1, straight lines along which u is constant.

Theorem 11.4.1 *A divide, associated with the admissible solution u, with initial data u_0, along which u is constant $\bar{u}$, issues from the point $(\bar{x}, 0)$ of the x-axis if and only if*

(11.4.2)
$$\int_{\bar{x}}^z [u_0(y) - \bar{u}]dy \geq 0 , \quad -\infty < z < \infty .$$

Proof. Assume first (11.4.2) holds. Apply (11.4.1) with $u^* \equiv \bar{u}$, $t \in (0, \infty)$, $x = \bar{x} + tf'(\bar{u})$. In particular, $\xi^*(\tau) = \bar{x} + \tau f'(\bar{u})$ and $\xi^*(0) = \bar{x}$. Hence the right-hand side of (11.4.1) is nonnegative, by account of (11.4.2). But then both integrals on the left-hand side must vanish, so that $u(x\pm, t) = \bar{u}$. We have thus established that the straight line $x = \bar{x} + tf'(\bar{u})$ is a shock free characteristic on $[0, \infty)$, that is a divide associated with u.

Conversely, assume the straight line $x = \bar{x} + tf'(\bar{u})$ is a divide associated with u. Take any $z \in (-\infty, \infty)$ and fix $\tilde{u}$ such that $\tilde{u} < \bar{u}$ if $z > \bar{x}$ and $\tilde{u} > \bar{u}$ if $z < \bar{x}$. The straight lines $z + tf'(\tilde{u})$ and $\bar{x} + tf'(\bar{u})$ will then intersect at a point (x, t) with $t > 0$. We apply (11.4.1) with $u^* \equiv \tilde{u}$, in which case $\xi(0) = \bar{x}, \xi^*(0) = z$. The left-hand side is nonpositive and so

(11.4.3)
$$\int_z^{\bar{x}} [u_0(y) - \tilde{u}]dy \leq 0 .$$

Letting $\tilde{u} \to \bar{u}$ we arrive at (11.4.2). This completes the proof.

The above proposition has implications on the existence of important time invariants of solutions:

Theorem 11.4.2 *Assume u_0 is integrable over $(-\infty, \infty)$ and the maxima*

$$(11.4.4) \qquad \max_x \int_x^{-\infty} u_0(y)dy = q_- , \qquad \max_x \int_x^{\infty} u_0(y)dy = q_+$$

exist. If u is the admissible solution with initial data u_0, then, for any $t > 0$,

$$(11.4.5) \qquad \max_x \int_x^{-\infty} u(y,t)dy = q_- , \qquad \max_x \int_x^{\infty} u(y,t)dy = q_+ .$$

Proof. Notice that q_- exists if and only if q_+ exists and in fact, by virtue of Theorem 11.4.1, both maxima are attained on the set of $\bar{x}$ with the property that the straight line $x = \bar{x} + tf'(0)$ is a divide associated with u, along which u is constant, equal to zero. But then, again by Theorem 11.4.1, both maxima in (11.4.5) will be attained at $\hat{x} = \bar{x} + tf'(0)$.

We now normalize f by $f(0) = 0$ and integrate (11.1.1) over the sets $\{(y, \tau) : 0 < \tau < t, -\infty < y < \bar{x} + \tau f'(0)\}$ and $\{(y, \tau) : 0 < \tau < t, \bar{x} + \tau f'(0) < y < \infty\}$. Applying Green's theorem, and since u vanishes along the straight line $x = \bar{x} + \tau f'(0)$,

$$(11.4.6) \qquad \int_{\hat{x}}^{-\infty} u(y,t)dy = \int_{\bar{x}}^{-\infty} u_0(y)dy , \qquad \int_{\hat{x}}^{\infty} u(y,t)dy = \int_{\bar{x}}^{\infty} u_0(y)dy ,$$

which verifies (11.4.5). The proof is complete.

One of the most striking features of genuinely nonlinear scalar conservation laws is that admissible solutions may be determined explicitly from the initial data through the following procedure. We start out with the Legendre transform

$$(11.4.7) \qquad g(v) = \max_u [uv - f(u)] ,$$

noting that the maximum is attained at $u = [f']^{-1}(v)$. With given initial data $u_0(\cdot)$ we then associate the *Lax function*

$$(11.4.8) \qquad G(y; x, t) = \int_0^y u_0(z)dz + tg\left(\frac{x-y}{t}\right) ,$$

defined for $(x, t) \in (-\infty, \infty) \times (0, \infty)$ and $y \in (-\infty, \infty)$.

Theorem 11.4.3 *For fixed $(x, t) \in (-\infty, \infty) \times (0, \infty)$, the Lax function $G(y; x, t)$ is minimized at a point $\bar{y} \in (-\infty, \infty)$ if and only if the straight line segment that connects the points (x, t) and $(\bar{y}, 0)$ is a generalized characteristic associated with the admissible solution u with initial data u_0, which is shock free.*

Proof. We fix y and $\bar{y}$ in $(-\infty, \infty)$, integrate (11.1.1) over the triangle with vertices $(x, t), (y, 0), (\bar{y}, 0)$ and apply Green's theorem to get

(11.4.9)

$$\int_0^{\bar{y}} u_0(z)dz + \int_0^t \left[\frac{x-\bar{y}}{t} u(\bar{y} + \tau\frac{x-\bar{y}}{t}\pm, \tau) - f(u(\bar{y} + \tau\frac{x-\bar{y}}{t}\pm, \tau)) \right] d\tau$$

$$= \int_0^y u_0(z)dz + \int_0^t \left[\frac{x-y}{t} u(y + \tau\frac{x-y}{t}\pm, \tau) - f(u(y + \tau\frac{x-y}{t}\pm, \tau)) \right] d\tau .$$

By virtue of (11.4.7) and (11.4.8), the left-hand side of (11.4.9) is less than, or equal to $G(\bar{y}; x, t)$, with equality if and only if $f'(u(\bar{y} + \tau\frac{x-\bar{y}}{t}\pm, \tau)) = \frac{x-\bar{y}}{t}$, almost everywhere on $(0, t)$, i.e., if and only if the straight line segment that connects the points (x, t) and $(\bar{y}, 0)$ is a shock free characteristic. Similarly, the right-hand side of (11.4.9) is less than, or equal to $G(y; x, t)$, with equality if and only if the straight line segment that connects the points (x, t) and $(y, 0)$ is a shock free characteristic. Assuming then that the straight line segment connecting (x, t) with $(\bar{y}, 0)$ is indeed a shock free characteristic, we deduce from (11.4.9) that $G(\bar{y}; x, t) \leq G(y; x, t)$ for any $y \in (-\infty, \infty)$.

Conversely, assume $G(\bar{y}; x, t) \leq G(y; x, t)$, for all $y \in (-\infty, \infty)$. In particular, pick y so that $(y, 0)$ is the intercept by the x-axis of the minimal backward characteristic emanating from (x, t). As shown above, y is a minimizer of $G(\cdot; x, t)$ and so $G(y; x, t) = G(\bar{y}; x, t)$. Moreover, the right-hand side of (11.4.9) equals $G(y; x, t)$ and hence so does the left-hand side. As explained above, this implies that the straight line segment connecting (x, t) with $(\bar{y}, 0)$ is a shock free characteristic. The proof is complete.

The above proposition may be used to determine the admissible solution u from its initial data u_0: For fixed $(x, t) \in (-\infty, \infty) \times (0, \infty)$, we let y_- and y_+ denote the smallest and the largest minimizer of $G(\cdot; x, t)$ over $(-\infty, \infty)$. We then have

(11.4.10) $$u(x\pm, t) = [f']^{-1}\left(\frac{x - y_\pm}{t}\right) .$$

By account of Theorems 11.3.2, 11.3.3 and 11.3.4, we conclude that (x, t) is a point of continuity of u if and only if $y_- = y_+$; a point of jump discontinuity of u if and only if $y_- < y_+$ and y_-, y_+ are the only minimizers of $G(\cdot; x, t)$; or an irregular point of u if and only if $y_- < y_+$ and there exist additional minimizers of $G(\cdot; x, t)$ in the interval (y_-, y_+). One may develop the entire theory of genuinely nonlinear scalar conservation laws on the basis of the above construction of admissible solutions, in lieu of the approach via generalized characteristics.

The change of variables $u = \partial_x v$, reduces the conservation law (11.1.1) to the *Hamilton-Jacobi equation*

(11.4.11) $$\partial_t v(x, t) + f(\partial_x v(x, t)) = 0 .$$

In that context, u is an admissible weak solution of (11.1.1) if and only if v is a *viscosity solution* of (11.4.11); (references in Section 11.9). In fact, Theorems 11.4.2 and 11.4.3 reflect properties of solutions of Hamilton-Jacobi equations rather than of hyperbolic conservation laws, in that they readily extend to the multi-space dimensional versions of the former though not of the latter.

11.5 Decay of Solutions Induced by Entropy Dissipation

Genuine nonlinearity gives rise to a multitude of dissipative mechanisms which, acting individually or collectively, affect the large time behavior of solutions. In this section we shall get acquainted with examples in which the principal agent of damping is entropy dissipation.

Theorem 11.5.1 *Let u be the admissible solution with initial data u_0 such that*

$$(11.5.1) \qquad \int_x^{x+\ell} u_0(y)dy = O(\ell^r) , \quad \text{as } \ell \to \infty ,$$

for some $r \in [0, 1)$, uniformly in x on $(-\infty, \infty)$. Then

$$(11.5.2) \qquad u(x\pm, t) = O\left(t^{-\frac{1-r}{2-r}}\right) , \quad \text{as } t \to \infty ,$$

uniformly in x on $(-\infty, \infty)$.

Proof. We fix $(x, t) \in (-\infty, \infty) \times (0, \infty)$ and write (11.4.1) for $u^* \equiv 0$. Notice that $\xi(0) - \xi^*(0) = t[f'(u(x\pm, t)) - f'(0)]$. Also recall that both integrals on the left-hand side are nonpositive. Consequently, using (11.5.1), we deduce

$$(11.5.3) \qquad \Phi(u(x\pm, t)) = O(t^{r-1}) , \quad \text{as } t \to \infty ,$$

uniformly in x on $(-\infty, \infty)$, where we have set

$$(11.5.4) \qquad \Phi(u) = \frac{f(0) - f(u) + uf'(u)}{|f'(u) - f'(0)|^r} = \frac{\int_0^u vf''(v)dv}{|\int_0^u f''(v)dv|^r} .$$

A simple estimation yields $\Phi(u) \geq K|u|^{2-r}$, with $K > 0$, and so (11.5.3) implies (11.5.2). This completes the proof.

In particular, when $u_0 \in L^p$, by virtue of Hölder's inequality, (11.5.1) holds with $r = 1 - \frac{1}{p}$. Therefore, Theorem 11.5.1 has the following corollary:

Theorem 11.5.2 *Let u be the admissible solution with initial data $u_0 \in L^p(-\infty, \infty)$, $1 \leq p < \infty$. Then*

$$(11.5.5) \qquad u(x\pm, t) = O\left(t^{-\frac{p}{p+1}}\right) , \quad \text{as } t \to \infty ,$$

uniformly in x on $(-\infty, \infty)$.

In the above examples the comparison function was the trivial solution $u^* \equiv 0$. Next we consider the case where the comparison function is the solution of a Riemann problem comprising two constant states u_- and u_+, $u_- > u_+$, joined by a shock, namely,

$$(11.5.6) \qquad u^*(x,t) = \begin{cases} u_-, & x < st \\ u_+, & x > st \,, \end{cases}$$

where

$$(11.5.7) \qquad s = \frac{f(u_+) - f(u_-)}{u_+ - u_-} \,.$$

Theorem 11.5.3 *Let u denote the admissible solution with intial data u_0 such that the improper integrals $\int_{-\infty}^{0} [u_0(y) - u_-]dy$ and $\int_{0}^{\infty} [u_0(y) - u_+]dy$ exist, with $u_- > u_+$. Normalize the origin $x = 0$ so that*

$$(11.5.8) \qquad \int_{-\infty}^{0} [u_0(y) - u_-]dy + \int_{0}^{\infty} [u_0(y) - u_+]dy = 0 \,.$$

Consider any forward characteristic $\chi(\cdot)$ issuing from $(0,0)$. Then, as $t \to \infty$,

$$(11.5.9) \qquad \chi(t) = st + o(1) \,,$$

with s given by (11.5.7), and

$$(11.5.10) \qquad u(x\pm, t) = \begin{cases} u_- + o(t^{-1/2}) \,, & \text{uniformly for } x < \chi(t) \,, \\ u_+ + o(t^{-1/2}), & \text{uniformly for } x > \chi(t) \,. \end{cases}$$

Proof. Fix any $(x,t) \in (-\infty, \infty) \times (0, \infty)$ and write (11.4.1) for the solution u, with initial data u_0, and the comparison solution u^* given by (11.5.6). Since $f'(u_-) > s > f'(u_+)$, as $t \to \infty$, $\xi^*(0) \to -\infty$, uniformly in x on $(-\infty, st)$, and $\xi^*(0) \to \infty$, uniformly in x on (st, ∞). Observe that, similarly, as $t \to \infty$, $\xi(0) \to -\infty$, uniformly in x on $(-\infty, \chi(t))$, and $\xi(0) \to \infty$ uniformly in x on $(\chi(t), \infty)$. Indeed, in the opposite case one would be able to find a sequence $\{(x_n, t_n)\}$, with $t_n \to \infty$ as $n \to \infty$, such that the intercepts $\xi_n(0)$ of the minimal backward characteristics $\xi_n(\cdot)$ emanating from (x_n, t_n) are confined in a bounded set. But then some subsequence of $\{\xi_n(\cdot)\}$ would converge to a divide issuing from some point $(\bar{x}, 0)$. However, this is impossible, because, since $u_- > u_+$, (11.5.8) is incompatible with (11.4.2), for any $\bar{x} \in (-\infty, \infty)$ and every $\bar{u} \in (-\infty, \infty)$.

In view of the above, (11.5.8) implies that the right-hand side of (11.4.1) is $o(1)$, as $t \to \infty$, uniformly in x on $(-\infty, \infty)$. The same will then be true for each integral on the left-hand side of (11.4.1), because they are of the same sign (nonpositive).

Consider first points $(x, t) \in (-\infty, \infty) \times (0, \infty)$ with $x < \min\{\chi(t), st\}$. Then $\xi(\tau) < s\tau$, $0 < \tau < t$, and so the first integral on the left-hand side of (11.4.1) yields

$$(11.5.11) \qquad t\{f(u(x\pm, t)) - f(u_-) - f'(u(x\pm, t))[u(x\pm, t) - u_-]\} = o(1) \,.$$

Since f is uniformly convex, (11.5.11) implies $u(x\pm, t) - u_- = o(t^{-1/2})$.

A similar argument demonstrates that for points $(x, t) \in (-\infty, \infty) \times (0, \infty)$ with $x > \max\{\chi(t), st\}$ we have $u(x\pm, t) - u_+ = o(t^{-1/2})$.

Next, consider points $(x, t) \in (-\infty, \infty) \times (0, \infty)$ with $st \leq x < \chi(t)$. Then $\xi(\cdot)$ will have to intersect the straight line $x = st$, say at $\tau = r$, $r \in [0, t]$, in which case the first integral on the left-hand side of (11.4.1) gives

$$(11.5.12) \quad (t - r)\{f(u(x\pm, t)) - f(u_+) - f'(u(x\pm, t))[u(x\pm, t) - u_+]\} = o(1) ,$$

$$(11.5.13) \quad r\{f(u(x\pm, t)) - f(u_-) - f'(u(x\pm, t))[u(x\pm, t) - u_-]\} = o(1) .$$

For $x < \chi(t)$, it was shown above that $\xi(0) \to -\infty$, as $t \to \infty$, and this in turn implies $r \to \infty$. It then follows from (11.5.13) and the convexity of f that $u(x\pm, t) = u_- + o(1)$. Then (11.5.12) implies that $t - r = o(1)$ so that $\chi(t) - st = o(1)$ and (11.5.13) yields (11.5.11). From (11.5.11) and the convexity of f we deduce, as before, $u(x\pm, t) - u_- = o(t^{-1/2})$.

A similar argument establishes that for points $(x, t) \in (-\infty, \infty) \times (0, \infty)$ with $\chi(t) < x \leq st$ we have $u(x\pm, t) - u_+ = o(t^{-1/2})$ and also $\chi(t) - st = o(1)$. This completes the proof.

11.6 Spreading of Characteristics and Development of N-Waves

Another feature of genuine nonlinearity, affecting the large time behavior of solutions, is spreading of characteristics. In order to see the effects of this mechanism, we shall study the asymptotic behavior of solutions with initial data of compact support. We already know, by account of Theorem 11.5.2, that the amplitude decays to zero as $O(t^{-1/2})$. The closer examination here will reveal that asymptotically the solution attains the profile of an *N-wave*, namely, a centered rarefaction wave flanked from both sides by shocks whose amplitudes decay like $O(t^{-1/2})$.

Theorem 11.6.1 *Let u be the admissible solution with initial data u_0, such that $u_0(x) = 0$ for $|x| > \ell$. Consider the minimal forward characteristic $\chi_-(\cdot)$ issuing from $(-\ell, 0)$ and the maximal forward characteristic $\chi_+(\cdot)$ issuing from $(\ell, 0)$. Then*

$$(11.6.1) \qquad u(x\pm, t) = 0 , \quad \text{for } t > 0 \text{ and } x < \chi_-(t) \text{ or } x > \chi_+(t) .$$

As $t \to \infty$,

$$(11.6.2) \quad f'(u(x\pm, t)) = \frac{x}{t} + O\left(\frac{1}{t}\right) , \quad \text{uniformly for } \chi_-(t) < x < \chi_+(t) ,$$

$(11.6.3)$

$$u(x\pm, t) = \frac{1}{f''(0)}\left[\frac{x}{t} - f'(0)\right] + O\left(\frac{1}{t}\right) , \quad \text{uniformly for } \chi_-(t) < x < \chi_+(t) ,$$

$$(11.6.4) \quad \begin{cases} \chi_-(t) = t f'(0) - [2q_- t f''(0)]^{1/2} + O(1) \,, \\ \chi_+(t) = t f'(0) + [2q_+ t f''(0)]^{1/2} + O(1) \,, \end{cases}$$

with q_- and q_+ given by (11.4.4).

Proof. Since $\chi_-(\cdot)$ is minimal and $\chi_+(\cdot)$ is maximal, the extremal backward characteristics emanating from any point (x, t) with $t > 0$ and $x < \chi_-(t)$ or $x > \chi_+(t)$ will be intercepted by the x-axis outside the support of u_0. This establishes (11.6.1).

On the other hand, the minimal or maximal backward characteristic $\xi(\cdot)$ emanating from a point (x, t) with $t > 0$ and $\chi_-(t) < x < \chi_+(t)$ will be intercepted by the x-axis inside the interval $[-\ell, \ell]$, i.e., $\xi(0) \in [-\ell, \ell]$. Consequently, as $t \to \infty$, $x - t f'(u(x\pm, t)) = \xi(0) = O(1)$, which yields (11.6.2).[1]

By account of Theorem 11.5.2, u is $O(t^{-1/2})$, as $t \to \infty$, and thus, assuming f is C^3, $f'(u) = f'(0) + f''(0)u + O(t^{-1})$. Therefore, (11.6.3) follows from (11.6.2).

To derive the asymptotics of $\chi_\pm(t)$, as $t \to \infty$, we first note that $0 \ge \dot{\chi}_-(t) - f'(0) \ge O(t^{-1/2})$, $0 \le \dot{\chi}_+(t) - f'(0) \le O(t^{-1/2})$ and so $0 \ge \chi_-(t) - t f'(0) \ge O(t^{1/2})$, $0 \le \chi_+(t) - t f'(0) \le O(t^{1/2})$. Next we appeal to Theorem 11.4.2: A divide $x = \overline{x} + t f'(0)$ originates from some point $(\overline{x}, 0)$, with $\overline{x} \in [-\ell, \ell]$, along which u is zero, and for any $t > 0$,

$$(11.6.5) \quad \int_{\overline{x} + t f'(0)}^{\chi_-(t)} u(y, t) dy = q_- \,, \quad \int_{\overline{x} + t f'(0)}^{\chi_+(t)} u(y, t) dy = q_+ \,.$$

In (11.6.5) we insert u from its asymptotic form (11.6.3) and after performing the simple integration we deduce

$$(11.6.6) \quad \frac{1}{2q_\pm t f''(0)} [\chi_\pm(t) - t f'(0)]^2 = 1 + O(t^{-1/2})$$

whence (11.6.4) follows. The proof is complete.

11.7 Confinement of Characteristics and Formation of Sawtoothed Profiles

The confinement of the intercepts of extremal backward characteristics in a bounded interval of the x-axis induces bounds on the decreasing variation of characteristic speeds and thereby, by virtue of genuine nonlinearity, on the decreasing variation of the solution itself.

Theorem 11.7.1 Let $\chi_-(\cdot)$ and $\chi_+(\cdot)$ be generalized characteristics on $[0, \infty)$, associated with an admissible solution u, and $\chi_-(t) < \chi_+(t)$ for $t \in [0, \infty)$. Then,

---

[1] As $t \to \infty$, the $\xi(0)$ accumulate at the set of points from which divides originate. In the generic case where (11.4.2) holds, with $\overline{u} = 0$, at a single point $\overline{x}$, which we normalize so that $\overline{x} = 0$, the $\xi(0)$ accumulate at the origin and hence in (11.6.2) $O(t^{-1})$ is upgraded to $o(t^{-1})$. When, in addition, u_0 is C^1 and $u_0'(0) > 0$, then in (11.6.2) $O(t^{-1})$ is improved to $O(t^{-2})$ and, for t large, the profile $u(\cdot, t)$ is C^1 on the interval $(\chi_-(t), \chi_+(t))$.

for any $t > 0$, the decreasing variation of the function $f'(u(\cdot, t))$ over the interval $(\chi_-(t), \chi_+(t))$ cannot exceed $[\chi_+(0) - \chi_-(0)]t^{-1}$. Thus the decreasing variation of $u(\cdot, t)$ over the interval $(\chi_-(t), \chi_+(t))$ is $O(t^{-1})$ as $t \to \infty$.

Proof. Fix $t > 0$ and consider any mesh $\chi_-(t) < x_1 < x_2 < \cdots < x_{2m-1} < x_{2m} < \chi_+(t)$ such that (x_i, t) is a point of continuity of u and $u(x_{2k-1}, t) > u(x_{2k}, t)$, $k = 1, \cdots, m$. Let $\xi_i(\cdot)$ denote the (unique) backward characteristic emanating from (x_i, t). Then $\chi_-(0) \le \xi_1(0) \le \cdots \le \xi_{2m}(0) \le \chi_+(0)$. Furthermore, $\xi_i(0) = x_i - tf'(u(x_i, t))$ and so

$$(11.7.1) \qquad \sum_{k=1}^{m} t[f'(u(x_{2k-1}, t)) - f'(u(x_{2k}, t))] \le \chi_+(0) - \chi_-(0)$$

whence the assertion of the theorem follows. This completes the proof.

In particular, referring to the setting of Theorem 11.6.1, we deduce that the decreasing variation of the N-wave profile $u(\cdot, t)$ over the interval $(\chi_-(t), \chi_+(t))$ is $O(t^{-1})$, as $t \to \infty$.

Another corollary of Theorem 11.7.1 is that when the initial data u_0, and thereby the solution u, are periodic in x, then the decreasing variation, and hence also the total variation, of $u(\cdot, t)$ over any period interval is $O(t^{-1})$ as $t \to \infty$. We may achieve finer resolution than $O(t^{-1})$ by paying closer attention to the initial data:

Theorem 11.7.2 *Let u be an admissible solution with initial data u_0. Assume $\chi_-(t) = x_- + tf'(\bar{u})$ and $\chi_+(t) = x_+ + tf'(\bar{u})$, $x_- < x_+$, are adjacent divides associated with u, that is (11.4.2) holds for $\bar{x} = x_-$ and $\bar{x} = x_+$ but for no other $\bar{x}$ in the interval (x_-, x_+). Then*

$$(11.7.2) \qquad \int_{\chi_-(t)}^{\chi_+(t)} u(x, t)dx = \int_{x_-}^{x_+} u_0(y)dy = (x_+ - x_-)\bar{u} , \quad t \in [0, \infty) .$$

Consider any forward characteristic $\psi(\cdot)$ issuing from the point $(\frac{x_- + x_+}{2}, 0)$. Then, as $t \to \infty$,

$$(11.7.3) \qquad \psi(t) = \frac{1}{2}[\chi_-(t) + \chi_+(t)] + o(1) ,$$

$(11.7.4)$

$$u(x\pm, t) = \begin{cases} \bar{u} + \dfrac{1}{f''(\bar{u})} \dfrac{x - \chi_-(t)}{t} + o\left(\dfrac{1}{t}\right) , & \textit{uniformly for } \chi_-(t) < x < \psi(t) , \\[3ex] \bar{u} + \dfrac{1}{f''(\bar{u})} \dfrac{x - \chi_+(t)}{t} + o\left(\dfrac{1}{t}\right) , & \textit{uniformly for } \psi(t) < x < \chi_+(t) . \end{cases}$$

Moreover, the decreasing variation of $u(\cdot, t)$ over the intervals $(\chi_-(t), \psi(t))$ and $(\psi(t), \chi_+(t))$ is $o(t^{-1})$ as $t \to \infty$.

Proof. To verify the first equality in (11.7.2), it suffices to integrate (11.1.1) over the parallelogram $\{(x, \tau) : 0 < \tau < t, \chi_-(\tau) < x < \chi_+(\tau)\}$ and then apply Green's theorem. The second equality in (11.7.2) follows because (11.4.2) holds for both $\overline{x} = x_-$ and $\overline{x} = x_+$.

For $t > 0$, we let $\xi_-^t(\cdot)$ and $\xi_+^t(\cdot)$ denote the minimal and the maximal backward characteristics emanating from the point $(\chi(t), t)$. As $t \uparrow \infty$, $\xi_-^t(0) \downarrow x_-$ and $\xi_+^t(0) \uparrow x_+$, because otherwise there would exist divides originating at points $(\overline{x}, 0)$ with $\overline{x} \in (x_-, x_+)$, contrary to our assumptions. It then follows from Theorem 11.7.1 that the decreasing variation of $f'(u(\cdot, t))$, and thereby also the decreasing variation of $u(\cdot, t)$ itself, over the intervals $(\chi_-(t), \psi(t))$ and $(\psi(t), \chi_+(t))$ is $o(t^{-1})$ as $t \to \infty$.

The extremal backward characteristics emanating from any point (x, t) with $\chi_-(t) < x < \psi(t)$ (or $\psi(t) < x < \chi_+(t)$) will be intercepted by the x-axis inside the interval $[x_-, \xi_-^t(0)]$ (or $[\xi_+^t(0), x_+]$) and thus

(11.7.5)
$$x - tf'(u(x\pm, t)) = \begin{cases} x_- + o(t^{-1}) \,, & \text{uniformly for } \chi_-(t) < x < \psi(t) \,, \\ x_+ + o(t^{-1}) \,, & \text{uniformly for } \psi(t) < x < \chi_+(t) \,. \end{cases}$$

Since $u(\chi_-(t), t) = u(\chi_+(t), t) = \overline{u}$, Theorem 11.7.1 implies $u - \overline{u} = O(t^{-1})$ and so, as $t \to \infty$, $f'(u) = f'(\overline{u}) + f''(\overline{u})(u - \overline{u}) + O(t^{-2})$. This together with (11.7.5) yield (11.7.4).

Finally, introducing u from (11.7.4) into (11.7.2) we arrive at (11.7.3). The proof is complete.

We shall employ the above proposition to describe the asymptotics of periodic solutions:

Theorem 11.7.3 *When the initial data u_0 are periodic, with mean $\overline{u}$, then, as $t \to \infty$, the admissible solution u tends, at the rate $o(t^{-1})$, to a periodic serrated profile consisting of wavelets of the form (11.7.4). The number of wavelets (or teeth) per period equals the number of divides per period or, equivalently, the number of points on any interval of the x-axis of period length at which the primitive of $u_0 - \overline{u}$ attains its minimum. In particular, in the generic case where the minimum of the primitive of $u_0 - \overline{u}$ is attained at a unique point on each period interval, u tends to a saw-tooth shaped profile with a single tooth per period.*

Proof. It is an immediate corollary of Theorems 11.4.1 and 11.7.2. If u_0 is periodic, (11.4.2) may hold only when $\overline{u}$ is the mean of u_0 and is attained at points $\overline{x}$ where the primitive of $u_0 - \overline{u}$ is minimized. The set of such points is obviously invariant under period translations and contains at least one (generically precisely one) point in each interval of period length.

11.8 Comparison Theorems and L^1 Stability

The assertions of Theorem 6.2.2 will be reestablished here, in sharper form, for the special case of genuinely nonlinear scalar conservation laws (11.1.1), in one-space dimension. The key factor will be the properties of the function

$$(11.8.1) \quad Q(u, v, w) = \begin{cases} f(v) - f(u) - \dfrac{f(u) - f(w)}{u - w}[v - u] , & \text{if } u \neq w , \\[2mm] f(v) - f(u) - f'(u)[v - u] , & \text{if } u = w , \end{cases}$$

defined for u, v and w in $\mathbb{R}$. Clearly, $Q(u, v, w) = Q(w, v, u)$. Since f is uniformly convex, $Q(u, v, w)$ will be negative when v lies between u and w, and positive when v lies outside the interval with endpoints u and w. In particular, for the Burgers equation (4.2.1), $Q(u, v, w) = \frac{1}{2}(v - u)(v - w)$.

The first step is to refine the ordering property:

Theorem 11.8.1 *Let u and $\bar{u}$ be admissible solutions of (11.1.1), on the upper half-plane, with respective initial data u_0 and $\bar{u}_0$ such that*

$$(11.8.2) \qquad u_0(x) \leq \bar{u}_0(x) , \quad \text{for all } x \in (y, \bar{y}) .$$

Let $\psi(\cdot)$ be any forward characteristic, associated with the solution u, issuing from the point $(y, 0)$, and $\bar{\psi}(\cdot)$ be any forward characteristic, associated with $\bar{u}$, issuing from $(\bar{y}, 0)$. Then, for any $t > 0$ with $\psi(t) < \bar{\psi}(t)$,

$$(11.8.3) \qquad u(x, t) \leq \bar{u}(x, t) , \quad \text{for all } x \in (\psi(t), \bar{\psi}(t)) .$$

Proof. We fix any interval $(z, \bar{z})$ with $\psi(t) < z < \bar{z} < \bar{\psi}(t)$ and consider the maximal backward characteristic $\xi(\cdot)$, associated with the solution u, emanating from the point (z, t), and the minimal backward characteristic $\bar{\zeta}(\cdot)$, associated with $\bar{u}$, emanating from the point $(\bar{z}, t)$. Thus, $\xi(0) \geq y$ and $\bar{\zeta}(0) \leq \bar{y}$.

Suppose first $\xi(0) < \bar{\zeta}(0)$. We integrate the equation

$$(11.8.4) \qquad \partial_t[u - \bar{u}] + \partial_x[f(u) - f(\bar{u})] = 0$$

over the trapezoid $\{(x, \tau) : 0 < \tau < t, \, \xi(\tau) < x < \bar{\zeta}(\tau)\}$ and apply Green's theorem to get

$$(11.8.5) \begin{aligned} & \int_z^{\bar{z}} [u(x, t) - \bar{u}(x, t)]dx - \int_{\xi(0)}^{\bar{\zeta}(0)} [u_0(x) - \bar{u}_0(x)]dx \\ &= - \int_0^t Q(u(\xi(\tau), \tau), \bar{u}(\xi(\tau), \tau), u(\xi(\tau), \tau))d\tau \\ & \quad - \int_0^t Q(\bar{u}(\bar{\zeta}(\tau), \tau), u(\bar{\zeta}(\tau), \tau), \bar{u}(\bar{\zeta}(\tau), \tau))d\tau . \end{aligned}$$

Both integrals on the right-hand side of (11.8.5) are nonnegative. Hence, by virtue of (11.8.2), the integral of $u(\cdot, t) - \bar{u}(\cdot, t)$ over $(z, \bar{z})$ is nonpositive.

Supppose now $\xi(0) \geq \overline{\zeta}(0)$. Then the straight lines $\xi(\cdot)$ and $\overline{\zeta}(\cdot)$ must intersect at some time $s \in [0, t)$. In that case we integrate (11.8.4) over the triangle $\{(x, \tau) : s < \tau < t, \xi(\tau) < x < \overline{\zeta}(\tau)\}$ and employ the same argument as above to deduce that the integral of $u(\cdot, t) - \overline{u}(\cdot, t)$ over $(z, \overline{z})$ is again nonpositive.

Since $(z, \overline{z})$ is an arbitrary subinterval of $(\psi(t), \overline{\psi}(t))$, we conclude (11.8.2). The proof is complete.

As a corollary of the above theorem, we infer that the number of sign changes of the function $u(\cdot, t) - \overline{u}(\cdot, t)$ over $(-\infty, \infty)$ is nonincreasing with time. Indeed, assume there are points $-\infty = y_0 < y_1 < \cdots < y_n < y_{n+1} = \infty$ such that, on each interval (y_i, y_{i+1}), $u_0(\cdot) - \overline{u}_0(\cdot)$ is nonnegative when i is even and nonpositive when i is odd. Let $\psi_i(\cdot)$ be any forward characteristic, associated with the solution u, issuing from the point $(y_i, 0)$ with i odd, and $\overline{\psi}_i(\cdot)$ any forward characteristic, associated with $\overline{u}$, issuing from $(y_i, 0)$ with i even. These curves are generally assigned finite life spans, according to the following prescription. At the time t_1 of the earliest collision between some ψ_i and some $\overline{\psi}_j$, these two curves are terminated. Then, at the time t_2 of the next collision between any (surviving) ψ_k and $\overline{\psi}_\ell$, these two curves are likewise terminated; and so on. By virtue of Theorem 11.8.1, the function $u(\cdot, t) - \overline{u}(\cdot, t)$ undergoes n sign changes for any $t \in [0, t_1)$, $n - 2$ sign changes for any $t \in [t_1, t_2)$, and so on. In particular, the so called *lap number*, which counts the crossings of the graph of the solution $u(\cdot, t)$ with any fixed constant $\overline{u}$, is nonincreasing with time.

By Theorem 6.2.2, the spatial L^1 distance of any pair of admissible solutions of a scalar conservation law is nonincreasing with time. In the present setting, it will be shown that it is actually possible to determine under what conditions is the L^1 distance strictly decreasing and at what rate:

Theorem 11.8.2 *Let u and $\overline{u}$ be admissible solutions of* (11.1.1) *with initial data u_0 and $\overline{u}_0$ in $L^1(-\infty, \infty)$. Thus $\|u(\cdot, t) - \overline{u}(\cdot, t)\|_{L^1(-\infty, \infty)}$ is a nonincreasing function of t which is locally Lipschitz on $(0, \infty)$. For any fixed $t \in (0, \infty)$, consider the (possibly empty and at most countable) sets*

$$(11.8.6) \quad \begin{cases} \mathscr{J} = \{y \in (-\infty, \infty) : u_+ < \overline{u}_+ \leq \overline{u}_- < u_-\}, \\ \overline{\mathscr{J}} = \{y \in (-\infty, \infty) : \overline{u}_+ < u_+ \leq u_- < \overline{u}_-\}, \end{cases}$$

where $u_\pm$ and $\overline{u}_\pm$ stand for $u(y\pm, t)$ and $\overline{u}(y\pm, t)$, respectively. Let

$$(11.8.7)_1 \quad u_* = \begin{cases} u_\pm \text{ if } u_+ = u_-, \\ u_- \text{ if } u_+ < u_- \text{ and } \dfrac{f(u_+) - f(u_-)}{u_+ - u_-} \geq \dfrac{f(\overline{u}_+) - f(\overline{u}_-)}{\overline{u}_+ - \overline{u}_-}, \\ u_+ \text{ if } u_+ < u_- \text{ and } \dfrac{f(u_+) - f(u_-)}{u_+ - u_-} < \dfrac{f(\overline{u}_+) - f(\overline{u}_-)}{\overline{u}_+ - \overline{u}_-}, \end{cases}$$

$$(11.8.7)_2 \quad \bar{u}_* = \begin{cases} \bar{u}_\pm \text{ if } \bar{u}_+ = \bar{u}_- , \\[2mm] \bar{u}_- \text{ if } \bar{u}_+ < \bar{u}_- \text{ and } \dfrac{f(\bar{u}_+) - f(\bar{u}_-)}{\bar{u}_+ - \bar{u}_-} \geq \dfrac{f(u_+) - f(u_-)}{u_+ - u_-} , \\[3mm] \bar{u}_+ \text{ if } \bar{u}_+ < \bar{u}_- \text{ and } \dfrac{f(\bar{u}_+) - f(\bar{u}_-)}{\bar{u}_+ - \bar{u}_-} < \dfrac{f(u_+) - f(u_-)}{u_+ - u_-} . \end{cases}$$

Then

(11.8.8)

$$\frac{d^+}{dt} \|u(\cdot, t) - \bar{u}(\cdot, t)\|_{L^1(-\infty, \infty)} = 2 \sum_{y \in \mathscr{I}} Q(u_-, \bar{u}_*, u_+) + 2 \sum_{y \in \overline{\mathscr{I}}} Q(\bar{u}_-, u_*, \bar{u}_+) .$$

Proof. First we establish (11.8.8) for the special case where $u(\cdot, t) - \bar{u}(\cdot, t)$ undergoes a finite number of sign changes on $(-\infty, \infty)$, i.e., there are points $-\infty = y_0 < y_1 < \cdots < y_n < y_{n+1} = \infty$ such that, on each interval (y_i, y_{i+1}), $u(\cdot, t) - \bar{u}(\cdot, t)$ is nonnegative when i is even and nonpositive when i is odd. In particular, any $y \in \mathscr{I}$ must be one of the y_i, with i odd, and any $y \in \overline{\mathscr{I}}$ must be one of the y_i, with i even.

Let $\psi_i(\cdot)$ be the (unique) forward characteristic, associated with the solution u, issuing from the point (y_i, t) with i odd, and $\overline{\psi}_i(\cdot)$ be the forward characteristic, associated with $\bar{u}$, issuing from (y_i, t) with i even. We fix $s > t$ with $s - t$ so small that no collisions of the above curves may occur on $[t, s]$, and integrate (11.8.4) over the domains $\{(x, \tau) : t < \tau < s, \psi_i(\tau) < x < \overline{\psi}_{i+1}(\tau)\}$, for i odd, and $\{(x, \tau) : t < \tau < s, \overline{\psi}_i(\tau) < x < \psi_{i+1}(\tau)\}$, for i even. We apply Green's theorem and employ Theorem 11.8.1, to deduce

(11.8.9)

$$\|u(\cdot, s) - \bar{u}(\cdot, s)\|_{L^1(-\infty, \infty)} - \|u(\cdot, t) - \bar{u}(\cdot, t)\|_{L^1(-\infty, \infty)}$$

$$= \sum_{i \text{ even}} \int_{\overline{\psi}_i(s)}^{\psi_{i+1}(s)} [u(x, s) - \bar{u}(x, s)]dx + \sum_{i \text{ odd}} \int_{\psi_i(s)}^{\overline{\psi}_{i+1}(s)} [\bar{u}(x, s) - u(x, s)]dx$$

$$- \sum_{i \text{ even}} \int_{y_i}^{y_{i+1}} [u(x, t) - \bar{u}(x, t)]dx - \sum_{i \text{ odd}} \int_{y_i}^{y_{i+1}} [\bar{u}(x, t) - u(x, t)]dx$$

$$= \sum_{i \text{ odd}} \int_t^s \{Q(u(\psi_i(\tau)-, \tau), \bar{u}(\psi_i(\tau)-, \tau), u(\psi_i(\tau)+, \tau))$$

$$+ Q(u(\psi_i(\tau)+, \tau), \bar{u}(\psi_i(\tau)+, \tau), u(\psi_i(\tau)-, \tau))\}d\tau$$

$$+ \sum_{i \text{ even}} \int_t^s \{Q(\bar{u}(\overline{\psi}_i(\tau)-, \tau), u(\overline{\psi}_i(\tau)-, \tau), \bar{u}(\overline{\psi}_i(\tau)+, \tau))$$

$$+ Q(\bar{u}(\overline{\psi}_i(\tau)+, \tau), u(\overline{\psi}_i(\tau)+, \tau), \bar{u}(\overline{\psi}_i(\tau)-, \tau))\}d\tau .$$

By virtue of Theorem 11.3.1, as $s \downarrow t$ the integrand in the first integral on the right-hand side of (11.8.9) tends to zero, if $y_i \notin \mathscr{J}$, or to $2Q(u_-, \overline{u}_*, u_+)$, if $y_i \in \mathscr{J}$. Similarly, the integrand in the second integral on the right-hand side of (11.8.9) tends to zero, if $y_i \notin \overline{\mathscr{J}}$, or to $2Q(\overline{u}_-, u_*, \overline{u}_+)$, if $y_i \in \overline{\mathscr{J}}$. Therefore, upon dividing (11.8.9) by $s - t$ and letting $s \downarrow t$, we arrive at (11.8.8).

We now turn to the general situation, where $u(\cdot, t) - \overline{u}(\cdot, t)$ may undergo infinitely many sign changes over $(-\infty, \infty)$. In that case, the open set $\{x \in (-\infty, \infty) : u(x\pm, t) - \overline{u}(x\pm, t) < 0\}$ is the countable union of disjoint open intervals $(y_i, \overline{y}_i)$. For $m = 1, 2, \cdots$, we let u_m denote the admissible solution of our conservation law (11.1.1) on $(-\infty, \infty) \times [t, \infty)$, with

$$(11.8.10) \qquad u_m(x, t) = \begin{cases} \overline{u}(x, t) , & x \in \bigcup_{i=m}^{\infty} (y_i, \overline{y}_i) \\ u(x, t) , & \text{otherwise} . \end{cases}$$

Thus $u_m(\cdot, t) - \overline{u}(\cdot, t)$ undergoes a finite number of sign changes over $(-\infty, \infty)$ and so, for $\tau \geq t$, $\frac{d^+}{d\tau} \|u_m(\cdot, \tau) - \overline{u}(\cdot, \tau)\|_{L^1}$ is evaluated by the analog of (11.8.8). Moreover, the function $\tau \mapsto \frac{d^+}{d\tau} \|u_m(\cdot, \tau) - \overline{u}(\cdot, \tau)\|_{L^1}$ is right-continuous at t and the modulus of right continuity is independent of m. To verify this, note that the total contribution of small jumps to the rate of change of $\|u_m(\cdot, \tau) - \overline{u}(\cdot, \tau)\|_{L^1}$ is small, controlled by the total variation of $u(\cdot, t)$ and $\overline{u}(\cdot, t)$ over $(-\infty, \infty)$, while the contribution of the (finite number of) large jumps is right-continuous, by account of Theorem 11.3.1. Therefore, by passing to the limit, as $m \to \infty$, we establish (11.8.8) for general solutions u and $\overline{u}$. The proof is complete.

According to the above theorem, the L^1 distance of $u(\cdot, t)$ and $\overline{u}(\cdot, t)$ may decrease only when the graph of either one of these functions happens to cross the graph of the other at a point of jump discontinuity. More robust contraction is realized in terms of a new metric which weighs the L^1 distance of two solutions by a weight specially tailored to them.

For v and $\overline{v}$ in $BV(-\infty, \infty)$, let

$(11.8.11)$

$$\rho(v, \overline{v}) = \int_{-\infty}^{\infty} \{(V(x) + \overline{V}(\infty) - \overline{V}(x))[v(x) - \overline{v}(x)]^+ + (\overline{V}(x) + V(\infty) - V(x))[\overline{v}(x) - v(x)]^+\} dx ,$$

where the superscript $+$ denotes "positive part", $w^+ = \max\{w, 0\}$, and V or $\overline{V}$ denotes the variation function of v or $\overline{v}$, i.e., $V(x) = TV_{(-\infty, x)}v(\cdot)$, $\overline{V}(x) = TV_{(-\infty, x)}\overline{v}(\cdot)$.

Theorem 11.8.3 *Let u and $\overline{u}$ be admissible solutions of* (11.1.1) *with initial data u_0 and $\overline{u}_0$ in $BV(-\infty, \infty)$. Then, for any fixed $t \in (0, \infty)$,*

$$\frac{d^+}{dt}\rho(u(\cdot,t),\overline{u}(\cdot,t))$$

$$\leq -\int_{-\infty}^{\infty} Q(u(x,t),\overline{u}(x,t),u(x,t))dV_t^c(x)$$

(11.8.12)
$$-\int_{-\infty}^{\infty} Q(\overline{u}(x,t),u(x,t),\overline{u}(x,t))d\overline{V}_t^c(x)$$

$$-\sum_{y\in\mathscr{H}}(u_- - u_+)Q(u_-,\overline{u}_*,u_+) - \sum_{y\in\overline{\mathscr{H}}}(\overline{u}_- - \overline{u}_+)Q(\overline{u}_-,u_*,\overline{u}_+)$$

$$+(V_t(\infty)+\overline{V}_t(\infty))\left\{\sum_{y\in\mathscr{J}} Q(u_-,\overline{u}_*,u_+) + \sum_{y\in\overline{\mathscr{J}}} Q(\overline{u}_-,u_*,\overline{u}_+)\right\},$$

where V_t or $\overline{V}_t$ is the variation function of $u(\cdot,t)$ or $\overline{u}(\cdot,t)$; V_t^c or $\overline{V}_t^c$ denotes the continuous part of V_t or $\overline{V}_t$; $u_\pm$ or $\overline{u}_\pm$ stand for $u(y\pm,t)$ or $\overline{u}(y\pm,t)$ while u_* and $\overline{u}_*$ are again determined through $(11.8.7)_1$ and $(11.8.7)_2$; the sets $\mathscr{J}$ and $\overline{\mathscr{J}}$ are defined by (11.8.6) and $\mathscr{H}$ or $\overline{\mathscr{H}}$ denotes the set of jump points of $u(\cdot,t)$ or $\overline{u}(\cdot,t)$:

(11.8.13)
$$\begin{cases} \mathscr{H} = \{y \in (-\infty,\infty) : u_+ < u_-\}, \\ \overline{\mathscr{H}} = \{y \in (-\infty,\infty) : \overline{u}_+ < \overline{u}_-\}. \end{cases}$$

Proof. We begin as in the proof of Theorem 11.8.2: We assume there are points $-\infty = y_0 < y_1 < \cdots < y_n < y_{n+1} = \infty$ such that, on each interval (y_i, y_{i+1}), $u(\cdot,t) - \overline{u}(\cdot,t)$ is nonnegative when i is even and nonpositive when i is odd. We consider the forward characteristic $\psi_i(\cdot)$, associated with u, issuing from each point (y_i, t), with i odd, and the forward characteristic $\overline{\psi}_i(\cdot)$, associated with $\overline{u}$, issuing from each (y_i, t), with i even.

We focus our attention on some (y_i, y_{i+1}) with i even. Let us assume $-\infty < y_i < y_{i+1} < \infty$, as the other cases are simpler. With the exception of $\overline{\psi}_i(\cdot)$, all characteristics to be considered below will be associated with the solution u. The argument varies soomewhat, depending on whether the forward characteristic χ_0 issuing from (y_i, t) lies to the left or to the right of $\overline{\psi}_i(\cdot)$; for definiteness, we shall treat the latter case, which is slightly more complicated.

We fix ε positive and small and identify all $z_1, \cdots, z_N$, $y_i < z_1 < \cdots < z_N < y_{i+1}$, such that $u(z_I-, t) - u(z_I+, t) \geq \varepsilon$, $I = 1, \cdots, N$. We consider the forward characteristic $\chi_I(\cdot)$ issuing from the point (z_I, t), $I = 1, \cdots, N$. Then we select $s > t$ with $s - t$ so small that the following hold: (a) No intersection of any two of the characteristics $\chi_0, \chi_1, \cdots, \chi_N$ and ψ_{i+1} may occur on the time interval $[t, s]$. (b) For $I = 1, \cdots, N$, if $\zeta_I(\cdot)$ and $\xi_I(\cdot)$ denote the minimal and the maximal backward characteristics emanating from the point $(\chi_I(s), s)$, then the total variation of $u(\cdot,t)$ over the intervals $(\zeta_I(t), z_I)$ and $(z_I, \xi_I(t))$ does not exceed ε/N. (c) If $\zeta(\cdot)$ denotes the minimal backward characteristic emanating from $(\psi_{i+1}(s), s)$, then the total variation of $u(\cdot,t)$ over the interval $(\zeta(t), y_{i+1})$ does not exceed ε. (d) If $\zeta_0(\cdot)$ is the minimal backward characteristic emanating from $(\psi_i(s), s)$

and $\xi_0(\cdot)$ is the maximal backward characteristic emanating from $(\chi_0(s), s)$, then the total variation of $u(\cdot, t)$ over the intervals $(\zeta_0(t), y_i)$ and $(y_i, \xi_0(t))$ does not exceed ε.

For $I = 0, \cdots, N - 1$, and some k to be fixed later, we set a mesh on the interval $[\chi_I(s), \chi_{I+1}(s)] : \chi_I(s) = x_I^0 < x_I^1 < \cdots < x_I^k < x_I^{k+1} = \chi_{I+1}(s)$; and likewise for the interval $[\chi_N(s), \psi_{i+1}(s)] : \chi_N(s) = x_N^0 < x_N^1 < \cdots < x_N^k < x_N^{k+1} = \psi_{i+1}(s)$. For $I = 0, \cdots, N$ and $j = 1, \cdots, k$, we consider the maximal backward characteristic $\xi_I^j(\cdot)$ emanating from the point (x_I^j, s) and identify its intercept $z_I^j = \xi_I^j(t)$ by the t-time line. We also set $z_0^0 = y_i, z_N^{k+1} = y_{i+1}$ and $z_{I-1}^{k+1} = z_I^0 = z_I, I = 1, \cdots, N$.

We now note the identity

$$(11.8.14) \qquad\qquad R - S = -D ,$$

where

$$(11.8.15) \qquad \begin{aligned} R &= \int_{\overline{\psi}_i(s)}^{\chi_0(s)} V_t(y_i)[u(x, s) - \overline{u}(x, s)]dx \\ &+ \sum_{I=0}^{N} \sum_{j=0}^{k} \int_{x_I^j}^{x_I^{j+1}} V_t(z_I^i +)[u(x, s) - \overline{u}(x, s)]dx , \end{aligned}$$

$$(11.8.16) \qquad S = \sum_{I=0}^{N} \sum_{j=0}^{k} \int_{z_I^j}^{z_I^{j+1}} V_t(z_I^j +)[u(x, t) - \overline{u}(x, t)]dx ,$$

$(11.8.17)$

$$\begin{aligned} D &= \sum_{I=0}^{N} \sum_{j=1}^{k} \int_t^s [V_t(z_I^j +) - V_t(z_I^{j-1} +)] \\ &\qquad\qquad \times Q(u(\xi_I^j(\tau), \tau), \overline{u}(\xi_I^j(\tau)-, \tau), u(\xi_I^j(\tau), \tau))d\tau \\ &+ \sum_{I=1}^{N} \int_t^s [V_t(z_I +) - V_t(z_{I-1}^k +)] \\ &\qquad\qquad \times Q(u(\chi_I(\tau)-, \tau), \overline{u}(\chi_I(\tau)-, \tau), u(\chi_I(\tau)+, \tau))d\tau \\ &+ \int_t^s [V_t(y_i +) - V_t(y_i)]Q(u(\chi_0(\tau)-, \tau), \overline{u}(\chi_0(\tau)-, \tau), u(\chi_0(\tau)+, \tau))d\tau \\ &- \int_t^s V_t(y_i)Q(\overline{u}(\overline{\psi}_i(\tau)-, \tau), u(\overline{\psi}_i(\tau)+, \tau), \overline{u}(\overline{\psi}_i(\tau)+, \tau))d\tau \\ &- \int_t^s V_t(z_N^k +)Q(u(\psi_{i+1}(\tau)-, \tau), \overline{u}(\psi_{i+1}(\tau)-, \tau), u(\psi_{i+1}(\tau)+, \tau))d\tau . \end{aligned}$$

To verify (11.8.14), one first integrates (11.8.4) over the domains $\{(x, \tau) : t < \tau < s, \overline{\psi}_i(\tau) < x < \chi_0(\tau)\}$, $\{(x, \tau) : t < \tau < s, \xi_I^j(\tau) < x < \xi_I^{j+1}(\tau)\}$, $\{(x, \tau) : t < \tau < s, \chi_I(\tau) < x < \xi_I^1(\tau)\}$, $\{(x, \tau) : t < \tau < s, \xi_I^k(\tau) < x < \chi_{I+1}(\tau)\}$, $\{(x, \tau) : t < \tau < s, \xi_N^k(\tau) < x < \psi_{i+1}(\tau)\}$ and applies Green's theorem; then forms the weighted sum of the resulting equations, with respective weights $V_t(y_i), V_t(z_I^j +), V_t(z_I +), V_t(z_I^k +), V_t(z_N^k +)$.

To estimate R, we note that $V_t(y_i) \geq V_s(\chi_0(s))$, and $V_t(z_I^j+) \geq V_s(x_I^j+)$, $I = 0, \cdots, N, j = 0, \cdots, k$. Hence, if we pick the $x_I^{j+1} - x_I^j$ sufficiently small, we can guarantee

$$(11.8.18) \qquad R \geq \int_{\overline{\psi}_i(s)}^{\psi_{i+1}(s)} V_s(x)[u(x,s) - \overline{u}(x,s)]dx - (s-t)\varepsilon .$$

To estimate S, it suffices to observe that $V_t(\cdot)$ is nondecreasing, and so

$$(11.8.19) \qquad S \leq \int_{\overline{\psi}_i(t)}^{\psi_{i+1}(t)} V_t(x)[u(x,t) - \overline{u}(x,t)]dx .$$

To estimate D, the first remark is that, due to the properties of Q, all five terms are nonnegative. For $I = 0, \cdots, N$ and $j = 1, \cdots, k$, $V_t(z_I^j+) - V_t(z_I^{j-1}+) \geq V_t^c(z_I^j) - V_t^c(z_I^{j-1})$. Furthermore,

$$(11.8.20) \qquad \begin{aligned} &Q(u(\xi_I^j(\tau),\tau), \overline{u}(\xi_I^j(\tau)-,\tau), u(\xi_I^j(\tau),\tau)) \\ &= Q(u(z_I^j,t), \overline{u}(p_\tau(z_I^j),t), u(z_I^j,t)) , \end{aligned}$$

where the monotone increasing function p_τ is determined through

$$(11.8.21) \qquad p_\tau(x) = x + (\tau - t)[f'(u(x,t)) - f'(\overline{u}(p_\tau(x),t))] .$$

Upon choosing the $x_I^{j+1} - x_I^j$ so small that the oscillation of $V_t^c(\cdot)$ over each one of the intervals (z_I^j, z_I^{j+1}) does not exceed ε, the standard estimates on Stieltjes integral imply

$(11.8.22)$

$$\begin{aligned} &\sum_{I=0}^{N} \sum_{j=1}^{k} [V_t(z_I^j+) - V_t(z_I^{j-1}+)]Q(u(\xi_I^j(\tau),\tau), \overline{u}(\xi_I^j(\tau)-,\tau), u(\xi_I^j(\tau),\tau)) \\ &\geq \int_{y_i}^{y_{i+1}} Q(u(x,t), \overline{u}(p_\tau(x),t), u(x,t))dV_t^c(x) - c\varepsilon . \end{aligned}$$

We now combine (11.8.14) with (11.8.18), (11.8.19), (11.8.17) and (11.8.22), then we divide the resulting inequality by $s - t$, we let $s \downarrow t$; and finally we let $\varepsilon \downarrow 0$. This yields

$(11.8.23)$

$$\begin{aligned} &\frac{d^+}{dt} \int_{\overline{\psi}_i(t)}^{\psi_{i+1}(t)} V_t(x)[u(x,t) - \overline{u}(x,t)]dx \\ &\leq -\int_{y_i}^{y_{i+1}} Q(u(x,t), \overline{u}(x,t), u(x,t))dV_t^c(x) \\ &\quad - \sum (u_- - u_+)Q(u_-, \overline{u}_*, u_+) \\ &\quad + V_t(y_i)Q(\overline{u}_-, u_*, \overline{u}_+) + V_t(y_{i+1})Q(u_-, \overline{u}_*, u_+) , \end{aligned}$$

where the summation runs over all y in $\mathcal{H} \cap (y_i, y_{i+1})$, plus y_i if $y_i \in \mathcal{H}$ and χ_0 lies to the right of $\overline{\psi}_i$, as assumed above. The $u_\pm, \overline{u}_\pm, u_*$ and $\overline{u}_*$ are of course evaluated at the corresponding y.

Next we focus on intervals (y_i, y_{i+1}) with i odd. A completely symmetrical argument yields, in the place of (11.8.23),

$$
\frac{d^+}{dt} \int_{\psi_i(t)}^{\overline{\psi}_{i+1}(t)} (V_t(\infty) - V_t(x))[\overline{u}(x, t) - u(x, t)]dx
$$

(11.8.24)
$$
\leq - \int_{y_i}^{y_{i+1}} Q(u(x, t), \overline{u}(x, t), u(x, t))dV_t^c(x)
$$

$$
- \sum (u_- - u_+)Q(u_-, \overline{u}_*, u_+)
$$

$$
+ (V_t(\infty) - V_t(y_i+))Q(u_-, \overline{u}_*, u_+)
$$

$$
+ (V_t(\infty) - V_t(y_{i+1}))Q(\overline{u}_-, u_*, \overline{u}_+) \, ,
$$

where the summation runs over all y in $\mathcal{H} \cap (y_i, y_{i+1})$, plus y_{i+1} if $y_{i+1} \in \mathcal{H}$ and the forward characteristic, associated with u, issuing from the point (y_{i+1}, t) lies to the left of $\overline{\psi}_{i+1}$.

We thus write (11.8.23), for all i even, then (11.8.24), for all i odd, and sum over $i = 0, \cdots, n$. This yields

(11.8.25)
$$
\frac{d^+}{dt} \int_{-\infty}^{\infty} \{V_t(x)[u(x, t) - \overline{u}(x, t)]^+ + (V_t(\infty) - V_t(x))[\overline{u}(x, t) - u(x, t)]^+\}dx
$$

$$
\leq - \int_{-\infty}^{\infty} Q(u(x, t), \overline{u}(x, t), u(x, t))dV_t^c(x) - \sum_{y \in \mathcal{H}} (u_- - u_+)Q(u_-, \overline{u}_*, u_+)
$$

$$
+ V_t(\infty)\left\{ \sum_{y \in \mathcal{J}} Q(u_-, \overline{u}_*, u_+) + \sum_{y \in \overline{\mathcal{J}}} Q(\overline{u}_-, u_*, \overline{u}_+) \right\} .
$$

By employing a technical argument, as in the proof of Theorem 11.8.2, one shows that (11.8.25) remains valid even when $u(\cdot, t) - \overline{u}(\cdot, t)$ is allowed to undergo infinitely many sign changes on $(-\infty, \infty)$.

Upon writing the inequality resulting from (11.8.25) by interchanging the roles of u and $\overline{u}$, and combining it with (11.8.25), one arrives at (11.8.12). The proof is complete.

The estimate (11.8.12) is sharp, in that it holds as equality, at least for piecewise smooth solutions. All terms on the right-hand side of (11.8.12) are negative, with the exception of $-(u_- - u_+)Q(u_-, \overline{u}_*, u_+)$, for $y \in \mathcal{J}$, and $-(\overline{u}_- - \overline{u}_+)Q(\overline{u}_-, u_*, \overline{u}_+)$, for $y \in \overline{\mathcal{J}}$. However, even these positive terms are offset by the negative terms $V_t(\infty)Q(u_-, \overline{u}_*, u_+)$ and $\overline{V}_t(\infty)Q(\overline{u}_-, u_*, \overline{u}_+)$. Thus, $\rho(u(\cdot, t), \overline{u}(\cdot, t))$ is generally strictly decreasing.

An analog of the functional ρ will be employed in Chapter XIV for establishing L^1 stability of solutions for systems of conservation laws.

11.9 Notes

There is voluminous literature on the scalar conservation law in one-space dimension, especially the genuinely nonlinear case, beginning with the seminal paper of Hopf [1], on the Burgers equation, already cited in earlier chapters.

In the 1950's, the qualitative theory was developed by the Russian school, headed by Oleinik [1,2,4], based on the vanishing viscosity approach as well as on the Lax-Friedrichs finite difference scheme (Lax [1]). It is in that context that Theorem 11.2.2 was originally established. The reader may find an exposition in the text of Smoller [1]. The culmination of that approach was the development of the theory of scalar conservation laws in several spaces dimensions, discussed in Chapter VI.

In a different direction, Lax [2] discovered the explicit representation (11.4.10) for solutions and employed it to establish the existence of invariants (Theorem 11.4.2), the development of N-wave under initial data of compact support (Theorem 11.6.1) as well as the formation of saw-tooth profiles under periodic initial data (Theorem 11.7.3). The original proof, by Schaeffer [1], that generically solutions are piecewise smooth was also based on the same method. This approach readily extends (Oleinik [1]) to inhomogeneous, genuinely nonlinear scalar conservation laws, which may also be casted as Hamilton-Jacobi equations. A thorough presentation of the theory of viscosity solutions for Hamilton-Jacobi equations is found in the monograph by Lions [1].

The approach via generalized characteristics, pursued in this chapter, is taken from Dafermos [7]. One of its advantages is that it may be readily extended not only to inhomogeneous conservation laws but even to inhomogeneous balance laws (Dafermos [8]) as well as to conservation laws that are not genuinely nonlinear (Dafermos [11], Jenssen [2]).

The property that the lap number of solutions of conservation laws (8.6.2) with viscosity is nonincreasing with time was discovered independently by Nickel [1] and Matano [1]. The L^1 contraction property for piecewise smooth solutions in one-space dimension was noted by Quinn [1]. The functional (11.8.11), in alternative, albeit completely equivalent form, was designed by Liu and Yang [3], who employ it to establish Theorem 11.8.3, for piecewise smooth solutions.

So much is known about the scalar conservation and balance law in one-space dimension that it would be pointless to attempt to provide comprehensive coverage. What follows is just a sample of relevant results.

Let us begin with the genuinely nonlinear case. For a probabilistic interpretation of generalized characteristics, see Rezakhanlou [1]. Regularity and generic regularity for inhomogeneous balance laws is investigated, by the method of generalized characteristics, in Dafermos [8]. The same method is used to study the effects of inhomogeneity and source terms on the large time behavior of solutions, in Dafermos [14], Lyberopoulos [1,2], Fan and Hale [1,2], Härterich [1], Mascia and Sinestrari [1] and Fan, Jin and Teng [1]. Problems of this type are also treated by different methods in Liu [15] and Sinestrari [1]. For an interesting application of the method of generalized characteristics in elastostatics, under incompressibility and inextensibility constraints, see Choksi [1].

An explicit representation of admissible solutions on the quarter-plane, analogous to Lax's formula for the upper half-plane, is presented in LeFloch [1]. An analog of Lax's formula has also been derived for the special systems with coinciding shock and rarefaction wave curves; see Benzoni-Gavage [1].

The analog of (11.2.1) holds for scalar conservation laws (6.1.1), in several space variables, if $g_\alpha(u) = f(u)v_\alpha$, where v is a constant vector (Hoff [1]).

For a Chapman-Enskog type regularization of the scalar conservation law, see Shochet and Tadmor [1].

A kinetic formulation, different from the one discussed in Section 6.6, is presented in Brenier and Corrias [1].

The connection of the scalar conservation law with the system of "pressureless gas," that is (7.1.7) with $p \equiv 0$, and the related model of "sticky particles" is investigated in E, Rykov and Sinai [1], Brenier and Grenier [1] and Bouchut and James [1]. The interesting theory of the pressureless gas is developed in Wang and Ding [1] and Wang, Huang and Ding [1].

Homogenization effects under random periodic forcing are demonstrated in E [2,3], E and Serre [1] and E, Khanin, Mazel and Sinai [1].

The case where $f(u, x)$ is piecewise constant in x is discussed in Lyons [1], Klingenberg and Risebro [1] and Diehl [1].

Regularity of solutions in Besov spaces is established in Lucier [2]. For the rate of convergence of numerical schemes see e.g. Nessyahu and Tadmor [1] and Osher and Tadmor [1].

When f has inflection points, the structure of solutions is considerably more intricate, due to the formation of contact discontinuities, which become sources of signals propagating into the future. For the construction of solutions, see Ballou [1]. Regularity is discussed in Ballou [2], Guckenheimer [1], Dafermos [11] and Cheverry [4]. The large time behavior is investigated in Dafermos [1,11], Greenberg and Tong [1], Conlon [1], Cheng [1,2,3], Weinberger [1], Sinestrari [2], Cheverry [4] and Mascia [1].

In the special case $f(u) = u^m$, the properties of solutions may be studied effectively with the help of the induced self-similarity transformation; see Bénilan and Crandall [1] and Liu and Pierre [1]. This last paper also considers initial data that are merely measures. The limit behavior as $m \to \infty$ is discussed in Xu [1].

Chapter XII. Genuinely Nonlinear Systems of Two Conservation Laws

The theory of solutions of genuinely nonlinear, strictly hyperbolic systems of two conservation laws will be developed in this chapter at a level of precision comparable to that for genuinely nonlinear scalar conservation laws, expounded in Chapter XI. This will be achieved by exploiting the presence of coordinate systems of Riemann invariants and the induced rich family of entropy-entropy flux pairs. The principal tools in the investigation will be generalized characteristics and entropy estimates.

The analysis will reveal a close similarity in the structure of solutions of scalar conservation laws and pairs of conservation laws. Thus, as in the scalar case, jump discontinuities are generally generated by the collision of shocks and/or the focussing of compression waves, and are then resolved into wave fans approximated locally by the solution of associated Riemann problems.

The total variation of the trace of solutions along space-like curves is controlled by the total variation of the initial data, and spreading of rarefaction waves affects total variation, as in the scalar case.

The dissipative mechanisms encountered in the scalar case are here at work as well, and have similar effects on the large time behavior of solutions. Entropy dissipation induces $O(t^{-1/2})$ decay of solutions with initial data in $L^1(-\infty, \infty)$. When the initial data have compact support, the two characteristic families asymptotically decouple, the characteristics spread and form a single N-wave profile for each family. Finally, as in the scalar case, confinement of characteristics under periodic initial data induces $O(t^{-1})$ decay in the total variation per period and formation of sawtoothed profiles, one for each characteristic family.

12.1 Notation and Assumptions

We consider a genuinely nonlinear, strictly hyperbolic system of two conservation laws,

$$(12.1.1) \qquad \partial_t U(x, t) + \partial_x F(U(x, t)) = 0 ,$$

on some disk $\mathscr{O}$ centered at the origin. The eigenvalues of DF (characteristic speeds) will here be denoted by λ and μ, with $\lambda(U) < 0 < \mu(U)$ for $U \in \mathscr{O}$, and the associated eigenvectors will be denoted by R and S.

The system is endowed with a coordinate system (z, w) of Riemann invariants, vanishing at the origin $U = 0$, and normalized according to (7.3.8):

$$(12.1.2) \qquad DzR = 1 , \quad DzS = 0 , \quad DwR = 0 , \quad DwS = 1 .$$

The condition of genuine nonlinearity is now expressed by (7.5.3), which here reads

$$(12.1.3) \qquad \lambda_z < 0 , \quad \mu_w > 0 .$$

The direction in the inequalities (12.1.3) has been selected so that z increases across admissible weak 1-shocks while w decreases across admissible weak 2-shocks.

For definiteness, we will consider systems with the property that the interaction of any two shocks of the same characteristic family produces a shock of the same family and a rarefaction wave of the opposite family. Note that this condition is here expressed by

$$(12.1.4) \qquad S^T D^2 zS > 0 , \quad R^T D^2 wR > 0 .$$

Indeed, in conjunction with (8.2.19), (12.1.3) and Theorem 8.3.1, the inequalities (12.1.4) imply that z increases across admissible weak 2-shocks while w decreases across admissible weak 1-shocks. Therefore, the admissible shock and rarefaction wave curves emanating from the state $(\bar{z}, \bar{w})$ have the shape depicted in Fig. 12.1.1. Consequently, as seen in Fig. 12.1.2(a), a 2-shock that joins the state (z_ℓ, w_ℓ), on the left, with the state (z_m, w_m), on the right, interacts with a 2-shock that joins (z_m, w_m), on the left, with the state (z_r, w_r), on the right, to produce a 1-rarefaction wave, joining (z_ℓ, w_ℓ), on the left, with a state (z_0, w_ℓ), on the right, and a 2-shock joining (z_0, w_ℓ), on the left, with (z_r, w_r), on the right, as depicted in Fig. 12.1.2(b). Similarly, the interaction of two 1-shocks produces a 1-shock and a 2-rarefaction wave.

Also for definiteness, we assume

$$(12.1.5) \qquad \lambda_w < 0 , \quad \mu_z > 0 ,$$

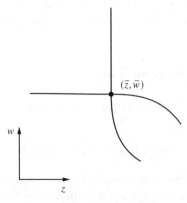

Fig. 12.1.1.

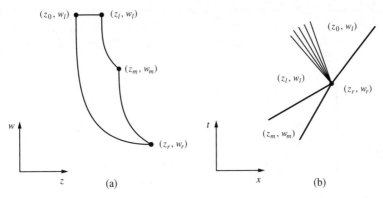

Fig. 12.1.2a,b.

or equivalently, by virtue of (7.3.14) and (7.4.15),

$$(12.1.6) \qquad R^T D^2 z S > 0 , \quad S^T D^2 w R > 0 .$$

The prototypical example is the system (7.1.6) of isentropic thermoelasticity, which satisfies all three assumptions (12.1.3), (12.1.4) and (12.1.6), with Riemann invariants (7.3.2), provided $\sigma''(u) < 0$, i.e., the elastic medium is a soft spring or a gas. When the medium is a hard spring, i.e., $\sigma''(u) > 0$, the sign of the Riemann invariants in (7.3.2) has to be reversed.

12.2 Entropy-Entropy Flux Pairs

As explained in Section 7.4, our system is endowed with a rich family of entropy-entropy flux pairs (η, q), which may be determined as functions of the Riemann invariants (z, w) by solving the system (7.4.12), namely

$$(12.2.1) \qquad q_z = \lambda \eta_z , \quad q_w = \mu \eta_w .$$

The integrability condition (7.4.13) now takes the form

$$(12.2.2) \qquad \eta_{zw} + \frac{\lambda_w}{\lambda - \mu} \eta_z + \frac{\mu_z}{\mu - \lambda} \eta_w = 0 .$$

The entropy $\eta(z, w)$ will be a convex function of the original state variable U when the inequalities (7.4.16) hold, that is,

$$(12.2.3) \qquad \begin{cases} \eta_{zz} + (R^T D^2 z R)\eta_z + (R^T D^2 w R)\eta_w \geq 0 , \\ \eta_{ww} + (S^T D^2 z S)\eta_z + (S^T D^2 w S)\eta_w \geq 0 . \end{cases}$$

In the course of our investigation, we shall face the need to construct entropy-entropy flux pairs with prescribed specifications, by solving (12.2.1) or (12.2.2) under assigned side conditions. To verify that the constructed entropy satisfies the condition (12.2.3), for convexity, it usually becomes necessary to estimate the second derivatives η_{zz} and η_{ww} in terms of the first derivatives η_z and η_w. For that purpose, one may employ the equations obtained by differentiating (12.2.2) with respect to z and w:

$$(12.2.4) \quad \begin{cases} \eta_{zzw} + \dfrac{\lambda_w}{\lambda - \mu}\eta_{zz} = \dfrac{(\mu - \lambda)\lambda_{zw} + \lambda_z\lambda_w - 2\lambda_w\mu_z}{(\lambda - \mu)^2}\eta_z \\ \qquad\qquad\qquad + \dfrac{(\lambda - \mu)\mu_{zz} - \lambda_z\mu_z + 2\mu_z^2}{(\lambda - \mu)^2}\eta_w \,, \\[3mm] \eta_{wwz} + \dfrac{\mu_z}{\mu - \lambda}\eta_{ww} = \dfrac{(\mu - \lambda)\lambda_{ww} - \lambda_w\mu_w + 2\lambda_w^2}{(\mu - \lambda)^2}\eta_z \\ \qquad\qquad\qquad + \dfrac{(\lambda - \mu)\mu_{zw} + \mu_z\mu_w - 2\lambda_w\mu_z}{(\mu - \lambda)^2}\eta_w \,. \end{cases}$$

As an illustration, we consider the important family of *Lax entropy-entropy flux pairs*

$$(12.2.5) \quad \begin{cases} \eta(z, w) = e^{kz}[\phi(z, w) + \frac{1}{k}\chi(z, w) + O(\frac{1}{k^2})] \,, \\ q(z, w) = e^{kz}\lambda(z, w)[\psi(z, w) + \frac{1}{k}\theta(z, w) + O(\frac{1}{k^2})] \,, \end{cases}$$

$$(12.2.6) \quad \begin{cases} \eta(z, w) = e^{kw}[\alpha(z, w) + \frac{1}{k}\beta(z, w) + O(\frac{1}{k^2})] \,, \\ q(z, w) = e^{kw}\mu(z, w)[\gamma(z, w) + \frac{1}{k}\delta(z, w) + O(\frac{1}{k^2})] \,, \end{cases}$$

where k is a parameter. These are designed to vary stiffly with one of the two Riemann invariants so as to be employed for decoupling the two characteristic families. To construct them, one substitutes η and q from (12.2.5) or (12.2.6) into the system (12.2.1), thus deriving recurrence relations for the coefficients, and then shows that the remainder is $O(k^{-2})$. The recurrence relations for the coefficients of the family (12.2.5) read as follows:

$$(12.2.7) \qquad\qquad \psi = \phi \,,$$

$$(12.2.8) \qquad\qquad \lambda\theta + (\lambda\psi)_z = \lambda\chi + \lambda\phi_z \,,$$

$$(12.2.9) \qquad\qquad (\lambda\psi)_w = \mu\phi_w \,.$$

Combining (12.2.7) with (12.2.9) yields

$$(12.2.10) \qquad\qquad (\mu - \lambda)\phi_w = \lambda_w\phi \,,$$

which may be satisfied by selecting

$$(12.2.11) \qquad \phi(z, w) = \exp\int_0^w \frac{\lambda_w(z, \omega)}{\mu(z, \omega) - \lambda(z, \omega)}d\omega \,.$$

In particular, this ϕ is positive, uniformly bounded away from zero on compact sets. Hence, for k sufficiently large, the inequalities (12.2.3) will hold, the second one by virtue of (12.1.4). Consequently, for k large the Lax entropy is a convex function of U.

Important implications of (12.2.7) and (12.2.8) are the estimates

$$(12.2.12) \qquad q - \lambda\eta = \frac{1}{k}e^{kz}\left[-\lambda_z\phi + O\left(\frac{1}{k}\right)\right],$$

$$(12.2.13) \qquad q - (\lambda + \varepsilon)\eta = -e^{kz}\left[\varepsilon\phi + O\left(\frac{1}{k}\right)\right],$$

whose usefulness will become clear later.

12.3 Local Structure of Solutions

Throughout this chapter, U will denote a function of locally bounded variation, defined on $(-\infty, \infty) \times [0, \infty)$ and taking values in a disk of small radius, centered at the origin, which is a weak solution of (12.1.1) satisfying the Lax E-condition, in the sense described in Section 10.1. In particular,

$$(12.3.1) \qquad \partial_t \eta(U(x, t)) + \partial_x q(U(x, t)) \leq 0$$

will hold, in the sense of measures, for any entropy-entropy flux pair (η, q), with η convex.

The notion of generalized characteristic, developed in Chapter X, will play a pivotal role in the discussion.

Definition 12.3.1 A Lipschitz curve, with graph $\mathscr{A}$ embedded in the upper half-plane, is called *space-like* relative to U when every point $(\bar{x}, \bar{t}) \in \mathscr{A}$ has the following property: The set $\{(x, t) : 0 \leq t < \bar{t}, \zeta(t) < x < \xi(t)\}$ of points confined between the graphs of the maximal backward 2-characteristic $\zeta(\cdot)$ and the minimal backward 1-characteristic $\xi(\cdot)$, emanating from $(\bar{x}, \bar{t})$, has empty intersection with $\mathscr{A}$.

Clearly, any generalized characteristic, of either family, associated with U, is space-like relative to U. Similarly, all time lines, $t = $ constant, are space-like.

The solution U will be conveniently monitored through its induced Riemann invariant coordinates (z, w). In Section 12.5, it is shown that the total variation of the trace of z and w along space-like curves is controlled by the total variation of their initial data. In anticipation of that result, we shall be assuming henceforth that, for any space-like curve $t = t^*(x)$, $z(x\pm, t^*(x))$ and $w(x\pm, t^*(x))$ are functions of bounded variation, with total variation bounded by a positive constant θ. Since the oscillation of the solution is small and arguments will be local, we may assume without further loss of generality that θ is small.

In order to describe the local structure of the solution, we associate with the generic point $(\bar{x}, \bar{t})$ of the upper half-plane eight, not necessarily distinct, curves (see Fig. 12.3.1) determined as follows:

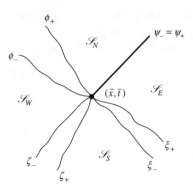

Fig. 12.3.1.

For $t < \bar{t}$: $\xi_-(\cdot)$ and $\xi_+(\cdot)$ are the minimal and the maximal backward 1-characteristics emanating from $(\bar{x}, \bar{t})$; similarly, $\zeta_-(\cdot)$ and $\zeta_+(\cdot)$ are the minimal and the maximal backward 2-characteristics emanating from $(\bar{x}, \bar{t})$.

For $t > \bar{t}$: $\phi_+(\cdot)$ is the maximal forward 1-characteristic and $\psi_-(\cdot)$ is the minimal forward 2-characteristic issuing from $(\bar{x}, \bar{t})$. To determine the remaining two curves $\phi_-(\cdot)$ and $\psi_+(\cdot)$, we consider the minimal backward 1-characteristic $\xi(\cdot)$ and the maximal backward 2-characteristic $\zeta(\cdot)$ emanating from the generic point (x, t) and locate the points $\xi(\bar{t})$ and $\zeta(\bar{t})$ where these characteristics are intercepted by the $\bar{t}$-time line. Then $\phi_-(t)$ is determined by the property that $\xi(\bar{t}) < \bar{x}$ when $x < \phi_-(t)$ and $\xi(\bar{t}) \geq \bar{x}$ when $x > \phi_-(t)$. Similarly, $\psi_+(t)$ is characterized by the property that $\zeta(\bar{t}) \leq \bar{x}$ when $x < \psi_+(t)$ and $\zeta(\bar{t}) > \bar{x}$ when $x > \psi_+(t)$. In particular, $\phi_-(t) \leq \phi_+(t)$ and if $\phi_-(t) < x < \phi_+(t)$ then $\xi(\bar{t}) = \bar{x}$. Similarly, $\psi_-(t) \leq \psi_+(t)$ and $\psi_-(t) < x < \psi_+(t)$ implies $\zeta(\bar{t}) = \bar{x}$.

We fix $\tau > \bar{t}$ and let $\xi_\tau(\cdot)$ denote the minimal backward 1-characteristic emanating from the point $(\phi_-(\tau), \tau)$. We also consider any sequence $\{x_m\}$ converging from above to $\phi_-(\tau)$ and let $\xi_m(\cdot)$ denote the minimal backward 1-characteristic emanating from (x_m, τ). Then the sequence $\{\xi_m(\cdot)\}$, or some subsequence thereof, will converge to some backward 1-characteristic $\hat{\xi}_\tau(\cdot)$ emanating from $(\phi_-(\tau), \tau)$. Moreover, for any $\bar{t} \leq t \leq \tau$, it is $\xi_\tau(t) \leq \phi_-(t) \leq \hat{\xi}_\tau(t)$. In particular, this implies that $\phi_-(\cdot)$ is a Lipschitz continuous space-like curve, with slope in the range of λ. Similarly, $\psi_+(\cdot)$ is a Lipschitz continuous space-like curve, with slope in the range of μ.

Referring again to Fig. 12.3.1, we see that the aforementioned curves border regions:

$$(12.3.2) \qquad \mathscr{S}_W = \{(x, t) : x < \bar{x} , \quad \zeta_-^{-1}(x) < t < \phi_-^{-1}(x)\} ,$$

$$(12.3.3) \qquad \mathscr{S}_E = \{(x, t) : x > \bar{x} , \quad \xi_+^{-1}(x) < t < \psi_+^{-1}(x)\} ,$$

$$(12.3.4) \qquad \mathscr{S}_N = \{(x, t) : t > \bar{t} , \quad \phi_+(t) < x < \psi_-(t)\} ,$$

$$(12.3.5) \qquad \mathscr{S}_S = \{(x, t) : t < \bar{t} , \quad \zeta_+(t) < x < \xi_-(t)\} .$$

Definition 12.3.2 The solution is called *locally regular* at the point $(\bar{x}, \bar{t})$ of the upper half-plane when the following hold:

(a) As (x, t) tends to $(\bar{x}, \bar{t})$ through any one of the regions $\mathscr{S}_W, \mathscr{S}_E, \mathscr{S}_N$ or $\mathscr{S}_S$, $(z(x\pm, t), w(x\pm, t))$ tend to respective limits (z_W, w_W), (z_E, w_E), (z_N, w_N) or (z_S, w_S). In particular, $z_W = z(\bar{x}-, \bar{t})$, $w_W = w(\bar{x}-, \bar{t})$, $z_E = z(\bar{x}+, \bar{t})$, $w_E = w(\bar{x}+, \bar{t})$.

(b)$_1$ If $p_\ell(\cdot)$ and $p_r(\cdot)$ are any two backward 1-characteristics emanating from $(\bar{x}, \bar{t})$, with $\xi_-(t) \le p_\ell(t) < p_r(t) \le \xi_+(t)$, for $t < \bar{t}$, then

$$(12.3.6)_1$$
$$z_S = \lim_{t\uparrow\bar{t}} z(\xi_-(t)\pm, t) \le \lim_{t\uparrow\bar{t}} z(p_\ell(t)-, t) \le \lim_{t\uparrow\bar{t}} z(p_\ell(t)+, t)$$
$$\le \lim_{t\uparrow\bar{t}} z(p_r(t)-, t) \le \lim_{t\uparrow\bar{t}} z(p_r(t)+, t) \le \lim_{t\uparrow\bar{t}} z(\xi_+(t)\pm, t) = z_E ,$$

$$(12.3.7)_1$$
$$w_S = \lim_{t\uparrow\bar{t}} w(\xi_-(t)\pm, t) \ge \lim_{t\uparrow\bar{t}} w(p_\ell(t)-, t) \ge \lim_{t\uparrow\bar{t}} w(p_\ell(t)+, t)$$
$$\ge \lim_{t\uparrow\bar{t}} w(p_r(t)-, t) \ge \lim_{t\uparrow\bar{t}} w(p_r(t)+, t) \ge \lim_{t\uparrow\bar{t}} w(\xi_+(t)\pm, t) = w_E .$$

(b)$_2$ If $q_\ell(\cdot)$ and $q_r(\cdot)$ are any two backward 2-characteristics emanating from $(\bar{x}, \bar{t})$, with $\zeta_-(t) \le q_\ell(t) < q_r(t) \le \zeta_+(t)$, for $t < \bar{t}$, then

$$(12.3.6)_2$$
$$w_W = \lim_{t\uparrow\bar{t}} w(\zeta_-(t)\pm, t) \ge \lim_{t\uparrow\bar{t}} w(q_\ell(t)-, t) \ge \lim_{t\uparrow\bar{t}} w(q_\ell(t)+, t)$$
$$\ge \lim_{t\uparrow\bar{t}} w(q_r(t)-, t) \ge \lim_{t\uparrow\bar{t}} w(q_r(t)+, t) \ge \lim_{t\uparrow\bar{t}} w(\zeta_+(t)\pm, t) = w_S ,$$

$$(12.3.7)_2$$
$$z_W = \lim_{t\uparrow\bar{t}} z(\zeta_-(t)\pm, t) \le \lim_{t\uparrow\bar{t}} z(q_\ell(t)-, t) \le \lim_{t\uparrow\bar{t}} z(q_\ell(t)+, t)$$
$$\le \lim_{t\uparrow\bar{t}} z(q_r(t)-, t) \le \lim_{t\uparrow\bar{t}} z(q_r(t)+, t) \le \lim_{t\uparrow\bar{t}} z(\zeta_+(t)\pm, t) = z_S .$$

(c)$_1$ If $\phi_-(t) = \phi_+(t)$, for $\bar{t} < t < \bar{t} + s$, then $z_W \le z_N$, $w_W \ge w_N$. On the other hand, if $\phi_-(t) < \phi_+(t)$, for $\bar{t} < t < \bar{t}+s$, then $w_W = w_N$ and as (x, t) tends to $(\bar{x}, \bar{t})$ through the region $\{(x, t) : t > \bar{t}, \phi_-(t) < x < \phi_+(t)\}$, $w(x\pm, t)$ tends to w_W. Furthermore, if $p_\ell(\cdot)$ and $p_r(\cdot)$ are any two forward 1-characteristics issuing from $(\bar{x}, \bar{t})$, with $\phi_-(t) \le p_\ell(t) \le p_r(t) \le \phi_+(t)$, for $\bar{t} < t < \bar{t} + s$, then

$$(12.3.8)_1$$
$$z_W = \lim_{t\downarrow\bar{t}} z(\phi_-(t)\pm, t) \ge \lim_{t\downarrow\bar{t}} z(p_\ell(t)-, t) = \lim_{t\downarrow\bar{t}} z(p_\ell(t)+, t)$$
$$\ge \lim_{t\downarrow\bar{t}} z(p_r(t)-, t) = \lim_{t\downarrow\bar{t}} z(p_r(t)+, t) \ge \lim_{t\downarrow\bar{t}} z(\phi_+(t)\pm, t) = z_N .$$

(c)$_2$ If $\psi_-(t) = \psi_+(t)$, for $\bar{t} < t < \bar{t} + s$, then $w_N \ge w_E$, $z_N \le z_E$. On the other hand, if $\psi_-(t) < \psi_+(t)$, for $\bar{t} < t < \bar{t}+s$, then $z_N = z_E$ and as (x, t) tends to $(\bar{x}, \bar{t})$ through the region $\{(x, t) : t > \bar{t}, \psi_-(t) < x < \psi_+(t)\}$, $z(x\pm, t)$ tends to z_E. Furthermore, if $q_\ell(\cdot)$ and $q_r(\cdot)$ are any two forward 2-characteristics issuing from $(\bar{x}, \bar{t})$, with $\psi_-(t) \le q_\ell(t) \le q_r(t) \le \psi_+(t)$, for $\bar{t} < t < \bar{t} + s$, then

$$(12.3.8)_2$$
$$w_N = \lim_{t\downarrow\bar{t}} w(\psi_-(t)\pm, t) \le \lim_{t\downarrow\bar{t}} w(q_\ell(t)-, t) = \lim_{t\downarrow\bar{t}} w(q_\ell(t)+, t)$$
$$\le \lim_{t\downarrow\bar{t}} w(q_r(t)-, t) = \lim_{t\downarrow\bar{t}} w(q_r(t)+, t) \le \lim_{t\downarrow\bar{t}} w(\psi_+(t)\pm, t) = w_E .$$

The justification of the above definition lies in

Theorem 12.3.1 *For θ sufficiently small, the solution is locally regular at any point of the upper half-plane.*

The proof will be provided in the next section. However, the following remarks are here in order. Definition 12.3.2 is motivated by experience with piecewise smooth solutions. Indeed, at points of local regularity incoming waves of the two characteristic families collide to generate a jump discontinuity, which is then resolved into an outgoing wave fan. Statements $(b)_1$ and $(b)_2$ regulate the incoming waves, allowing for any combination of admissible shocks and focussing compression waves. Statements $(c)_1$ and $(c)_2$ characterize the outgoing wave fan. In particular, $(c)_1$ implies that the state (z_W, w_W), on the left, may be joined with the state (z_N, w_N), on the right, by a 1-rarefaction wave or admissible 1-shock; while $(c)_2$ implies that the state (z_N, w_N), on the left, may be joined with the state (z_E, w_E), on the right, by a 2-rarefaction wave or admissible 2-shock. Thus, the outgoing wave fan is locally approximated by the solution of the Riemann problem with end-states $(z(\bar{x}-, \bar{t}), w(\bar{x}-, \bar{t}))$ and $(z(\bar{x}+, \bar{t}), w(\bar{x}+, \bar{t}))$.

A simple corollary of Theorem 12.3.1 is that $\phi_-(\cdot)$ is a 1-characteristic while $\psi_+(\cdot)$ is a 2-characteristic.

Definition 12.3.2 and Theorem 12.3.1 apply even to points on the initial line, $\bar{t} = 0$, after discarding the irrelevant parts of the statements, pertaining to $t < \bar{t}$. It should be noted, however, that there is an important difference between $\bar{t} = 0$ and $\bar{t} > 0$. In the former case, $(z(\bar{x}\pm, 0), w(\bar{x}\pm, 0))$ are unrestricted, being induced arbitrarily by the initial data, and hence the outgoing wave fan may comprise any combination of shocks and rarefaction waves. By contrast, when $\bar{t} > 0$, statements $(b)_1$ and $(b)_2$ in Definition 12.3.2 induce the restrictions $z_W \leq z_E$ and $w_W \geq w_E$. This, combined with statements $(c)_1$ and $(c)_2$, rules out the possibility that both outgoing waves may be rarefactions.

12.4 Propagation of Riemann Invariants Along Extremal Backward Characteristics

The theory of the genuinely nonlinear scalar conservation law, expounded in Chapter XI, owes its simplicity to the observation that extremal backward generalized characteristics are essentially classical characteristics, namely straight lines along which the solution stays constant. It is thus natural to investigate whether solutions U of systems (12.1.1) exhibit similar behavior. When U is Lipschitz continuous, the Riemann invariants z and w stay constant along 1-characteristics and 2-characteristics, respectively, by virtue of Theorem 7.3.2. One should not expect, however, that this will hold for weak solutions, because Riemann invariants generally jump across shocks of both characteristic families. In the context of piecewise smooth solutions, Theorem 8.2.3 implies that, under the current normalization conditions, the trace of z (or w) along shock free 1-characteristics (or

2-characteristics) is a nonincreasing step function. The jumps of z (or w) occur at the points where the characteristic crosses a shock of the opposite family, and are of cubic order in the strength of the crossed shock. It is remarkable that this property essentially carries over to general weak solutions:

Theorem 12.4.1 *Let $\xi(\cdot)$ be the minimal (or maximal) backward 1-characteristic (or 2-characteristic) emanating from any fixed point $(\overline{x}, \overline{t})$ of the upper half-plane. Set*

$$(12.4.1) \qquad \overline{z}(t) = z(\xi(t)-, t) , \quad \overline{w}(t) = w(\xi(t)+, t) , \quad 0 \le t \le \overline{t} .$$

Then $\overline{z}(\cdot)$ (or $\overline{w}(\cdot)$) is a nonincreasing saltus function whose variation is concentrated in the set of points of jump discontinuity of $\overline{w}(\cdot)$ (or $\overline{z}(\cdot)$). Furthermore, if $\tau \in (0, \overline{t})$ is any point of jump discontinuity of $\overline{z}(\cdot)$ (or $\overline{w}(\cdot)$), then

$$(12.4.2)_1 \qquad \overline{z}(\tau-) - \overline{z}(\tau+) \le a[\overline{w}(\tau+) - \overline{w}(\tau)]^3 ,$$

or

$$(12.4.2)_2 \qquad \overline{w}(\tau-) - \overline{w}(\tau+) \le a[\overline{z}(\tau+) - \overline{z}(\tau)]^3 ,$$

where a is a positive constant depending solely on F.

The proof of the above proposition will be intermingled with the proof of Theorem 12.3.1, on local regularity of the solution, and will be partitioned into several steps. The assumption that the trace of (z, w) along space-like curves has bounded variation will be employed only for special space-like curves, namely, generalized characteristics and time lines, $t = $ constant.

Proposition 12.4.1 *When $\xi(\cdot)$ is the minimal (or maximal) backward 1-characteristic (or 2-characteristic) emanating from $(\overline{x}, \overline{t})$, $\overline{z}(\cdot)$ (or $\overline{w}(\cdot)$) is nonincreasing on $[0, \overline{t}]$.*

Proof. The two cases are quite similar, so it will suffice to discuss the first one, namely where $\xi(\cdot)$ is a 1-characteristic. Then, by virtue of Theorem 10.3.2, $\xi(\cdot)$ is shock free and hence

$$(12.4.3) \qquad \dot{\xi}(t) = \lambda(U(\xi(t)\pm, t)) , \quad \text{a.e. on } [0, \overline{t}] .$$

We fix numbers τ and s, with $0 \le \tau < s \le \overline{t}$. For ε positive and small, we let $\xi_\varepsilon(\cdot)$ denote the minimal Filippov solution of the ordinary differential equation

$$(12.4.4) \qquad \frac{dx}{dt} = \lambda(U(x, t)) + \varepsilon ,$$

on $[\tau, s]$, with initial condition $\xi_\varepsilon(s) = \xi(s) - \varepsilon$. Applying (12.1.1), as equality of measures, to arcs of the graph of $\xi_\varepsilon(\cdot)$ and using Theorem 1.7.5, we deduce

$$(12.4.5)$$
$$F(U(\xi_\varepsilon(t)+, t)) - F(U(\xi_\varepsilon(t)-, t)) - \dot{\xi}_\varepsilon(t)[U(\xi_\varepsilon(t)+, t) - U(\xi_\varepsilon(t)-, t)] = 0 ,$$

almost everywhere on $[\tau, s]$. Thus, $\xi_\varepsilon(\cdot)$ propagates with speed $\lambda(U(\xi_\varepsilon(t)\pm, t))+\varepsilon$, at points of approximate continuity, or with 1-shock speed, at points of approximate jump discontinuity. In particular, $\lambda(U(\xi_\varepsilon(t)+, t)) \leq \lambda(U(\xi_\varepsilon(t)-, t))$, almost everywhere on $[\tau, s]$, and so, by the definition of Filippov solutions of (12.4.4),

$$(12.4.6) \qquad \dot\xi_\varepsilon(t) \geq \lambda(U(\xi_\varepsilon(t)+, t)) + \varepsilon , \quad \text{a.e. on } [\tau, s] .$$

For any entropy-entropy flux pair (η, q), with η convex, integrating (12.3.1) over the region $\{(x, t) : \tau < t < s, \xi_\varepsilon(t) < x < \xi(t)\}$ and applying Green's theorem yields

$$(12.4.7)$$
$$\int_{\xi_\varepsilon(s)}^{\xi(s)} \eta(U(x, s))dx - \int_{\xi_\varepsilon(\tau)}^{\xi(\tau)} \eta(U(x, \tau))dx$$
$$\leq - \int_\tau^s \{q(U(\xi(t)-, t)) - \dot\xi(t)\eta(U(\xi(t)-, t))\}dt$$
$$+ \int_\tau^s \{q(U(\xi_\varepsilon(t)+, t)) - \dot\xi_\varepsilon(t)\eta(U(\xi_\varepsilon(t)+, t))\}dt .$$

In particular, we write (12.4.7) for the Lax entropy-entropy flux pair (12.2.5). For k large, the right-hand side of (12.4.7) is nonpositive, by virtue of (12.4.3), (12.4.6), (12.2.12), (12.1.3) and (12.2.13). Hence

$$(12.4.8) \qquad \int_{\xi_\varepsilon(s)}^{\xi(s)} \eta(z(x, s), w(x, s))dx \leq \int_{\xi_\varepsilon(\tau)}^{\xi(\tau)} \eta(z(x, \tau), w(x, \tau))dx .$$

We raise (12.4.8) to the power $1/k$ and then let $k \to \infty$. This yields

$$(12.4.9) \qquad \operatorname*{ess\,sup}_{(\xi_\varepsilon(s),\xi(s))} z(\cdot, s) \leq \operatorname*{ess\,sup}_{(\xi_\varepsilon(\tau),\xi(\tau))} z(\cdot, \tau) .$$

Finally, we let $\varepsilon \downarrow 0$. By standard theory of Filippov solutions, the family $\{\xi_\varepsilon(\cdot)\}$ contains a sequence which converges, uniformly on $[\tau, s]$, to some Filippov solution $\xi_0(\cdot)$ of the differential equation $dx/dt = \lambda(U(x, t))$, with initial condition $\xi_0(s) = \xi(s)$. But then $\xi_0(\cdot)$ is a backward 1-characteristic emanating from the point $(\xi(s), s)$. Moreover, $\xi_0(t) \leq \xi(t)$, for $\tau \leq t \leq s$. Since $\xi(\cdot)$ is minimal, $\xi_0(\cdot)$ must coincide with $\xi(\cdot)$ on $[\tau, s]$. Thus (12.4.9) implies $\bar z(s) \leq \bar z(\tau)$ and so $\bar z(\cdot)$ is nonincreasing on $[\tau, s]$. The proof is complete.

Propositon 12.4.2 *Let $\xi(\cdot)$ be the minimal (or maximal) backward 1-characteristic (or 2-characteristic) emanating from $(\bar x, \bar t)$. Then, for any $\tau \in (0, \bar t]$,*

$$(12.4.10)_1 \qquad z(\xi(\tau)-, \tau) \leq \bar z(\tau-) \leq z(\xi(\tau)+, \tau) ,$$

or

$$(12.4.10)_2 \qquad w(\xi(\tau)-, \tau) \geq \bar w(\tau-) \geq w(\xi(\tau)+, \tau) .$$

In particular,

(12.4.11)
$$z(x-,t) \le z(x+,t) \,, \quad w(x-,t) \ge w(x+,t) \,, \quad -\infty < x < \infty \,, \quad 0 < t < \infty \,.$$

This will be established in conjunction with

Proposition 12.4.3 *Let $\xi(\cdot)$ be the minimal (or maximal) backward 1-characteristic (or 2-characteristic) emanating from $(\overline{x}, \overline{t})$. For any τ and s with $0 < \tau < s \le \overline{t}$,*

$$(12.4.12)_1 \qquad z(\xi(\tau)+, \tau) - z(\xi(s)+, s) \le b\, osc_{[\tau,s]}\overline{w}(\cdot)\, TV_{[\tau,s]}\overline{w}(\cdot) \,,$$

or

$$(12.4.12)_2 \qquad w(\xi(\tau)-, \tau) - w(\xi(s)-, s) \le b\, osc_{[\tau,s]}\overline{z}(\cdot)\, TV_{[\tau,s]}\overline{z}(\cdot) \,,$$

where b is a positive constant depending solely on F. Further, if $\overline{w}(\tau+) > \overline{w}(\tau)$ (or $\overline{z}(\tau+) > \overline{z}(\tau)$), then $(12.4.2)_1$ (or $(12.4.2)_2$) holds.

Proof. It suffices to discuss the case where $\xi(\cdot)$ is a 1-characteristic. Consider any convex entropy η with associated entropy flux q. We fix ε positive and small and integrate (12.3.1) over the region $\{(x, t) : \tau < t < s, \xi(t) < x < \xi(t) + \varepsilon\}$. Notice that both curves $x = \xi(t)$ and $x = \xi(t)+\varepsilon$ have slope $\lambda(\overline{z}(t), \overline{w}(t))$, almost everywhere on (τ, s). Therefore, Green's theorem yields

$$(12.4.13) \qquad \begin{aligned} &\int_{\xi(s)}^{\xi(s)+\varepsilon} \eta(z(x, s), w(x, s))dx - \int_{\xi(\tau)}^{\xi(\tau)+\varepsilon} \eta(z(x, \tau), w(x, \tau))dx \\ &\le - \int_\tau^s H(z(\xi(t)+\varepsilon+, t), w(\xi(t)+\varepsilon+, t), \overline{z}(t), \overline{w}(t))dt \,, \end{aligned}$$

under the notation

$$(12.4.14) \quad H(z, w, \overline{z}, \overline{w}) = q(z, w) - q(\overline{z}, \overline{w}) - \lambda(\overline{z}, \overline{w})[\eta(z, w) - \eta(\overline{z}, \overline{w})] \,.$$

One easily verifies, with the help of (12.2.1), that

$$(12.4.15) \qquad H_z(z, w, \overline{z}, \overline{w}) = [\lambda(z, w) - \lambda(\overline{z}, \overline{w})]\eta_z(z, w) \,,$$

$$(12.4.16) \qquad H_w(z, w, \overline{z}, \overline{w}) = [\mu(z, w) - \lambda(\overline{z}, \overline{w})]\eta_w(z, w) \,,$$

$$(12.4.17) \quad H_{zz}(z, w, \overline{z}, \overline{w}) = \lambda_z(z, w)\eta_z(z, w) + [\lambda(z, w) - \lambda(\overline{z}, \overline{w})]\eta_{zz}(z, w),$$

$$(12.4.18) \quad H_{zw}(z, w, \overline{z}, \overline{w}) = \lambda_w(z, w)\eta_z(z, w) + [\lambda(z, w) - \lambda(\overline{z}, \overline{w})]\eta_{zw}(z, w) \,,$$

$$(12.4.19) \quad H_{ww}(z, w, \overline{z}, \overline{w}) = \mu_w(z, w)\eta_w(z, w) + [\mu(z, w) - \lambda(\overline{z}, \overline{w})]\eta_{ww}(z, w).$$

We introduce the notation $z_0 = z(\xi(\tau)+, \tau)$, $w_0 = w(\xi(\tau)+, \tau) = \overline{w}(\tau)$, $z_1 = z(\xi(s)+, s)$, $w_1 = w(\xi(s)+, s) = \overline{w}(s)$ and set $\delta = osc_{[\tau,s]}\overline{w}(\cdot)$. We then apply (12.4.13) for the entropy η constructed by solving the Goursat problem for (12.2.2), with data

$$(12.4.20) \quad \begin{cases} \eta(z, w_0) = -(z - z_0) + \beta(z - z_0)^2 \, , \\ \eta(z_0, w) = -3\beta\delta(w - w_0) + \beta(w - w_0)^2 \, , \end{cases}$$

where β is a positive constant, sufficiently large for the following to hold on a small neighborhood of the point (z_0, w_0):

$$(12.4.21) \qquad\qquad \eta \text{ is a convex function of } U \, ,$$

$$(12.4.22) \qquad\qquad \eta(z, w) \text{ is a convex function of } (z, w) \, ,$$

$$(12.4.23) \qquad\qquad H(z, w, \overline{z}, \overline{w}) \text{ is a convex function of } (z, w) \, .$$

It is possible to satisfy the above requirements when $|z - z_0|$, $|w - w_0|$ and δ are sufficiently small. In particular, (12.4.21) will hold by virtue of (12.2.3), (12.1.4), (12.4.20), (12.2.2) and (12.2.4). Similarly, (12.4.22) follows from (12.4.20), (12.2.2) and (12.2.4). Finally, (12.4.23) is verified by combining (12.4.17), (12.4.18), (12.4.19), (12.4.20), (12.2.2) and (12.2.4).

By virtue of (12.4.23), (12.4.15) and (12.4.16),

$$(12.4.24) \qquad H(z, w, \overline{z}, \overline{w}) \geq [\mu(\overline{z}, \overline{w}) - \lambda(\overline{z}, \overline{w})]\eta_w(\overline{z}, \overline{w})[w - \overline{w}].$$

One may estimate $\eta_w(\overline{z}(t), \overline{w}(t))$ by integrating (12.2.2), as an ordinary differential equation for η_w, along the line $w = \overline{w}(t)$, starting out from the initial value $\eta_w(z_0, \overline{w}(t))$ at $z = z_0$. Because $|\overline{w}(t) - w_0| \leq \delta$, (12.4.20) gives $-5\beta\delta \leq \eta_w(z_0, \overline{w}(t)) \leq -\beta\delta < 0$. Since $\lambda_w < 0$ and $\eta_z < 0$, (12.2.2) then implies $\eta_w(z, \overline{w}(t)) < 0$, for $z \leq z_0$. In anticipation of (12.4.10)$_1$, we now assume $z_0 \geq \overline{z}(\tau)$, which we already know will apply for almost all choices of τ in $(0, s)$, namely when $z(\xi(\tau)-, \tau) = z(\xi(\tau)+, \tau)$. By Proposition 12.4.1, $\overline{z}(t) \leq \overline{z}(\tau)$ and so $\eta_w(\overline{z}(\tau), \overline{w}(\tau)) < 0$, for $\tau \leq t \leq s$.

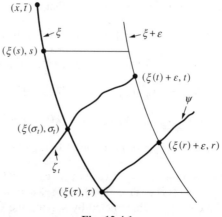

Fig. 12.4.1.

For $t \in [\tau, s]$, let $\zeta_t(\cdot)$ denote the maximal backward 2-characteristic emanating from the point $(\xi(t) + \varepsilon, t)$ (Fig. 12.4.1). We also draw the maximal forward 2-characteristic $\psi(\cdot)$, issuing from the point $(\xi(\tau), \tau)$, which collides with the curve $x = \xi(t) + \varepsilon$ at time r, where $0 < r - \tau < c_0 \varepsilon$.

For $t \in (r, s)$, the graph of $\zeta_t(\cdot)$ intersects the graph of $\xi(\cdot)$ at time σ_t. By Proposition 12.4.1,

(12.4.25)
$$w(\xi(t) + \varepsilon +, t) = w(\zeta_t(t)+, t) \leq w(\zeta_t(\sigma_t)+, \sigma_t) = w(\xi(\sigma_t)+, \sigma_t) = \overline{w}(\sigma_t) .$$

Since $\eta_w(\overline{z}(t), \overline{w}(t)) < 0$, (12.4.24) and (12.4.25) together imply

(12.4.26)
$$H(z(\xi(t) + \varepsilon +, t), w(\xi(t) + \varepsilon +, t), \overline{z}(t), \overline{w}(t))$$
$$\geq [\mu(\overline{z}(t), \overline{w}(t)) - \lambda(\overline{z}(t), \overline{w}(t))] \eta_w(\overline{z}(t), \overline{w}(t)) [\overline{w}(\sigma_t) - \overline{w}(t)] .$$

Because the two characteristic speeds λ and μ are strictly separated, we deduce $0 < t - \sigma_t < c_1 \varepsilon$ and so (12.4.26) yields

(12.4.27)
$$-\int_r^s H(z(\xi(t) + \varepsilon +, t), w(\xi(t) + \varepsilon +, t), \overline{z}(t), \overline{w}(t))dt$$
$$\leq c_2 \varepsilon \sup_{(\tau, s)} |\eta_w(\overline{z}(\cdot), \overline{w}(\cdot))| \, NV_{(\tau, s)} \overline{w}(\cdot) ,$$

with NV denoting negative (i.e., decreasing) variation.

Next, we restrict t to the interval (τ, r). Then, $\zeta_t(\cdot)$ is intercepted by the τ-time line at $\zeta_t(\tau) \in [\xi(\tau), \xi(\tau) + \varepsilon]$. By virtue of Proposition 12.4.1,

(12.4.28)
$$w(\xi(t) + \varepsilon +, t) = w(\zeta_t(t)+, t) \leq w(\zeta_t(\tau)+, \tau) = w_0 + o(1) , \quad \text{as } \varepsilon \downarrow 0 .$$

On the other hand, setting $z_+ = \overline{z}(\tau+)$, $w_+ = \overline{w}(\tau+)$, we have $\overline{z}(t) = z_+ + o(1)$, $\overline{w}(t) = w_+ + o(1)$, as $\varepsilon \downarrow 0$. Therefore, combining (12.4.24) with (12.4.28) yields

(12.4.29)
$$-\int_\tau^r H(z(\xi(t) + \varepsilon +, t), w(\xi(t) + \varepsilon +, t), \overline{z}(t), \overline{w}(t))dt$$
$$\leq -[\mu(z_+, w_+) - \lambda(z_+, w_+)] \eta_w(z_+, w_+) [w_0 - w_+](r - \tau) + o(\varepsilon) .$$

We now multiply (12.4.13) by $1/\varepsilon$ and then let $\varepsilon \downarrow 0$. Using (12.4.27), (12.4.28) and recalling that $0 < r - \tau < c_0 \varepsilon$, we obtain

(12.4.30) $\eta(z_1, w_1) - \eta(z_0, w_0) \leq c_3 \sup_{(\tau, s)} |\eta_w(\overline{z}(\cdot), \overline{w}(\cdot))| \, NV_{[\tau, s)} \overline{w}(\cdot) .$

In particular, s is the limit of an increasing sequence of τ with the property that $z(\xi(\tau)-, \tau) = z(\xi(\tau)+, \tau)$, for which (12.4.30) is valid. It follows that $\eta(z_1, w_1) \leq \eta(\overline{z}(s-), \overline{w}(s-))$. Now applying (12.4.25) for $t = s$, and letting $\varepsilon \downarrow 0$, yields $w_1 \leq \overline{w}(s-)$. Also, $\eta_w < 0$, $\eta_z < 0$. Hence, $\overline{z}(s-) \leq z_1$. By Proposition 12.4.1, $\overline{z}(s) \leq \overline{z}(s-)$ and so $z(\xi(s)-, s) = \overline{z}(s) \leq \overline{z}(s-) \leq z_1 = z(\xi(s)+, s)$. Since s is arbitrary, we may write these inequalities for $s = \tau$ and this verifies (12.4.10)$_1$. Proposition 12.4.2 has now been proved. Furthermore, $z_0 \geq \overline{z}(\tau)$ has been established and hence (12.4.30) is valid for all τ and s with $0 < \tau < s \leq \overline{t}$.

From (12.4.22) and (12.4.20) follows

(12.4.31) $\eta(z_1, w_1) - \eta(z_0, w_0) \geq z_0 - z_1 - 3\beta\delta(w_1 - w_0)$.

Combining (12.4.30) with (12.4.31),

(12.4.32) $z_0 - z_1 \leq 3\beta\delta(w_1 - w_0) + c_3 \sup_{(\tau,s)} |\eta_w(\overline{z}(\cdot), \overline{w}(\cdot))| \, NV_{[\tau,s)}\overline{w}(\cdot)$.

To establish $(12.4.12)_1$ for general τ and s, it would suffice to verify it just for τ and s with $s - \tau$ so small that $TV_{[\tau,s]}\overline{w}(\cdot) < 2\delta$. For such τ and s, (12.4.32) gives the preliminary estimate $z_0 - z_1 \leq c_4\delta$, and in fact $z_0 - \overline{z}(t) \leq c_4\delta$, for all $t \in (\tau, s)$. Then, since $|\eta_w(z_0, \overline{w}(t))| \leq 5\beta\delta$, (12.2.2) implies $\sup_{(\tau,s)} |\eta_w(\overline{z}(\cdot), \overline{w}(\cdot))| \leq c_5\delta$. Inserting this estimate into (12.4.32), we arrive at $(12.4.12)_1$, with $b = 3\beta + c_3 c_5$.

Finally, we assume $\overline{w}(\tau+) > \overline{w}(\tau)$, say $w_+ - w_0 = \delta_0 > 0$, and proceed to verify $(12.4.2)_1$. Keeping τ fixed, we choose $s - \tau$ so small that $TV_{[\tau,s]}\overline{w}(\cdot) < 2\delta_0$ and hence $\delta < 2\delta_0$. We need to improve the estimate (12.4.29) and thus we restrict t to the interval $[\tau, r]$.

By account of (12.4.22), (12.4.15) and (12.4.16),

(12.4.33)
$$H(z(\xi(t) + \varepsilon+, t), w(\xi(t) + \varepsilon+, t), \overline{z}(t), \overline{w}(t))$$
$$\geq H(z_+, w_0, \overline{z}(t), \overline{w}(t))$$
$$- [\lambda(z_+, w_0) - \lambda(\overline{z}(t), \overline{w}(t))][z(\xi(t) + \varepsilon+, t) - z_+]$$
$$- 3\beta\delta[\mu(z_+, w_0) - \lambda(\overline{z}(t), \overline{w}(t))][w(\xi(t) + \varepsilon+, t) - w_0] .$$

We have already seen that, as $\varepsilon \downarrow 0$, $\overline{z}(t) = z_+ + o(1)$, $\overline{w}(t) = w_+ + o(1)$. In particular, for ε small, $\lambda(z_+, w_0) - \lambda(\overline{z}(t), \overline{w}(t)) > 0$, by virtue of (12.1.5). Furthermore, if $\hat{\xi}(\cdot)$ denotes the minimal backward 1-characteristic emanating from any point (x, t) with $\xi(t) < x < \xi(t) + 2\varepsilon$, Proposition 12.4.1 implies that $z(x-, t) \leq z(\hat{\xi}(\tau)-, \tau) = z_0 + o(1)$, as $\varepsilon \downarrow 0$. On the other hand, $(12.4.12)_1$, with $s \downarrow \tau$, implies $z_0 - z_+ \leq b\delta_0^2$. Therefore, as $\varepsilon \downarrow 0$, $z(\xi(t) + \varepsilon+, t) \leq z_+ + b\delta_0^2 + o(1)$. Finally, we recall (12.4.28). Collecting the above, we deduce from (12.4.33):

(12.4.34)
$$H(z(\xi(t) + \varepsilon+, t), w(\xi(t) + \varepsilon+, t), \overline{z}(t), \overline{w}(t))$$
$$\geq H(z_+, w_0, z_+, w_+) - c_6\delta_0^3 + o(1) , \quad \text{as } \varepsilon \downarrow 0 .$$

To estimate the right-hand side of (12.4.34), let us visualize q as a function of (z, η). By the chain rule and (12.2.1), we deduce $q_\eta = \mu$, $q_{\eta\eta} = \mu_w/\eta_w$. For $w \in [w_0, w_+]$, $q_{\eta\eta} < 0$. Hence

(12.4.35)
$$H(z_+, w_0, z_+, w_+) \geq [\mu(z_+, w_0) - \lambda(z_+, w_+)][\eta(z_+, w_0) - \eta(z_+, w_+)] .$$

The next step is to show

(12.4.36) $\dfrac{r - \tau}{\varepsilon} \geq \dfrac{1}{\mu(z_0, w_+) - \lambda(z_+, w_+)} + o(1) , \quad \text{as } \varepsilon \downarrow 0$.

To see this, let us begin with

$$\varepsilon = \psi(r) - \xi(r) = \int_\tau^r [\dot{\psi}(t) - \dot{\xi}(t)]dt$$

(12.4.37)

$$\leq \int_\tau^r [\mu(z(\psi(t)-,t), w(\psi(t)-,t)) - \lambda(\bar{z}(t), \overline{w}(t))]dt \;.$$

As shown above, $z(\psi(t)-,t) \leq z_0 + o(1)$, as $\varepsilon \downarrow 0$. On the other hand, the maximal backward 2-characteristic $\zeta(\cdot)$, emanating from any (x,t) with $\xi(t) < x < \psi(t)$, will intersect the graph of $\xi(\cdot)$ at time $\sigma \in (\tau, r]$ and so, by Proposition 12.4.1, $w(x+,t) \leq \overline{w}(\sigma)$. In particular, $w(\psi(t)-,t) \leq w_+ + o(1)$, as $\varepsilon \downarrow 0$. Since $\mu_z > 0$ and $\mu_w > 0$, (12.4.37) implies $\varepsilon \leq (r-\tau)[\mu(z_0, w_+) - \lambda(z_+, w_+) + o(1)]$ whence (12.4.36) immediately follows.

Once again we multiply (12.4.13) by $1/\varepsilon$, let $\varepsilon \downarrow 0$ and then also let $s \downarrow \tau$. Combining (12.4.27), (12.4.34), (12.4.35) and (12.4.36), we conclude:

(12.4.38)

$$\eta(z_+, w_0) - \eta(z_0, w_0) \leq \frac{\mu(z_0, w_+) - \mu(z_+, w_0)}{\mu(z_0, w_+) - \lambda(z_+, w_+)}[\eta(z_+, w_0) - \eta(z_+, w_+)] + c_7\delta_0^3 \;.$$

By virtue of (12.4.20), $\eta(z_+, w_0) - \eta(z_0, w_0) \geq z_0 - z_+$. The right-hand side of (12.4.38) is bounded by $a\delta_0^3$, because $\eta_w = O(\delta_0)$. Therefore, $z_0 - z_+ \leq a\delta_0^3$. Now $\bar{z}(\tau-) \leq z_0$, by account of (12.4.10)$_1$. Hence $\bar{z}(\tau-) - \bar{z}(\tau+) \leq a\delta_0^3$, which establishes (12.4.2)$_1$.

Since total variation is additive, we deduce immediately

Corollary 12.4.1 *In* (12.4.12)$_1$ *(or* (12.4.12)$_2$*), $osc_{[\tau,s]}\overline{w}(\cdot)$ (or $osc_{[\tau,s]}\bar{z}(\cdot)$) may be replaced by the local oscillation of $\overline{w}(\cdot)$ (or $\bar{z}(\cdot)$) in the interval $[\tau, s]$, which is measured by the maximum jump of $\overline{w}(\cdot)$ (or $\bar{z}(\cdot)$) in $[\tau, s]$. In particular, $\bar{z}(\cdot)$ (or $\overline{w}(\cdot)$) is a saltus function whose variation is concentrated in the set of points of jump discontinuity of $\overline{w}(\cdot)$ (or $\bar{z}(\cdot)$).*

We have thus verified all the assertions of Theorem 12.4.1, except that (12.4.2) has been established under the extraneous assumption $\overline{w}(\tau) < \overline{w}(\tau+)$. By Proposition 12.4.2, $\overline{w}(\tau) = w(\xi(\tau)+, \tau) \leq w(\xi(\tau)-, \tau)$. On the other hand, when $(\xi(\tau), \tau)$ is a point of local regularity of the solution, Condition $(c)_1$ of Definition 12.3.2 implies $\overline{w}(\tau+) = w(\xi(\tau)-, \tau)$. Hence, by establishing Theorem 12.3.1, we will justify, in particular, the assumption $\overline{w}(\tau) < \overline{w}(\tau+)$.

We thus turn to the proof of Theorem 12.3.1. Our main tool will be the estimate (12.4.12). In what follows, δ will denote an upper bound of the oscillation of z and w on the upper half-plane. We fix any point $(\bar{x}, \bar{t})$ of the upper half-plane and construct the curves $\xi_\pm(\cdot)$, $\zeta_\pm(\cdot)$, $\phi_\pm(\cdot)$ and $\psi_\pm(\cdot)$, as described in Section 12.3 and sketched in Fig. 12.3.1. The first step is to verify the part of Condition (a) of Definition 12.3.2 pertaining to the "western" sector $\mathscr{S}_W$.

Proposition 12.4.4 *For θ sufficiently small, as (x,t) tends to $(\bar{x}, \bar{t})$ through the region $\mathscr{S}_W$, defined by (12.3.2), $(z(x\pm, t), w(x\pm, t))$ tend to (z_W, w_W), where $z_W = z(\bar{x}-, \bar{t})$, $w_W = w(\bar{x}-, \bar{t})$.*

Proof. We shall construct a sequence $x_0 < x_1 < x_2 < \cdots < \bar{x}$ such that, for $m = 0, 1, 2, \cdots$,

$$(12.4.39) \qquad osc._{\mathscr{S}_W \cap \{x > x_m\}} z \leq (3b\theta)^m \delta , \qquad osc._{\mathscr{S}_W \cap \{x > x_m\}} w \leq (3b\theta)^m \delta ,$$

where b is the constant appearing in (12.4.12). Clearly, (12.4.39) will readily imply the assertion of the proposition, provided $3b\theta < 1$.

For $m = 0$, (12.4.39) is satisfied with $x_0 = -\infty$. Arguing by induction, let us assume $x_0 < x_1 < \cdots < x_{k-1} < \bar{x}$ have already been fixed so that (12.4.39) holds for $m = 0, \cdots, k - 1$. We proceed to determine x_k. We fix $\hat{t} \in (0, \bar{t})$ with $\bar{t} - \hat{t}$ so small that $\zeta_-(\hat{t}) > x_{k-1}$ and the oscillation of $z(\zeta_-(\tau)\pm, \tau)$ over the interval $[\hat{t}, \bar{t})$ does not exceed $\frac{1}{3}(3b\theta)^k \delta$. Next we locate $\hat{x} \in (x_{k-1}, \zeta_-(\hat{t}))$ with $\zeta_-(\hat{t}) - \hat{x}$ so small that the oscillation of $w(y-, \hat{t})$ over the interval $(\hat{x}, \zeta_-(\hat{t})]$ is similarly bounded by $\frac{1}{3}(3b\theta)^k \delta$.

By the construction of $\phi_-(\cdot)$, the minimal backward 1-characteristic $\xi(\cdot)$ emanating from any point (x, t) in $\mathscr{S}_W \cap \{x > x_k\}$ stays to the left of the graph of $\phi_-(\cdot)$. At the same time, as (x, t) tends to $(\bar{x}, \bar{t})$ through $\mathscr{S}_W$, the maximal backward 2-characteristic $\zeta(\cdot)$ emanating from it will tend to some backward 2-characteristic emanating from $(\bar{x}, \bar{t})$, which necessarily lies to the right of the minimal characteristic $\zeta_-(\cdot)$ or coincides with $\zeta_-(\cdot)$. It follows that when $\bar{x} - x_k$ is sufficiently small, $\xi(\cdot)$ will have to cross the graph of $\zeta_-(\cdot)$ at some time $t^* \in (\hat{t}, \bar{t})$, while $\zeta(\cdot)$ must intersect either the graph of $\zeta_-(\cdot)$ at some time $\tilde{t} \in (\hat{t}, \bar{t})$ or the $\hat{t}$-time line at some $\tilde{x} \in (\hat{x}, \zeta_-(\hat{t})]$.

By virtue of Propositions 12.4.1 and 12.4.2,

$$(12.4.40) \qquad z(x-, t) \leq z(\xi(t^*)-, t^*) = z(\zeta_-(t^*)-, t^*) \leq z(\zeta_-(t^*)+, t^*) .$$

By account of (12.4.39), for $m = k - 1$, and the construction of $\hat{t}$, the oscillation of $w(\xi(\tau)+, \tau)$ over the interval $[t^*, t]$ does not exceed $(3b\theta)^{k-1}\delta + \frac{1}{3}(3b\theta)^k\delta$, which in turn is majorized by $2(3b\theta)^{k-1}\delta$. Then $(12.4.12)_1$ yields

$$(12.4.41)$$
$$z(x+, t) \geq z(\xi(t^*)+, t^*) - 2b\theta(3b\theta)^{k-1}\delta = z(\zeta_-(t^*)+, t^*) - \frac{2}{3}(3b\theta)^k\delta .$$

Recalling that the oscillation of $z(\zeta_-(\tau)+, \tau)$ over $[\hat{t}, \bar{t})$ is bounded by $\frac{1}{3}(3b\theta)^k\delta$, (12.4.40) and (12.4.41) together imply the bound (12.4.39) on the oscillation of z, for $m = k$.

The argument for w is similar: Assume, for example, that $\xi(\cdot)$ intersects the $\hat{t}$-time line, rather than the graph of $\zeta_-(\cdot)$. By virtue of Propositions 12.4.1 and 12.4.2,

$$(12.4.42) \qquad w(x+, t) \leq w(\zeta(\hat{t})+, \hat{t}) = w(\tilde{x}+, \hat{t}) \leq w(\tilde{x}-, \hat{t}) .$$

The oscillation of $z(\zeta(\tau)-, \tau)$ over the interval $[\hat{t}, t]$ does not exceed $(3b\theta)^{k-1}\delta$, by account of (12.4.39), for $m = k - 1$. Then $(12.4.12)_2$ implies

$$(12.4.43) \quad w(x-, t) \geq w(\zeta(\hat{t})-, \hat{t}) - b\theta(3b\theta)^{k-1}\delta = w(\tilde{x}-, \hat{t}) - \frac{1}{3}(3b\theta)^k\delta .$$

The bound (12.4.39) on the oscillation of w, for $m = k$, now easily follows from (12.4.42), (12.4.43) and the construction of $\hat{t}$ and $\hat{x}$. The proof is complete.

The part of Condition (a) of Definition 12.3.2 pertaining to the "eastern" sector $\mathscr{S}_E$ is validated by a completely symmetrical argument. The next step is to check the part of Condition (a) that pertains to the "southern" sector $\mathscr{S}_S$.

Proposition 12.4.5 *For θ sufficiently small, as (x, t) tends to $(\bar{x}, \bar{t})$ through the region $\mathscr{S}_S$, defined by (12.3.5), $(z(x\pm, t), w(x\pm, t))$ tend to a constant state (z_S, w_S).*

Proof. Similar to the proof of Proposition 12.4.4, the aim is here to determine $t_0 < t_1 < \cdots < \bar{t}$ such that

$$(12.4.44) \qquad osc_{\mathscr{S}_S \cap \{t > t_m\}} z \leq (4b\theta)^m \delta , \qquad osc_{\mathscr{S}_S \cap \{t > t_m\}} w \leq (4b\theta)^m \delta ,$$

for $m = 0, 1, 2, \cdots$. For $m = 0$, (12.4.44) is satisfied with $t_0 = 0$. Arguing by induction, we assume $t_0 < t_1 < \cdots < t_{k-1} < \bar{t}$ have already been fixed so that (12.4.44) holds for $m = 0, \cdots, k - 1$, and proceed to determine t_k. We fix $\hat{t} \in (t_{k-1}, \bar{t})$ with $\bar{t} - \hat{t}$ so small that the oscillation of $z(\zeta_+(\tau)-, \tau)$, $w(\zeta_+(\tau)+, \tau)$, $z(\xi_-(\tau)-, \tau)$, $w(\xi_-(\tau)+, \tau)$ over the interval $[\hat{t}, \bar{t}]$ does not exceed $\frac{1}{4}(4b\theta)^k \delta$. Next we fix $\hat{x}$ and $\tilde{x}$ in the interval $(\zeta_+(\hat{t}), \xi_-(\hat{t}))$ with $\hat{x} - \zeta_+(\hat{t})$ and $\xi_-(\hat{t}) - \tilde{x}$ so small that the oscillation of $z(y-, \hat{t})$ over the interval $(\tilde{x}, \xi_-(\hat{t})]$ and the oscillation of $w(y+, \hat{t})$ over the interval $[\zeta_+(\hat{t}), \hat{x})$ do not exceed $\frac{1}{4}(4b\theta)^k \delta$.

Since $\xi_-(\cdot)$ is the minimal backward 1-characteristic and $\zeta_+(\cdot)$ is the maximal backward 2-characteristic emanating from $(\bar{x}, \bar{t})$, we can find $t_k \in (\hat{t}, \bar{t})$ with $\bar{t} - t_k$ so small that the following hold for any (x, t) in $\mathscr{S}_S \cap \{t > t_k\}$: (a) the minimal backward 1-characteristic $\xi(\cdot)$ emanating from (x, t) must intersect either the $\hat{t}$-time line at $x' \in (\tilde{x}, \xi_-(\hat{t})]$ or the graph of $\xi_-(\cdot)$ at time $t' \in (\hat{t}, \bar{t})$; and (b) the maximal backward 2-characteristic $\zeta(\cdot)$ emanating from (x, t) must intersect either the $\hat{t}$-time line at $x^* \in [\zeta_+(\hat{t}), \hat{x})$ or the graph of $\zeta_+(\cdot)$ at some time $t^* \in (\hat{t}, \bar{t})$. One then repeats the argument employed in the proof of Proposition 12.4.4 to verify the (12.4.44) is indeed satisfied for $m = k$, with t_k determined as above. The proof is complete.

To conclude the validation of Condition (a) of Definition 12.3.2, it remains to check the part pertaining to the "northern" sector $\mathscr{S}_N$.

Proposition 12.4.6 *For θ sufficiently small, as (x, t) tends to $(\bar{x}, \bar{t})$ through the region $\mathscr{S}_N$, defined by (12.3.4), $(z(x\pm, t), w(x\pm, t))$ tend to a constant state (z_N, w_N).*

Proof. For definiteness, we treat the configuration depicted in Fig. 12.3.1, where $\psi_- \equiv \psi_+$, so that $\psi_-(\cdot)$ is a 2-shock of positive strength at $t = \bar{t}$, while $\phi_-(t) < \phi_+(t)$, for $t > \bar{t}$, in which case, as we shall see in Proposition 12.4.8, it is
$$\lim_{t \downarrow \bar{t}} z(\phi_+(t)-, t) = \lim_{t \downarrow \bar{t}} z(\phi_+(t)+, t) \text{ and } \lim_{t \downarrow \bar{t}} w(\phi_+(t)-, t) = \lim_{t \downarrow \bar{t}} w(\phi_-(t)+, t).$$

Only slight modifications in the argument are needed for the case of alternative feasible configurations.

The aim is to find $t_0 > t_1 > \cdots > \bar{t}$ such that

$$(12.4.45) \qquad osc_{\mathscr{S}_N \cap \{t < t_m\}} z \leq a(ab\theta)^m \delta , \quad osc_{\mathscr{S}_N \cap \{t < t_m\}} w \leq 3(ab\theta)^m \delta ,$$

for $m = 0, 1, 2, \cdots$, where a is a constant, $a \geq 1$, independent of m and θ, to be specified below. Clearly, (12.4.45) is satisfied for $m = 0$, with $t_0 = \infty$. Arguing by induction, we assume $t_0 > t_1 > \cdots > t_{k-1} > \bar{t}$ have already been fixed so that (12.4.45) holds for $m = 0, \cdots, k - 1$, and proceed to determine t_k.

We select $t_k \in (\bar{t}, t_{k-1})$ with $t_k - \bar{t}$ so small that the oscillation of $z(\phi_+(\tau)-, \tau)$ over the interval $(\bar{t}, t_k)$ does not exceed $a(ab\theta)^{k-1}\delta$, the oscillation of $w(\phi_+(\tau)-, \tau)$ over $(\bar{t}, t_k)$ is bounded by $(ab\theta)^k \delta$, and the oscillation of $U(\psi_-(\tau)-, \tau)$ over $(\bar{t}, t_k)$ is majorized by $(ab\theta)^{2k} \delta^2$.

The bound (12.4.45) on the oscillation of w, for $m = k$, will be established by the procedure employed in the proof of Proposition 12.4.4 and 12.4.5. We thus fix any (x, t) in $\mathscr{S}_N \cap \{t < t_k\}$ and consider the maximal backward 2-characteristic $\zeta(\cdot)$ emanating from it, which intersects the graph of $\phi_+(\cdot)$ at some time $\tilde{t} \in (\bar{t}, t_k)$. By virtue of Propositions 12.4.1 and 12.4.2:

$$(12.4.46) \qquad w(x+, t) \leq w(\zeta(\tilde{t})+, \tilde{t}) = w(\phi_+(\tilde{t})+, \tilde{t}) \leq w(\phi_+(\tilde{t})-, \tilde{t}) .$$

By account of (12.4.45), for $m = k - 1$, and the construction of t_k, the oscillation of $z(\zeta(\tau)-, \tau)$ over the interval $[\tilde{t}, t]$ does not exceed $2a(ab\theta)^{k-1}\delta$. Then $(12.4.12)_2$ implies

$$(12.4.47) \quad w(x-, t) \geq w(\zeta(\tilde{t})-, \tilde{t}) - 2(ab\theta)^k \delta = w(\phi_+(\tilde{t})-, \tilde{t}) - 2(ab\theta)^k \delta .$$

The inequalities (12.4.46), (12.4.47) coupled with the condition that the oscillation of $w(\phi_+(\tau)-, \tau)$ over $(\bar{t}, t_k)$ is majorized by $(ab\theta)^k \delta$ readily yield the bound (12.4.45) on the oscillation of w, for $m = k$.

To derive the corresponding bound on the oscillation of z requires an entirely different argument. Let us define $\overline{U} = \lim_{t \downarrow \bar{t}} U(\psi_-(t)-, t)$, with induced values $(\bar{z}, \bar{w})$ for the Riemann invariants, and then set $\Delta z = z - \bar{z}$, $\Delta w = w - \bar{w}$. On $\mathscr{S}_N \cap \{t < t_k\}$, as shown above,

$$(12.4.48) \qquad |\Delta w| \leq 3(ab\theta)^k \delta .$$

We construct the minimal backward 1-characteristic $\xi(\cdot)$, emanating from any point (y, t) of approximate continuity in $\mathscr{S}_N \cap \{t < t_k\}$, which is intercepted by the graph of $\psi_-(\cdot)$ at time $t^* \in (\bar{t}, t_k)$. Proposition 12.4.1 implies that $z(y, t) \leq z(\xi(t^*)-, t^*) = z(\psi_-(t^*)-, t^*)$ and this in conjunction with the selection of t_k yields

$$(12.4.49) \qquad \Delta z(y, t) \leq c_1 (ab\theta)^{2k} \delta^2 ,$$

for some constant c_1 independent of k and θ.

We now fix any point of approximate continuity (x, t) in $\mathscr{S}_N \cap \{t < t_k\}$. We consider, as above, the minimal backward 1-characteristic $\xi(\cdot)$ emanating from (x, t), which is intercepted by the graph of $\psi_-(\cdot)$ at time $t^* \in (\bar{t}, t_k)$, and integrate (12.1.1) over $\{(y, \tau) : t^* < \tau < t, \xi(\tau) < y < \psi_-(\tau)\}$. By Green's theorem,

(12.4.50)

$$\int_x^{\psi_-(t)} [U(y, t) - \overline{U}] dy$$

$$+ \int_{t^*}^t \{F(U(\psi_-(\tau)-, \tau)) - F(\overline{U}) - \dot{\psi}_-(\tau)[U(\psi_-(\tau)-, \tau) - \overline{U}]\} d\tau$$

$$- \int_{t^*}^t \{F(U(\xi(\tau)+, \tau)) - F(\overline{U}) - \lambda(U(\xi(\tau)+, \tau))[U(\xi(\tau)+, \tau) - \overline{U}]\} d\tau = 0 .$$

Applying repeatedly (7.3.12), we obtain, for $U = U(z, w)$,

(12.4.51) $$U = \overline{U} + \Delta z R(\overline{U}) + \Delta w S(\overline{U}) + O(\Delta z^2 + \Delta w^2) ,$$

(12.4.52)
$$F(U) - F(\overline{U}) - \lambda(U)[U - \overline{U}] = \Delta w[\mu(\overline{U}) - \lambda(\overline{U})]S(\overline{U})$$
$$- \frac{1}{2} \Delta z^2 \lambda_z(\overline{U}) R(\overline{U}) - \Delta z \Delta w \lambda_z(\overline{U}) S(\overline{U}) + O(\Delta w^2 + |\Delta z|^3) .$$

We also note that the oscillation of $w(\xi(\tau)+, \tau)$ over the interval $(t^*, t]$ is bounded by $3(ab\theta)^k \delta$ and so, by account of (12.4.12)$_1$ and Proposition 12.4.2, we have

(12.4.53) $$0 \leq \Delta z(\xi(\tau)+, \tau) - \Delta z(x, t) \leq 3b\theta(ab\theta)^k \delta \leq 3(ab\theta)^k \delta ,$$

for any $\tau \in (t^*, t)$.

We substitute from (12.4.51), (12.4.52) into (12.4.50) and then multiply the resulting equation, from the left, by $Dz(\overline{U})$. By using (12.1.2), (12.4.49), (12.4.48), (12.1.3), (12.4.53) and the properties of t_k, we end up with

(12.4.54) $$\Delta z^2(x, t) \leq c(ab\theta)^{2k} \delta^2 ,$$

where c is a constant independent of (x, t), k and θ. Consequently, upon selecting $a = \max\{1, 2\sqrt{c}\}$, we arrive at the desired bound (12.4.45) on the oscillation of z, for $m = k$. This completes the proof.

To establish Condition (b) of Definition 12.3.2, we demonstrate

Proposition 12.4.7 *Let $p_\ell(\cdot)$ and $p_r(\cdot)$ be any backward 1-characteristics emanating from $(\bar{x}, \bar{t})$, with $p_\ell(t) < p_r(t)$, for $t < \bar{t}$. If θ is sufficiently small, then*

(12.4.55) $$\lim_{t \uparrow \bar{t}} z(p_\ell(t)+, t) \leq \lim_{t \uparrow \bar{t}} z(p_r(t)-, t) ,$$

(12.4.56) $$\lim_{t \uparrow \bar{t}} w(p_\ell(t)+, t) \geq \lim_{t \uparrow \bar{t}} w(p_r(t)-, t) .$$

Proof. Consider any sequence $\{(x_n, t_n)\}$ with $t_n \uparrow \bar{t}$, as $n \to \infty$, and $x_n \in (p_\ell(t_n), p_r(t_n))$ so close to $p_r(t_n)$ that $\lim_{n\to\infty} [w(x_n+, t_n) - w(p_r(t_n)-, t_n)] = 0$. Let $\zeta_n(\cdot)$ denote the maximal backward 2-characteristic emanating from (x_n, t_n), which intersects the graph of $p_\ell(\cdot)$ at time t_n^*. By virtue of Proposition 12.4.1, $w(x_n+, t_n) \leq w(\zeta_n(t_n^*)+, t_n^*) = w(p_\ell(t_n^*)+, t_n^*)$. Since $t_n^* \uparrow \bar{t}$, as $n \to \infty$, this establishes (12.4.56).

To verify (12.4.55), we begin with another sequence $\{(x_n, t_n)\}$, where $t_n \uparrow \bar{t}$, as $n \to \infty$, and $x_n \in (p_\ell(t_n), p_r(t_n))$ is so close to $p_\ell(t_n)$ that we now have $\lim_{n\to\infty} [z(x_n-, t_n) - z(p_\ell(t_n)+, t_n)] = 0$. We construct the minimal backward 1-characteristics $\xi_n(\cdot)$ and $\xi_n^*(\cdot)$, emanating from the points (x_n, t_n) and $(p_r(t_n), t_n)$, respectively. By minimality, $\xi_n(t) \leq \xi_n^*(t) \leq p_r(t)$, for $t \leq t_n$. As $n \to \infty$, $\{\xi_n(\cdot)\}$ and $\{\xi_n^*(\cdot)\}$ will converge, uniformly, to shock free minimal 1-separatrices (in the sense of Definition 10.3.1) $\chi(\cdot)$ and $\chi^*(\cdot)$, emanating from $(\bar{x}, \bar{t})$, such that $\chi(t) \leq \chi^*(t) \leq p_r(t)$, for $t \leq \bar{t}$. In particular, $\dot{\chi}(\bar{t}-) \geq \dot{\chi}^*(\bar{t}-)$ and so

$$(12.4.57) \quad \lim_{t\uparrow\bar{t}} \lambda(z(\chi(t)\pm, t), w(\chi(t)\pm, t)) \geq \lim_{t\uparrow\bar{t}} \lambda(z(\chi^*(t)\pm, t), w(\chi^*(t)\pm, t)) \,.$$

Applying (12.4.56) with $\chi(\cdot)$ and $\chi^*(\cdot)$ in the roles of $p_\ell(\cdot)$ and $p_r(\cdot)$ yields

$$(12.4.58) \qquad \lim_{t\uparrow\bar{t}} w(\chi(t)+, t) \geq \lim_{t\uparrow\bar{t}} w(\chi^*(t)-, t) \,.$$

Since $\lambda_z < 0$ and $\lambda_w < 0$, (12.4.57) and (12.4.58) together imply

$$(12.4.59) \qquad \lim_{t\uparrow\bar{t}} z(\chi(t)\pm, t) \leq \lim_{t\uparrow\bar{t}} z(\chi^*(t)\pm, t) \,.$$

By virtue of Proposition 12.4.1, $z(\xi_n(t)-, t)$ and $z(\xi_n^*(t)-, t)$ are nonincreasing functions on $[0, t_n]$ and so

$$(12.4.60) \qquad \lim_{t\uparrow\bar{t}} z(\chi(t)\pm, t) \geq \lim_{t\uparrow\bar{t}} z(p_\ell(t)+, t) \,,$$

$$(12.4.61) \qquad \lim_{t\uparrow\bar{t}} z(\chi^*(t)\pm, t) \geq \lim_{t\uparrow\bar{t}} z(p_r(t)-, t) \,.$$

Thus, to complete the proof of (12.4.55), one has to show

$$(12.4.62) \qquad \lim_{t\uparrow\bar{t}} z(\chi^*(t)\pm, t) = \lim_{t\uparrow\bar{t}} z(p_r(t)-, t) \,.$$

Since (12.4.62) is trivially true when $\chi^* \equiv p_r$, we take up the case where $\chi^*(t) < p_r(t)$, for $t < \bar{t}$. We set $\mathscr{S} = \{(x, t) : 0 \leq t < \bar{t}, \chi^*(t) < x < p_r(t)\}$. We shall verify (12.4.62) by constructing $t_0 < t_1 < \cdots < \bar{t}$ such that

$$(12.4.63) \qquad osc_{\mathscr{S}\cap\{t>t_m\}}z \leq (3b\theta)^m \delta \,, \quad osc_{\mathscr{S}\cap\{t>t_m\}}w \leq (3b\theta)^m \delta \,,$$

for $m = 0, 1, 2, \cdots$.

For $m = 0$, (12.4.63) is satisfied with $t_0 = 0$. Arguing by induction, we assume $t_0 < t_1 < \cdots < t_{k-1} < \bar{t}$ have already been fixed so that (12.4.63) holds for $m = 0, \cdots, k-1$, and proceed to determine t_k. We fix $\hat{t} \in (t_{k-1}, \bar{t})$ with $\bar{t} - \hat{t}$

so small that the oscillation of $z(\chi^*(\tau)\pm, \tau)$ and $w(\chi^*(\tau)-, \tau)$ over the interval $[\hat{t}, \bar{t})$ does not exceed $\frac{1}{3}(3b\theta)^k\delta$. Next we locate $\hat{x} \in (\chi^*(\hat{t}), p_r(\hat{t}))$ with $\hat{x} - \chi^*(\hat{t})$ so small that the oscillation of $z(y+, \hat{t})$ over the interval $[\chi^*(\hat{t}), \hat{x})$ is similarly bounded by $\frac{1}{3}(3b\theta)^k\delta$.

By the construction of $\chi^*(\cdot)$, if we fix $t_k \in (\hat{t}, \bar{t})$ with $\bar{t} - t_k$ sufficiently small, then the minimal backward 1-characteristic $\xi(\cdot)$, emanating from any point (x, t) in $\mathscr{S} \cap \{t > t_k\}$, will intersect either the graph of $\chi^*(\cdot)$ at some time $t^* \in (\hat{t}, \bar{t})$ or the $\hat{t}$-time line at some $x^* \in (\chi^*(\hat{t}), \hat{x})$; while the maximal backward 2-characteristic $\zeta(\cdot)$, emanating from (x, t), will intersect the graph of $\chi^*(\cdot)$ at some time $\tilde{t} \in (\hat{t}, \bar{t})$.

Assume, for definiteness, that $\xi(\cdot)$ intersects the $\hat{t}$-time line. By virtue of Propositions 12.4.1 and 12.4.2,

$$(12.4.64) \qquad z(x-, t) \leq z(\xi(\hat{t})-, \hat{t}) = z(x^*-, \hat{t}) \leq z(x^*+, \hat{t}) .$$

By account of (12.4.63), for $m = k - 1$, the oscillation of $w(\xi(\tau)+, \tau)$ over the interval $[\hat{t}, t]$ does not exceed $(3b\theta)^{k-1}\delta$. It then follows from $(12.4.12)_1$ that

$$(12.4.65) \qquad z(x+, t) \geq z(\xi(\hat{t})+, \hat{t}) - b\theta(3b\theta)^{k-1}\delta = z(x^*+, \hat{t}) - \frac{1}{3}(3b\theta)^k\delta .$$

Recalling that the oscillation of $z(y+, \hat{t})$ over $[\chi^*(\hat{t}), \hat{x})$ and the oscillation of $z(\chi^*(\tau)+, \tau)$ over $[\hat{t}, \bar{t})$ are bounded by $\frac{1}{3}(3b\theta)^k\delta$, (12.4.64) and (12.4.65) together imply the bound (12.4.63) on the oscillation of z, for $m = k$.

The argument for w is similar: On the one hand, Propositions 12.4.1 and 12.4.2 give

$$(12.4.66) \qquad w(x+, t) \leq w(\zeta(\tilde{t})+, \tilde{t}) = w(\chi^*(\tilde{t})+, \tilde{t}) \leq w(\chi^*(\tilde{t})-, \tilde{t}) .$$

On the other hand, considering that the oscillation of $z(\zeta(\tau)-, \tau)$ over the interval $[\tilde{t}, t]$ is bounded by $(3b\theta)^{k-1}\delta + \frac{1}{3}(3b\theta)^k\delta$, which in turn is smaller than $2(3b\theta)^{k-1}\delta$, $(12.4.12)_2$ yields

$$(12.4.67) \quad w(x-, t) \geq w(\zeta(\tilde{t})-, \tilde{t}) - 2b\theta(3b\theta)^{k-1}\delta = w(\chi^*(\tilde{t})-, \tilde{t}) - \frac{2}{3}(3b\theta)^k\delta .$$

Since the oscillation of $w(\chi^*(\tau)-, \tau)$ over $[\hat{t}, \bar{t})$ does not exceed $\frac{1}{3}(3b\theta)^k\delta$, the inequalities (12.4.66) and (12.4.67) together imply the bound (12.4.63) on the oscillation of w, for $m = k$. The proof of the proposition is now complete.

In particular, one may apply Proposition 12.4.7 with $\xi(\cdot)$ and/or $\xi^*(\cdot)$ in the role of $p_\ell(\cdot)$ or $p_r(\cdot)$, so that, by virtue of Proposition 12.4.2, the inequalities $(12.3.6)_1$ and $(12.3.7)_1$ follow from (12.4.55) and (12.4.56). We have thus verified condition $(b)_1$ of Definition 12.3.2. Condition $(b)_2$ may be validated by a completely symmetrical argument.

It remains to check Condition (c) of Definition 12.3.2. It will suffice to verify $(c)_1$, because then $(c)_2$ will readily follow by a similar argument. In the shock case, $\phi_- \equiv \phi_+$, the required inequalities $z_W \leq z_N$ and $w_W \geq w_N$ are immediate corollaries of Proposition 12.4.2. Thus, one need only consider the rarefaction wave case.

Proposition 12.4.8 *Suppose $\phi_-(t) < \phi_+(t)$, for $t > \bar{t}$. For θ sufficiently small, as (x, t) tends to $(\bar{x}, \bar{t})$ through the region $\mathscr{W} = \{(x, t) : t > \bar{t}, \phi_-(t) < x < \phi_+(t)\}$, $w(x\pm, t)$ tend to w_W. Furthermore, $(12.3.8)_1$ holds for any 1-characteristics $p_\ell(\cdot)$ and $p_r(\cdot)$, with $\phi_-(t) \leq p_\ell(t) \leq p_r(t) \leq \phi_+(t)$, for $t > \bar{t}$.*

Proof. Consider (x, t) that tend to $(\bar{x}, \bar{t})$ through $\mathscr{W}$. The maximal backward 2-characteristic $\zeta(\cdot)$ emanating from (x, t) is intercepted by the $\bar{t}$-time line at $\zeta(\bar{t})$, which tends from below to $\bar{x}$. It then follows from Proposition 12.4.1 that $\lim \sup w(x\pm, t) \leq w_W$. To verify the assertion of the proposition, one needs to show that $\lim \inf w(x\pm, t) = w_W$. The plan is to argue by contradiction and so we make the hypothesis $\lim \inf w(x\pm, t) = w_W - \beta$, with $\beta > 0$.

We fix $\hat{t} > \bar{t}$ with $\hat{t} - \bar{t}$ so small that

$$(12.4.68) \quad w_W - 2\beta < w(x\pm, t) \leq w_W + \beta , \quad \bar{t} < t < \hat{t} , \quad \phi_-(t) < x < \phi_+(t)$$

and also the oscillation of the functions $z(\phi_-(t)\pm, t)$ and $w(\phi_-(t)\pm, t)$ over the interval $(\bar{t}, \hat{t})$ does not exceed $\frac{1}{2}\beta$.

We consider the maximal backward 2-characteristic $\zeta(\cdot)$ emanating from any point $(\tilde{x}, \tilde{t})$, with $\bar{t} < \tilde{t} < \hat{t}$, $\phi_-(\tilde{t}) < x < \phi_+(\tilde{t})$, and intersecting the graph of $\phi_-(\cdot)$ at time $t^* \in (\bar{t}, \hat{t})$. We demonstrate that, when θ is sufficiently small, independent of β, then

$$(12.4.69) \quad w(\zeta(t)-, t) - w(\tilde{x}-, \tilde{t}) \leq \frac{1}{4}\beta , \quad t^* < t < \tilde{t} .$$

Indeed, if (12.4.69) were false, one may find t_1, t_2, with $t^* < t_1 < t_2 \leq \tilde{t}$ and $t_2 - t_1$ arbitrarily small, such that

$$(12.4.70) \quad |z(\zeta(t_1)\pm, t_1) - z(\zeta(t_2)\pm, t_2)| > \frac{\beta}{4b\theta} .$$

In particular, if $\xi_1(\cdot)$ and $\xi_2(\cdot)$ denote the minimal backward 1- characteristics, which emanate from the points $(\zeta(t_1), t_1)$ and $(\zeta(t_2), t_2)$, respectively, and thus necessarily pass through the point $(\bar{x}, \bar{t})$, then t_1 and t_2 may be fixed so close that

$$(12.4.71) \quad \begin{aligned} 0 \leq &\int_0^{t_1} \lambda(z(\xi_2(t)-, t), w(\xi_2(t)-, t))dt \\ &- \int_0^{t_1} \lambda(z(\xi_1(t)-, t), w(\xi_1(t)-, t))dt \leq \beta t_0 . \end{aligned}$$

By virtue of (12.4.68), $|w(\xi_2(t)-, t) - w(\xi_1(t)-, t)| < 3\beta$, for all t in $(\bar{t}, t_1)$. Also, by account of Proposition 12.4.1, $(12.4.12)_1$ and (12.4.68), we have

$$(12.4.72) \quad \begin{cases} z(\zeta(t_1)-, t_1) \leq z(\xi_1(t)-, t) = z(\xi_1(t)+, t) \leq z(\zeta(t_1)+, t_1) + 3\beta b\theta , \\ z(\zeta(t_2)-, t_2) \leq z(\xi_2(t)-, t) = z(\xi_2(t)+, t) \leq z(\zeta(t_2)+, t_2) + 3\beta b\theta , \end{cases}$$

for almost all t in $(\bar{t}, t_1)$. It is now clear that, for θ sufficiently small, (12.4.72) renders the inequalities (12.4.70) and (12.4.71) incompatible. This provides the desired contradiction that verifies (12.4.69).

By Proposition 12.4.4, and the construction of $\hat{t}$,

$$(12.4.73) \qquad \lim_{t \downarrow \bar{t}} z(\phi_-(t)-, t) = z_W , \qquad \lim_{t \downarrow \bar{t}} w(\phi_-(t)-, t) = w_W ,$$

$$(12.4.74) \qquad |z(\phi_-(t^*)-, t^*) - z_W| \le \frac{1}{2}\beta , \qquad |w(\phi_-(t^*)-, t^*) - w_W| \le \frac{1}{2}\beta .$$

The next step is to establish an estimate

$$(12.4.75) \qquad |z(\phi_-(t^*)-, t^*) - \lim_{t \downarrow t^*} z(\zeta(t)-, t)| \le a\beta ,$$

for some constant a independent of θ and β. Let

$$(12.4.76) \qquad \lim_{t \downarrow \bar{t}} z(\phi_-(t)+, t) = z_W + \gamma ,$$

with $\gamma \ge 0$. We fix $t_3 \in (t^*, \hat{t})$ and $x_3 \in (\phi_-(t_3), \phi_+(t_3))$, with $x_3 - \phi_-(t_3)$ so small that

$$(12.4.77) \qquad |z(x_3\pm, t_3) - z_W - \gamma| \le \beta .$$

By also choosing $t_3 - t^*$ sufficiently small, we can guarantee that the minimal backward 1-characteristic $\xi(\cdot)$, emanating from the point (x_3, t_3), will intersect the graph of $\zeta(\cdot)$ at time t_4, arbitrarily close to t^*. By Proposition 12.4.1, $z(\zeta(t_4)-, t_4) \ge z(x_3-, t_3)$. On the other hand, by virtue of (12.4.68), Proposition 12.4.3 implies $z(\zeta(t_4)+, t_4) \le z(x_3+, t_3) + 3b\theta\beta$. Hence, for θ so small that $6b\theta \le 1$, we have $|z_W + \gamma - \lim_{t \downarrow t^*} z(\zeta(t)-, t)| \le \frac{3}{2}\beta$. In conjunction with (12.4.74), this yields

$$(12.4.78) \qquad |z(\phi_-(t^*)-, t^*) - \lim_{t \downarrow t^*} z(\zeta(t)-, t)| \le 2\beta + \gamma .$$

Thus, to verify (12.4.75), one has to show $\gamma \le c\beta$.

The characteristic $\xi(\cdot)$ lies to the right of $\phi_-(\cdot)$ and passes through the point $(\bar{x}, \bar{t})$, so $\dot\phi_-(\bar{t}+) \le \dot\xi(\bar{t}+)$. By account of (12.4.73), (12.4.76), (8.2.1), (8.2.2), (7.3.12), (8.2.3), and (12.1.2), we conclude

$$(12.4.79) \qquad \dot\phi_-(\bar{t}+) = \lambda(z_W, w_W) + \frac{1}{2}\lambda_z(z_W, w_W)\gamma + O(\gamma^2) .$$

To estimate $\dot\xi(\bar{t}+) = \lim_{t \downarrow \bar{t}} \lambda(z(\xi(t)-, t), w(\xi(t)-, t))$, recall that $\lambda_z < 0$, $\lambda_w < 0$, $z(\xi(t)-, t) \ge z(x_3-, t_3) \ge z_W + \gamma - \beta$, $w(\xi(t)-, t) \ge w_W - 2\beta$, and so

$$(12.4.80)$$
$$\dot\xi(\bar{t}+) \le \lambda(z_W + \gamma - \beta, w_W - 2\beta) = \lambda(z_W, w_W) + \lambda_z(z_W, w_W)\gamma + O(\beta + \gamma^2) .$$

Therefore, $\gamma = O(\beta)$ and (12.4.75) follows from (12.4.78).

By virtue of Proposition 12.4.3, (12.4.75) yields

$$(12.4.81) \qquad w(\phi_-(t^*)-, t^*) - \lim_{t \downarrow t^*} w(\zeta(t)-, t) \leq ab\theta\beta \ .$$

Hence, if $\theta \leq (8ab)^{-1}$, then (12.4.69), (12.4.81) and (12.4.74) together imply $w_W - w(\tilde{x}-, \tilde{t}) \leq \frac{7}{8}\beta$, for all $(\tilde{x}, \tilde{t})$ in $\mathscr{W} \cap \{t < \hat{t}\}$. This provides the desired contradiction to the hypothesis $\lim \inf w(x\pm, t) = w_W - \beta$, with $\beta > 0$, thus verifying the assertion that, as (x, t) tends to $(\overline{x}, \overline{t})$ through $\mathscr{W}$, $w(x\pm, t)$ tend to w_W.

We now focus attention on $\phi_+(\cdot)$. We already have $\lim_{t \downarrow \overline{t}} w(\phi_+(t)-, t) = w_W$, $\lim_{t \downarrow \overline{t}} z(\phi_+(t)+, t) = z_N$, $\lim_{t \downarrow \overline{t}} w(\phi_+(t)+, t) = w_N$. We set $z_0 = \lim_{t \downarrow \overline{t}} z(\phi_+(t)-, t)$. Then $\lambda(z_0, w_W) \geq \dot{\phi}_+(\overline{t}+) \geq \lambda(z_N, w_N)$. We shall show $\dot{\phi}_+(\overline{t}+) = \lambda(z_0, w_W)$ so as to infer $z_N = z_0$, $w_N = w_W$. We consider the minimal backward 1-characteristic $\xi(\cdot)$ emanating from the point $(\phi_+(t_5), t_5)$, where $t_5 - \overline{t}$ is very small. The assertion $z_N = z_0$, $w_N = w_W$ is clearly true when $\xi \equiv \phi_+$, so let us assume $\xi(t) < \phi_+(t)$ for $t \in (\overline{t}, t_5)$. Then $|w(\xi(t)+, t) - w_W|$ is very small on $(\overline{t}, t_5)$. Moreover, by Proposition 12.4.3, the oscillation of $z(\xi(t)+, t)$ over the interval $(\overline{t}, t_5)$ is very small so this function takes values near z_0. Hence, $t_5 - \overline{t}$ sufficiently small renders $\dot{\xi}(\overline{t}+)$ arbitrarily close to $\lambda(z_0, w_W)$. Since $\dot{\xi}(\overline{t}+) \leq \dot{\phi}_+(\overline{t}+)$, we conclude that $\dot{\phi}_+(\overline{t}+) \geq \lambda(z_0, w_W)$ and thus necessarily $\dot{\phi}_+(\overline{t}_+) = \lambda(z_0, w_W)$.

Consider now any forward 1-characteristic $\chi(\cdot)$ issuing from $(\overline{x}, \overline{t})$, with $\phi_-(t) \leq \chi(t) \leq \phi_+(t)$, for $t > \overline{t}$. Since $\lim_{t \downarrow \overline{t}} w(\chi(t)-, t)$ and $\lim_{t \downarrow \overline{t}} w(\chi(t)+, t)$ take the same value, namely w_W, we infer that $\lim_{t \downarrow \overline{t}} z(\chi(t)-, t)$ and $\lim_{t \downarrow \overline{t}} z(\chi(t)+, t)$ must also take the same value, say z_χ. In particular, $\dot{\chi}(\overline{t}+) = \lambda(z_\chi, w_W)$. Hence, if $p_\ell(\cdot)$ and $p_r(\cdot)$ are any 1-characteristics, with $\phi_-(t) \leq p_\ell(t) \leq p_r(t) \leq \phi_+(t)$, for $t > \overline{t}$, the inequalities $\dot{\phi}_-(\overline{t}+) \leq \dot{p}_\ell(\overline{t}+) \leq \dot{p}_r(\overline{t}+) \leq \dot{\phi}_+(\overline{t}+)$, ordering the speeds of propagation at $\overline{t}$, together with $\lambda_z < 0$, imply $(12.3.8)_1$. The proof is complete.

We have now completed the proof of Theorem 12.3.1, on local regularity, as well as of Theorem 12.4.1, on the laws of propagation of Riemann invariants along extremal backward characteristics. These will serve as the principal tools for deriving a priori estimates leading to a description of the long time behavior of solutions.

Henceforth, our solutions will be normalized on $(-\infty, \infty) \times (0, \infty)$ by defining $(z(x, t), w(x, t)) = (z_S, w_S)$, namely the "southern" limit at (x, t). The trace of the solution on any space-like curve is then defined as the restriction of the normalized (z, w) to this curve. In particular, this renders the trace of (z, w) along the minimal backward 1-characteristic and the maximal backward 2-characteristic, emanating from any point $(\overline{x}, \overline{t})$, continuous from the left on $(0, \overline{t}]$.

12.5 Bounds on Solutions

We consider a solution, normalized as above, bounded by

$$(12.5.1) \qquad |z(x,t)| + |w(x,t)| < 2\delta , \quad -\infty < x < \infty , \quad 0 < t < \infty ,$$

where δ is a small positive constant. It is convenient to regard the initial data as multi-valued functions, allowing $(z(x,0), w(x,0))$ to take as values any state in the range of the solution of the Riemann problem with end-states $(z(x\pm, 0), w(x\pm, 0))$. The supremum and total variation are measured for the selection that maximizes these quantities. We then assume

$$(12.5.2) \qquad \sup_{(-\infty,\infty)} |z(\cdot,0)| + \sup_{(-\infty,\infty)} |w(\cdot,0)| \le \delta ,$$

$$(12.5.3) \qquad TV_{(-\infty,\infty)} z(\cdot, 0) + TV_{(-\infty,\infty)} w(\cdot, 0) < a\delta^{-1} ,$$

where a is a small constant, to be fixed later, independently of δ. Thus, there is a tradeoff, allowing for arbitrarily large total variation at the expense of keeping the oscillation sufficiently small. The aim is to establish bounds on the solution. In what follows, c will stand for a generic constant that depends solely on F. The principal result is

Theorem 12.5.1 *Consider any space-like curve* $t = t^*(x)$, $x_\ell \le x \le x_r$, *in the upper half-plane, along which the trace of* (z, w) *is denoted by* (z^*, w^*). *Then*

$$(12.5.4)_1 \quad \begin{aligned} TV_{[x_\ell, x_r]} z^*(\cdot) &\le TV_{[\xi_\ell(0), \xi_r(0)]} z(\cdot, 0) \\ &\quad + c\delta^2 \{ TV_{[\zeta_\ell(0), \xi_r(0)]} z(\cdot, 0) + TV_{[\zeta_\ell(0), \xi_r(0)]} w(\cdot, 0) \} , \end{aligned}$$

$$(12.5.4)_2 \quad \begin{aligned} TV_{[x_\ell, x_r]} w^*(\cdot) &\le TV_{[\zeta_\ell(0), \zeta_r(0)]} w(\cdot, 0) \\ &\quad + c\delta^2 \{ TV_{[\zeta_\ell(0), \xi_r(0)]} z(\cdot, 0) + TV_{[\zeta_\ell(0), \xi_r(0)]} w(\cdot, 0) \} , \end{aligned}$$

where $\xi_\ell(\cdot), \xi_r(\cdot)$ *are the minimal backward 1-characteristics and* $\zeta_\ell(\cdot), \zeta_r(\cdot)$ *are the maximal backward 2-characteristics emanating from the endpoints* (x_ℓ, t_ℓ) *and* (x_r, t_r) *of the graph of* $t^*(\cdot)$.

Since generalized characteristics are space-like curves, one may combine the above proposition with Theorem 12.4.1 and the assumptions (12.5.1), (12.5.3) to deduce the following corollary:

Theorem 12.5.2 *For any point* (x, t) *of the upper half-plane:*

$$(12.5.5)_1 \qquad \sup_{(-\infty,\infty)} z(\cdot, 0) \ge z(x, t) \ge \inf_{(-\infty,\infty)} z(\cdot, 0) - ca\delta ,$$

$$(12.5.5)_2 \qquad \sup_{(-\infty,\infty)} w(\cdot, 0) \ge w(x, t) \ge \inf_{(-\infty,\infty)} w(\cdot, 0) - ca\delta .$$

Thus, on account of our assumption (12.5.2) and by selecting a sufficiently small, we secure a posteriori that the solution will satisfy (12.5.1).

The task of proving Theorem 12.5.1 is quite arduous and will require extensive preparation. In the course of the proof we shall verify that certain quantities measuring the total amount of wave interaction are also bounded.

Consider a 1-shock joining the state (z_-, w_-), on the left, with the state (z_+, w_+), on the right. The jumps $\Delta z = z_+ - z_-$ and $\Delta w = w_+ - w_-$ are related through an equation

$$(12.5.6)_1 \qquad \Delta w = f(\Delta z; z_-, w_-)$$

resulting from the reparametrization of the 1-shock curve emanating from the state (z_-, w_-). In particular, f and its first two derivatives with respect to Δz vanish at $\Delta z = 0$ and hence f as well as $\partial f/\partial z_-$ and $\partial f/\partial w_-$ are $O(\Delta z^3)$ as $\Delta z \to 0$.

Similarly, the jumps $\Delta w = w_+ - w_-$ and $\Delta z = z_+ - z_-$ of the Riemann invariants across a 2-shock joining the state (z_-, w_-), on the left, with the state (z_+, w_+), on the right, are related through an equation

$$(12.5.6)_2 \qquad \Delta z = g(\Delta w; z_+, w_+)$$

resulting from the reparametrization of the backward 2-shock curve (see Section 9.3) that emanates from the state (z_+, w_+). Furthermore, g together with $\partial g/\partial z_+$ and $\partial g/\partial w_+$ are $O(\Delta w^3)$ as $\Delta w \to 0$.

For convenience, points of the upper half-plane will be labelled by single capital letters I, J, etc. With any point $I = (\bar{x}, \bar{t})$ we associate the special characteristics $\phi_\pm^I$, $\psi_\pm^I$, $\xi_\pm^I$, $\zeta_\pm^I$ emanating from it, as discussed in Section 12.3 and depicted in Fig. 12.3.1, and identify the limits (z_W^I, w_W^I), (z_E^I, w_E^I), (z_N^I, w_N^I), (z_S^I, w_S^I) as I is approached through the sectors $\mathscr{S}_W^I$, $\mathscr{S}_E^I$, $\mathscr{S}_N^I$, $\mathscr{S}_S^I$. From I emanate minimal 1-separatrices $p_\pm^I$ and maximal 2-separatrices $q_\pm^I$ constructed as follows: p_-^I (or q_+^I) is simply the minimal (or maximal) backward 1-characteristic ξ_-^I (or 2-characteristic ζ_+^I) emanating from I; while p_+^I (or q_-^I) is the limit of a sequence of minimal (or maximal) backward 1-characteristics ξ_n (or 2-characteristics ζ_n) emanating from points (x_n, t_n) in $\mathscr{S}_E^I$ (or $\mathscr{S}_W^I$), where $(x_n, t_n) \to (\bar{x}, \bar{t})$, as $n \to \infty$. We introduce the notation

$$(12.5.7)_1 \qquad \mathscr{F}_I = \{(x, t) : 0 \le t < \bar{t}, \quad p_-^I(t) \le x \le p_+^I(t)\},$$

$$(12.5.7)_2 \qquad \mathscr{G}_I = \{(x, t) : 0 \le t < \bar{t}, \quad q_-^I(t) \le x \le q_+^I(t)\}.$$

By virtue of Theorems 12.3.1 and 12.4.1,

$$(12.5.8)_1 \qquad \lim_{t \uparrow \bar{t}} z(p_-^I(t), t) = z_S^I, \quad \lim_{t \uparrow \bar{t}} z(p_+^I(t), t) = z_E^I,$$

$$(12.5.8)_2 \qquad \lim_{t \uparrow \bar{t}} w(q_-^I(t), t) = w_W^I, \quad \lim_{t \uparrow \bar{t}} w(q_+^I(t), t) = w_S^I.$$

The cumulative strength of 1-waves and 2-waves, incoming at I, is respectively measured by

$$(12.5.9) \qquad \Delta z^I = z_E^I - z_S^I, \quad \Delta w^I = w_S^I - w_W^I.$$

If the incoming 1-waves alone were allowed to interact, they would produce an outgoing 1-shock with w-amplitude

$$(12.5.10)_1 \qquad \Delta w_*^I = f(\Delta z^I; z_S^I, w_S^I) \,,$$

together with an outgoing 2-rarefaction wave. Consequently, $|\Delta w_*^I|$ exceeds the cumulative w-strength $|w_E^I - w_S^I|$ of incoming 1-waves. Similarly, the interaction of incoming 2-waves alone would produce an outgoing 2-shock with z-amplitude

$$(12.5.10)_2 \qquad \Delta z_*^I = g(\Delta w^I; z_S^I, w_S^I) \,,$$

exceeding their cumulative z-strength $z_S^I - z_W^I$. Note that if $z_S^I = z_W^I$, $w_S^I = w_W^I$ then $\Delta w_*^I = w_N^I - w_W^I$ while if $z_S^I = z_E^I$, $w_S^I = w_E^I$ then $\Delta z_*^I = z_E^I - z_N^I$.

We visualize the upper half-plane as a partially ordered set under the relation induced by the rule $I < J$ whenever J is confined between the graphs of the minimal 1-separatrices p_-^I and p_+^I emanating from I. In particular, when J lies strictly to the right of the graph of p_-^I, then I lies on the graph of the 1-characteristic ϕ_-^J emanating from J. Thus $I < J$ implies that I always lies on the graph of a forward 1-characteristic issuing from J, that is either ϕ_-^J or p_-^J. This special characteristic will be denoted by χ_-^J.

We consider 1-*characteristic trees* $\mathcal{M}$ consisting of a finite set of points of the upper half-plane, called *nodes*, with the following properties: $\mathcal{M}$ contains a unique minimal node I_0, namely the *root* of the tree. Furthermore, if J and K are any two nodes, then the point I of confluence of the forward 1-characteristics χ_-^J and χ_-^K, which pass through the root I_0, is also a node of $\mathcal{M}$. In general, $\mathcal{M}$ will contain several maximal nodes (Fig. 12.5.1).

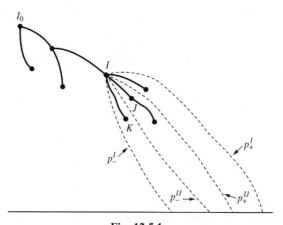

Fig. 12.5.1.

Every node $J \neq I_0$ is *consecutive* to some node I, namely, its strict greatest lower bound relative to $\mathcal{M}$. The set of nodes that are consecutive to a node I is denoted by $\mathcal{C}_I$. When J is consecutive to I, the pair (I, J) is called a *link*.

A finite sequence $\{I_0, I_1, \cdots, I_m\}$ of nodes such that I_{j+1} is consecutive to I_j, for $j = 0, \cdots, m - 1$, which connects the root I_0 with some maximal node I_m, constitutes a *chain* of $\mathcal{M}$.

If (I, J) is a link of $\mathcal{M}$, so that $I = (\chi^J_-(\bar{t}), \bar{t})$, we set

$$(12.5.11)_1 \qquad z^{IJ}_\pm = \lim_{t \uparrow \bar{t}} z(\chi^J_-(t) \pm, t) , \quad w^{IJ}_\pm = \lim_{t \uparrow \bar{t}} w(\chi^J_-(t) \pm, t) ,$$

$$(12.5.12)_1 \qquad \Delta z^{IJ} = z^{IJ}_+ - z^{IJ}_- , \quad \Delta w^{IJ} = w^{IJ}_+ - w^{IJ}_- .$$

In particular,

$$(12.5.13)_1 \qquad \Delta w^{IJ} = f(\Delta z^{IJ}; z^{IJ}_-, w^{IJ}_-) .$$

With (I, J) we associate minimal 1-separatrices $p^{IJ}_\pm$, emanating from I, constructed as follows: p^{IJ}_- is the $t \uparrow \bar{t}$ limit of the family ξ_t of minimal backward 1-characteristics emanating from the point $(\chi^J_-(t), t)$; while p^{IJ}_+ is the limit of a sequence of minimal backward 1-characteristics ξ_n emanating from points (x_n, t_n) such that, as $n \to \infty$, $t_n \uparrow \bar{t}$, $x_n - \chi^J_-(t_n) \downarrow 0$, $z(x_n-, t_n) \to z^{IJ}_+$, $w(x_n-, t_n) \to w^{IJ}_+$. Notice that the graphs of $p^{IJ}_\pm$ are confined between the graph of p^I_- and the graph of p^I_+; see Fig. 12.5.1. In turn, the graphs of $p^J_\pm$, as well as the graphs of $p^K_\pm$, for any $K > J$, are confined between the graph of p^{IJ}_- and the graph of p^{IJ}_+. Furthermore,

$$(12.5.14)_1 \qquad \lim_{t \uparrow \bar{t}} z(p^{IJ}_-(t), t) = z^{IJ}_- , \quad \lim_{t \uparrow \bar{t}} z(p^{IJ}_+(t), t) = z^{IJ}_+ .$$

Indeed, the first of the above two equations has already been established in the context of the proof of Proposition 12.4.7 (under different notation; see (12.4.62)); while the second may be verified by a similar argument.

We now set

$$(12.5.15)_1 \qquad \mathcal{R}_1(\mathcal{M}) = - \sum_{I \in \mathcal{M}} \left[\Delta w^I_* - \sum_{J \in \mathcal{C}_I} \Delta w^{IJ} \right] ,$$

$$(12.5.16)_1 \qquad \mathcal{Q}_1(\mathcal{M}) = \sum_{I \in \mathcal{M}} \sum_{J \in \mathcal{C}_I} |\Delta w^{IJ} - \Delta w^J_*| .$$

By virtue of $(12.3.7)_1$,

$$(12.5.17)_1 \qquad \sum_{J \in \mathcal{C}_I} \Delta w^{IJ} \geq w^I_E - w^I_S \geq \Delta w^I_* ,$$

so that both $\mathcal{R}_1$ and $\mathcal{Q}_1$ are nonnegative.

With subsets $\mathcal{F}$ of the upper half-plane, we associate functionals

$$(12.5.18)_1 \qquad \mathcal{R}_1(\mathcal{F}) = \sup_{\mathcal{F}} \sum_{\mathcal{M} \in \mathcal{F}} \mathcal{R}_1(\mathcal{M}) ,$$

$$(12.5.19)_1 \qquad \mathcal{Q}_1(\mathcal{F}) = \sup_{\mathcal{F}} \sum_{\mathcal{M} \in \mathcal{F}} \mathcal{Q}_1(\mathcal{M}) ,$$

where $\mathscr{I}$ denotes any (finite) collection of 1-characteristic trees $\mathscr{M}$ contained in $\mathscr{F}$, which are disjoint, in the sense that the roots of any pair of them are non-comparable. One may view $\mathscr{R}_1(\mathscr{F})$ as a measure of the amount of 1-wave interactions inside $\mathscr{F}$, and $\mathscr{Q}_1(\mathscr{F})$ as a measure of strengthening of 1-shocks induced by interaction with 2-waves.

We introduce corresponding notions for the 2-characteristic family: $I < J$ whenever J is confined between the graphs of the maximal 2-separatrices q_-^I and q_+^I emanating from I. In that case, I lies on the graph of a forward 2-characteristic χ_+^J issuing from J, namely either ψ_+^J or q_+^J. One may then construct 2-*characteristic* trees $\mathscr{N}$, with *nodes*, *root*, *links* and *chains* defined as above. In the place of $(12.5.11)_1$, $(12.5.12)_1$ and $(12.5.13)_1$, we now have

$(12.5.11)_2$
$$z_\pm^{IJ} = \lim_{t\uparrow\bar{t}} z(\chi_+^J(t)\pm, t) \ , \quad w_\pm^{IJ} = \lim_{t\uparrow\bar{t}} w(\chi_+^J(t)\pm, t) \ ,$$

$(12.5.12)_2$
$$\Delta z^{IJ} = z_+^{IJ} - z_-^{IJ} \ , \quad \Delta w^{IJ} = w_+^{IJ} - w_-^{IJ} \ ,$$

$(12.5.13)_2$
$$\Delta z^{IJ} = g(\Delta w^{IJ}; z_+^{IJ}, w_+^{IJ}) \ .$$

With links (I, J) we associate maximal 2-separatrices $q_\pm^{IJ}$, emanating from I, in analogy to $p_\pm^{IJ}$. The graphs of $q_\pm^{IJ}$ are confined between the graphs of q_-^I and q_+^I. On the other hand, the graphs of $q_\pm^J$ are confined between the graphs of q_-^{IJ} and q_+^{IJ}. In the place of $(12.5.14)_1$,

$(12.5.14)_2$
$$\lim_{t\uparrow\bar{t}} w(q_-^{IJ}(t), t) = w_-^{IJ} \ , \quad \lim_{t\uparrow\bar{t}} w(q_+^{IJ}(t), t) = w_+^{IJ} \ .$$

Analogs of $(12.5.15)_1$ and $(12.5.16)_1$ are also defined:

$(12.5.15)_2$
$$\mathscr{R}_2(\mathscr{N}) = \sum_{I\in\mathscr{N}}\left[\Delta z_*^I - \sum_{J\in\mathscr{C}_I}\Delta z^{IJ}\right] \ ,$$

$(12.5.16)_2$
$$\mathscr{Q}_2(\mathscr{N}) = \sum_{I\in\mathscr{N}}\sum_{J\in\mathscr{C}_I}|\Delta z^{IJ} - \Delta z_*^J| \ ,$$

which are nonnegative since

$(12.5.17)_2$
$$\sum_{J\in\mathscr{C}_I}\Delta z^{IJ} \le z_S^I - z_W^I \le \Delta z_*^I \ .$$

This induces functionals analogous to $\mathscr{R}_1$ and $\mathscr{Q}_1$:

$(12.5.18)_2$
$$\mathscr{R}_2(\mathscr{F}) = \sup_{\mathscr{I}}\sum_{\mathscr{N}\in\mathscr{I}}\mathscr{R}_2(\mathscr{N}) \ ,$$

$(12.5.19)_2$
$$\mathscr{Q}_2(\mathscr{F}) = \sup_{\mathscr{I}}\sum_{\mathscr{N}\in\mathscr{I}}\mathscr{Q}_2(\mathscr{N}) \ .$$

Proposition 12.5.1 *Let $\mathscr{F}_1, \cdots, \mathscr{F}_m$ be a collection of subsets of a set $\mathscr{F}$ contained in the upper half-plane. Suppose that for any $I \in \mathscr{F}_i$ and $J \in \mathscr{F}_j$ which are* *comparable, say* $I < J$, *the arc of the characteristic* χ^J_- *(or* χ^J_+*) which connects* J *to* I *is contained in* $\mathscr{F}$. *Then*

$$(12.5.20)_1 \qquad \sum_{i=1}^{m}\{\mathscr{R}_1(\mathscr{F}_i) + \mathscr{Q}_1(\mathscr{F}_i)\} \le k\{\mathscr{R}_1(\mathscr{F}) + \mathscr{Q}_1(\mathscr{F})\} \,,$$

or

$$(12.5.20)_2 \qquad \sum_{i=1}^{m}\{\mathscr{R}_2(\mathscr{F}_i) + \mathscr{Q}_2(\mathscr{F}_i)\} \le k\{\mathscr{R}_2(\mathscr{F}) + \mathscr{Q}_2(\mathscr{F})\} \,,$$

where k *is the smallest positive integer with the property that any* $k + 1$ *of* $\mathscr{F}_1, \cdots, \mathscr{F}_m$ *have empty intersection.*

Proof. It will suffice to verify $(12.5.20)_1$. With each $i = 1, \cdots, m$, we associate a family $\mathscr{J}_i$ of disjoint 1-characteristic trees $\mathscr{M}$ contained in $\mathscr{F}_i$. Clearly, by adjoining if necessary additional nodes contained in $\mathscr{F}$, one may extend the collection of the $\mathscr{J}_i$ into a single family $\mathscr{J}$ of disjoint trees contained in $\mathscr{F}$. The contribution of the additional nodes may only increase the value of $\mathscr{R}_1$ and $\mathscr{Q}_1$. Therefore,

$$(12.5.21) \qquad \sum_{i=1}^{m}\sum_{\mathscr{M}\in\mathscr{J}_i}\{\mathscr{R}_1(\mathscr{M}) + \mathscr{Q}_1(\mathscr{M})\} \le k \sum_{\mathscr{M}\in\mathscr{J}}\{\mathscr{R}_1(\mathscr{M}) + \mathscr{Q}_1(\mathscr{M})\} \,,$$

where the factor k appears on the right-hand side because the same node or link may be counted up to k times on the left-hand side. Recalling $(12.5.18)_1$ and $(12.5.19)_1$, we arrive at $(12.5.20)_1$. The proof is complete.

Proposition 12.5.2 *Consider a space-like curve* $t = \bar{t}(x), \hat{x} \le x \le \tilde{x}$, *in the upper half-plane. The trace of* (z, w) *along* $\bar{t}$ *is denoted by* $(\bar{z}, \bar{w})$. *Let* $\hat{p}(\cdot)$ *and* $\tilde{p}(\cdot)$ *(or* $\hat{q}(\cdot)$ *and* $\tilde{q}(\cdot)$*) be minimal (or maximal) 1-separatrices (or 2-separatrices) emanating from the left endpoint* $(\hat{x}, \hat{t})$ *and the right endpoint* $(\tilde{x}, \tilde{t})$ *of the graph of* $\bar{t}$. *The trace of* z *(or* w*) along* $\hat{p}$ *and* $\tilde{p}$ *(or* $\hat{q}$ *and* $\tilde{q}$*) is denoted by* $\hat{z}$ *and* $\tilde{z}$ *(or* $\hat{w}$ *and* $\tilde{w}$*). Let* $\mathscr{F}$ *(or* $\mathscr{G}$*) stand for the region bordered by the graphs of* $\hat{p}, \tilde{p}$ *(or* $\hat{q}, \tilde{q}$*),* $\bar{t}$ *and the x-axis. Then*

$$(12.5.22)_1 \qquad \begin{aligned} |\tilde{z}(\tilde{t}-) - \hat{z}(\hat{t}-)| &\le |\tilde{z}(0+) - \hat{z}(0+)| + c\delta^2 T V_{[\hat{x},\tilde{x}]}\overline{w}(\cdot) \\ &\quad + \mathscr{R}_2(\mathscr{F}) + \mathscr{Q}_2(\mathscr{F}) \,, \end{aligned}$$

or

$$(12.5.22)_2 \qquad \begin{aligned} |\tilde{w}(\tilde{t}-) - \hat{w}(\hat{t}-)| &\le |\tilde{w}(0+) - \hat{w}(0+)| + c\delta^2 T V_{[\hat{x},\tilde{x}]}\overline{z}(\cdot) \\ &\quad + \mathscr{R}_1(\mathscr{G}) + \mathscr{Q}_1(\mathscr{G}) \,. \end{aligned}$$

Proof. It will suffice to verify $(12.5.22)_1$. We write

$$(12.5.23) \quad \tilde{z}(\tilde{t}-) - \hat{z}(\hat{t}-) = [\tilde{z}(0+) - \hat{z}(0+)] + [\tilde{z}(\tilde{t}-) - \tilde{z}(0+)] - [\hat{z}(\hat{t}-) - \hat{z}(0+)] \,.$$

By virtue of Theorem 12.4.1,

$$(12.5.24) \quad \begin{cases} \hat{z}(\hat{\imath}-) - \hat{z}(0+) = \sum[\hat{z}(\tau+) - \hat{z}(\tau-)] \,, \\ \tilde{z}(\tilde{\imath}-) - \tilde{z}(0+) = \sum[\tilde{z}(\tau+) - \tilde{z}(\tau-)] \,, \end{cases}$$

where the summations run over the countable set of jump discontinuities of $\hat{z}(\cdot)$ and $\tilde{z}(\cdot)$.

By account of Theorem 12.3.1, if $\overline{z}(\cdot)$ is the trace of z along any minimal 1-separatrix which passes through some point $K = (x, \tau)$, then

$$(12.5.25) \quad z_S^K - z_W^K \leq \overline{z}(\tau-) - \overline{z}(\tau+) \leq \Delta z_*^K \,.$$

Starting out from points K of jump discontinuity of $\hat{z}(\cdot)$ on the graph of $\hat{p}$, we construct the characteristic ϕ_-^K and extend it until it intersects the graph of either $\tilde{p}$ or $\overline{\imath}$. This generates families of disjoint 2-characteristic trees $\mathcal{N}$, with maximal nodes $K_1 = (x_1, \tau_1), \cdots, K_m = (x_m, \tau_m)$ lying on the graph of $\hat{p}$ and root $K_0 = (x_0, \tau_0)$ lying on the graph of either $\tilde{p}$ or $\overline{\imath}$. In the former case, by account of (12.5.25), (12.5.15)$_2$ and (12.5.16)$_2$,

$$(12.5.26) \quad \left| \tilde{z}(\tau_0+) - \tilde{z}(\tau_0-) - \sum_{\ell=1}^{m}[\hat{z}(\tau_\ell+) - \hat{z}(\tau_\ell-)] \right| \leq \mathcal{B}(\mathcal{N}) + \mathcal{Q}(\mathcal{N}) \,.$$

On the other hand, if K_0 lies on the graph of $\overline{\imath}$,

$$(12.5.27) \quad \sum_{J \in \mathcal{C}_{K_0}} \Delta z^{K_0 J} \leq z_S^{K_0} - z_W^{K_0} \leq c\delta^2 |w_S^{K_0} - w_W^{K_0}| \,,$$

and so

$$(12.5.28) \quad \left| -\sum_{\ell=1}^{m}[\hat{z}(\tau_\ell+) - \hat{z}(\tau_\ell-)] \right| \leq c\delta^2 |w_S^{K_0} - w_W^{K_0}| + \mathcal{B}(\mathcal{N}) + \mathcal{Q}(\mathcal{N}) \,.$$

Suppose that on the graph of $\tilde{p}$ still remain points K_0 of jump discontinuity of $\tilde{z}(\cdot)$ which cannot be realized as roots of trees with maximal nodes on the graph of $\hat{p}$. We then adjoin (trivial) 2-characteristic trees $\mathcal{N}$ which contain a single node, namely such a $K_0 = (x_0, \tau_0)$, in which case

$$(12.5.29) \quad |\tilde{z}(\tau_0+) - \tilde{z}(\tau_0-)| \leq \mathcal{B}(\mathcal{N}) + \mathcal{Q}(\mathcal{N}) \,.$$

Recalling (12.5.23) and tallying the jump discontinuities of $\overline{z}_1(\cdot)$ and $\overline{z}_2(\cdot)$, as indicated in (12.5.24), according to (12.5.26), (12.5.28) or (12.5.29), we arrive at (12.5.22)$_1$. The proof is complete.

Proposition 12.5.3 *Under the assumptions of Theorem 12.5.1,*

$$(12.5.30)_1$$

$$TV_{[x_\ell, x_r]}z^*(\cdot) \leq TV_{[\xi_\ell(0), \xi_r(0)]}z(\cdot, 0) + c\delta^2 TV_{[x_\ell, x_r]}w^*(\cdot) + 2\{\mathcal{B}(\mathcal{F}) + \mathcal{Q}(\mathcal{F})\} \,,$$

$(12.5.30)_2$

$$TV_{[x_\ell,x_r]}w^*(\cdot) \le TV_{[\zeta_\ell(0),\zeta_r(0)]}w(\cdot,0) + c\delta^2 TV_{[x_\ell,x_r]}z^*(\cdot) + 2\{\mathscr{R}(\mathscr{G}) + \mathscr{Q}(\mathscr{G})\},$$

where $\mathscr{F}$ denotes the region bordered by the graphs of ξ_ℓ, ξ_r, t^*, and the x-axis while $\mathscr{G}$ stands for the region bordered by the graphs of ζ_ℓ, ζ_r, t^*, and the x-axis.

Proof. It will suffice to establish $(12.5.30)_1$. We have to estimate

$$(12.5.31) \qquad TV_{[x_\ell,x_r]}z^*(\cdot) = \sup \sum_{i=1}^{m} |z_S^{L_i} - z_S^{L_{i-1}}|,$$

where the supremum is taken over all finite sequences $\{L_0, \cdots, L_m\}$ of points along t^* (Fig. 12.5.2).

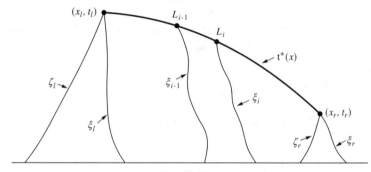

Fig. 12.5.2.

We construct the minimal backward 1-characteristics ξ_i emanating from $L_i = (x_i, t_i)$, $i = 0, \cdots, m$, and let $z_i(\cdot)$ denote the trace of z along $\xi_i(\cdot)$. We apply Proposition 12.5.2 with $\bar{t}$ the arc of t^* with endpoints L_{i-1} and L_i; $\hat{x} = x_{i-1}$; $\tilde{x} = x_i$; $\hat{p} = \xi_{i-1}$; $\tilde{p} = \xi_i$; and $\mathscr{F} = \mathscr{F}_i$, namely the region bordered by the graphs of ξ_{i-1}, ξ_i, t^*, and the x-axis. The estimate $(12.5.22)_1$ then yields

$(12.5.32)$

$$|z_S^{L_i} - z_S^{L_{i-1}}| \le |z_i(0+) - z_{i-1}(0+)| + c\delta^2 TV_{[x_{i-1},x_i]}w^*(\cdot) + \mathscr{R}_2(\mathscr{F}_i) + \mathscr{Q}_2(\mathscr{F}_i).$$

Combining (12.5.31), (12.5.32) and Proposition 12.5.1, we arrive at $(12.5.30)_1$. The proof is complete.

Proposition 12.5.4 *Let $\mathscr{M}$ (or $\mathscr{N}$) be a 1-characteristic (or 2-characteristic) tree rooted at I_0. Then*

$(12.5.33)_1$
$$\mathscr{R}(\mathscr{M}) + \mathscr{Q}(\mathscr{M}) \le c\delta^2(1 + V_{\mathscr{M}}) \{TV_{[p_-^{I_0}(0), p_+^{I_0}(0)]}z(\cdot, 0)$$
$$+ \mathscr{R}_2(\mathscr{F}_{I_0}) + \mathscr{Q}_2(\mathscr{F}_{I_0})\},$$

or

$$(12.5.33)_2 \quad \mathscr{R}_2(\mathscr{N}) + \mathscr{Q}_2(\mathscr{N}) \leq c\delta^2(1 + W_{\mathscr{N}})\{TV_{[q_-^{I_0}(0),q_+^{I_0}(0)]}w(\cdot,0) \\ + \mathscr{R}_1(\mathscr{G}_{I_0}) + \mathscr{Q}_1(\mathscr{G}_{I_0})\},$$

where $V_{\mathscr{M}}$ (or $W_{\mathscr{N}}$) denotes the maximum of

$$(12.5.34)_1 \quad \sum_{i=0}^{m-1}\{|z_-^{I_iI_{i+1}} - z_S^{I_{i+1}}| + |w_-^{I_iI_{i+1}} - w_S^{I_{i+1}}|\}$$

or

$$(12.5.34)_2 \quad \sum_{i=0}^{m-1}\{|z_+^{I_iI_{i+1}} - z_S^{I_{i+1}}| + |w_+^{I_iI_{i+1}} - w_S^{I_{i+1}}|\}$$

over all chains $\{I_0, \cdots, I_m\}$ of $\mathscr{M}$ (or $\mathscr{N}$).

Proof. It will suffice to validate $(12.5.33)_1$, the other case being completely analogous. By virtue of $(12.5.15)_1$ and $(12.5.16)_1$,

$$(12.5.35) \quad \begin{aligned} \mathscr{R}_1(\mathscr{M}) &\leq -\sum_{I\in\mathscr{M}}[\Delta w_*^I - \sum_{J\in\mathscr{C}_I}\Delta w_*^J] + \mathscr{Q}_1(\mathscr{M}) \\ &= \sum_{\substack{\text{maximal}\\\text{nodes}}}\Delta w_*^K - \Delta w_*^{I_0} + \mathscr{Q}_1(\mathscr{M}). \end{aligned}$$

Since $\Delta w_*^K \leq 0$, to establish $(12.5.33)_1$ it is sufficient to show

$$(12.5.36) \quad -\Delta w_*^{I_0} \leq c\delta^2\{TV_{[p_-^{I_0}(0),p_+^{I_0}(0)]}z(\cdot,0) + \mathscr{R}_2(\mathscr{F}_{I_0}) + \mathscr{Q}_2(\mathscr{F}_{I_0})\},$$

$$(12.5.37) \quad \mathscr{Q}_1(\mathscr{M}) \leq c\delta^2(1 + V_{\mathscr{M}})\{TV_{[p_-^{I_0}(0),p_+^{I_0}(0)]}z(\cdot,0) + \mathscr{R}_2(\mathscr{F}_{I_0}) + \mathscr{Q}_2(\mathscr{F}_{I_0})\}.$$

To demonstrate (12.5.36), we first employ $(12.5.10)_1$ to get

$$(12.5.38) \quad -\Delta w_*^{I_0} = -f(\Delta z^{I_0}; z_S^{I_0}, w_S^{I_0}) \leq c\delta^2\Delta z^{I_0},$$

and then, to estimate Δz^{I_0}, we apply Proposition 12.5.2, with $(\hat{x},\hat{t}) = (\tilde{x},\tilde{t}) = I_0$, $\hat{p} = p_-^{I_0}$ and $\tilde{p} = p_+^{I_0}$.

We now turn to the proof of (12.5.37), recalling the definition $(12.5.16)_1$ of $\mathscr{Q}_1(\mathscr{M})$. For any nodes $I \in \mathscr{M}$ and $J \in \mathscr{C}_I$, we use $(12.5.10)_1$ and $(12.5.13)_1$ to get

$$(12.5.39) \quad \begin{aligned} \Delta w^{IJ} - \Delta w_*^J &= f(\Delta z^{IJ}; z_-^{IJ}, w_-^{IJ}) - f(\Delta z^{IJ}; z_S^J, w_S^J) \\ &\quad + f(\Delta z^{IJ}; z_S^J, w_S^J) - f(\Delta z^J; z_S^J, w_S^J). \end{aligned}$$

By account of the properties of the function f,

$$(12.5.40) \quad \begin{aligned} |f(\Delta z^{IJ}; z_-^{IJ}, w_-^{IJ}) &- f(\Delta z^{IJ}; z_S^J, w_S^J)| \\ &\leq c\delta^2\Delta z^{IJ}\{|z_-^{IJ} - z_S^J| + |w_-^{IJ} - w_S^J|\}, \end{aligned}$$

(12.5.41) $|f(\Delta z^{IJ}; z_S^J, w_S^J) - f(\Delta z^J; z_S^J, w_S^J)| \le c\delta^2 |\Delta z^{IJ} - \Delta z^J|$.

Thus, to verify (12.5.37) we have to show

(12.5.42)
$$\sum_{I \in \mathcal{M}} \sum_{J \in \mathcal{C}_I} \Delta z^{IJ} \{|z_-^{IJ} - z_S^J| + |w_-^{IJ} - w_S^J|\}$$
$$\le V_{\mathcal{M}} \{\Delta z^{I_0} + \sum_{I \in \mathcal{M}} \sum_{J \in \mathcal{C}_I} |\Delta z^{IJ} - \Delta z^J|\} \,,$$

(12.5.43)
$$\sum_{I \in \mathcal{M}} \sum_{J \in \mathcal{C}_I} |\Delta z^{IJ} - \Delta z^J| \le c \{TV_{[p_-^{I_0}(0), p_+^{I_0}(0)]} z(\cdot, 0) + \mathcal{P}_2(\mathcal{F}_{I_0}) + \mathcal{Q}_2(\mathcal{F}_{I_0})\} \,.$$

We tackle (12.5.42) first. We perform the summation starting out from the maximal nodes and moving down towards the root of $\mathcal{M}$. For $L \in \mathcal{M}$, we let $\mathcal{M}_L$ denote the subtree of $\mathcal{M}$ which is rooted at L and contains all $I \in \mathcal{M}$ with $L < I$. For some $K \in \mathcal{M}$, assume

(12.5.44)
$$\sum_{I \in \mathcal{M}_L} \sum_{J \in \mathcal{C}_I} \Delta z^{IJ} \{|z_-^{IJ} - z_S^J| + |w_-^{IJ} - w_S^J|\}$$
$$\le V_{\mathcal{M}_L} \left\{ \Delta z^L + \sum_{I \in \mathcal{M}_L} \sum_{J \in \mathcal{C}_I} |\Delta z^{IJ} - \Delta z^J| \right\}$$

holds for every $L \in \mathcal{C}_K$. Since $\Delta z^L \le \Delta z^{KL} + |\Delta z^{KL} - \Delta z^L|$ and

(12.5.45)
$$\sum_{L \in \mathcal{C}_K} \Delta z^{KL} \le \Delta z^K \,,$$

(12.5.44) implies

(12.5.46)
$$\sum_{I \in \mathcal{M}_K} \sum_{J \in \mathcal{C}_I} \Delta z^{IJ} \{|z_-^{IJ} - z_S^J| + |w_-^{IJ} - w_S^J|\}$$
$$\le \sum_{L \in \mathcal{C}_K} \Delta z^{KL} \{|z_-^{KL} - z_S^L| + |w_-^{KL} - w_S^L| + V_{\mathcal{M}_L}\}$$
$$+ \sum_{L \in \mathcal{C}_K} V_{\mathcal{M}_L} \{|\Delta z^{KL} - \Delta z^L| + \sum_{I \in \mathcal{M}_L} \sum_{J \in \mathcal{C}_I} |\Delta z^{IJ} - \Delta z^J|\}$$
$$\le V_{\mathcal{M}_K} \{\Delta z^K + \sum_{I \in \mathcal{M}_K} \sum_{J \in \mathcal{C}_I} |\Delta z^{IJ} - \Delta z^J|\} \,.$$

Thus, proceeding step by step, we arrive at (12.5.42).

It remains to show (12.5.43). We note that

(12.5.47) $\Delta z^{IJ} - \Delta z^J = [z_+^{IJ} - z_E^J] + [z_S^J - z_-^{IJ}]$.

We estimate the right-hand side by applying Proposition 12.5.2 twice: First with $(\hat{x}, \hat{t}) = J$, $(\tilde{x}, \tilde{t}) = I$, $\hat{p} = p_+^J$, $\tilde{p} = p_+^{IJ}$, and then with $(\hat{x}, \hat{t}) = I$, $(\tilde{x}, \tilde{t}) = J$, $\hat{p} = p_-^{IJ}$, $\tilde{p} = p_-^J$. In either case, the arc of χ_J^- joining I to J serves as $\bar{t}$. We combine the derivation of $(12.5.22)_1$ for the two cases: The characteristic ϕ_-^K

issuing from any point K on the graph of p_+^J is always intercepted by the graph of p_+^{IJ}; never by the graph of χ_J^-. On the other hand, ϕ_-^K issuing from points K on the graph of p_-^{IJ} and crossing the graph of χ_J^-, may be prolonged until they intersect the graph of p_+^{IJ}. Consequently, the contribution of the common $\bar{t}$ cancels out and we are left with the estimate

$$(12.5.48) \qquad \begin{aligned} |\Delta z^{IJ} - \Delta z^J| &\le TV_{[p_-^{IJ}(0),\, p_-^J(0)]} z(\cdot, 0) + TV_{[p_+^J(0),\, p_+^{IJ}(0)]} z(\cdot, 0) \\ &\quad + \mathscr{B}_2(\mathscr{F}_{IJ}) + \mathscr{Q}_2(\mathscr{F}_{IJ}) \,, \end{aligned}$$

with $\mathscr{F}_{IJ}$ defined through

$$(12.5.49) \qquad \mathscr{F}_{IJ} = \{(x, t) : 0 \le t < t_I,\ p_-^{IJ}(t) \le x \le p_+^{IJ}\} \cap \overline{\mathscr{F}_J^C} \,.$$

When (I, J) and (K, L) are any two distinct links (possibly with $I = K$), then the intervals $(p_-^{IJ}(0), p_-^J(0))$, $(p_+^J(0), p_+^{IJ}(0))$, $(p_-^{KL}(0), p_-^L(0))$ and $(p_+^L(0), p_+^{KL}(0))$ are pairwise disjoint; likewise, the interiors of the sets $\mathscr{F}_{IJ}$ and $\mathscr{F}_{KL}$ are disjoint. Therefore, by virtue of Proposition 12.5.1, tallying (12.5.48) over $J \in \mathscr{C}_I$ and then over $I \in \mathscr{M}$ yields (12.5.43). The proof is complete.

Proposition 12.5.5 *Under the assumptions of Theorem 12.5.1, if $\mathscr{H}$ denotes the region bordered by the graphs of ζ_ℓ, ξ_r, t^*, and the x-axis, then*

$$(12.5.50) \qquad \begin{aligned} &\mathscr{R}(\mathscr{H}) + \mathscr{Q}_1(\mathscr{H}) + \mathscr{B}_2(\mathscr{H}) + \mathscr{Q}_2(\mathscr{H}) \\ &\le c\delta^2 \{ TV_{[\zeta_\ell(0), \xi_r(0)]} z(\cdot, 0) + TV_{[\zeta_\ell(0), \xi_r(0)]} w(\cdot, 0) \} \,. \end{aligned}$$

Proof. Consider any family $\mathscr{J}$ of disjoint 1-characteristic trees $\mathscr{M}$ contained in $\mathscr{H}$. If I and J are the roots of any two trees in $\mathscr{J}$, $(p_-^I(0), p_+^I(0))$ and $(p_-^J(0), p_+^J(0))$ are disjoint intervals contained in $(\zeta_\ell(0), \xi_\ell(0))$; also $\mathscr{F}_I$ and $\mathscr{F}_J$ are subsets of $\mathscr{H}$ with disjoint interiors. Consequently, by combining Propositions 12.5.1 and 12.5.4 we deduce

$$(12.5.51)_1 \qquad \begin{aligned} &\mathscr{R}(\mathscr{H}) + \mathscr{Q}_1(\mathscr{H}) \\ &\le c\delta^2(1 + V_\mathscr{H}) \{ TV_{[\zeta_\ell(0), \xi_r(0)]} z(\cdot, 0) + \mathscr{B}_2(\mathscr{H}) + \mathscr{Q}_2(\mathscr{H}) \} \,, \end{aligned}$$

where $V_\mathscr{H}$ denotes the supremum of the total variation of the trace of (z, w) over all 2-characteristics with graph contained in $\mathscr{H}$.

Similarly,

$$(12.5.51)_2 \qquad \begin{aligned} &\mathscr{B}_2(\mathscr{H}) + \mathscr{Q}_2(\mathscr{H}) \\ &\le c\delta^2(1 + W_\mathscr{H}) \{ TV_{[\zeta_\ell(0), \xi_r(0)]} w(\cdot, 0) + \mathscr{R}(\mathscr{H}) + \mathscr{Q}_1(\mathscr{H}) \} \,, \end{aligned}$$

where $W_\mathscr{H}$ stands for the supremum of the total variation of the trace of (z, w) over all 2-characteristics with graph contained in $\mathscr{H}$.

The constants in $(12.5.30)_1$ and $(12.5.30)_2$ do not depend on the particular t^*, so long as $\mathscr{H}$ remains fixed. In particular, we may apply these estimates for t^* any 1-characteristic or 2-characteristic, contained in $\mathscr{H}$. Therefore,

$(12.5.52)_1$
$$(1 - c\delta^2)V_{\mathscr{H}} \leq T V_{[\zeta_\ell(0),\xi_r(0)]}z(\cdot, 0) + T V_{[\zeta_\ell(0),\xi_r(0)]}w(\cdot, 0)$$
$$+ 2\{\mathscr{R}(\mathscr{H}) + \mathscr{Q}_1(\mathscr{H}) + \mathscr{R}_2(\mathscr{H}) + \mathscr{Q}_2(\mathscr{H})\} ,$$

$(12.5.52)_2$
$$(1 - c\delta^2)W_{\mathscr{H}} \leq T V_{[\zeta_\ell(0),\xi_r(0)]}z(\cdot, 0) + T V_{[\zeta_\ell(0),\xi_r(0)]}w(\cdot, 0)$$
$$+ 2\{\mathscr{R}(\mathscr{H}) + \mathscr{Q}_1(\mathscr{H}) + \mathscr{R}_2(\mathscr{H}) + \mathscr{Q}_2(\mathscr{H})\} .$$

Combining $(12.5.51)_1$, $(12.5.51)_2$, $(12.5.52)_1$, $(12.5.52)_2$ and recalling (12.5.3), we deduce (12.5.50), provided δ is sufficiently small. This completes the proof.

We now combine Propositions 12.5.3 and 12.5.5. Since $\mathscr{F}$ and $\mathscr{G}$ are subsets of $\mathscr{H}$, $(12.5.30)_1$, $(12.5.30)_2$ and (12.5.50) together imply $(12.5.4)_1$ and $(12.5.4)_2$. The assertion of Theorem 12.5.1 has thus been established.

In addition to serving as a stepping stone in the proof of Theorem 12.5.1, Proposition 12.5.5 reveals that the amount of self-interaction of waves of the first and second characteristic family, measured by $\mathscr{R}$ and $\mathscr{R}_2$, respectively, as well as the amount of mutual interaction of waves of opposite families, measured by $\mathscr{Q}_1$ and $\mathscr{Q}_2$, are bounded and controlled by the total variation of the initial data.

In our derivation of (12.5.4), the initial data were regarded as multi-valued and their total variation was evaluated for the "most unfavorable" selection of allowable values. According to this convention, the set of values of $z(x, 0)$ is either confined between $z(x-, 0)$ and $z(x+, 0)$ or else it lies within $c|w(x+, 0) - w(x-, 0)|^3$ distance from $z(x+, 0)$; and an analogous property holds for $w(x, 0)$. Consequently, (12.5.4) will still hold, with readjusted constant c, when $(z(\cdot, 0), w(\cdot, 0))$ are renormalized to be single-valued, for example continuous from the right at $\xi_r(0)$ and at $\zeta_r(0)$ and continuous from the left at any other point.

12.6 Spreading of Rarefaction Waves

In Section 11.2 we saw that the spreading of rarefaction waves induces one-sided Lipschitz conditions on solutions of genuinely nonlinear scalar conservation laws. Here we shall encounter a similar effect in the context of our system (12.1.1) of two conservation laws. We shall see that the spreading of 1- (or 2-) rarefaction waves acts to reduce the falling (or rising) slope of the corresponding Riemann invariant z (or w). Due to intervening wave interactions, this mechanism is no longer capable to sustain one-sided Lipschitz conditions, as in the scalar case; it still manages, however, to induce bounds on the total variation of solutions, independently of the initial data.

Let us consider again the solution (z, w) discussed in the previous section, with small oscillation (12.5.1). The principal result is

Theorem 12.6.1 *For any* $-\infty < x < y < \infty$ *and* $t > 0$,

$$(12.6.1) \qquad TV_{[x,y]}z(\cdot,t) + TV_{[x,y]}w(\cdot,t) \le b\frac{y-x}{t} + \beta\delta \,,$$

where b *and* β *are constants that may depend on* F *but are independent of the initial data.*

The proof of the above theorem will be partitioned into several steps. The notation introduced in Section 12.5 will be used here freely. In particular, as before, c will stand for a generic constant that may depend on F but is independent of δ.

Proposition 12.6.1 *Fix* $\bar{t} > 0$ *and consider any* $-\infty < x_\ell < x_r < \infty$, *with* $x_r - x_\ell$ *small compared to* $\bar{t}$. *Construct the minimal (or maximal) backward 1- (or 2-) characteristics* $\xi_\ell(\cdot), \xi_r(\cdot)$ *(or* $\zeta_\ell(\cdot), \zeta_r(\cdot)$*) emanating from* $(x_\ell, \bar{t})$, $(x_r, \bar{t})$, *and let* $\mathscr{F}$ *(or* $\mathscr{G}$*) denote the region bordered by the graphs of* ξ_ℓ, ξ_r *(or* ζ_ℓ, ζ_r*) and the time lines* $t = \bar{t}$ *and* $t = \bar{t}/2$. *Then*

$$(12.6.2)_1 \qquad z(x_\ell, \bar{t}) - z(x_r, \bar{t}) \le \hat{c}\exp(\overline{c}\delta\overline{V})\frac{x_r - x_\ell}{\bar{t}} + \mathscr{B}_2(\mathscr{F}) + \mathscr{C}_2(\mathscr{F}) \,,$$

or

$$(12.6.2)_2 \qquad w(x_r, \bar{t}) - w(x_\ell, \bar{t}) \le \hat{c}\exp(\overline{c}\delta\overline{V})\frac{x_r - x_\ell}{\bar{t}} + \mathscr{R}_1(\mathscr{G}) + \mathscr{C}_1(\mathscr{G}) \,,$$

where $\overline{V}$ *denotes the total variation of the trace of* w *(or* z*) along* $\xi_\ell(\cdot)$ *(or* $\zeta_r(\cdot)$*) over the interval* $[\frac{1}{2}\bar{t}, \bar{t}]$.

Proof. It will suffice to establish $(12.6.2)_1$. We let $(z_\ell(\cdot), w_\ell(\cdot))$ and $(z_r(\cdot), w_r(\cdot))$ denote the trace of (z, w) along $\xi_\ell(\cdot)$ and $\xi_r(\cdot)$, respectively.

We consider the infimum $\tilde{\mu}$ and the supremum $\overline{\mu}$ of the characteristic speed $\mu(z, w)$ over the range of the solution. The straight lines with slope $\tilde{\mu}$ and $\overline{\mu}$ emanating from the point $(\xi_r(t), t)$, $t \in [\frac{1}{2}\bar{t}, \bar{t}]$, are intercepted by $\xi_\ell(\cdot)$ at time $f(t)$ and $g(t)$, respectively. Both functions f and g are Lipschitz with slope $1 + O(\delta)$ and

$$(12.6.3) \qquad 0 \le g(t) - f(t) \le c_1\delta[\xi_r(t) - \xi_\ell(f(t))] \,.$$

The map that carries $(\xi_r(t), t)$ to $(\xi_\ell(f(t)), f(t))$ induces a pairing of points of the graphs of ξ_ℓ and ξ_r. From

$$(12.6.4) \qquad \xi_r(t) - \xi_\ell(f(t)) = \tilde{\mu}[t - f(t)] \,,$$

we obtain

$$(12.6.5) \qquad \dot{f}(t) = 1 - \frac{1}{\tilde{\mu} - \dot{\xi}_\ell(f(t))}[\dot{\xi}_r(t) - \dot{\xi}_\ell(f(t))] \,,$$

$$(12.6.6) \qquad \frac{d}{dt}[\xi_r(t) - \xi_\ell(f(t))] = \frac{\tilde{\mu}}{\tilde{\mu} - \dot{\xi}_\ell(f(t))}[\dot{\xi}_r(t) - \dot{\xi}_\ell(f(t))] \,,$$

almost everywhere on $[\frac{1}{2}\bar{t}, \bar{t}]$. In order to bound from below the right-hand side of (12.6.6), we begin with

$$(12.6.7) \quad \begin{aligned} \dot{\xi}_r(t) - \dot{\xi}_\ell(f(t)) &= \lambda(z_r(t), w_r(t)) - \lambda(z_\ell(f(t)), w_\ell(f(t))) \\ &= \bar{\lambda}_z[z_r(t) - z_\ell(f(t))] + \bar{\lambda}_w[w_r(t) - w_\ell(f(t))] . \end{aligned}$$

By virtue of Theorem 12.4.1,

$$(12.6.8) \quad z_r(t) - z_\ell(f(t)) \le z(x_r, \bar{t}) - z(x_\ell, \bar{t}) - \sum [z_r(\tau+) - z_r(\tau-)] ,$$

where the summation runs over the set of jump points of $z_r(\cdot)$ inside the interval $(t, \bar{t})$. As in the proof of Proposition 12.5.2, with each one of these jump points τ one may associate the trivial 2-characteristic tree $\mathcal{N}$ which consists of the single node $(\xi_r(\tau), \tau)$ so as to deduce

$$(12.6.9) \quad -\sum [z_r(\tau+) - z_r(\tau-)] \le \mathcal{B}_2(\mathcal{F}) + \mathcal{Q}_2(\mathcal{F}) .$$

For $t \in [\frac{1}{2}\bar{t}, \bar{t}]$, we construct the maximal backward 2-characteristic emanating from $(\xi_r(t), t)$, which is intercepted by $\xi_\ell(\cdot)$ at time $h(t)$, $f(t) \le h(t) \le g(t)$. By account of Theorem 12.4.1, $w_\ell(h(t)) \ge w_r(t)$ and so

$$(12.6.10) \quad w_r(t) - w_\ell(f(t)) \le w_\ell(h(t)) - w_\ell(f(t)) \le V(f(t)) - V(g(t)) ,$$

where $V(\tau)$ measures the total variation of $w_\ell(\cdot)$ over the interval $[\tau, \bar{t})$.

We now integrate (12.6.6) over the interval $(s, \bar{t})$. Recalling that $\bar{\lambda}_z < 0$, $\bar{\lambda}_w < 0$, upon combining (12.6.7), (12.6.8), (12.6.9) and (12.6.10), we deduce

(12.6.11)

$$\begin{aligned} \xi_r(s) - \xi_\ell(f(s)) &\le \xi_r(\bar{t}) - \xi_\ell(f(\bar{t})) \\ &\quad + c_2^{-1}(\bar{t} - s)[z(x_r, \bar{t}) - z(x_\ell, \bar{t}) + \mathcal{B}_2(\mathcal{F}) + \mathcal{Q}_2(\mathcal{F})] \\ &\quad + c_3 \int_s^{\bar{t}} [V(f(t)) - V(g(t))]dt . \end{aligned}$$

By interchanging the order of integration,

$$\begin{aligned} \int_s^{\bar{t}} [V(f(t)) - V(g(t))]dt &= -\int_s^{\bar{t}} \int_{f(t)}^{g(t)} dV(\tau)dt \\ &\le -\int_{f(s)}^{f(\bar{t})} [f^{-1}(\tau) - g^{-1}(\tau)]dV(\tau) - \int_{f(\bar{t})}^{g(\bar{t})} [\bar{t} - g^{-1}(\tau)]dV(\tau) \\ &= -\int_s^{\bar{t}} [t - g^{-1}(f(t))]dV(f(t)) - \int_{f(\bar{t})}^{g(\bar{t})} [\bar{t} - g^{-1}(\tau)]dV(\tau) . \end{aligned}$$

(12.6.12)

By account of (12.6.3),

$$(12.6.13) \quad t - g^{-1}(f(t)) \le c_4\delta[\xi_r(t) - \xi_\ell(f(t))] , \quad \tfrac{1}{2}\bar{t} \le t \le \bar{t} ,$$

(12.6.14) $\bar{t} - g^{-1}(\tau) \le c_4 \delta[\xi_r(\bar{t}) - \xi_\ell(f(\bar{t}))]$, $f(\bar{t}) \le \tau \le g(\bar{t})$,

and hence (12.6.11) yields

(12.6.15)
$$\xi_r(s) - \xi_\ell(f(s)) \le \exp(c_3 c_4 \delta \overline{V})[\xi_r(\bar{t}) - \xi_\ell(f(\bar{t}))]$$
$$+ c_2^{-1}(\bar{t} - s)[z(x_r, \bar{t}) - z(x_\ell, \bar{t}) + \mathscr{P}_2(\mathscr{F}) + \mathscr{Q}_2(\mathscr{F})]$$
$$- c_3 c_4 \delta \int_s^{\bar{t}} [\xi_r(t) - \xi_\ell(f(t))]dV(f(t)) ,$$

for any $s \in [\frac{3}{4}\bar{t}, \bar{t}]$. Integrating the above, Gronwall-type, inequality, we obtain

$$\xi_r(s) - \xi_\ell(f(s)) \le \exp(2c_3 c_4 \delta \overline{V})[\xi_r(\bar{t}) - \xi_\ell(f(\bar{t}))]$$
(12.6.16)
$$+ c_2^{-1}\left[\int_s^{\bar{t}} \exp\{c_3 c_4 \delta[V(f(s)) - V(f(\tau))]\}d\tau\right]$$
$$\times [z(x_r, \bar{t}) - z(x_\ell, \bar{t}) + \mathscr{P}_2(\mathscr{F}) + \mathscr{Q}_2(\mathscr{F})] .$$

We apply (12.6.16) for $s = \frac{3}{4}\bar{t}$. The left-hand side of (12.6.16) is nonnegative. Also, $\xi_r(\bar{t}) - \xi_\ell(f(\bar{t})) \le c_5(x_r - x_\ell)$. Therefore, (12.6.16) implies (12.6.2)$_1$ with $\bar{c} = 2c_3 c_4$, $\hat{c} = 4c_2 c_5$. The proof is complete.

In what follows, we shall be operating under the assumption that the constants $\overline{V}$ appearing in (12.6.2)$_1$ and (12.6.2)$_2$ satisfy

(12.6.17) $\bar{c}\delta\overline{V} \le \ell n 2$.

This will certainly be the case, by virtue of Theorem 12.5.1, when the initial data satisfy (12.5.3) with a sufficiently small. Furthermore, because of the finite domain of dependence property, (12.6.17) shall hold for $\bar{t}$ sufficiently small, even when the initial data have only locally bounded variation and satisfy (12.5.2) with δ sufficiently small. It will be shown below that (12.6.17) actually holds for any $\bar{t} > 0$, provided only the initial data have sufficiently small oscillation, i.e., δ is small.

Proposition 12.6.2 *For any* $-\infty < \bar{x} < \bar{y} < \infty$ *and* $\bar{t} > 0$,

$$N V_{[\bar{x},\bar{y}]} z(\cdot, \bar{t}) + P V_{[\bar{x},\bar{y}]} w(\cdot, \bar{t}) \le 4\hat{c}\frac{\bar{y} - \bar{x}}{\bar{t}}$$
(12.6.18)
$$+ c\delta^2 \left\{ T V_{[\bar{x} - \frac{1}{2}\bar{\mu}\bar{t}, \bar{y} - \frac{1}{2}\bar{\lambda}\bar{t}]} z(\cdot, \frac{1}{2}\bar{t}) + T V_{[\bar{x} - \frac{1}{2}\bar{\mu}\bar{t}, \bar{y} - \frac{1}{2}\bar{\lambda}\bar{t}]} w(\cdot, \frac{1}{2}\bar{t}) \right\} ,$$

$$T V_{[\bar{x},\bar{y}]} z(\cdot, \bar{t}) + T V_{[\bar{x},\bar{y}]} w(\cdot, \bar{t}) \le 8\hat{c}\frac{\bar{y} - \bar{x}}{\bar{t}} + 8\delta$$
(12.6.19)
$$+ c\delta^2 \left\{ T V_{[\bar{x} - \frac{1}{2}\bar{\mu}\bar{t}, \bar{y} - \frac{1}{2}\bar{\lambda}\bar{t}]} z(\cdot, \frac{1}{2}\bar{t}) + T V_{[\bar{x} - \frac{1}{2}\bar{\mu}\bar{t}, \bar{y} - \frac{1}{2}\bar{\lambda}\bar{t}]} w(\cdot, \frac{1}{2}\bar{t}) \right\} ,$$

where $\bar{\lambda}$ is the infimum of $\lambda(z, w)$ and $\bar{\mu}$ is the supremum of $\mu(z, w)$ over the range of the solution.

Proof. By combining $(12.6.2)_1$, $(12.6.2)_2$, $(12.6.17)$ and Proposition 12.5.1, we immediately infer

(12.6.20)
$$NV_{[\bar{x},\bar{y}]}z(\cdot, \bar{t}) + PV_{[\bar{x},\bar{y}]}w(\cdot, \bar{t})$$
$$\leq 4\hat{c}\frac{\bar{y} - \bar{x}}{\bar{t}} + 2[\mathscr{R}_1(\mathscr{H}) + \mathscr{Q}_1(\mathscr{H}) + \mathscr{R}_2(\mathscr{H}) + \mathscr{Q}_2(\mathscr{H})] ,$$

where $\mathscr{H}$ denotes the region bordered by the graph of the minimal backward 1-characteristic $\xi(\cdot)$ emanating from $(\bar{y}, \bar{t})$, the graph of the maximal backward 2-characteristic $\zeta(\cdot)$ emanating from $(\bar{x}, \bar{t})$, and the time lines $t = \bar{t}$ and $t = \bar{t}/2$.

We estimate $\mathscr{R}_1(\mathscr{H}) + \mathscr{Q}_1(\mathscr{H}) + \mathscr{R}_2(\mathscr{H}) + \mathscr{Q}_2(\mathscr{H})$ by applying Proposition 12.5.5, with the time origin shifted from $t = 0$ to $t = \bar{t}/2$. This yields (12.6.18).

Since total variation is the sum of negative variation and positive variation, while the difference of negative variation and positive variation is majorized by the oscillation, (12.6.18) together with (12.5.1) yield (12.6.19). The proof is complete.

Proof of Theorem 12.6.1 To verify (12.6.1), we first write (12.6.19) with $\bar{t} = t$, $\bar{x} = x$ and $\bar{y} = y$. To estimate the right-hand side of the resulting inequality, we reapply (12.6.19), for $\bar{t} = \frac{1}{2}t$, $\bar{x} = x - \frac{1}{2}\bar{\mu}t$ and $\bar{y} = x - \frac{1}{2}\bar{\lambda}t$. This yields

(12.6.21)
$$TV_{[x-\frac{1}{2}\bar{\mu}t, y-\frac{1}{2}\bar{\lambda}t]}z(\cdot, \tfrac{1}{2}t) + TV_{[x-\frac{1}{2}\bar{\mu}t, y-\frac{1}{2}\bar{\lambda}t]}w(\cdot, \tfrac{1}{2}t)$$
$$\leq 16\hat{c}\frac{y - x}{t} + 8\hat{c}(\bar{\mu} - \bar{\lambda}) + 8\delta$$
$$+ c\delta^2 \left\{ TV_{[x-\frac{3}{4}\bar{\mu}t, y-\frac{3}{4}\bar{\lambda}t]}z(\cdot, \tfrac{1}{4}t) + TV_{[x-\frac{3}{4}\bar{\mu}t, y-\frac{3}{4}\bar{\lambda}t]}w(\cdot, \tfrac{1}{4}t) \right\} .$$

Similarly, to estimate the right-hand side of (12.6.21), we apply (12.6.19) with $\bar{t} = \frac{1}{4}t$, $\bar{x} = x - \frac{3}{4}\bar{\mu}t$ and $\bar{y} = y - \frac{3}{4}\bar{\lambda}t$. We thus obtain

(12.6.22)
$$TV_{[x-\frac{3}{4}\bar{\mu}t, y-\frac{3}{4}\bar{\lambda}t]}z(\cdot, \tfrac{1}{4}t) + TV_{[x-\frac{3}{4}\bar{\mu}t, y-\frac{3}{4}\bar{\lambda}t]}w(\cdot, \tfrac{1}{4}t)$$
$$\leq 32\hat{c}\frac{y - x}{t} + 24\hat{c}(\bar{\mu} - \bar{\lambda}) + 8\delta$$
$$+ c\delta^2 \left\{ TV_{[x-\frac{7}{8}\bar{\mu}t, y-\frac{7}{8}\bar{\lambda}t]}z(\cdot, \tfrac{1}{8}t) + TV_{[x-\frac{7}{8}\bar{\mu}t, y-\frac{7}{8}\bar{\lambda}t]}w(\cdot, \tfrac{1}{8}t) \right\} .$$

Continuing on and passing to the limit, we arrive at (12.6.1) with

(12.6.23)
$$b = \frac{8\hat{c}}{1 - 2c\delta^2} , \quad \beta = \frac{8}{1 - c\delta^2} + \frac{8c\hat{c}\delta(\bar{\mu} - \bar{\lambda})}{(1 - c\delta^2)(1 - 2c\delta^2)} .$$

The above derivations hinge on the assumption that (12.6.17) holds; hence, in order to complete the proof, we now have to verify this condition. Recalling the definition of $\overline{V}$ in Proposition 12.6.1 and applying Theorem 12.5.1, with time origin shifted from 0 to $\frac{1}{2}t$, we deduce

$$(12.6.24) \qquad \overline{V} \leq c \sup_{\overline{x}} \left\{ TV_{[\overline{x}-\frac{1}{2}\overline{\mu}t,\overline{x}-\frac{1}{2}\overline{\lambda}t]} z(\cdot,\tfrac{1}{2}t) + TV_{[\overline{x}-\frac{1}{2}\overline{\mu}t,\overline{x}-\frac{1}{2}\overline{\lambda}t]} w(\cdot,\tfrac{1}{2}t) \right\} .$$

We estimate the right-hand side of (12.6.24) by means of (12.6.1), which yields

$$(12.6.25) \qquad \overline{V} \leq cb(\overline{\mu} - \overline{\lambda}) + c\beta\delta ,$$

so that (12.6.17) is indeed satisfied, provided δ is sufficiently small. The proof is complete.

We now show that the L^∞ bound (12.5.1), which has been assumed throughout this section, is induced by initial data of sufficiently small oscillation, regardless of the size of their total variation.

Theorem 12.6.2 *There is a positive constant γ, depending solely on F, such that solutions generated by initial data with small oscillation*

$$(12.6.26) \qquad |z(x,0)| + |w(x,0)| < \gamma\delta^2 , \qquad -\infty < x < \infty ,$$

but unrestricted total variation, satisfy (12.5.1).

Proof. Assuming (12.6.26) holds, with γ small, we shall demonstrate that $-\delta < z(x,t) < \delta$ and $-\delta < w(x,t) < \delta$ on the upper half-plane. Arguing by contradiction, suppose any one of the above four inequalities is violated at some point, say for example $z(\overline{x},\overline{t}) \geq \delta$.

We determine $\overline{y}$ through $8\hat{c}(\overline{y} - \overline{x}) = \delta\overline{t}$, where $\hat{c}$ is the constant appearing in (12.6.2)$_1$, and apply (12.6.18). The first term on the right-hand side of (12.6.18) is here bounded by $\frac{1}{4}\delta$; the second term is bounded by $\hat{c}\delta^2$, on account of (12.6.1). Consequently, for δ sufficiently small, the negative (decreasing) variation of $z(\cdot,\overline{t})$ over the interval $[\overline{x},\overline{y}]$ does not exceed $\frac{1}{2}\delta$. It follows that $z(x,\overline{t}) \geq \frac{1}{2}\delta$, for all $x \in [\overline{x},\overline{y}]$. In particular,

$$(12.6.27) \qquad \int_{\overline{x}}^{\overline{y}} [|z(x,\overline{t})| + |w(x,\overline{t})|]dx \geq (\overline{y} - \overline{x})\frac{\delta}{2} = \frac{1}{16\hat{c}}\delta^2\overline{t} .$$

We now appeal to the L^1 estimate (12.8.3), which will be established in Section 12.8, Proposition 12.8.1, and combine it with (12.6.26) to deduce

$$(12.6.28) \qquad \int_{\overline{x}}^{\overline{y}} [|z(x,\overline{t})| + |w(x,\overline{t})|]dx \leq 4[(\overline{y} - \overline{x}) + 2c\overline{t}]\gamma\delta^2 = \gamma\left[\frac{\delta}{2\hat{c}} + 8c\right]\delta^2\overline{t} .$$

It is clear that for γ sufficiently small (12.6.27) is inconsistent with (12.6.28), and this provides the desired contradiction. The proof is complete.

In conjunction with the compactness properties of BV functions, recounted in Section 1.7, the estimate (12.6.1) indicates that, starting out with solutions with initial data of locally bounded variation, one may construct, via completion, BV_{loc} solutions under initial data that are merely in L^∞, with sufficiently small oscillation. Thus, the solution operator of genuinely nonlinear systems of two consesrvation laws regularizes the initial data by the mechanism already encountered in the context of the genuinely nonlinear scalar conservation law (Theorem 11.2.2).

12.7 Regularity of Solutions

The information collected thus far paints the following picture for the regularity of solutions:

Theorem 12.7.1 *Let* $U(x,t)$ *be an admissible BV solution of the genuinely nonlinear system* (12.1.1) *of two conservation laws, with the properties recounted in the previous sections. Then*

(a) *Any point* $(\bar{x}, \bar{t})$ *of approximate continuity is a point of continuity of* U.

(b) *Any point* $(\bar{x}, \bar{t})$ *of approximate jump discontinuity is a point of (classical) jump discontinuity of* U.

(c) *Any irregular point* $(\bar{x}, \bar{t})$ *is the focus of a centered compression wave of either, or both, characteristic families, and/or a point of interaction of shocks of the same or opposite characteristic families.*

(d) *The set of irregular points is (at most) countable.*

Proof. Assertions (a), (b) and (c) are immediate corollaries of Theorem 12.3.1. In particular, $(\bar{x}, \bar{t})$ is a point of approximate continuity if and only if $(z_W, w_W) = (z_E, w_E)$, in which case all four limits (z_W, w_W), (z_E, w_E), (z_N, w_N) and (z_S, w_S) coincide. When $(z_W, w_W) \neq (z_E, w_E)$, then $(\bar{x}, \bar{t})$ is a point of approximate jump discontinuity in the 1-shock set if $(z_W, w_W) = (z_S, w_S)$, $(z_E, w_E) = (z_N, w_N)$; or a point of approximate jump discontinuity in the 2-shock set if $(z_W, w_W) = (z_N, w_N)$, $(z_E, w_E) = (z_S, w_S)$; and an irregular point in all other cases.

To verify asssertion (d), assume the irregular point $I = (\bar{x}, \bar{t})$ is a node of some 1-characteristic tree $\mathcal{M}$ or a 2-characteristic tree $\mathcal{N}$. If I is the focussing point of a centered 1-compression wave and/or point of interaction of 1-shocks, then, by virtue of (12.5.15)$_1$, I will register a positive contribution to $\mathscr{R}(\mathcal{M})$. Similarly, if I is the focussing point of a centered 2-compression wave and/or point of interaction of 2-shocks, then, by acocunt of (12.5.15)$_2$, I will register a positive contribution to $\mathscr{R}(\mathcal{N})$. Finally, suppose I is a point of interaction of a 1-shock with a 2-shock. We adjoin to $\mathcal{M}$ an additional node K lying on the graph of χ_-^I very close to I. Then $|\Delta w^{KI} - \Delta w_*^I| > 0$ and so, by (12.5.16)$_1$, we get a positive contribution to $\mathcal{Q}(\mathcal{M})$. Since the total amount of wave interaction is bounded, by virtue of Proposition 12.5.5, we conclude that the set of irregular points is necessarily (at most) countable. This completes the proof.

An analog of Theorem 11.3.5 is also in force here:

Theorem 12.7.2 *Assume the set $\mathscr{C}$ of points of continuity of the solution U has nonempty interior $\mathscr{C}^0$. Then U is locally Lipschitz continuous on $\mathscr{C}^0$.*

Proof. We verify that z is locally Lipschitz continuous on $\mathscr{C}^0$. Assume $(\bar{x}, \bar{t}) \in \mathscr{C}^0$ and $\mathscr{C}$ contains a rectangle $\{(x, t) : |x - \bar{x}| < kp, |t - \bar{t}| < p\}$, with $p > 0$ and k large compared to $|\lambda|$ and μ. By shifting the axes, we may assume, without loss of generality, that $\bar{t} = p$. We fix $\bar{y} > \bar{x}$, where $\bar{y} - \bar{x}$ is small compared to p, and apply $(12.6.2)_1$, with $x_\ell = \bar{x}$, $x_r = \bar{y}$. Since the solution is continuous in the rectangle, both $\mathscr{B}_2(\mathscr{F})$ and $\mathscr{Q}_2(\mathscr{F})$ vanish and so, recalling (12.6.17),

$$(12.7.1) \qquad z(\bar{x}, \bar{t}) - z(\bar{y}, \bar{t}) \leq \frac{2\hat{c}}{p}(\bar{y} - \bar{x}) .$$

The functions $(\hat{z}, \hat{w})(x, t) = (z, w)(\bar{x} + \bar{y} - x, 2p - t)$ are Riemann invariants of another solution $\hat{U}$ which is continuous, and thereby admissible, on the rectangle $\{(x, t) : |x - \bar{y}| < kp, |t - \bar{t}| < p\}$. Applying (12.7.1) to $\hat{z}$ yields

$$(12.7.2) \qquad z(\bar{y}, \bar{t}) - z(\bar{x}, \bar{t}) = \hat{z}(\bar{x}, \bar{t}) - \hat{z}(\bar{y}, \bar{t}) \leq \frac{2\hat{c}}{p}(\bar{y} - \bar{x}) .$$

We now fix $\bar{s} > \bar{t}$, with $\bar{s} - \bar{t}$ small compared to p. We construct the minimal backward 1-characteristic ξ emanating from $(\bar{x}, \bar{s})$, which is intercepted by the $\bar{t}$-time line at the point $\bar{y} = \xi(\bar{t})$, where $0 < \bar{y} - \bar{x} \leq -\bar{\lambda}(\bar{s} - \bar{t})$. By Theorem 12.4.1, $z(\bar{x}, s) = z(\bar{y}, \bar{t})$ and so, by virtue of (12.7.1) and (12.7.2),

$$(12.7.3) \qquad |z(\bar{x}, \bar{s}) - z(\bar{x}, \bar{t})| \leq \frac{2\hat{c}}{p}(\bar{y} - \bar{x}) \leq -\frac{2\bar{\lambda}\hat{c}}{p}(\bar{s} - \bar{t}) .$$

Thus z is Lipschitz.

A similar argument shows that w is also Lipschitz in $\mathscr{C}^0$. This completes the proof. $\qquad \square$

12.8 Initial Data in L^1

Recall that, by virtue of Theorem 11.5.2, initial data in L^1 induce decay of solutions of genuinely nonlinear scalar conservation laws, as $t \to \infty$, at the rate $O(t^{-\frac{1}{2}})$. The aim here is to establish an analogous result in the context of genuinely nonlinear systems of two conservation laws. Accordingly, we consider a solution $(z(x, t), w(x, t))$ of small oscillation (12.5.1), with initial data of unrestricted total variation, which lie in $L^1(-\infty, \infty)$:

$$(12.8.1) \qquad L = \int_{-\infty}^{\infty} [|z(x, 0)| + |w(x, 0)|] dx < \infty .$$

The principal result is

Theorem 12.8.1 *As* $t \to \infty$,

$$(12.8.2) \qquad\qquad (z(x,t), w(x,t)) = O(t^{-\frac{1}{2}}) ,$$

uniformly in x *on* $(-\infty, \infty)$.

The proof will be partitioned into several steps.

Proposition 12.8.1 *For any* $\bar{t} \in [0, \infty)$, *and* $-\infty < \bar{x} < \bar{y} < \infty$,

$$(12.8.3) \qquad \int_{\bar{x}}^{\bar{y}} [|z(x,\bar{t})| + |w(x,\bar{t})|]dx \le 4 \int_{\bar{x}-c\bar{t}}^{\bar{y}+c\bar{t}} [|z(x,0)| + |w(x,0)|]dx .$$

In particular, $(z(\cdot,\bar{t}), w(\cdot,\bar{t}))$ *are in* $L^1(-\infty, \infty)$.

Proof. We construct a Lipschitz continuous entropy η by solving the Goursat problem for (12.2.2) with prescribed data

$$(12.8.4) \qquad \begin{cases} \eta(z, 0) = |z| + \alpha z^2, & -\infty < z < \infty , \\ \eta(0, w) = |w| + \alpha z^2, & -\infty < w < \infty , \end{cases}$$

where α is a positive constant. From (12.2.3) it follows that, for α sufficiently large, η is a convex function of U on some neighborhood of the origin containing the range of the solution.

Combining (12.2.2) and (12.8.4), one easily deduces, for δ small,

$$(12.8.5)$$
$$\frac{1}{2}(|z| + |w|) \le \eta(z, w) \le 2(|z| + |w|) , \quad -2\delta < z < 2\delta , \quad -2\delta < w < 2\delta .$$

Furthermore, if q is the entropy flux associated with η, normalized by $q(0, 0) = 0$, (12.2.1) and (12.8.5) imply

$$(12.8.6) \qquad |q(z, w)| \le c\eta(z, w) , \quad -2\delta < z < 2\delta , \quad -2\delta < w < 2\delta .$$

We now fix $\bar{t} > 0$, $-\infty < \bar{x} < \bar{y} < \infty$ and integrate (12.3.1), for the entropy-entropy flux pair (η, q) constructed above, over the trapezoid $\{(x,t) : 0 < t < \bar{t}, \bar{x} - c(\bar{t} - t) < x < \bar{y} + c(\bar{t} - t)\}$. Upon using (12.8.6), this yields

$$(12.8.7) \qquad \int_{\bar{x}}^{\bar{y}} \eta(z(x,\bar{t}), w(x,\bar{t}))dx \le \int_{\bar{x}-c\bar{t}}^{\bar{y}+c\bar{t}} \eta(z(x,0), w(x,0))dx .$$

By virtue of (12.8.5), (12.8.7) implies (12.8.3). The proof is complete.

Proposition 12.8.2 *Let* $(\bar{z}(\cdot), \bar{w}(\cdot))$ *denote the trace of* (z, w) *along the minimal (or maximal) backward 1- (or 2-) characteristic* $\xi(\cdot)$ *(or* $\zeta(\cdot)$*) emanating from any point* $(\bar{y}, \bar{t})$ *of the upper half-plane. Then*

$(12.8.8)_1$
$$\int_0^{\bar{t}} [\bar{z}^2(t) + |\overline{w}(t)|]dt \leq \tilde{c}L ,$$

or

$(12.8.8)_2$
$$\int_0^{\bar{t}} [|\bar{z}(t)| + \overline{w}^2(t)]dt \leq \tilde{c}L .$$

Proof. It will suffice to verify $(12.8.8)_1$. Suppose η is any Lipschitz continuous convex entropy associated with entropy flux q and $\eta(0, 0) = 0, q(0, 0) = 0$. We fix $\bar{x} < \bar{y}$ and integrate (12.3.1) over $\{(x, t) : 0 < t < \bar{t}, \bar{x} < x < \xi(t)\}$ to get

(12.8.9)
$$\int_{\bar{x}}^{\bar{y}} \eta(z(x, \bar{t}), w(x, \bar{t}))dx - \int_{\bar{x}}^{\xi(0)} \eta(z(x, 0), w(x, 0))dx$$
$$+ \int_0^{\bar{t}} G(\bar{z}(t), \overline{w}(t))dt - \int_0^{\bar{t}} q(z(\bar{x}+, t), w(\bar{x}+, t))dt \leq 0 ,$$

where G is defined by

(12.8.10)
$$G(z, w) = q(z, w) - \lambda(z, w)\eta(z, w) .$$

We seek an entropy-entropy flux pair that renders $G(z, w)$ positive definite on $(-2\delta, 2\delta) \times (-2\delta, 2\delta)$. By account of (12.2.1),

(12.8.11)
$$G_z = -\lambda_z \eta ,$$

(12.8.12)
$$G_w = [(\mu - \lambda)\eta]_w - \mu_w \eta ,$$

which indicate that G decays fast, at least quadratically, as $z \to 0$, but it may decay more slowly, even linearly, as $w \to 0$.

We construct an entropy η by solving the Goursat problem for (12.2.2) with data

(12.8.13)
$$\begin{cases} \eta(z, 0) = 2z + \alpha z^2 , & -\infty < z < \infty , \\ \eta(0, w) = |w| + \alpha w^2 , & -\infty < w < \infty . \end{cases}$$

For α sufficiently large, it follows from (12.2.3) that η is a convex function of U on some neighborhood of the origin containing the range of the solution. From (12.8.12), (12.2.2) and (12.8.13) we deduce

(12.8.14)
$$G(0, w) = [\mu(0, 0) - \lambda(0, 0)]|w| + O(w^2) ,$$

(12.8.15)
$$\eta(z, w) = 2z + |w| + O(z^2 + w^2) ,$$

for (z, w) near the origin. Combining (12.8.14) with (12.8.11) and (12.8.15), we conclude

(12.8.16) $G(z, w) = [\mu(0, 0) - \lambda(0, 0)]|w| - \lambda_z(0, 0)z^2 + O(w^2 + |zw| + |z|^3) .$

We now return to (12.8.9). By account of Proposition 12.8.1, $(z(\cdot, t), w(\cdot, t))$ are in $L^1(-\infty, \infty)$, for all $t \in [0, \bar{t}]$, and hence

$$(12.8.17) \qquad \liminf_{\bar{x} \to -\infty} | \int_0^{\bar{t}} q(z(\bar{x}+, t), w(\bar{x}+, t))dt | = 0 .$$

Therefore, (12.8.9), (12.8.17), (12.8.15), (12.8.3) and (12.8.1) together imply

$$(12.8.18) \qquad \int_0^{\bar{t}} G(\bar{z}(t), \overline{w}(t))dt \leq 12L ,$$

provided (12.5.1) holds, with δ sufficiently small. The assertion $(12.8.8)_1$ now follows easily from (12.8.18), (12.8.16) and (12.1.3). This completes the proof.

Proposition 12.8.2 indicates that along minimal backward 1-characteristics z is $O(t^{-\frac{1}{2}})$ and w is $O(t^{-1})$, while along maximal backward 2-characteristics z is $O(t^{-1})$ and w is $O(t^{-\frac{1}{2}})$. In fact, recalling that $\bar{z}(\cdot)$ and $\overline{w}(\cdot)$ are nonincreasing along minimal and maximal backward 1- and 2-characteristics, respectively, we infer directly from $(12.8.8)_1$ and $(12.8.8)_2$ that the positive parts of $z(x, t)$ and $w(x, t)$ are $O(t^{-\frac{1}{2}})$, as $t \to \infty$. The proof of Theorem 12.8.1 will now be completed by establishing $O(t^{-\frac{1}{2}})$ decay on both sides:

Proposition 12.8.3 *For δ sufficiently small,*

$$(12.8.19)_1 \qquad z^2(x, t) \leq \frac{8\tilde{c}L}{t} ,$$

$$(12.8.19)_2 \qquad w^2(x, t) \leq \frac{8\tilde{c}L}{t} ,$$

hold, for all $-\infty < x < \infty, 0 < t < \infty$, where $\tilde{c}$ is the constant in $(12.8.8)_1$ and $(12.8.8)_2$.

Proof. Arguing by contradiction, suppose the assertion is false and let $\bar{t} > 0$ be the greatest lower bound of the set of points t on which $(12.8.19)_1$ and/or $(12.8.19)_2$ is violated for some x. According to Theorem 12.3.1, the continuation of the solution beyond $\bar{t}$ is initiated by resolution of Riemann problems along the $\bar{t}$-time line. Consequently, since $(12.8.19)_1$ and/or $(12.8.19)_2$ fail for $t > \bar{t}$, one can find $\bar{y} \in (-\infty, \infty)$ such that

$$(12.8.20)_1 \qquad z^2(\bar{y}, \bar{t}) > \frac{4\tilde{c}L}{\bar{t}} ,$$

and/or

$$(12.8.20)_2 \qquad w^2(\bar{y}, \bar{t}) > \frac{4\tilde{c}L}{\bar{t}} .$$

For definiteness, assume $(12.8.20)_1$ holds.

Let $(\bar{z}(\cdot), \bar{w}(\cdot))$ denote the trace of (z, w) along the minimal backward 1-characteristic $\xi(\cdot)$ emanating from $(\bar{y}, \bar{t})$. By applying Theorem 12.5.1, with the time origin shifted from $t = 0$ to $t = \bar{t}/2$, we deduce

(12.8.21)
$$TV_{[\frac{1}{2}\bar{t},\bar{t}]}\bar{w}(\cdot) \leq \hat{c}\left\{TV_{[\bar{y}-\frac{1}{2}\bar{\mu}\bar{t},\bar{y}-\frac{1}{2}\bar{\lambda}\bar{t}]}\, z(\cdot, \tfrac{1}{2}\bar{t}) + TV_{[\bar{y}-\frac{1}{2}\bar{\mu}\bar{t},\bar{y}-\frac{1}{2}\bar{\lambda}\bar{t}]}\, w(\cdot, \tfrac{1}{2}\bar{t})\right\} ,$$

where $\bar{\lambda}$ stands for the infimum of $\lambda(z, w)$ and $\bar{\mu}$ denotes the supremum of $\mu(z, w)$ over the range of the solution. We estimate the right-hand side of (12.8.21) with the help of Theorem 12.6.1 thus obtaining

(12.8.22)
$$TV_{[\frac{1}{2}\bar{t},\bar{t}]}\bar{w}(\cdot) \leq \hat{c}[b(\bar{\mu} - \bar{\lambda}) + \beta\delta].$$

By hypothesis,

(12.8.23)
$$\bar{w}^2(t) \leq \frac{16\tilde{c}L}{\bar{t}} , \qquad \frac{\bar{t}}{2} \leq t < \bar{t} .$$

We also have $|\bar{z}(t)| \leq 2\delta$. Therefore, by applying $(12.4.2)_1$ we deduce

(12.8.24)
$$\bar{z}^2(\bar{t}-) - \bar{z}^2(t) \leq \bar{c}\delta\frac{4\tilde{c}L}{\bar{t}} ,$$

with $\bar{c} = 64a\hat{c}[b(\bar{\mu} - \bar{\lambda}) + \beta\delta]$.

Since $\bar{z}(\bar{t}-) = z(\bar{y}, \bar{t})$, combining $(12.8.20)_1$ with (12.8.24) yields

(12.8.25)
$$\bar{z}^2(t) \geq \frac{4\tilde{c}L}{\bar{t}}(1 - \bar{c}\delta) , \qquad \frac{\bar{t}}{2} \leq t < \bar{t} .$$

From (12.8.25),

(12.8.26)
$$\int_{\frac{\bar{t}}{2}}^{\bar{t}} \bar{z}^2(t)dt \geq 2\tilde{c}L(1 - \bar{c}\delta) ,$$

which provides the desired contradiction to $(12.8.8)_1$, when δ is sufficiently small. The proof is complete.

12.9 Initial Data with Compact Support

Here we consider the large time behavior of solutions, with small oscillation (12.5.1), to our genuinely nonlinear system (12.1.1) of two conservation laws under initial data $(z(x, 0), w(x, 0))$ which vanish outside a bounded interval $[-\ell, \ell]$. We already know, from Section 12.8, that $(z(x, t), w(x, t)) = O(t^{-\frac{1}{2}})$. The aim is to examine the asymptotics in finer scale, establishing the analog of Theorem 11.6.1 on the genuinely nonlinear scalar conservation law.

Theorem 12.9.1 *Employing the notation introduced in Section* 12.3, *consider the special forward characteristics* $\phi_-(\cdot)$, $\psi_-(\cdot)$ *issuing from* $(-\ell, 0)$ *and* $\phi_+(\cdot)$, $\psi_+(\cdot)$ *issuing from* $(\ell, 0)$. *Then*

(a) *For t large,* ϕ_-, ψ_-, ϕ_+ *and* ψ_+ *propagate according to*

$$(12.9.1)_1 \qquad \phi_-(t) = \lambda(0,0)t - (p_-t)^{\frac{1}{2}} + O(1) \,,$$

$$(12.9.1)_2 \qquad \psi_+(t) = \mu(0,0)t + (q_+t)^{\frac{1}{2}} + O(1) \,,$$

$$(12.9.2)_1 \qquad \phi_+(t) = \lambda(0,0)t + (p_+t)^{\frac{1}{2}} + O(t^{\frac{1}{4}}) \,,$$

$$(12.9.2)_2 \qquad \psi_-(t) = \mu(0,0)t - (q_-t)^{\frac{1}{2}} + O(t^{\frac{1}{4}}) \,,$$

where p_-, p_+, q_- *and* q_+ *are nonnegative constants.*

(b) *For* $t > 0$ *and either* $x < \phi_-(t)$ *or* $x > \psi_+(t)$,

$$(12.9.3) \qquad z(x,t) = 0 \,, \quad w(x,t) = 0 \,.$$

(c) *For t large,*

$$(12.9.4) \qquad TV_{[\phi_-(t),\psi_+(t)]} \, z(\cdot,t) + TV_{[\phi_-(t),\psi_+(t)]} \, w(\cdot,t) = O(t^{-\frac{1}{2}}) \,.$$

(d) *For t large and* $\phi_-(t) < x < \phi_+(t)$,

$$(12.9.5)_1 \qquad \lambda(z(x,t),0) = \frac{x}{t} + O\left(\frac{1}{t}\right) \,,$$

while for $\psi_-(t) < x < \psi_+(t)$,

$$(12.9.5)_2 \qquad \mu(0, w(x,t)) = \frac{x}{t} + O\left(\frac{1}{t}\right) \,.$$

(e) *For t large and* $x > \phi_+(t)$, *if* $p_+ > 0$ *then*

$$(12.9.6)_1 \qquad 0 \le -z(x,t) \le c[x - \lambda(0,0)t]^{-\frac{3}{2}} \,,$$

while for $x < \psi_-(t)$, *if* $q_- > 0$ *then*

$$(12.9.6)_2 \qquad 0 \le -w(x,t) \le c[\mu(0,0)t - x]^{-\frac{3}{2}} \,.$$

According to the above proposition, as $t \to \infty$ the two characteristic families decouple and each one develops a N-wave profile, of width $O(t^{\frac{1}{2}})$ and strength $O(t^{-\frac{1}{2}})$, which propagates into the rest state at characteristic speed. When one of p_-, p_+ (or q_-, q_+) vanishes, the 1- (or 2-) N-wave is one-sided, of triangular profile. If both p_-, p_+ (or q_-, q_+) vanish, the 1- (or 2-) N-wave is absent altogether. In the wake of the N-waves, the solution decays at the rate $O(t^{-\frac{3}{4}})$, so long as $p_+ > 0$ and $q_- > 0$. In cones properly contained in the wake, the decay is even faster, $O(t^{-\frac{3}{2}})$.

Statement (b) of Theorem 12.9.1 is an immediate corollary of Theorem 12.5.1. The remaining assertions will be established in several steps.

Proposition 12.9.1 *As* $t \to \infty$, *the total variation decays according to* (12.9.4).

Proof. We fix t large and construct the maximal forward 1-characteristic $\chi_-(\cdot)$ issuing from $(\psi_+(t^{\frac{1}{2}}), t^{\frac{1}{2}})$ and the minimal forward 2-characteristic $\chi_+(\cdot)$ issuing from $(\phi_-(t^{\frac{1}{2}}), t^{\frac{1}{2}})$.

To estimate the total variation over the interval $(\chi_-(t), \chi_+(t))$, we apply Theorem 12.5.1, shifting the time origin from 0 to $t^{\frac{1}{2}}$. The minimal backward 1-characteristics as well as the maximal backward 2-characteristics emanating from points (x, t) with $\chi_-(t) < x < \chi_+(t)$ are intercepted by the $t^{\frac{1}{2}}$-time line outside the support of the solution. Furthermore, the oscillation of (z, w) along the $t^{\frac{1}{2}}$-time line is $O(t^{-\frac{1}{4}})$ so that in (12.5.4)$_1$ and (12.5.4)$_2$ one may take $\delta = O(t^{-\frac{1}{4}})$. Therefore,

$$(12.9.7) \qquad T V_{(\chi_-(t),\chi_+(t))} z(\cdot, t) + T V_{(\chi_-(t),\chi_+(t))} w(\cdot, t) = O(t^{-\frac{1}{2}}) .$$

To estimate the total variation over the intervals $[\phi_-(t), \chi_-(t)]$ and $[\chi_+(t), \psi_+(t)]$, we apply Theorem 12.6.1, shifting the time origin from 0 to $\frac{1}{2}t$. The oscillation of (z, w) along the $\frac{1}{2}t$-time line is $O(t^{-\frac{1}{2}})$ so that in (12.6.1) we may take $\delta = O(t^{-\frac{1}{2}})$. Since $\chi_-(t) - \phi_-(t)$ and $\psi_+(t) - \chi_+(t)$ are $O(t^{\frac{1}{2}})$, this yields

$$(12.9.8) \qquad \begin{cases} T V_{[\phi_-(t),\chi_-(t)]} z(\cdot, t) + T V_{[\phi_-(t),\chi_-(t)]} w(\cdot, t) = O(t^{-\frac{1}{2}}) , \\ T V_{[\chi_+(t),\psi_+(t)]} z(\cdot, t) + T V_{[\chi_+(t),\psi_+(t)]} w(\cdot, t) = O(t^{-\frac{1}{2}}) . \end{cases}$$

Combining (12.9.7) with (12.9.8), we arrive at (12.9.4). This completes the proof.

Proposition 12.9.2 *Let* $\bar{\lambda}$ *be any fixed strict upper bound of* $\lambda(z, w)$ *and* $\bar{\mu}$ *any fixed strict lower bound of* $\mu(z, w)$, *over the range of the solution. Then, for* t *large and* $x > \bar{\lambda}t$,

$$(12.9.9)_1 \qquad z(x, t) = O(t^{-\frac{3}{2}}) ,$$

while for $x < \bar{\mu}t$,

$$(12.9.9)_2 \qquad w(x, t) = O(t^{-\frac{3}{2}}) .$$

Proof. We fix t large and $x > \bar{\lambda}t$. Since $\bar{\lambda}$ is a strict upper bound of $\lambda(z, w)$, the minimal backward 1-characteristic $\xi(\cdot)$ emanating from (x, t) will be intercepted by the graph of ψ_+ at time $t_1 \geq \kappa t$, where κ is a positive constant depending solely on $\bar{\lambda}$. If $(\bar{z}(\cdot), \bar{w}(\cdot))$ denotes the trace of (z, w) along $\xi(\cdot)$, then the oscillation of $\bar{w}(\cdot)$ over $[t_1, t]$ is $O(t^{-\frac{1}{2}})$. Applying Theorem 12.5.1, with time origin shifted to t_1, and using Proposition 12.9.1, we deduce that the total variation of $\bar{w}(\cdot)$ over $[t_1, t]$

is likewise $O(t^{-\frac{1}{2}})$. It then follows from Theorem 12.4.1 that $\bar{z}(t-) = O(t^{-\frac{3}{2}})$. Since $z(x, t) = \bar{z}(t-)$, we arrive at $(12.9.9)_1$.

In a similar fashion, one establishes $(12.9.9)_2$, for $x < \bar{\mu}t$. The proof is complete.

Proposition 12.9.3 *Assertion* (d) *of Theorem* 12.9.1 *holds.*

Proof. By the construction of ϕ_- and ϕ_+, the minimal backward 1-characteristic $\xi(\cdot)$ emanating from any point (x, t) with $\phi_-(t) < x < \phi_+(t)$ will be intercepted by the x-axis on the interval $[-\ell, \ell]$. Therefore, if $(\bar{z}(\cdot), \bar{w}(\cdot))$ denotes the trace of (z, w) along $\xi(\cdot)$,

$$(12.9.10) \quad \begin{aligned} x &= \int_1^t \lambda(\bar{z}(\tau), \bar{w}(\tau))d\tau + \xi(1) \\ &= t\lambda(z(x, t), 0) + \int_1^t \{\bar{\lambda}_z[\bar{z}(\tau) - \bar{z}(t-)] + \bar{\lambda}_w \bar{w}(\tau)\}d\tau + O(1) \ . \end{aligned}$$

By account of Proposition 12.9.2, $\bar{w}(\tau) = O(\tau^{-\frac{3}{2}})$. Applying Theorem 12.5.1, with time origin shifted to τ, and using Proposition 12.9.1, we deduce that the total variation of $\bar{w}(\cdot)$ over $[\tau, t]$ is $O(\tau^{-\frac{1}{2}})$. It then follows from Theorem 12.4.1 that $\bar{z}(\tau) - \bar{z}(t-)$ is $O(\tau^{-\frac{7}{2}})$. In particular, the integral on the right-hand side of (12.9.10) is $O(1)$ and this establishes $(12.9.5)_1$.

A similar argument shows $(12.9.5)_2$. The proof is complete.

Proposition 12.9.4 *For t large, $\phi_-(t)$ and $\psi_+(t)$ satisfy $(12.9.1)_1$ and $(12.9.1)_2$.*

Proof. For t large, $\phi_-(t)$ joins the state $(z(\phi_-(t)-, t), w(\phi_-(t)-, t)) = (0, 0)$, on the left, with the state $(z(\phi_-(t)+, t), w(\phi_-(t)+, t))$, on the right, where $w(\phi_-(t)+, t) = O(t^{-\frac{3}{2}})$, while $z(\phi_-(t)+, t)$ satisfies $(12.9.5)_1$ for $x = \phi_-(t)$. The jump accross $\phi_-(t)$ is $O(t^{-\frac{1}{2}})$. Consequently, by use of (8.1.9) we infer

$$(12.9.11) \quad \dot{\phi}_-(t) = \frac{1}{2}\lambda(0, 0) + \frac{1}{2t}\phi_-(t) + O\left(\frac{1}{t}\right) ,$$

almost everywhere.

We set $\phi_-(t) = \lambda(0, 0)t - v(t)$. By the admissibility condition $\dot{\phi}_-(t) \leq \lambda(0, 0)$, we deduce that $\dot{v}(t) \geq 0$. Substituting into (12.9.11), yields

$$(12.9.12) \quad \dot{v}(t) = \frac{1}{2t}v(t) + O\left(\frac{1}{t}\right) .$$

If $v(t) = O(1)$, as $t \to \infty$, we obtain $(12.9.1)_1$ with $p_- = 0$. On the other hand, if $v(t) \uparrow \infty$, as $t \to \infty$, then (12.9.12) implies $v(t) = (p_- t)^{\frac{1}{2}} + O(1)$, which establishes $(12.9.1)_1$ with $p_- > 0$.

One validates $(12.9.1)_2$ by a similar argument. The proof is complete.

Proposition 12.9.5 *For t large, $\phi_+(t)$ and $\psi_-(t)$ satisfy (12.9.2)$_1$ and (12.9.2)$_2$. Furthermore, Assertion* (e) *of Theorem* 12.9.1 *holds.*

Proof. For t large, $\phi_+(t)$ joins the state $(z(\phi_+(t)-, t), w(\phi_+(t)-, t))$, on the left, with the state $(z(\phi_+(t)+, t), w(\phi_+(t)+, t))$, on the right, where both $w(\phi_+(t)\pm, t)$ are $O(t^{-\frac{3}{2}})$, while $z(\phi_+(t)-, t)$ satisfies (12.9.5)$_1$ for $x = \phi_+(t)$. The jump across $\phi_+(t)$ is $O(t^{-\frac{1}{2}})$. Hence, by use of (8.1.9) we obtain

$$(12.9.13) \qquad \dot{\phi}_+(t) = \frac{1}{2}\lambda(z(\phi_+(t)+, t), 0) + \frac{1}{2t}\phi_+(t) + O\left(\frac{1}{t}\right) .$$

Since $\phi_+(\cdot)$ is maximal, minimal backward 1-characteristics $\zeta(\cdot)$ emanating from points (x, t) with $x > \phi_+(t)$ stay strictly to the right of $\phi_+(\cdot)$ on $[0, t]$ and are thus intercepted by the x-axis at $\zeta(0) > \ell$. By virtue of Theorem 12.4.1, it follows that $z(\phi_+(t)+, t) \le 0$ and so $\lambda(z(\phi_+(t)+, t), 0) \ge \lambda(0, 0)$.

We now set $\phi_+(t) = \lambda(0, 0)t + v(t)$, $\lambda(z(\phi_+(t)+, t), 0) = \lambda(0, 0) + g(t)$. As shown above, $g(t) \ge 0$. Furthermore, by virtue of the admissibility condition $\dot{\phi}_+(t) \ge \lambda(z(\phi_+(t)+, t), w(\phi_+(t)+, t))$ we deduce $\dot{v}(t) \ge g(t) + O(t^{-\frac{3}{2}})$. When $v(t) = O(1)$, as $t \to \infty$, we obtain (12.9.2)$_1$, with $p_+ = 0$, corresponding to the case of one-sided N-wave. This case is delicate and shall not be discussed here, so let us assume $v(t) \to \infty$, as $t \to \infty$.

Substituting $\phi_+(t)$ into (12.9.13), we obtain

$$(12.9.14) \qquad \dot{v}(t) = \frac{1}{2t}v(t) + \frac{1}{2}g(t) + O\left(\frac{1}{t}\right) .$$

Since $g(t) \ge 0$, (12.9.14) yields $v(t) \ge \alpha t^{\frac{1}{2}}$, with $\alpha > 0$. On the other hand, we know that $v(t) = O(t^{\frac{1}{2}})$ and so (12.9.14) implies

$$(12.9.15) \qquad \frac{\dot{v}}{v} \ge \frac{1}{2t} + \beta g(t)t^{-\frac{1}{2}} + O(t^{-\frac{3}{2}}) .$$

It is clear that (12.9.15) induces a contradiction to $v(t) = O(t^{\frac{1}{2}})$ unless

$$(12.9.16) \qquad \int_1^\infty g(\tau)\tau^{-\frac{1}{2}}d\tau < \infty .$$

We now demonstrate that, in consequence of (12.9.16), there is $T > 0$ with the property that

$$(12.9.17) \qquad \inf\left\{\tau^{\frac{1}{2}}g(\tau) : \frac{t}{2} \le \tau \le t\right\} < \frac{1}{2}\alpha , \qquad \text{for all } t > T .$$

Indeed, if this assertion were false, we could find a sequence $\{t_m\}$, with $t_{m+1} \ge 2t_m$, $m = 1, 2, \cdots$, along which (12.9.17) is violated. But then

$$(12.9.18) \qquad \int_1^\infty g(\tau)\tau^{-\frac{1}{2}}d\tau \ge \frac{1}{2}\alpha \sum_m \int_{\frac{1}{2}t_m}^{t_m} \frac{dt}{t} = \infty ,$$

in contradiction to (12.9.16).

Let us fix (x, t), with $t > T$ and $x > \phi_+(t)$. The minimal backward 1-characteristic $\zeta(\cdot)$ emanating from (x, t) stays strictly to the right of $\phi_+(\cdot)$. We locate $\bar{t} \in [\frac{1}{2}t, t]$ such that

$$(12.9.19) \qquad \lambda(z(\phi_+(\bar{t})+, \bar{t}), 0) - \lambda(0, 0) = g(\bar{t}) < \frac{1}{4}\alpha\bar{t}^{-\frac{1}{2}}$$

and consider the minimal backward 1-characteristic $\xi(\cdot)$ emanating from a point $(\bar{x}, \bar{t})$, where $\bar{x}$ lies between $\phi_+(\bar{t})$ and $\zeta(\bar{t})$ and is so close to $\phi_+(\bar{t})$ that

$$(12.9.20) \qquad \lambda(z(\bar{x}, \bar{t}), 0) - \lambda(0, 0) < \frac{1}{4}\alpha\bar{t}^{-\frac{1}{2}} .$$

Let $(\bar{z}(\cdot), \bar{w}(\cdot))$ denote the trace of (z, w) along $\xi(\cdot)$. By virtue of Theorem 12.4.1, $\bar{z}(\cdot)$ is a nonincreasing function on $(0, \bar{t})$ so that $\bar{z}(\tau) \leq \bar{z}(\bar{t}-) = z(\bar{x}, \bar{t})$. Consequently, by account of (12.9.20),

$$(12.9.21) \qquad \begin{aligned} \dot{\xi}(\tau) &= \lambda(\bar{z}(\tau), \bar{w}(\tau)) \leq \lambda(z(\bar{x}, \bar{t}), \bar{w}(\tau)) \\ &\leq \lambda(0, 0) + \frac{1}{2}\alpha\bar{t}^{-\frac{1}{2}} + \bar{c}|\bar{w}(\tau)| . \end{aligned}$$

The integral of $|\bar{w}(\cdot)|$ over $[0, \bar{t}]$ is $O(1)$, by virtue of Proposition 12.8.2. Moreover,

$$(12.9.22) \qquad \xi(\bar{t}) = \bar{x} > \phi_+(\bar{t}) \geq \lambda(0, 0)\bar{t} + \alpha\bar{t}^{\frac{1}{2}} .$$

Therefore, integrating (12.9.21) over $[0, \bar{t}]$ yields

$$(12.9.23) \qquad \xi(0) \geq \frac{1}{2}\alpha\bar{t}^{\frac{1}{2}} + O(1) \geq \frac{\sqrt{2}}{4}\alpha t^{\frac{1}{2}} + O(1) .$$

Since $\zeta(\cdot)$ stays to the right of $\xi(\cdot)$, (12.9.23) implies, in particular, that the graph of $\zeta(\cdot)$ will intersect the graph of $\psi_+(\cdot)$ at time $\hat{t} = O(t^{\frac{1}{2}})$.

Let $(\hat{z}(\cdot), \hat{w}(\cdot))$ denote the trace of (z, w) along $\zeta(\cdot)$. The oscillation of $\hat{w}(\cdot)$ over $[\hat{t}, t)$ is $O(t^{-\frac{1}{4}})$. Furthermore, by account of Theorem 12.5.1, with time origin shifted to $\hat{t}$, and Proposition 12.9.1, we deduce that the total variation of $\hat{w}(\cdot)$ over $[\hat{t}, t)$ is also $O(t^{-\frac{1}{4}})$. It then follows from Theorem 12.4.1 that $\hat{z}(t-) = O(t^{-\frac{3}{4}})$.

By virtue of the above result, (12.9.21) now implies

$$(12.9.24) \qquad \dot{\xi}(\tau) \leq \lambda(0, 0) + O(t^{-\frac{3}{4}}) + \bar{c}|\bar{w}(\tau)| ,$$

which, upon integrating over $[0, t]$, yields

$$(12.9.25) \qquad \xi(0) \geq x - \lambda(0, 0)t + O(t^{\frac{1}{4}}) \geq \frac{1}{2}[x - \lambda(0, 0)t] .$$

Thus, $\hat{t} \geq c'[x - \lambda(0, 0)t]$. But then the oscillation and total variation of $\hat{w}(\cdot)$ over $[\hat{t}, t]$ is bounded by $\hat{c}[x - \lambda(0, 0)t]^{-\frac{1}{2}}$, in which case $(12.9.6)_1$ follows from Theorem 12.4.1.

Finally, we return to (12.9.14). Since $z(\phi_+(t)+,t)$ is $O(t^{-\frac{3}{4}})$, we get that $g(t) = O(t^{-\frac{3}{4}})$ and this in turn yields $v(t) = (p_+ t)^{\frac{1}{2}} + O(t^{\frac{1}{4}})$, with $p_+ > 0$. We have thus verified $(12.9.2)_1$.

A similar argument establishes $(12.9.6)_2$, for $x < \psi_-(t)$, and validates $(12.9.2)_2$. This completes the proof of Proposition 12.9.5 and thereby the proof of Theorem 12.9.1.

It is now easy to determine the large time asymptotics of the solution $U(x,t)$ in $L^1(-\infty, \infty)$. Starting out from the (finite) Taylor expansion

$$(12.9.26) \qquad U(z, w) = z R(0, 0) + w S(0, 0) + O(z^2 + w^2) ,$$

and using Theorem 12.9.1, we conclude

Theorem 12.9.2 *Assume $p_+ > 0$ and $q_- > 0$. Then, as $t \to \infty$,*

$$(12.9.27)$$
$$\|U(x, t) - M(x, t; p_-, p_+) R(0, 0) - N(x, t; q_-, q_+) S(0, 0)\|_{L^1(-\infty,\infty)} = O(t^{-\frac{1}{4}}) ,$$

where M and N denote the N-wave profiles:

$(12.9.28)_1$

$$M(x, t; p_-, p_+) = \begin{cases} \dfrac{x - \lambda(0, 0)t}{\lambda_z(0, 0)t} , & \text{for } -(p_- t)^{\frac{1}{2}} \le x - \lambda(0, 0)t \le (p_+ t)^{\frac{1}{2}} \\ 0 & \text{otherwise} , \end{cases}$$

$(12.9.28)_2$

$$N(x, t; q_-, q_+) = \begin{cases} \dfrac{x - \mu(0, 0)t}{\mu_w(0, 0)t} , & \text{for } -(q_- t)^{\frac{1}{2}} \le x - \mu(0, 0)t \le (q_+ t)^{\frac{1}{2}} \\ 0 & \text{otherwise} . \end{cases}$$

12.10 Periodic Solutions

The study of genuinely nonlinear hyperbolic systems (12.1.1) of two conservation laws will be completed with a discussion of the large time behavior of solutions with small oscillation, which are periodic,

$$(12.10.1) \qquad U(x + \ell, t) = U(x, t) , \qquad -\infty < x < \infty , \qquad t > 0 ,$$

and have zero mean[1]:

$$(12.10.2) \qquad \int_y^{y+\ell} U(x, t)dx = 0 , \qquad -\infty < y < \infty , \qquad t > 0 .$$

---

[1] If the initial value problem has unique solution, initial data that are periodic with zero mean necessarily generate solutions with the same property.

The confinement of waves resulting from periodicity induces active interactions and cancellation. As a result, the total variation per period decays at the rate $O(t^{-1})$:

Theorem 12.10.1 *For any* $x \in (\infty, \infty)$, *as* $t \to \infty$,

(12.10.3)
$$TV_{[x,x+\ell]} z(\cdot, t) + TV_{[x,x+\ell]} w(\cdot, t) \leq \frac{b\ell}{t} .$$

Proof. We apply (12.6.1) with $y = x + n\ell$; then we divide by n and we let $n \to \infty$. This completes the proof.

We now investigate the asymptotics at the scale $O(t^{-1})$. The mechanism encountered in Section 11.7, in the context of genuinely nonlinear scalar conservation laws, namely the confinement of the intercepts of extremal backward characteristics in intervals of the x-axis of period length, is here in force as well and generates similar, serrated asymptotic profiles. The nodes of the profiles are again tracked by divides, in the sense of Definition 10.3.1.

Theorem 12.10.2 *The upper half-plane is partitioned by minimal (or maximal) 1- (or 2-) divides along which* z *(or* w) *decays rapidly to zero,* $O(t^{-2})$, *as* $t \to \infty$. *Let* $\chi_-(\cdot)$ *and* $\chi_+(\cdot)$ *be any two adjacent 1- (or 2-) divides, with* $\chi_-(t) < \chi_+(t)$. *Then* $\chi_+(t) - \chi_-(t)$ *approaches a constant at the rate* $O(t^{-1})$, *as* $t \to \infty$. *Furthermore, between* χ_- *and* χ_+ *lies a 1- (or 2-) characteristic* ψ *such that, as* $t \to \infty$,

(12.10.4)
$$\psi(t) = \frac{1}{2}[\chi_-(t) + \chi_+(t)] + o(1) ,$$

(12.10.5)₁ $\lambda_z(0,0)z(x,t) =$
$$\begin{cases} \dfrac{x - \chi_-(t)}{t} + o\left(\dfrac{1}{t}\right) , & \chi_-(t) < x < \psi(t) , \\[2ex] \dfrac{x - \chi_+(t)}{t} + o\left(\dfrac{1}{t}\right) , & \psi(t) < x < \chi_+(t) , \end{cases}$$

or

(12.10.5)₂ $\mu_w(0,0)w(x,t) =$
$$\begin{cases} \dfrac{x - \chi_-(t)}{t} + o\left(\dfrac{1}{t}\right) , & \chi_-(t) < x < \psi(t) , \\[2ex] \dfrac{x - \chi_+(t)}{t} + o\left(\dfrac{1}{t}\right) , & \psi(t) < x < \chi_+(t) . \end{cases}$$

The first step towards proving the above proposition is to investigate the large time behavior of divides:

Proposition 12.10.1 *Along minimal (or maximal) 1- (or 2-) divides,* z *(or* w) *decays at the rate* $O(t^{-2})$, *as* $t \to \infty$. *Furthermore, if* $\chi_-(\cdot)$ *and* $\chi_+(\cdot)$ *are any two minimal (or maximal) 1- (or 2-) divides, then, as* $t \to \infty$,

(12.10.6)
$$\chi_+(t) - \chi_-(t) = h_\infty + O\left(\frac{1}{t}\right) ,$$

(12.10.7)$_1$
$$\int_{\chi_-(t)}^{\chi_+(t)} z(x,t)dx = O\left(\frac{1}{t^2}\right) ,$$

or

(12.10.7)$_2$
$$\int_{\chi_-(t)}^{\chi_+(t)} w(x,t)dx = O\left(\frac{1}{t^2}\right) .$$

Proof. Assume $\chi(\cdot)$ is a minimal 1-divide, say the limit of a sequence $\{\xi_n(\cdot)\}$ of minimal backward 1-characteristics emanating from points $\{(x_n, t_n)\}$, with $t_n \to \infty$, as $n \to \infty$. Let $(z_n(\cdot), w_n(\cdot))$ denote the trace of (z, w) along $\xi_n(\cdot)$. Applying Theorem 12.5.1, with time origin shifted to τ, and using Theorem 12.10.1, we deduce that the total variation of $w_n(\cdot)$ over any interval $[\tau, \tau + 1] \subset [0, t_n]$ is $O(\tau^{-1})$, uniformly in n. Therefore, by virtue of Theorem 12.4.1, $z_n(\cdot)$ is a nonincreasing function on $[0, t_n]$ whose oscillation over $[\tau, \tau + 1]$ is $O(\tau^{-3})$, uniformly in n. It follows that the trace $\bar{z}(\cdot)$ of z along $\chi(\cdot)$ is likewise a nonincreasing function with $O(\tau^{-3})$ oscillation over $[\tau, \tau + 1]$. By tallying the oscillation of $\bar{z}(\cdot)$ over intervals of unit length, from t to infinity, we verify the assertion $\bar{z}(t) = O(t^{-2})$.

A similar argument shows that the trace $\bar{w}(\cdot)$ of w along maximal 2-divides is likewise $O(t^{-2})$, as $t \to \infty$.

Assume now $\chi_-(\cdot)$ and $\chi_+(\cdot)$ are two minimal 1-divides with $\chi_+(t) - \chi_-(t) = h(t) \geq 0$, for $0 \leq t < \infty$. Note that, because of periodicity, $h(0) < k\ell$, for some integer k, implies $h(t) \leq k\ell, 0 \leq t < \infty$. Letting $(z_-(\cdot), w_-(\cdot))$ and $(z_+(\cdot), w_+(\cdot))$ denote the trace of (z, w) along $\chi_-(\cdot)$ and $\chi_+(\cdot)$, respectively, we have

(12.10.8)
$$\dot{h}(\tau) = \lambda(z_+(\tau), w_+(\tau)) - \lambda(z_-(\tau), w_-(\tau)) ,$$

for almost all τ in $[0, \infty)$.

The maximal backward 2-characteristic $\zeta_\tau(\cdot)$ emanating from the point $(\chi_+(\tau), \tau)$ is intercepted by the graph of $\chi_-(\cdot)$ at time $\tau - f(\tau)$. If $(\hat{z}(\cdot), \hat{w}(\cdot))$ denotes the trace of (z, w) along $\zeta_\tau(\cdot)$, Theorems 12.5.1 and 12.10.1 together imply that the total variation of $\hat{z}(\cdot)$ over the interval $[\tau - f(\tau), \tau]$ is $O(\tau^{-1})$, as $\tau \to \infty$. It then follows from Theorem 12.4.1 that the oscillation of $\hat{w}(\cdot)$ over $[\tau - f(\tau), \tau]$ is $O(\tau^{-3})$. Hence

(12.10.9)
$$w_+(\tau) = w_-(\tau - f(\tau)) + O(\tau^{-3}) .$$

Since $z_\pm(\tau) = O(\tau^{-2})$, (12.10.8) yields

(12.10.10)
$$\dot{h}(\tau) = \lambda(0, w_-(\tau - f(\tau)) - \lambda(0, w_-(\tau)) + O\left(\frac{1}{\tau^2}\right) .$$

From $\dot{h}(\tau) = O(\tau^{-1})$ and $\dot{\zeta}_\tau = \mu(0,0) + O(\tau^{-1})$, we infer that the oscillation of $f(\cdot)$ over the interval $[\tau, \tau + 1]$ is $O(\tau^{-1})$. The total variation of $w_-(\cdot)$ over $[\tau, \tau + 1]$ is likewise $O(\tau^{-1})$. Then, for any $t < t' < \infty$,

$$(12.10.11) \qquad \left| \int_t^{t'} \{\lambda(0, w_-(\tau - f(\tau))) - \lambda(0, w_-(\tau))\} d\tau \right| \le \frac{c}{t} \, .$$

Upon combining (12.10.10) with (12.10.11), one arrives at (12.10.6).

Let $U_-(\cdot)$ and $U_+(\cdot)$ denote the trace of U along $\chi_-(\cdot)$ and $\chi_+(\cdot)$, respectively. Integration of (12.1.1) over $\{(x, \tau) : t < \tau < \infty, \chi_-(\tau) < x < \chi_+(\tau)\}$ gives

$$(12.10.12) \qquad \int_{\chi_-(t)}^{\chi_+(t)} U(x, t) dx = \int_t^{\infty} \{F(U_+(\tau)) - \lambda(U_+(\tau))U_+(\tau)$$
$$- F(U_-(\tau)) + \lambda(U_-(\tau))U_-(\tau)\} d\tau \, .$$

We multiply (12.10.12), from the left, by the row vector $Dz(0)$. By account of (7.3.12), $U_z = R$ and $U_w = S$ so that, using (12.1.2), we deduce

$$(12.10.13) \qquad Dz(0)U = z + O(z^2 + w^2) \, ,$$

$$(12.10.14) \quad Dz(0)[F(U) - \lambda(U)U] = Dz(0)F(0) + aw^2 + O(z^2 + |zw| + |w|^3) \, ,$$

where the constant a is the value of $\frac{1}{2}(\lambda - \mu)S^T D^2 zS$ at $U = 0$. By virtue of $z_{\pm}(\tau) = O(\tau^{-2})$, $w_{\pm}(\tau) = O(\tau^{-1})$ and (12.10.9), we conclude

$$(12.10.15) \qquad \int_{\chi_-(t)}^{\chi_+(t)} z(x, t) dx = a \int_t^{\infty} [w_-^2(\tau - f(\tau)) - w_-^2(\tau)] d\tau + O\left(\frac{1}{t^2}\right) \, .$$

As explained above, over the interval $[\tau, \tau+1]$ the oscillation of $f(\cdot)$ is $O(\tau^{-1})$ and the total variation of $w_-^2(\cdot)$ is $O(\tau^{-2})$. Then, the integral on the right-hand side of (12.10.15) is $O(t^{-2})$, as $t \to \infty$, which establishes (12.10.7)$_1$.

When $\chi_-(\cdot)$ and $\chi_+(\cdot)$ are maximal 2-divides, a similar argument verifies (12.10.6) and (12.10.7)$_2$. The proof is complete.

The remaining assertions of Theorem 12.10.2 will be established through the following

Proposition 12.10.2 *Consider any two adjacent minimal (or maximal) 1- (or 2-) divides $\chi_-(\cdot)$, $\chi_+(\cdot)$, with $\chi_-(t) < \chi_+(t)$, $0 \le t < \infty$. The special forward 1- (or 2-) characteristic $\phi_-(\cdot)$ (or $\psi_+(\cdot)$), in the notation of Section 12.3, issuing from any fixed point $(\bar{x}, 0)$, where $\chi_-(0) < \bar{x} < \chi_+(0)$, is denoted by $\psi(\cdot)$. Then $\psi(\cdot)$ satisfies (12.10.4). Furthermore, (12.10.5)$_1$ (or (12.10.5)$_2$) holds.*

Proof. It will suffice to discuss the case χ_-, χ_+ are 1-divides. We consider minimal backward 1-characteristics $\xi(\cdot)$ emanating from points (x, t), with $t > 0$ and $\chi_-(t) < x < \chi_+(t)$. Their graphs are trapped between the graphs of χ_- and χ_+. The intercepts $\xi(0)$ of such ξ, by the x-axis, cannot accumulate to any $\hat{x}$ in the open interval $(\chi_-(0), \chi_+(0))$, because in that case a minimal 1-divide would issue from the point $(\hat{x}, 0)$, contrary to our assumption that χ_-, χ_+ are adjacent. Therefore, by the construction of $\psi(\cdot)$ we infer that, as $t \to \infty, \xi(\tau) \to \chi_-(\tau)$, when

$x \in (\chi_-(t), \psi(t)]$, or $\xi(\tau) \to \chi_+(\tau)$, when $x \in (\psi(t), \chi_+(t)]$, the convergence being uniform on compact subsets of $[0, \infty)$.

We now fix $\xi(\cdot)$ emanating from some point (x, t), with $\chi_-(t) < x \leq \psi(t)$, and set $h(\tau) = \xi(\tau) - \chi_-(\tau), 0 \leq \tau \leq t$. Then, for almost all $\tau \in [0, t]$ we have

$$(12.10.16) \qquad \dot{h}(\tau) = \lambda(\bar{z}(\tau), \overline{w}(\tau)) - \lambda(z_-(\tau), w_-(\tau)),$$

where $(\bar{z}(\cdot), \overline{w}(\cdot))$ denotes the trace of (z, w) along $\xi(\cdot)$, while $(z_-(\cdot), w_-(\cdot))$ stands for the trace of (z, w) along $\chi_-(\cdot)$.

By virtue of Theorems 12.5.1 and 12.10.1, the total variation of $\overline{w}(\cdot)$ on any interval $[s, s+1] \subset [0, t]$ is $O(s^{-1})$. It then follows from Theorem 12.4.1 that the oscillation of $\bar{z}(\cdot)$ over $[s, s+1]$ is $O(s^{-3})$ and hence

$$(12.10.17) \qquad \bar{z}(\tau) = z(x, t) + O\left(\frac{1}{\tau^2}\right).$$

Furthermore, by Proposition 12.10.1, $z_-(\tau) = O(\tau^{-2})$. Also, $z(x, t) = O(t^{-1})$ so, *a fortiori*, $z(x, t) = O(\tau^{-1})$. By account of these observations, (12.10.16) yields

$$(12.10.18) \quad \dot{h}(\tau) = \lambda_z(0, 0)z(x, t) + \lambda(0, \overline{w}(\tau)) - \lambda(0, w_-(\tau)) + O\left(\frac{1}{\tau^2}\right).$$

For any fixed $\tau \gg 0$, we consider the maximal backward 2-characteristic $\zeta_\tau(\cdot)$ emanating from the point $(\xi(\tau), \tau)$, which is intercepted by the graph of $\chi_-(\cdot)$ at time $\tau - f(\tau)$. If $(\hat{z}(\cdot), \hat{w}(\cdot))$ denotes the trace of (z, w) along $\zeta_\tau(\cdot)$, Theorems 12.5.1 and 12.10.1 together imply that the total variation of $\hat{z}(\cdot)$ over the interval $[\tau - f(\tau), \tau]$ is $O(\tau^{-1})$. It then follows from Theorem 12.4.1 that the oscillation of $\hat{w}(\cdot)$ over $[\tau - f(\tau), \tau]$ is $O(\tau^{-3})$. Hence

$$(12.10.19) \qquad \overline{w}(\tau) = w_-(\tau - f(\tau)) + O\left(\frac{1}{\tau^3}\right),$$

and so (12.10.18) implies

$$(12.10.20)$$
$$h(\tau) = \lambda_z(0, 0)z(x, t) + \lambda(0, w_-(\tau - f(\tau))) - \lambda(0, w_-(\tau)) + O\left(\frac{1}{\tau^2}\right).$$

As in the proof of Proposition 12.10.1, on any interval $[\tau, \tau+1] \subset [0, t]$ the oscillation of $f(\cdot)$ is $O(\tau^{-1})$ and the total variation of $w_-(\cdot)$ is also $O(\tau^{-1})$. Therefore, upon integrating (12.10.20) over the interval $[s, t], 0 < s < t$, we deduce

$$(12.10.21) \quad x - \chi_-(t) - \lambda_z(0, 0)z(x, t)t = \xi(s) - \chi_-(s) + O\left(\frac{1}{s}\right) + sO\left(\frac{1}{t}\right).$$

With reference to the right-hand side of (12.10.21), given $\varepsilon > 0$, we first fix s so large that $O(s^{-1})$ is less than $\frac{1}{3}\varepsilon$. With s thus fixed, we determine $\hat{t}$ such that, for

$t \geq \hat{t}$, $sO(t^{-1})$ does not exceed $\frac{1}{3}\varepsilon$ while at the same time $\xi(s) - \chi_-(s) < \frac{1}{3}\varepsilon$, for all $x \in (\chi_-(t), \psi(t)]$. Clearly, this last condition need only be checked for $t = \hat{t}$, $x = \psi(\hat{t})$. We have thus verified that the left-hand side of (12.10.21) is $o(1)$, as $t \to \infty$, uniformly in x on $(\chi_-(t), \psi(t))$, which verifies the upper half of (12.10.5)$_1$. The lower half of (12.10.5)$_1$ is established by a similar argument. This completes the proof.

12.11 Notes

There is voluminous literature addressing various aspects of the theory of genuinely nonlinear systems of two conservation laws. The approach in this chapter, via the theory of generalized characteristics, is principally due to the author, and some of the proofs are recorded here in print for the first time. Most of the results were derived earlier in the framework of solutions constructed by the random choice method, which will be presented in Chapter XIII. The seminal contribution in that direction is Glimm and Lax [1].

The Lax entropies, discussed in Section 12.2, were first introduced in Lax [4].

A somewhat stronger version of Theorem 12.3.1 was established by DiPerna [1], for solutions constructed by the random choice method. Theorem 12.4.1 improves a proposition in Dafermos [16].

Theorems 12.5.1, 12.6.1 and 12.6.2 were originally established in Glimm and Lax [1], for solutions constructed by the random choice method, by use of the theory of approximate conservation laws and approximate characteristics, which will be outlined in Section 13.3. The treatment here employs and refines methodology developed by Dafermos and Geng [1,2], for special systems, and Trivisa [1], for general systems, albeit when solutions are "countably regular".

The results of Section 12.7 were established earlier by DiPerna [1], for solutions constructed by the random choice method.

For solutions with initial data in L^1, Temple [3] derives decay at the rate $O(1/\sqrt{\log t})$. The $O(t^{-\frac{1}{2}})$ decay rate established in Theorem 12.8.1, which is taken from Dafermos [16], is sharp.

The mechanism that generates N-wave profiles was understood quite early, through formal asymptotics (see Courant and Friedrichs [1]), even though a rigorous proof was lacking (Lax [2]). In a series of papers by DiPerna [2,4] and Liu [5,6,14], decay to N-waves of solutions with initial data of compact support, constructed by the random choice method, was established at progressively sharper rates, not only for genuinely nonlinear sytems of two conservation laws but even for systems of n conservation laws with characteristic families that are either genuinely nonlinear or linearly degenerate. The decay rates recorded in Theorem 12.9.1 are sharp. Relatively little is known for systems that are not genuinely nonlinear; see Zumbrun [1,2].

Theorem 12.10.1 is due to Glimm and Lax [1], while Theorem 12.10.2 is taken from Dafermos [18].

For applications of the theory of characteristics in investigating uniqueness, regularity and large time behavior of solutions of special systems with coinciding shock and rarefaction wave curves (Temple [1]), see Serre [7,9], Dafermos and Geng [1,2], Heibig [2], Heibig and Sahel [1] and Ostrov [1]. *BV* solutions for such systems have been constructed by the Godunov difference scheme (LeVeque and Temple [1]) as well as by the method of vanishing viscosity (Serre [1,9]).

Chapter XIII. The Random Choice Method

The endeavor of solving the initial-value problem for the scalar conservation law, in Chapter VI, owes its spectacular success to the L^1-contraction property, which applies not only to the solutions themselves but even to their approximations by means of vanishing viscosity, layering, relaxation, etc. For systems, however, the situation is different: L^1-contraction no longer applies; in its place, L^1-Lipschitz continuity will eventually be established, in Chapter XIV, albeit under substantial restrictions on the initial data. Furthermore, at the time of this writing, the requisite stability estimates have been established solely in the context of highly specialized approximating schemes that employ as building blocks the solution of the Riemann problem.

It is desirable to design approximating schemes that do not smear the shocks of the exact solution. It turns out, however, that this feature may be in conflict with the requirement of consistency of the algorithm. The random choice method, developed in this chapter, succeeds in striking the delicate balance of safeguarding consistency without smearing the sharpness of propagating shock fronts. This algorithm will be employed to establish the existence of admissible BV solutions under any initial data with sufficiently small total variation.

The important notions of approximate conservation laws and approximate characteristics will be introduced. The device of wave partitioning, which renders the issue of consistency of the construction scheme deterministic, will be briefly considered. A discussion will follow on how the algorithm shall be modified in order to handle inhomogeneity and source terms involved in hyperbolic systems of balance laws.

The chapter will close with a demonstration that, in systems of at least three conservation laws, when the total variation of the initial data is large, repeated collisions of shocks in resonance may drive the oscillation and/or total variation of solutions to infinity, in finite time.

13.1 The Construction Scheme

We consider the initial-value problem for a strictly hyperbolic system of conservation laws, defined on a ball $\mathscr{O}$ centered at the origin:

$$(13.1.1) \quad \begin{cases} \partial_t U(x,t) + \partial_x F(U(x,t)) = 0 , & -\infty < x < \infty , \quad 0 \le t < \infty , \\ U(x,0) = U_0(x) , & -\infty < x < \infty . \end{cases}$$

The initial data U_0 are functions of bounded variation on $(-\infty, \infty)$. The ultimate goal is to establish the following

Theorem 13.1.1 *There are positive constants δ_0 and δ_1 such that if*

$$(13.1.2) \qquad \sup_{(-\infty,\infty)} |U_0(\cdot)| < \delta_0 ,$$

$$(13.1.3) \qquad TV_{(-\infty,\infty)} U_0(\cdot) < \delta_1 ,$$

then there exists a solution U of (13.1.1), which is a function of locally bounded variation on $(-\infty, \infty) \times [0, \infty)$ taking values in $\mathcal{O}$. This solution satisfies the entropy admissibility criterion for any entropy-entropy flux pair (η, q) of the system, with $\eta(U)$ convex. Furthermore, for each fixed $t \in [0, \infty)$, $U(\cdot, t)$ is a function of bounded variation on $(-\infty, \infty)$ and

$$(13.1.4) \qquad \sup_{(-\infty,\infty)} |U(\cdot,t)| \le c_0 \sup_{(-\infty,\infty)} |U_0(\cdot)| , \quad 0 \le t < \infty ,$$

$$(13.1.5) \qquad TV_{(-\infty,\infty)} U(\cdot,t) \le c_1 TV_{(-\infty,\infty)} U_0(\cdot) , \quad 0 \le t < \infty ,$$

(13.1.6)
$$\int_{-\infty}^{\infty} |U(x,t) - U(x,\tau)| dx \le c_2 |t - \tau| TV_{(-\infty,\infty)} U_0(\cdot) , \quad 0 \le \tau < t < \infty ,$$

where c_0, c_1 and c_2 are constants depending solely on F. When the system is endowed with a coordinate system of Riemann invariants, δ_1 in (13.1.3) may be fixed arbitrarily large, so long as

$$(13.1.7) \qquad (\sup_{(-\infty,\infty)} |U_0(\cdot)|)(TV_{(-\infty,\infty)} U_0(\cdot)) < \delta_2 ,$$

with δ_2 sufficiently small, depending on δ_1.

The proof of the above proposition is quite lengthy and shall occupy the entire chapter. Even though the assertion holds at the level of generality stated above, certain steps in the proof (Sections 13.3, 13.4 and 13.6) will be carried out under the simplifying assumption that each characteristic family of the system is either genuinely nonlinear (7.6.13) or linearly degenerate (7.5.2). The solution U will be attained as the $h \downarrow 0$ limit of a family of approximate solutions U_h constructed by the following process.

We fix a spatial mesh-length h, which will serve as parameter, and a corresponding temporal mesh-length $\lambda^{-1}h$, where λ is a fixed upper bound of the characteristic speeds $|\lambda_i(U)|$, for $U \in \mathcal{O}$ and $i = 1, \cdots, n$. Setting $x_r = rh$, $r = 0, \pm 1, \pm 2, \cdots$ and $t_s = s\lambda^{-1}h$, $s = 0, 1, 2, \cdots$, we build the staggered grid of *mesh-points* (x_r, t_s), with $s = 0, 1, 2, \cdots$, and $r + s$ even.

Assuming now U_h has already been defined on $\{(x,t) : -\infty < x < \infty, 0 \le t < t_s\}$, we determine $U_h(\cdot, t_s)$ as a step function that is constant on intervals defined by neighboring mesh-points along the line $t = t_s$,

(13.1.8) $\qquad U_h(x, t_s) = U_s^r$, $\quad x_{r-1} < x < x_{r+1}$, $\quad r + s$ odd ,

and approximates the function $U_h(\cdot, t_s-)$. The major issue of selecting judiciously the constant states U_s^r will be addressed in Section 13.2.

Next we determine U_h on the strip $\{(x, t) : -\infty < x < \infty, t_s \leq t < t_{s+1}\}$ as a solution of our system, namely,

(13.1.9) $\quad \partial_t U_h(x, t) + \partial_x F(U_h(x, t)) = 0$, $\quad -\infty < x < \infty$, $\quad t_s \leq t < t_{s+1}$,

under the initial condition (13.1.8), along the line $t = t_s$. Notice that the solution of (13.1.9), (13.1.8) consists of centered wave fans issuing from the mesh-points lying on the t_s-time line (Fig. 13.1.1). The wave fan centered at the mesh point (x_r, t_s), $r + s$ even, is constructed by solving the Riemann problem for our system, with left state U_s^{r-1} and right state U_s^{r+1}. We employ admissible solutions, with shocks satisfying the viscous shock admissibility condition (cf. Chapter IX). The resulting outgoing waves from neighboring mesh-points do not interact on the time interval $[t_s, t_{s+1})$, due to our selection of the ratio λ of spatial and temporal mesh-lengths.

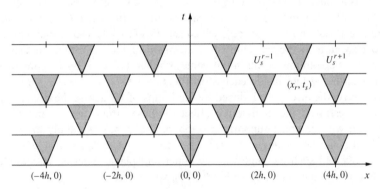

Fig. 13.1.1.

To initiate the algorithm, at $s = 0$, we employ the initial data:

(13.1.10) $\qquad U_h(x, 0-) = U_0(x)$, $\quad -\infty < x < \infty$.

The construction of U_h may proceed for as long as one may solve the resulting Riemann problems. As we saw in Chapter IX, this can be effected, in general, so long as the jumps $|U_s^{r+1} - U_s^{r-1}|$ stay sufficiently small.

After considerable preparation, we shall demonstrate, in Sections 13.5 and 13.6, that the U_h satisfy estimates

(13.1.11) $\qquad \sup_{(-\infty,\infty)} |U_h(\cdot, t)| \leq c_0 \sup_{(-\infty,\infty)} |U_0(\cdot)|$, $\quad 0 \leq t < \infty$;

(13.1.12) $\qquad TV_{(-\infty,\infty)} U_h(\cdot, t) \leq c_1 TV_{(-\infty,\infty)} U_0(\cdot)$, $\quad 0 \leq t < \infty$,

(13.1.13)

$$\int_{-\infty}^{\infty} |U_h(x,t) - U_h(x,\tau)| dx \le c_2(|t-\tau|+h)TV_{(-\infty,\infty)}U_0(\cdot), \quad 0 \le \tau < t < \infty.$$

In particular, (13.1.11) guarantees that when (13.1.2) holds with δ_0 sufficiently small, U_h may be constructed on the entire upper half-plane.

13.2 Compactness and Consistency

Deferring the proof of (13.1.11), (13.1.12) and (13.1.13) to Sections 13.5 and 13.6, we shall take here these stability estimates for granted and will examine their implications. By virtue of Theorem 1.7.1, (13.1.12) and (13.1.13) imply that U_h is in BV_{loc} and its total variation over any compact subset of $(-\infty, \infty) \times [0, \infty)$ is bounded, uniformly in h. It then follows from Theorem 1.7.2 that there is a sequence $\{h_m\}$, with $h_m \to 0$ as $m \to \infty$, such that

(13.2.1) $U_{h_m}(x,t) \to U(x,t)$,   as $m \to \infty$,   a.e. on $(-\infty, \infty) \times [0, \infty)$,

where U is a function in BV_{loc} on $(-\infty, \infty) \times [0, \infty)$. Furthermore, by account of (13.1.11), (13.1.12) and (13.1.13), for each fixed $t \in [0, \infty)$, $U(\cdot, t)$ is a function of bounded variation on $(-\infty, \infty)$, which satisfies (13.1.4), (13.1.5) and (13.1.6).

We now turn to the question of *consistency* of the algorithm, investigating whether U is a solution of the initial-value problem (13.1.1). By its construction, U_h satisfies the system inside each strip $\{(x,t) : -\infty < x < \infty, t_s \le t < t_{s+1}\}$. Consequently, the errors are induced by the jumps of U_h across the dividing lines $t = t_s$. To estimate the cumulative effect of these errors, we fix any C^∞ test function ϕ, with compact support on $(-\infty, \infty) \times [0, \infty)$, we apply the measure (13.1.9) to ϕ on the rectangle $\{(x,t) : x_{r-1} < x < x_{r+1}, t_s \le t < t_{s+1}, r+s \text{ odd}\}$ and sum over all such rectangles in the upper half-plane. After an integration by parts, and upon using (13.1.8) and (13.1.10), we obtain

(13.2.2)

$$\int_0^\infty \int_{-\infty}^\infty [\partial_t\phi U_h + \partial_x\phi F(U_h)]dxdt + \int_{-\infty}^\infty \phi(x,0)U_0(x)dx$$
$$= \sum_{s=0}^\infty \sum_{r+s \text{ odd}} \int_{x_{r-1}}^{x_{r+1}} \phi(x,t_s)[U_h(x,t_s-) - U_s^r]dx.$$

Therefore, U will be a weak solution of (13.1.1), i.e., the algorithm will be consistent, if U_s^r approximates the function $U_h(\cdot, t_s-)$, over the interval (x_{r-1}, x_{r+1}), in such a manner that the right-hand side of (13.2.2) tends to zero, as $h \downarrow 0$.

One may attain consistency via the *Lax-Friedrichs scheme*:

(13.2.3) $$U_s^r = \frac{1}{2h} \int_{x_{r-1}}^{x_{r+1}} U_h(x,t_s-)dx, \quad r+s \text{ odd}.$$

Indeed, with that choice, each integral on the right-hand side of (13.2.2) is majorized by $h^2 \max |\partial_x\phi| osc_{(x_{r-1},x_{r+1})}U_h(\cdot, t_s-)$. The sum of these integrals over r

is then majorized by $h^2 \max |\partial_x \phi| TV_{(-\infty,\infty)} U_h(\cdot, t_s-)$, which, in turn, is bounded by $c_1 \delta_1 h^2 \max |\partial_x \phi|$, on account of (13.1.12) and (13.1.3). The summation over s, within the support of ϕ, involves $O(h^{-1})$ terms, and so finally the right-hand side of (13.2.2) is $O(h)$, as $h \downarrow 0$.

Even though it passes the test of consistency, the Lax-Friedrichs scheme stumbles on the issue of stability: It is presently unknown whether estimates (13.1.12) and (13.1.13) hold within its framework. One of the drawbacks of this scheme is that it smears, through averaging, the shocks of the exact solution. This feature may be vividly illustrated in the context of the Riemann problem for the linear, scalar conservation law,

(13.2.4)
$$\begin{cases} \partial_t u(x,t) + a\lambda \partial_x u(x,t) = 0 , & -\infty < x < \infty , \quad 0 \le t < \infty \\ u(x,0) = \begin{cases} 0 , & -\infty < x < 0 \\ 1 , & 0 < x < \infty \end{cases} \end{cases}$$

where a is a constant in $(-1,1)$ (recall that λ denotes the ratio of the spatial and temporal mesh-lengths). The solution of (13.2.4) comprises, of course, the constant states $u = 0$, on the left, and $u = 1$, on the right, joined by the shock $x = a\lambda t$. The first four steps of the construction of the approximate solution u_h according to the Lax-Friedrichs scheme are depicted in Fig. 13.2.1. The smearing of the shock is clear.

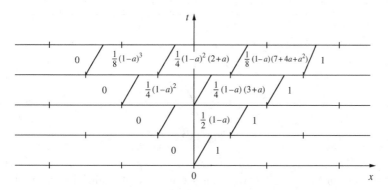

Fig. 13.2.1.

In order to prevent the smearing of shocks, we try a different policy for evaluating the U_s^r. We start out with some sequence $\wp = \{a_0, a_1, a_2, \cdots\}$, where $a_s \in (-1,1)$, we set $y_s^r = x_r + a_s h$, and build, on the upper half-plane, another staggered grid of points (y_s^r, t_s), with $s = 0,1,2,\cdots$ and $r+s$ odd. We employ (y_s^r, t_s) as a *sampling point* for the interval (x_{r-1}, x_{r+1}), on the t_s-time line, by selecting

(13.2.5)
$$U_s^r = \lim_{t \uparrow t_s} U_h(y_s^r -, t) , \quad r+s \text{ odd} .$$

To test this approach, we consider again the Riemann problem (13.2.4). The first few steps of the construction of the approximate solution u_h are depicted in Fig. 13.2.2. We observe that according to the rule (13.2.5), as one passes from $t = t_s$ to $t = t_{s+1}$, the shock is preserved but its location is shifted by h, to the left, when $a_s > a$, or to the right, when $a_s < a$. Consequently, in the limit $h \downarrow 0$ the shock will be thrown off course, unless the number m_- of indices $s \leq m$ with $a_s < a$ and the number m_+ of indices $s \leq m$ with $a_s > a$ are related through $m_- - m_+ \sim am$, as $m \to \infty$. Combining this with $m_- + m_+ = m$, we conclude that u_h will converge to the solution of (13.2.4) if and only if $m_-/m \to \frac{1}{2}(1+a)$ and $m_+/m \to \frac{1}{2}(1-a)$, as $m \to \infty$. For consistency of the algorithm, it will be necessary that the above condition hold for arbitrary $a \in (-1, 1)$. Clearly, this will be the case only when the sequence $\wp$ is *equidistributed* on the interval $(-1, 1)$, that is, for any subinterval $I \subset (-1, 1)$ of length $\mu(I)$:

$$(13.2.6) \qquad \lim_{m \to \infty} \frac{2}{m}[\text{number of indices } s \leq m \text{ with } a_s \in I] = \mu(I) ,$$

uniformly with respect to I.

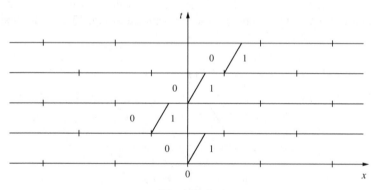

Fig. 13.2.2.

Later on, in Section 13.7, we shall see that the algorithm based on (13.2.5), with any sequence $\wp$ which is equidistributed in $(-1, 1)$, is indeed consistent, for the general initial-value problem (13.1.1); but this may only be established by paying the price of tracking the global wave pattern. The objective here is to demonstrate a slightly weaker result, whose proof however relies solely on the stability estimate (13.1.5). Roughly, it will be shown that if one picks the sequence $\wp$ at random, then the resulting algorithm will be consistent, with probability one. It is from this feature that the method derives its name: *random choice*.

We realize sequences $\wp$ as points in the Cartesian product space $\mathscr{A} = \prod_{s=0}^{\infty}(-1, 1)$. Each factor $(-1, 1)$ is regarded as a probability space, under Lebesgue measure rescaled by a factor $1/2$, and this induces a probability measure ν on $\mathscr{A}$ as well. In connection to our earlier discussions on consistency, it may be shown (references in Section 13.10) that almost all sequences $\wp \in \mathcal{A}$ are equidistributed in $(-1, 1)$. The main result is

Theorem 13.2.1 *There is a null subset $\mathcal{N}$ of $\mathcal{A}$ with the property that the algorithm induced by any sequence $\wp \in \mathcal{A}\backslash\mathcal{N}$ is consistent. That is, when the U_s^r are evaluated through* (13.2.5), *with $y_s^r = x_r + a_s h$, then the limit U in* (13.2.1) *is a solution of the initial-value problem* (13.1.1).

Proof. The right-hand side of (13.2.2) is completely determined by the spatial mesh-length h, the sequence $\wp$ and the test function ϕ, so it shall be denoted by $e(\wp; \phi, h)$. By virtue of (13.2.5),

$$(13.2.7) \qquad e(\wp; \phi, h) = \sum_{s=0}^{\infty} e_s(\wp; \phi, h) \; ,$$

where

$$(13.2.8) \qquad e_s(\wp; \phi, h) = \sum_{r+s \text{ odd}} \int_{x_{r-1}}^{x_{r+1}} \phi(x, t_s)[U_h(x, t_s-) - U_h(y_s^r, t_s-)]dx \; .$$

Note that the integral on the right-hand side of (13.2.8) is majorized by $2h \max |\phi| osc_{(x_{r-1}, x_{r+1})} U_h(\cdot, t_s-)$ and hence $e_s(\wp; \phi, h)$ itself is majorized by $2h \max |\phi| TV_{(-\infty, \infty)} U_h(\cdot, t_s-)$. By virtue of (13.1.12) and (13.1.3),

$$(13.2.9) \qquad |e_s(\wp; \phi, h)| \leq 2c_1 \delta_1 h \max |\phi| \; , \qquad s = 0, 1, 2, \cdots \; .$$

In the summation (13.2.7), the number of nonzero terms, lying inside the support of ϕ, is $O(h^{-1})$ and so the most one may extract from (13.2.9) is $e(\wp; \phi, h) = O(1)$, as $h \downarrow 0$. This again indicates that one should not expect consistency for an arbitrary sequence $\wp$. The success of the random choice method stems from the fact that, as $h \downarrow 0$, the average of $e_s(\wp; \phi, h)$ decays to zero faster than $e_s(\wp; \phi, h)$ itself. Indeed,

$$(13.2.10) \qquad \begin{aligned} &\int_{-1}^{1} \int_{x_{r-1}}^{x_{r+1}} \phi(x, t_s)[U_h(x, t_s-) - U_h(y_s^r, t_s-)]dx da_s \\ &= \frac{1}{h} \int_{x_{r-1}}^{x_{r+1}} \int_{x_{r-1}}^{x_{r+1}} \phi(x, t_s)[U_h(x, t_s-) - U_h(y, t_s-)]dx dy \end{aligned}$$

is majorized by $2h^2 \max |\partial_x \phi| osc_{(x_{r-1}, x_{r+1})} U_h(\cdot, t_s-)$. The sum over r of these integrals is then majorized by $2h^2 \max |\partial_x \phi| TV_{(-\infty, \infty)} U_h(\cdot, t_s-)$. Recalling (13.1.12) and (13.1.3), we finally conclude

$$(13.2.11) \qquad \left| \int_{-1}^{1} e_s(\wp; \phi, h) da_s \right| \leq 2c_1 \delta_1 h^2 \max |\partial_x \phi| \; , \qquad s = 0, 1, 2, \cdots \; .$$

Next we demonstrate that, for $0 \leq s < \sigma < \infty$, $e_s(\wp; \phi, h)$ and $e_\sigma(\wp; \phi, h)$ are "weakly correlated" in that their inner product in $\mathcal{A}$ decays to zero very rapidly,

$O(h^3)$, as $h \downarrow 0$. In the first place, $e_s(\wp; \phi, h)$ depends on $\wp$ solely through the first $s+1$ components $(a_0, \cdots, a_s)$ and, similarly, $e_\sigma(\wp; \phi, h)$ depends on $\wp$ only through $(a_0, \cdots, a_\sigma)$. Hence, upon using (13.2.9) and (13.2.11),

(13.2.12)
$$\left| \int_{\mathscr{A}} e_s(\wp; \phi, h) e_\sigma(\wp; \phi, h) dv(\wp) \right|$$
$$= \left| 2^{-\sigma-1} \int_{-1}^{1} \cdots \int_{-1}^{1} e_s \left(\int_{-1}^{1} e_\sigma da_\sigma \right) da_0 \cdots da_{\sigma-1} \right|$$
$$\leq 2c_1^2 \delta_1^2 h^3 \max |\phi| \max |\partial_x \phi| .$$

By virtue of (13.2.7),

(13.2.13)
$$|e|^2 = \sum_{s=0}^{\infty} |e_s|^2 + 2 \sum_{s=0}^{\infty} \sum_{\sigma=s+1}^{\infty} e_s e_\sigma .$$

Since ϕ has compact support, on the right-hand side of (13.2.13) the first summation contains $O(h^{-1})$ nonzero terms and the second summation contains $O(h^{-2})$ nonzero terms. Consequently, on account of (13.2.9) and (13.2.11),

(13.2.14)
$$\int_{\mathscr{A}} |e(\wp; \phi, h)|^2 dv(\wp) = O(h) , \quad \text{as } h \downarrow 0 .$$

This implies the existence of a null subset $\mathscr{N}_\phi$ of $\mathscr{A}$ such that $e(\wp; \phi, h_m) \to 0$, as $m \to \infty$, for any $\wp \in \mathscr{A} \setminus \mathscr{N}_\phi$. If $\{\phi_k\}$ is any countable set of test functions, which is dense in C^1 in the set of all test functions with compact support in $(-\infty, \infty) \times [0, \infty)$, the null subset $\mathscr{N} = \cup_k \mathscr{N}_{\phi_k}$ of $\mathscr{A}$ will obviously satisfy the asssertion of the theorem. The proof is complete.

To conclude this section, we discuss the admissibility of the constructed solution.

Theorem 13.2.2 *Assume the system is endowed with an entropy-entropy flux pair (η, q), where $\eta(U)$ is convex in $\mathscr{O}$. Then there is a null subset $\mathscr{N}$ of $\mathscr{A}$ with the following property: When the U_s^r are evaluated via (13.2.5), with $y_s^r = x_r + a_s h$, for any $\wp \in \mathscr{A} \setminus \mathscr{N}$, the limit U in (13.2.1) is a solution of (13.1.1) which satisfies the entropy admissibility criterion.*

Proof. Inside each strip $\{(x, t) : -\infty < x < \infty, t_s \leq t < t_{s+1}\}$, U_h is a solution of (13.1.9), with shocks that satisfy the viscous shock admissibility condition and thereby also the entropy shock admissibility criterion, relative to the entropy-entropy flux pair (η, q) (cf. Theorem 8.6.2). Therefore, we have

(13.2.15)
$$\partial_t \eta(U_h(x, t)) + \partial_x q(U_h(x, t)) \leq 0 , \quad -\infty < x < \infty , \quad t_s \leq t < t_{s+1} ,$$

in the sense of measures.

Consider any nonnegative C^∞ test function ϕ with compact support on $(-\infty, \infty) \times [0, \infty)$. We apply the measure (13.2.15) to ϕ on the rectangle $\{(x, t) : x_{r-1} < x < x_{r+1}, t_s \le t < t_{s+1}, r + s \text{ odd}\}$ and sum over all such rectangles in the upper half-plane. After an integration by parts, and upon using (13.1.8) and (13.1.10), this yields

$$
(13.2.16) \quad
\begin{aligned}
&\int_0^\infty \int_{-\infty}^\infty [\partial_t \phi \eta(U_h) + \partial_x \phi q(U_h)] dx dt + \int_{-\infty}^\infty \phi(x, 0) \eta(U_0(x)) dx \\
&\ge \sum_{s=0}^\infty \sum_{r+s \text{ odd}} \int_{x_{r-1}}^{x_{r+1}} \phi(x, t_s)[\eta(U_h(x, t_s-)) - \eta(U_s^r)] dx .
\end{aligned}
$$

Retracing the steps of the proof of Theorem 13.2.1, we deduce that there is a null subset $\mathcal{N}_\phi$ of $\mathcal{A}$ with the property that, when $\wp \in \mathcal{A} \backslash \mathcal{N}_\phi$, the right-hand side of (13.2.16) tends to zero, along the sequence $\{h_m\}$, as $m \to \infty$. Consequently, the limit U in (13.2.1) satisfies the inequality

$$
(13.2.17) \quad \int_0^\infty \int_{-\infty}^\infty [\partial_t \phi \eta(U) + \partial_x \phi q(U)] dx dt + \int_{-\infty}^\infty \phi(x, 0) \eta(U_0(x)) dx \ge 0 .
$$

We now consider any countable set $\{\phi_k\}$ of test functions, which is dense in C^1 in the set of all test functions with compact support in $(-\infty, \infty) \times [0, \infty)$, and define $\mathcal{N} = \cup_k \mathcal{N}_{\phi_k}$. It is clear that if one selects any $\wp \in \mathcal{A} \backslash \mathcal{N}$ then (13.2.17) will hold for all test functions ϕ and hence U will satisfy the entropy admissibility condition. This completes the proof.

In the absence of entropy-entropy flux pairs, or whenever the entropy admissibility criterion is not sufficiently discriminating to rule out all spurious solutions (cf. Chapter VIII), the question of admissibility of solutions constructed by the random choice method is subtle. It is plausible that the requisite shock admissibility conditions will hold at points of approximate jump discontinuity of the solution U, so long as they are satisfied by the shocks of the approximate solutions U_h. Proving this, however, requires a more refined treatment of the limit process that yields U from U_h. This may be attained by the method of wave partitioning which is outlined in Section 13.7.

13.3 Wave Interactions, Approximate Conservation Laws and Approximate Characteristics

We now embark on the long journey that will lead eventually to the stability estimates (13.1.11), (13.1.12) and (13.1.13). The first step is to estimate local changes in the total variation of the approximate solutions U_h. For simplicity, we limit the discussion to systems with characteristic families that are either genuinely nonlinear (7.6.13) or linearly degenerate (7.5.2). The general case is considerably more complicated; see Remark 13.4.1, in the next section.

According to the construction scheme, a portion of the wave fan emanating from the mesh-point (x_{r-1}, t_{s-1}), $r+s$ even, combines with a portion of the wave fan emanating from the mesh-point (x_{r+1}, t_{s-1}) to produce the wave fan that emanates from the mesh-point (x_r, t_s). This is conveniently illustrated by enclosing the mesh-point (x_r, t_s) in a diamond-shaped region Δ_s^r with vertices at the four surrounding sampling points, (y_s^{r-1}, t_s), (y_{s-1}^r, t_{s-1}), (y_s^{r+1}, t_s) and (y_{s+1}^r, t_{s+1}); see Fig. 13.3.1.

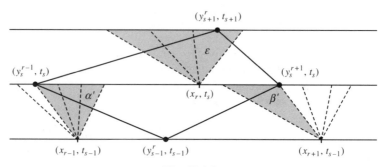

Fig. 13.3.1.

A wave fan emanating from (x_{r-1}, t_{s-1}) and joining the state U_s^{r-1}, on the left, with the state U_{s-1}^r, on the right, enters Δ_s^r through its "southwestern" edge. It may be represented, as explained in Sections 9.3 and 9.6, by the n-tuple $\alpha = (\alpha_1, \cdots, \alpha_n)$ of its wave amplitudes. A second wave fan, emanating from (x_{r+1}, t_{s-1}), joining the state U_{s-1}^r, on the left, with the state U_s^{r+1}, on the right, and similarly represented by the n-tuple $\beta = (\beta_1, \cdots, \beta_n)$ of wave amplitudes, enters Δ_s^r through its "southeastern" edge. Needless to say, depending on the relative location of mesh-points and sampling points, waves of certain families may be missing from the α or β fan, in which case the corresponding amplitude is set equal to zero.

The output from Δ_s^r consists of the full wave fan which emanates from (x_r, t_s), joins the state U_s^{r-1}, on the left, with the state U_s^{r+1}, on the right, and is represented by the n-tuple $\varepsilon = (\varepsilon_1, \cdots, \varepsilon_n)$ of wave amplitudes. A portion $\beta' = (\beta_1', \cdots, \beta_n')$ of ε exits through the "northwestern" edge of Δ_s^r and enters the diamond Δ_{s+1}^{r-1}, while the balance $\alpha' = (\alpha_1', \cdots, \alpha_n')$ exits through the "northeastern" edge of Δ_s^r and enters the diamond Δ_{s+1}^r. Clearly, $\varepsilon_i = \alpha_i' + \beta_i'$, $i = 1, \cdots, n$. Furthermore, there is $j = 1, \cdots, n$ such that $\alpha_i' = 0$ for $i = 1, \cdots, j-1$ and $\beta_i' = 0$ for $i = j+1, \cdots, n$. Both α_j' and β_j' may be nonzero, but then necessarily $\alpha_j' \beta_j' > 0$.

If the waves α and β were allowed to propagate freely, beyond the t_s-time line, the resulting wave interactions would generate a very intricate wave pattern. Nevertheless, following the discussion in Section 9.6, it should be expected that as $t \to \infty$ this wave pattern will reduce to a centered wave fan which is none other than ε. Thus the essense of our construction scheme is that it replaces real, complicated, wave patterns by their time-asymptotic, simpler, forms. In that connection, the role of "random choice" is to arrange the relative location of the wave fans in such a manner that "on the average" the law of "mass" conservation is realized.

According to the terminology of Section 9.6, the wave fan ε shall be regarded as the result of the interaction of the wave fan α, on the left, with the wave fan β, on the right. It is convenient to realize ε, α and β as n-vectors normed by the ℓ_1^n norm, in which case Theorem 9.6.1 yields the estimate

$$(13.3.1) \qquad |\varepsilon - (\alpha + \beta)| \leq [c_3 + c_4(|\alpha| + |\beta|)]\mathscr{D}(\Delta_s^r) \ ,$$

with c_3 and c_4 depending solely on F. In particular, $c_3 = 0$ when the system is endowed with a coordinate system of Riemann invariants. The symbol $\mathscr{D}(\Delta_s^r)$ is here being used, in the place of $D(\alpha, \beta)$ in Section 9.6, to denote the *amount of wave interaction in the diamond* Δ_s^r, namely,

$$(13.3.2) \qquad \mathscr{D}(\Delta_s^r) = \sum_{\text{app}} |\alpha_k||\beta_j| \ .$$

Formula (13.3.1) will serve as the vehicle for estimating how the total variation and the supremum of the approximate solutions U_h change with time, as a result of wave interactions.

By (13.3.1), when α_i and β_i have the same sign, the total strength $|\alpha_i'| + |\beta_i'|$ of i-waves leaving the diamond Δ_s^r approximately equals the total strength $|\alpha_i| + |\beta_i|$ of entering i-waves. However, when α_i and β_i have opposite signs, cancellation of i-waves takes place. To account for this phenomenon, which greatly affects the behavior of solutions, certain notions will now be introduced.

The *amount of i-wave cancellation in the diamond* Δ_s^r is conveniently measured by the quantity

$$(13.3.3) \qquad \mathscr{C}_i(\Delta_s^r) = \frac{1}{2}(|\alpha_i| + |\beta_i| - |\alpha_i + \beta_i|) \ .$$

In order to account separately for shocks and rarefaction waves, we rewrite (13.3.1) in the form

$$(13.3.4) \qquad \varepsilon_i^{\pm} = \alpha_i^{\pm} + \beta_i^{\pm} - \mathscr{C}_i(\Delta_s^r) + [c_3 O(1) + O(\tau)]\mathscr{D}(\Delta_s^r) \ ,$$

where the superscript plus or minus denotes positive or negative part of the amplitude, and τ is the oscillation of U_h.

Upon summing (13.3.4) over any collection of diamonds, whose union forms a domain Λ in the upper half-plane, we end up with equations

$$(13.3.5) \qquad L_i^{\pm}(\Lambda) = E_i^{\pm}(\Lambda) - \mathscr{C}_i(\Lambda) + [c_3 O(1) + O(\tau)]\mathscr{D}(\Lambda) \ ,$$

where E_i^- (or E_i^+) denotes the total amount of i-shock (or i-rarefaction wave) that enters Λ, L_i^- (or L_i^+) denotes the total amount of i-shock (or i-rarefaction wave) that leaves Λ, $\mathscr{C}_i(\Lambda)$ is the amount of i-wave cancellation inside Λ and $\mathscr{D}(\Lambda)$ is the amount of wave interaction inside Λ. The equations (13.3.5) express the balance of i-waves relative to Λ and, accordingly, are called *approximate conservation laws* for i-shocks (with minus sign) or i-rarefaction waves (with plus sign).

The total amount of wave cancellation in the diamond Δ_s^r is naturally measured by

$$(13.3.6) \qquad \mathscr{C}(\Delta_s^r) = \sum_{i=1}^{n} \mathscr{C}_i(\Delta_s^r) .$$

Notice that (13.3.1) implies

$$(13.3.7) \quad |\alpha'| + |\beta'| = |\varepsilon| \le |\alpha| + |\beta| - 2\mathscr{C}(\Delta_s^r) + [c_3 + c_4(|\alpha| + |\beta|)]\mathscr{D}(\Delta_s^r) .$$

An *approximate i-characteristic* associated with the approximate solution U_h, and defined on the time interval $[t_\ell, t_m)$, is a sequence $\chi^{(\ell)}, \cdots, \chi^{(m-1)}$ of straight line segments, such that, for $s = \ell, \cdots, m - 1$, $\chi^{(s)}$ is either a classical i-characteristic or an i-shock for U_h, emanating from some mesh-point (x_r, t_s), $r + s$ even, and defined on the time interval $[t_s, t_{s+1})$. Furthermore, for $s = \ell + 1, \cdots, m - 1$, $\chi^{(s)}$ is a proper sequel to $\chi^{(s-1)}$, according to the following rules: $\chi^{(s-1)}$ must enter the diamond Δ_s^r centered at (x_r, t_s). Whenever the interaction of i-waves entering Δ_s^r produces an i-shock, $\chi^{(s)}$ is that shock. On the other hand, when the interaction of the i-waves entering Δ_s^r produces an i-rarefaction wave, then $\chi^{(s)}$ is a classical i-characteristic identified by the requirement that the amount of i-rarefaction wave that leaves Δ_s^r on the left (right) of $\chi^{(s)}$ does not exceed the amount of i-rarefaction wave that enters Δ_s^r on the left (right) of $\chi^{(s-1)}$. In applying the above rule, we tacitly assume that $\varepsilon_i = \alpha_i + \beta_i$, disregarding the potential (small) contribution to i-rarefaction wave by wave interactions.

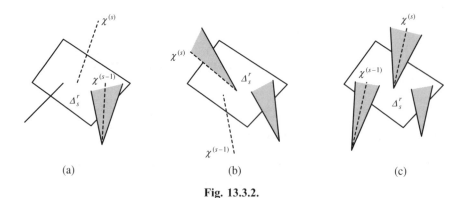

(a)   (b)   (c)

Fig. 13.3.2.

Figure 13.3.2 depicts three representative configurations. Only i-waves are illustrated and the approximate i-characteristic is drawn as a dotted line. In case (a), an i-shock interacts with an i-rarefaction wave to produce an i-shock. $\chi^{(s-1)}$ is a classical i-characteristic but $\chi^{(s)}$ will be the outgoing i-shock. In case (b), $\chi^{(s-1)}$ is an i-shock whose interaction with an i-rarefaction wave produces an

i-rarefaction wave. Since the amount of i-rarefaction wave that enters Δ_s^r on the left of $\chi^{(s-1)}$ is nil, $\chi^{(s)}$ must be the left edge of the outgoing rarefaction wave. Finally, in case (c) two i-rarefaction waves interact to produce an i-rarefaction wave. Then $\chi^{(s)}$ is selected so that the amount of i-rarefaction wave on its left equals the amount of i-rarefaction wave that enters Δ_s^r on the left of $\chi^{(s-1)}$. This will automatically assure that the amount of i-rarefaction wave that leaves Δ_s^r on the right of $\chi^{(s)}$ equals the amount of i-rarefaction wave that enters Δ_s^r on the right of $\chi^{(s-1)}$, provided one neglects potential contribution to i-rarefaction wave by wave interactions.

The above construction of approximate characteristics has been designed so that the following principle holds: Rarefaction waves cannot cross approximate characteristics of their own family. Consequently, approximate conservation laws

$$(13.3.8) \qquad L_i^+(\Lambda_\pm) = E_i^+(\Lambda_\pm) - \mathscr{C}_i(\Lambda_\pm) + [c_3 O(1) + O(\tau)]\mathscr{D}(\Lambda_\pm) ,$$

for i-rarefaction waves, hold for the domains $\Lambda_\pm$ in which the diamond Δ_s^r is divided by any approximate i-characteristic (Fig. 13.3.3).

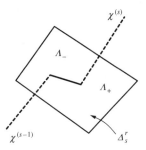

Fig. 13.3.3.

The corresponding approximate conservation laws for i-shocks assume a more complicated form, not recorded here, depending on how one apportions between Λ_- and Λ_+ the strength of i-shocks that lie on the dividing boundary of Λ_- and Λ_+.

One may immediately extend the approximate conservation laws for i-rarefaction waves from the single diamond to any domain Λ formed by the union of a collection of diamonds and thus write (13.3.8) for the domains $\Lambda_\pm$ into which Λ is divided by any approximate i-characteristic.

Approximate conservation laws may be employed to derive fine properties of approximate solutions, at least for systems of two conservation laws, which yield, in the limit, properties of solutions comparable to those established in Chapter XII by the method of generalized characteristics. Indeed, the $h \downarrow 0$ limit of any convergent sequence of approximate i-characteristics is necessarily a generalized i-characteristic, in the sense of Chapter X.

13.4 The Glimm Functional

The aim here is to establish bounds on the total variation of approximate solutions U_h along curves in a certain family. We are still operating under the assumption that each characteristic family is either genuinely nonlinear (7.6.13) or linearly degenerate (7.5.2).

A *mesh curve*, associated with U_h, is a polygonal graph with vertices that form a finite sequence of sample points $(y_{s_1}^{r_1}, t_{s_1}), \cdots, (y_{s_m}^{r_m}, t_{s_m})$, where $r_{\ell+1} = r_\ell + 1$ and $s_{\ell+1} = s_\ell - 1$ or $s_{\ell+1} = s_\ell + 1$ (Fig. 13.4.1). Thus the edges of any mesh curve I are also edges of diamond-shaped regions considered in the previous section. Any wave entering into a diamond through an edge shared with the mesh curve I is said to *cross* I.

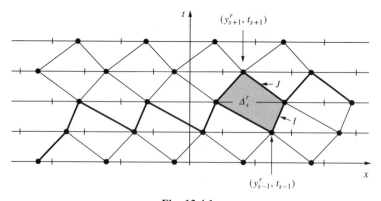

Fig. 13.4.1.

A mesh curve J is called an *immediate successor* of the mesh curve I when $J \backslash I$ is the upper (i.e., "northwestern" and "northeastern") boundary of some diamond, say Δ_s^r, and $I \backslash J$ is the lower (i.e., "southwestern" and "southeastern") boundary of Δ_s^r. Thus J has the same vertices as I, save for one, (y_{s-1}^r, t_{s-1}), which is replaced by (y_{s+1}^r, t_{s+1}). This induces a natural partial ordering in the family of mesh curves: J is a *successor* of I, $I < J$, whenever there is a finite sequence $I = I_0, I_1, \cdots, I_m = J$ of mesh curves such that I_ℓ is an immediate successor of $I_{\ell-1}$, for $\ell = 1, \cdots, m$.

With mesh curves I we associate the functionals

$$(13.4.1) \qquad\qquad \mathscr{S}(I) = \max |\gamma| \, ,$$

$$(13.4.2) \qquad\qquad \mathscr{L}(I) = \sum |\gamma| \, ,$$

where both the maximum and the summation are taken over all waves γ that cross I. Clearly, $\mathscr{S}(I)$ measures the oscillation and $\mathscr{L}(I)$ measures the total variation of U_h along the curve I. We shall estimate the supremum and total variation of U_h by monitoring how $\mathscr{S}$ and $\mathscr{L}$ change as one passes from I to its successors.

Assume J is an immediate successor of I, as depicted in Fig. 13.4.1. Wave fans $\alpha = (\alpha_1, \cdots, \alpha_n)$ and $\beta = (\beta_1, \cdots, \beta_n)$ enter the diamond Δ_s^r through its "southwestern" and "southeastern" edge, respectively, and interact to generate, as discussed in Section 13.3, the wave fan $\varepsilon = (\varepsilon_1, \cdots, \varepsilon_n)$. In turn, ε splits into two wave fans $\beta' = (\beta_1', \cdots, \beta_n')$ and $\alpha' = (\alpha_1', \cdots, \alpha_n')$, as explained in Section 13.3, which exit Δ_s^r through its "northwestern" and "northeastern" edge, respectively. Clearly, I and J are crossed by the same family of waves, with the exception of $(\alpha_1, \cdots, \alpha_n, \beta_1, \cdots, \beta_n)$, which cross I only, and $(\beta_1', \cdots, \beta_n', \alpha_1', \cdots, \alpha_n')$, which cross only J. Consequently, by virtue of (13.3.1) we deduce

$$(13.4.3) \qquad \mathscr{S}(J) \leq \mathscr{S}(I) + [c_3 + c_4 \mathscr{S}(I)]\mathscr{D}(\Delta_s^r) \, ,$$

$$(13.4.4) \qquad \mathscr{L}(J) \leq \mathscr{L}(I) + [c_3 + c_4 \mathscr{S}(I)]\mathscr{D}(\Delta_s^r) \, ,$$

where c_3 and c_4 are the constants that appear also in (13.3.1). In particular, when the system is endowed with a coordinate system of Riemann invariants, $c_3 = 0$. Clearly, $\mathscr{S}$ and $\mathscr{L}$ may increase as one passes from I to J and thus (13.4.3), (13.4.4) alone are insufficient to render the desired bounds (13.1.11), (13.1.12).

What saves the day is the realization that $\mathscr{L}$ may only increase as a result of interaction by approaching waves, which, after crossing paths, separate and move away from each other, never to meet again. Consequently, the potential for future interactions is embodied in the initial arrangement of waves and may thus be anticipated and estimated in advance. To formalize the above heuristic arguments, we shall associate with mesh curves I a functional $\mathscr{Q}(I)$ which measures the *potential for increase in the total variation due to future interactions by waves that cross I*.

Suppose that a k-wave ζ and a j-wave ξ are crossing the mesh curve I, with ζ lying to the left (i.e., "west") of ξ. The waves ζ and ξ are said to be *approaching* if either (a) $k > j$ or (b) $k = j$, the j-characteristic family is genuinely nonlinear and at least one of ζ and ξ is a shock. The reader should note the close analogy with the notion of approaching waves in two interacting wave fans, considered in Section 9.6. After this preparation, we set

$$(13.4.5) \qquad \mathscr{Q}(I) = \sum_{\text{app}} |\zeta||\xi| \, ,$$

where the summation runs over all pairs (ζ, ξ) of approaching waves that cross I. Clearly,

$$(13.4.6) \qquad \mathscr{Q}(I) \leq \frac{1}{2}[\mathscr{L}(I)]^2 \, .$$

In order to see how the potential for future wave interactions may change as one passes from I to its successors, let us estimate, from above, $\mathscr{Q}(J) - \mathscr{Q}(I)$, where, as before, J is the immediate successor of I depicted in Fig. 13.4.1. The contributions of different categories of pairs of approaching waves shall be tallied separately:

Pairs of waves ζ and ξ, both of which cross I as well as J, contribute equally to $Q(I)$ and $Q(J)$ and hence their net contribution to $Q(J) - Q(I)$ is nil.

By virtue of (13.3.2), the waves $(\alpha_1, \cdots, \alpha_n, \beta_1, \cdots, \beta_n)$, which cross I but not J, mutually interact to contribute $\mathscr{D}(\Delta_s^r)$ to $Q(I)$. By contrast, the family $(\beta_1', \cdots, \beta_n', \alpha_1', \cdot, \alpha_n')$ of waves, which cross J but not I, contributes nothing to $Q(J)$, through mutual interactions.

It remains to consider pairs of waves ζ and ξ where ζ crosses both I and J while ξ is only crossing one of I or J. Hence ξ must be one of $\alpha_j, \beta_j, \beta_j'$ or α_j'. Various possibilities arise, for instance, $(\zeta, \alpha_j), (\zeta, \beta_j), (\zeta, \beta_j')$ and (ζ, α_j') may all be approaching, or (ζ, α_j) and (ζ, β_j') may be approaching while $\alpha_j' = 0$ and (ζ, β_j) are not approaching, etc. By examining, however, all possible combinations, one easily deduces, with the help of (13.3.1), that the combined contribution to $Q(J) - Q(I)$ of $(\zeta, \alpha_j), (\zeta, \beta_j), (\zeta, \beta_j')$ and (ζ, α_j'), for $j = 1, \cdots, n$, cannot exceed $[c_3 + c_4 \mathscr{S}(I)]|\zeta|\mathscr{D}(\Delta_s^r)$. Summing up over all qualified ζ, we conclude that the overall contribution to $Q(J) - Q(I)$ by pairs of waves in the above category does not exceed $[c_3 + c_4 \mathscr{S}(I)]\mathscr{L}(I)\mathscr{D}(\Delta_s^r)$.

Collecting the above pieces together, we finally reach the conclusion

$$(13.4.7) \qquad Q(J) - Q(I) \leq \{[c_3 + c_4 \mathscr{S}(I)]\mathscr{L}(I) - 1\}\mathscr{D}(\Delta_s^r) ,$$

where c_3 and c_4 are the same constants appearing in (13.4.3) and (13.4.4). Recall that $c_3 = 0$ when the system is endowed with a coordinate system of Riemann invariants.

Remark 13.4.1 Similar estimates are valid for general systems, in which characteristic families are not necessarily genuinely nonlinear or linearly degenerate. However, $\mathscr{D}(\Delta_s^r)$ and $Q(I)$ have to be defined in a more delicate manner. Elementary i-waves are now i-wave fans, composed of i-rarefaction waves and i-shocks (possibly one-sided or two-sided contact discontinuities; cf. Section 9.3). Consider a k-wave ζ approaching from the left a j-wave ξ, crossing the mesh curve I. The contribution of the pair (ζ, ξ) to $Q(I)$ and/or $\mathscr{D}(\Delta_s^r)$ remains the same as in the genuinely nonlinear case, namely $|\zeta||\xi|$, when either (a) $k > j$ or (b) $k = j$, the j-characteristic family is not linearly degenerate and $\zeta\xi < 0$. However, when $k = j$ and $\zeta\xi > 0$, the contribution of (ζ, ξ) to $Q(I)$ and/or $\mathscr{D}(\Delta_s^r)$ is taken $\theta|\zeta||\xi|$, where θ denotes the angle (difference in wave speeds) between the fastest shock in the wave fan ζ and the slowest shock in the wave fan ξ. Employing the weighting factor θ is motivated by the following argument. Consider, as before, the mesh curve I and its immediate successor J, depicted in Fig. 13.4.1. Assume for simplicity that all waves crossing I are j-shocks. In particular, j-shocks α_j and β_j enter the diamond Δ_s^r and interact to produce the outgoing j-shock ε_j. To leading order in wave strength, we have $\varepsilon_j \sim \alpha_j + \beta_j$. Let σ_-, σ_+ and σ_0 denote, respectively, the speed of propagation of α_j, β_j and ε_j. A simple calculation shows that

$$(13.4.8) \qquad \varepsilon_j \sigma_0 \sim \alpha_j \sigma_- + \beta_j \sigma_+ ,$$

again to leading order in wave strength. Suppose now ζ is another j-shock, crossing I (and J) on the left of α_j and propagating with speed $\sigma, \sigma > \sigma_- > \sigma_0 > \sigma_+$. The contribution of the pair (ζ, ε_j) to $Q(J)$ is $(\sigma - \sigma_0)\zeta\varepsilon_j$. On the other hand, the combined contribution of the pairs (ζ, α_j) and (ζ, β_j) to $Q(I)$ is $(\sigma - \sigma_-)\zeta\alpha_j + (\sigma - \sigma_+)\zeta\beta_j$. With the help of (13.4.8), we now arrive at the desired conclusion that the net contribution of (ζ, α_j), (ζ, β_j) and (ζ, ε_j) to $Q(J) - Q(I)$ vanishes to linear order in wave strength. The detailed proof of (13.4.7) for general systems is too laborious to be included here; it may be found in the references cited in Section 13.10.

An important consequence of (13.4.7) is that when $\mathscr{L}(I)$ is suffiently small the potential Q for future wave interactions will decrease as one passes from the mesh curve I to its immediate successor J. We shall exploit this property to compensate for the possibility that $\mathscr{S}$ and $\mathscr{L}$ may be increasing, to the extent allowed by (13.4.3) and (13.4.4). Towards that end, we associate with mesh curves I the *Glimm functional*

$$(13.4.9) \qquad \mathscr{G}(I) = \mathscr{L}(I) + 2\kappa Q(I) \ ,$$

where κ is some fixed upper bound of $c_3 + c_4\mathscr{S}(I)$, independent of I and h. Even though $\mathscr{G}$ majorizes $\mathscr{L}$, it is actually equivalent to $\mathscr{L}$ by account of (13.4.6).

Theorem 13.4.1 *Let I be a mesh curve with $4\kappa\mathscr{L}(I) \leq 1$. Then, for any mesh curve J that is a successor of I,*

$$(13.4.10) \qquad \mathscr{G}(J) \leq \mathscr{G}(I) \ ,$$

$$(13.4.11) \qquad \mathscr{L}(J) \leq 2\mathscr{L}(I) \ .$$

Furthermore, the amount of wave interaction and the amount of wave cancellation in the diamonds confined between the curves I and J are bounded:

$$(13.4.12) \qquad \sum \mathscr{D}(\Delta_s^r) \leq [\mathscr{L}(I)]^2 \ ,$$

$$(13.4.13) \qquad \sum \mathscr{C}(\Delta_s^r) \leq \mathscr{L}(I) \ .$$

Proof. Assume first J is the immediate successor of I depicted in Fig. 13.4.1. Upon combining (13.4.9) with (13.4.4) and (13.4.7), we deduce

$$(13.4.14) \qquad \mathscr{G}(J) \leq \mathscr{G}(I) + \kappa[2\kappa\mathscr{G}(I) - 1]\mathscr{D}(\Delta_s^r) \ .$$

By virtue of (13.4.9), (13.4.6) and $4\kappa\mathscr{L}(I) \leq 1$, we obtain

$$(13.4.15) \qquad \mathscr{G}(I) \leq 2\mathscr{L}(I) \ ,$$

so that $2\kappa\mathscr{G}(I) \leq 1$, in which case (13.4.14) yields (13.4.10).

Assume now J is any successor of I. Iterating the above argument, we establish (13.4.10) for that case as well. Since $\mathscr{L}(J) \leq \mathscr{G}(J)$, (13.4.11) follows from

(13.4.10) and (13.4.15). Summing (13.4.7) over all diamonds confined between the curves I and J and using (13.4.11), we obtain

(13.4.16)
$$\frac{1}{2}\sum \mathscr{D}(\Delta_s^r) \le \mathcal{Q}(I) - \mathcal{Q}(J) ,$$

which yields (13.4.12), by virtue of (13.4.6).

We sum (13.3.7) over all the diamonds confined between the curves I and J, to get

(13.4.17)
$$2\sum \mathscr{C}(\Delta_s^r) \le \mathscr{L}(I) - \mathscr{L}(J) + \kappa \sum \mathscr{D}(\Delta_s^r) .$$

Combining (13.4.17) with (13.4.11) and (13.4.12) we arrive at (13.4.13). This completes the proof.

The above theorem is of fundamental importance. In particular, the estimates (13.4.10) and (13.4.11) shall provide the desired bounds on the total variation while (13.4.12) and (13.4.13) embody the dissipative effects of nonlinearity and have significant implications to regularity and large time behavior of solutions.

A first application of (13.4.12) is the following

Theorem 13.4.2 *Assume that the system is endowed with a coordinate system of Riemann invariants. Let I be a mesh curve with $4\kappa \mathscr{L}(I) \le 1$. Then, for any mesh curve J that is a successor of I,*

(13.4.18)
$$\mathscr{S}(J) \le \exp[c_4 \mathscr{L}(I)^2] \mathscr{S}(I) .$$

Proof. Assume first J is the immediate successor of I depicted in Fig. 13.4.1. Since $c_3 = 0$, (13.4.4) yields

(13.4.19)
$$\mathscr{S}(J) \le [1 + c_4 \mathscr{D}(\Delta_s^r)] \mathscr{S}(I) .$$

Iterating the above argument, we deduce that if J is any successor of I, then

(13.4.20)
$$\mathscr{S}(J) \le \prod [1 + c_4 \mathscr{D}(\Delta_s^r)] \mathscr{S}(I) ,$$

where the product runs over all the diamonds confined between the curves I and J. Combining (13.4.20) with (13.4.12), we arrive at (13.4.18). This completes the proof.

To apply Theorems 13.4.1 and 13.4.2, one needs to assume that $\kappa \mathscr{L}(I)$ is sufficiently small. For general systems this means that $\mathscr{L}(I)$ itself should be sufficiently small, while for systems endowed with a coordinate system of Riemann invariants $c_3 = 0$ and so it would suffice that $\mathscr{S}(I)\mathscr{L}(I)$ be sufficiently small.

There is a very special class of systems of two conservation laws in which, under proper normalization, $\mathscr{L}$ itself is decreasing as one passes from a mesh curve to its successors and hence we may bound $\mathscr{L}(J)$ even when $\mathscr{L}(I)$ is large. The only interesting representative of that class is the system

$$(13.4.21) \quad \begin{cases} \partial_t u - \partial_x v = 0 \\ \\ \partial_t v + \partial_x \left(\dfrac{1}{u} \right) = 0 \,, \end{cases}$$

namely the special case of (7.1.6) with $\sigma(u) = -u^{-1}$. In classical gas dynamics, this system governs the isothermal process of an ideal gas.

13.5 Bounds on the Total Variation

Here we prove the estimates (13.1.12) and (13.1.13), always operating under the assumption that the oscillation of U_h is bounded, uniformly in h. The vehicle will be the following corollary of Theorem 13.4.1:

Theorem 13.5.1 *Fix* $0 \le \tau < t < \infty$ *and* $-\infty < a < b < \infty$. *Assume that κ times the total variation of* $U_h(\cdot, t)$ *over the interval* $[a - \lambda(t - \tau) - 6h, b + \lambda(t - \tau) + 6h]$ *is sufficiently small.*[1] *Then*

$$(13.5.1) \quad TV_{[a,b]} U_h(\cdot, t) \le c_1 TV_{[a - \lambda(t - \tau) - 6h, b + \lambda(t - \tau) + 6h]} U_h(\cdot, \tau) \,,$$

where c_1 depends solely on F. *Furthermore, if x is a point of continuity of both* $U_h(\cdot, \tau)$ *and* $U_h(\cdot, t)$, *and κ times the total variation of* $U_h(\cdot, t)$ *over the interval* $[x - \lambda(t - \tau) - 6h, x + \lambda(t - \tau) + 6h]$ *is sufficiently small, then*

$$(13.5.2) \quad |U_h(x, t) - U_h(x, \tau)| \le c_5 TV_{[x - \lambda(t - \tau) - 6h, x + \lambda(t - \tau) + 6h]} U_h(\cdot, \tau) \,,$$

where c_5 depends solely on F.

Proof. First we determine nonnegative integers σ and s such that $t_\sigma \le \tau < t_{\sigma+1}$ and $t_s \le t < t_{s+1}$. Next we identify integers r_1 and r_2 such that $y_{s+1}^{r_1+1} < a \le y_{s+1}^{r_1+3}$ and $y_{s+1}^{r_2-3} \le b < y_{s+1}^{r_2-1}$. We then set $r_3 = r_1 - (s - \sigma)$ and $r_4 = r_2 + (s - \sigma)$.

We now construct two mesh curves I and J, as depicted in Fig. 13.5.1, by the following procedure: I originates at the sampling point $(y_\sigma^{r_3}, t_\sigma)$, and zig-zags between t_σ and $t_{\sigma+1}$ until it reaches the sampling point $(y_\sigma^{r_4}, t_\sigma)$ where it terminates. J also originates at $(y_\sigma^{r_3}, t_\sigma)$, takes $s - \sigma$ steps to the "northeast", reaching the sampling point $(y_s^{r_1}, t_s)$, then it zig-zags between t_s and t_{s+1} until it arrives at the sampling point $(y_s^{r_2}, t_s)$, and finally takes $s - \sigma$ steps to the "southeast" terminating at $(y_\sigma^{r_4}, t_\sigma)$.

Clearly,

$$(13.5.3) \quad TV_{[a,b]} U_h(\cdot, t) \le c_6 \mathscr{L}(J) \,.$$

It is easy to see that $y_\sigma^{r_3} \ge a - \lambda(t - \tau) - 6h$ and $y_\sigma^{r_4} \le b + \lambda(t - \tau) + 6h$. Therefore,

$$(13.5.4) \quad \mathscr{L}(I) \le c_7 TV_{[a - \lambda(t - \tau) - 6h, b + \lambda(t - \tau) + 6h]} U_h(\cdot, \tau) \,.$$

---

[1] As before, λ here denotes the ratio of spatial and temporal mesh-lengths.

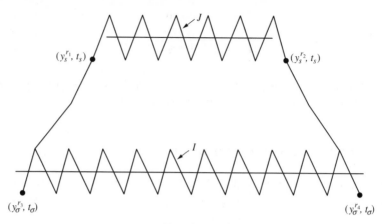

Fig. 13.5.1.

Also, J is a successor of I and hence, if $4\kappa \mathscr{L}(I) \leq 1$, Theorem 13.4.1 implies $\mathscr{L}(J) \leq 2\mathscr{L}(I)$. Combining this with (13.5.3) and (13.5.4), we arrive at (13.5.1), with $c_1 = 2c_6c_7$.

Given x, we repeat the above construction of I and J with $a = b = x$. We can identify a point (y', τ') on I with $U_h(y', \tau') = U_h(x, \tau)$ as well as a point (x', t') on J with $U_h(x', t') = U_h(x, t)$. Hence

(13.5.5) $|U_h(x, t) - U_h(x, \tau)| \leq c_8[\mathscr{L}(I) + \mathscr{L}(J)] \leq 3c_8\mathscr{L}(I)$.

From (13.5.5) and (13.5.4), with $a = b = x$, we deduce (13.5.2) with $c_5 = 3c_7c_8$. This completes the proof.

Upon applying (13.5.1) for $\tau = 0$, $a \to -\infty$, $b \to \infty$, and then using that $TV_{(-\infty,\infty)}U_h(\cdot, 0) \leq TV_{(-\infty,\infty)}U_0(\cdot)$, we verify (13.1.12).

Finally, we integrate (13.5.2) over $(-\infty, \infty)$, apply Fubini's theorem, and use (13.1.12) to get

(13.5.6)

$$\int_{-\infty}^{\infty} |U_h(x, t) - U_h(x, \tau)|dx \leq c_5 \int_{-\infty}^{\infty} TV_{[x-\lambda(t-\tau)-6h, x+\lambda(t-\tau)+6h]}U_h(\cdot, \tau)dx$$

$$= 2c_5[\lambda(t - \tau) + 6h]TV_{(-\infty,\infty)}U_h(\cdot, \tau)$$

$$\leq 2c_1c_5[\lambda(t - \tau) + 6h]TV_{(-\infty,\infty)}U_0(\cdot) ,$$

which establishes (13.1.13).

13.6 Bounds on the Supremum

One may readily obtain a bound on the L^∞ norm of U_h from (13.5.2), with $\tau = 0$:

(13.6.1) $\sup_{(-\infty,\infty)} |U_h(\cdot, t)| \leq \sup_{(-\infty,\infty)} |U_0(\cdot)| + c_5 TV_{(-\infty,\infty)}U_0(\cdot)$.

This estimate is not as strong as the asserted (13.1.11), because, in addition to the supremum, it involves the total variation of the initial data. Even so, combining (13.6.1) with the estimates (13.1.12) and (13.1.13), established in Section 13.5, allows us to invoke the results of Section 13.2 and thus infer the existence of a solution U to the initial-value problem (13.1.1), which is constructed as the limit of a sequence of approximate solutions; cf. (13.2.1). Clearly, U satisfies (13.1.5) and (13.1.6), by virtue of (13.1.12) and (13.1.13). We have thus verified all the assertions of Theorem 13.1.1, save (13.1.4). Despite the fact that it is inessential for demonstrating existence of solutions, (13.1.4) has intrinsic interest, as a statement of stability, and also plays a useful role in deriving other qualitative properties of solutions. It is thus important to discuss how the estimate (13.1.11), from which (13.1.4) derives, may be established.

We first note that for systems endowed with a coordinate system of Riemann invariants, (13.1.11) is an immediate corollary of Theorem 13.4.2 and thus δ_1 in (13.1.3) need not be small, so long as δ_2 in (13.1.7) is. The proof in this case is so simple because terms of quadratic order are missing in the interaction estimate (13.3.1), i.e., $c_3 = 0$. By contrast, in systems devoid of this special structure, the presence of interaction terms of quadratic order renders the situation much more complicated. The proof of (13.1.11) hinges on the special form of the quadratic terms, which, as seen in (9.6.13), involve the Lie brackets of the eigenvectors of DF. The analysis is too laborious to be reproduced here in its entirety, so only an outline of the main ideas shall be presented. The reader may find the details in the references cited in Section 13.10. At the outset, we limit our investigation to systems that are genuinely nonlinear.

The general strategy of the proof is motivated by the ideas expounded in Section 13.4, that culminated in the proof of Theorems 13.4.1 and 13.4.2. Two functionals, $\mathscr{S}$ and $\mathscr{P}$, will be associated with mesh curves I, where $\mathscr{S}(I)$ measures the oscillation of U_h over I while $\mathscr{P}(I)$ provides an estimate on how the oscillation may be affected by future wave interactions.

Recall that $\mathscr{S}(I)$ has already been defined in general terms by (13.4.1), which was adequate for the purposes of Section 13.4. Here, however, we need a more specific characterization of $\mathscr{S}(I)$, analogous to the definition (13.4.2) for $\mathscr{L}(I)$, namely in terms of the waves that cross the mesh curve I. Whereas total variation is estimated, as in (13.4.2), by tallying the strengths of the waves that cross I, in order to represent oscillation, one should account for the mutual cancellations of shocks and rarefaction waves of the same characteristic family. Accordingly, in the definition of $\mathscr{S}(I)$ one should consider each characteristic family separately and tally the signed amplitudes (rather than the strengths) of the waves.

A finite sequence $\xi = (\xi_1, \cdots, \xi_m)$ of waves that cross the mesh curve I is called *consecutive* when, for $\ell = 1, \cdots, m - 1$, ξ_ℓ and $\xi_{\ell+1}$ are separated by a constant state, i.e., the state on the right of ξ_ℓ coincides with the state on the left of $\xi_{\ell+1}$. With such ξ we associate the number

(13.6.2)
$$|\xi| = \sum_{j=1}^{n} \left| \sum_{j-\text{waves}} \xi_\ell \right| ,$$

where the second summation runs over the indices $\ell = 1, \cdots, m$ for which ξ_ℓ is a j-wave. We then define

(13.6.3)
$$\mathscr{S}(I) = \sup_\xi |\xi| ,$$

where the supremum is taken over all sequences of consecutive waves that cross I. After a little reflection, one sees that, as long as $\mathscr{L}(I)$ is sufficiently small, $\mathscr{S}(I)$ is equivalent to the oscillation of U_h over I.

As one passes from I to its successors, $\mathscr{S}$ changes for two reasons: First, as waves travel at different speeds, crossings occur and wave sequences are reordered (notice, however, that the relative order of waves of the same characteristic family is necessarily preserved). Secondly, the amplitude of waves changes in result of wave interactions, as indicated in (9.6.13). It turns out that the effect of wave interactions of third or higher order in wave strength may be estimated grossly, as in the proof of Theorem 13.4.2. However, the effect of wave interactions of quadratic order in wave strength is more significant and thus must be estimated with higher precision. This may be accomplished in an effective manner by realizing the quadratic terms in (9.6.13) as new *virtual waves* which should be accounted for, along with the actual waves.

The above may be formalized by introducing a functional $\mathscr{P}$ associated with mesh curves I, defined by

(13.6.4)
$$\mathscr{P}(I) = \sup_\xi \sup_{\gamma_\xi} |\gamma_\xi| .$$

In (13.6.4) the first supremum is taken over all sequences ξ of consecutive waves that cross I, while the second supremum applies over all wave sequences γ_ξ that may be generated through the following process: (a) by admissible reorderings of the waves in the sequence ξ, say the k-wave ξ_ρ and the j-wave ξ_σ exchange locations if $\rho < \sigma$ and $k > j$; and (b) by inserting any virtual waves that potentially may be generated from interactions of waves in the sequence ξ. Finally, the symbol $|\gamma_\xi|$ is understood in the sense of (13.6.2). The precise construction of γ_ξ entails a major technical endeavor not to be undertaken here (references in Section 13.10).

As long as the total variation is small, $\mathscr{P}$ is actually equivalent to $\mathscr{S}$:

(13.6.5)
$$\mathscr{S}(I) \leq \mathscr{P}(I) \leq [1 + c_9 \mathscr{L}(I)] \mathscr{S}(I) .$$

The idea of the proof of (13.6.5) is as follows. Recall that the principal difference between $\mathscr{L}(I)$ and $\mathscr{S}(I)$ is that in the former we tally the (positive) strengths of crossing waves while in the latter we sum the (signed) amplitudes of crossing waves, thus allowing for cancellation between waves in the same characteristic family but of opposite signs (i.e., shocks and rarefaction waves). Consider the interaction of a single j-wave, say ζ, with a number of k-waves. Since waves in the same characteristic family preserve their relative order, the interactions of the k-waves with ζ will occur consecutively and so the resulting virtual waves will also appear in the same order. Furthermore, whenever the amplitudes of the k-waves alternate in sign, then so do the corresponding Lie bracket terms. Consequently, the virtual waves undergo the same cancellation as their parent waves and thus the contribution to $\mathscr{P}(I)$ by the interaction of ζ with the k-waves will be like $O(1)|\zeta|\mathscr{S}(I)$. Summing over all ζ and $k = 1, \cdots, n$, we conclude that the total contribution to $\mathscr{P}(I)$ from interactions will be $O(1)\mathscr{L}(I).\mathscr{S}(I)$, whence (13.6.5) follows. The detailed proof is quite lengthy and may be found in the references.

The next step is to show that if J is the immediate successor of the mesh curve I depicted in Fig. 13.4.1, then

$$(13.6.6) \qquad \mathscr{P}(J) \le \mathscr{P}(I) + c_{10}.\mathscr{S}(I)\mathscr{D}(\Delta_s^r) \ .$$

The idea of the proof is as follows. Sequences of waves crossing J are reorderings of sequences that cross I, with the waves entering the diamond Δ_s^r through its "southwestern" and "southeastern" edges exchanging their relative locations as they exit Δ_s^r. Furthermore, as one passes from I to J the virtual waves produced by the interaction of the waves that enter Δ_s^r are converted into actual waves, embodied in the waves that exit Δ_s^r. Again, the detailed proof is quite lengthy and should be sought in the references.

By virtue of (13.6.5), we may substitute $\mathscr{P}(I)$ for $\mathscr{S}(I)$ on the right-hand side of (13.6.6), without violating the inequality. Therefore, upon iterating the argument, we conclude that if J is any successor of I, then

$$(13.6.7) \qquad \mathscr{P}(J) \le \prod [1 + c_{10}\mathscr{D}(\Delta_s^r)]\mathscr{P}(I) \ ,$$

where the product runs over all the diamonds Δ_s^r confined between the curves I and J.

We now assume $4\kappa\mathscr{L}(I) \le 1$ and appeal to Theorem 13.4.1. Combining (13.6.7), (13.4.12) and (13.6.5) yields

$$(13.6.8) \qquad \mathscr{S}(J) \le \exp[c_9\mathscr{L}(I) + c_{10}\mathscr{L}(I)^2].\mathscr{S}(I) \ ,$$

whence the desired estimate (13.1.11) readily follows.

13.7 Wave Partitioning

This section demonstrates that more refined information may be gleaned from the construction scheme by examining the global wave pattern. Following up on the discussion in Section 13.3, let us reconsider wave interactions in a diamond, assuming temporarily that we are dealing with scalar conservation laws, $n = 1$. The wave interaction estimate (13.3.1) now reduces to $\varepsilon = \alpha + \beta$. In one typical situation, distinct waves α and β enter the diamond through its "southwestern" and "southeastern" edge, respectively, and then merge into a single wave, say α', which exits through the "northeastern" edge. It is instructive to regard α' as a composite wave, partitioned into α and β, so that α and β retain their identities even after they merge. In the dual situation, a single rarefaction wave α enters the diamond through its "southwestern" edge and then splits into distinct rarefaction waves α' and β', where α' exits through the "northeastern" edge while β' exits through the "northwestern" edge. In that case it is again instructive to partition α

into α' and β' so as to exhibit that the seeds of these waves predated their birth. Of course, the speed of propagation of the composite wave should be assigned to the parts, in order to keep them together. By carrying out this process over several time steps, one may partition the wave ε issuing from the general mesh-point $(x_r, t_s), r + s$ even, into waves whose ancestry may be traced back all the way to $t = 0$. This construction renders an explicit picture of the global wave pattern, from which fine properties of solutions may be derived.

The same approach should work for systems as well, except that now one should take into account that each wave interaction generates new waves, in every characteristic family. The saving grace is that, in virtue of (13.3.1), the strength of these new waves generated in the diamond Δ_s^r is bounded by the amount of wave interaction $\mathscr{D}(\Delta_s^r)$ and hence their cumulative effect may be controlled with the help of (13.4.12). It is thus sufficient to perform approximate partitioning of waves, as described below. For simplicity, we only consider systems with characteristic families that are either genuinely nonlinear or linearly degenerate.

A *partitioning* of an i-shock, which joins the state U_-, on the left, with the state U_+, on the right, is induced by a finite sequence of states $U_- = U^0, U^1, \cdots, U^\nu = U_+$, such that, for $\mu = 1, \cdots, \nu$, U^μ lies on the i-shock curve emanating from U_-, and $\lambda_i(U^\mu) \leq \lambda_i(U^{\mu-1})$. Even though $U^{\mu-1}$ and U^μ are not joined, in general, by a shock, we visualize the pair $(U^{\mu-1}, U^\mu)$ as a *virtual wave*, with amplitude $V_i^\mu = U^\mu - U^{\mu-1}$, to which we assign speed of propagation λ_i^μ, equal to the speed of the i-shock (U_-, U_+).

In an analogous fashion, a *partitioning* of an i-rarefaction wave, which joins the state U_-, on the left, with the state U_+, on the right, is induced by a finite sequence of states $U_- = U^0, U^1, \cdots, U^\nu = U_+$, such that, for $\mu = 1, \cdots, \nu$, U^μ lies on the i-rarefaction wave curve emanating from U_-, and $\lambda_i(U^\mu) > \lambda_i(U^{\mu-1})$. Even though $U^{\mu-1}$ and U^μ are now joined by an i-rarefaction wave, we opt to regard, as in the shock case, $(U^{\mu-1}, U^\mu)$ as a *virtual wave* with amplitude $V_i^\mu = U^\mu - U^{\mu-1}$ and speed of propagation $\lambda_i^\mu = \lambda_i(U^{\mu-1})$.

After this preparation, let us consider the approximate solution U_h generated by our construction scheme for some sequence $\wp = \{a_0, a_1, \cdots\}$. We shall perform a special partitioning of the waves issuing from the mesh points, as described below.

We fix a positive integer σ. We consider the total amount of wave interaction

$$(13.7.1) \qquad\qquad D = \sum \mathscr{D}(\Delta_s^r)$$

and the total amount of wave cancellation

$$(13.7.2) \qquad\qquad C = \sum \mathscr{C}(\Delta_s^r) ,$$

over all diamonds Δ_s^r with $0 \leq s \leq \sigma$ and $r + s$ even.

Using the estimate (13.3.1) and after tedious examination of all possible sorts of interactions that may occur, it can be shown (references in Section 13.10) that for any (r, s) with $0 \leq s \leq \sigma$ and $r + s$ even, and each $i = 1, \cdots, n$, the i-wave emanating from the mesh-point (x_r, t_s) may be partitioned into virtual waves $\{V_i^\mu(r, s)\}$, with assigned speeds $\{\lambda_i^\mu(r, s)\}$, which satisfy the following conditions.

The range of the parameter μ is partitioned into two disjoint subsets M_- and M_+ such that

(a) for $\mu \in M_-$, the virtual waves $V_i^\mu(r, s)$ are weak, in the sense that their cumulative strength is bounded:

$$(13.7.3) \qquad \sum_{s=0}^{\sigma} \sum_{r+s \text{ even}} \sum_{i=1}^{n} \sum_{\mu \in M_-} |V_i^\mu(r, s)| \le c_{11}(C + D) \ ;$$

(b) for $\mu \in M_+$, the virtual waves $V_i^\mu(r, s)$ may have substantial strength but their amplitude and speed do not vary significantly over the time interval $[0, t_\sigma]$, in the sense that for any $s = 0, 1, \cdots, \sigma$,

$$(13.7.4) \qquad \sum_{r \text{ even}} \sum_{i=1}^{n} \sum_{\mu \in M_+} |V_i^\mu(r, 0) - V_i^\mu(\rho(s, r, i, \mu), s)| \le c_{12}D \ ,$$

$$(13.7.5) \qquad \sum_{r \text{ even}} \sum_{i=1}^{n} \sum_{\mu \in M_+} |V_i^\mu(r, 0)||\lambda_i^\mu(r, 0) - \lambda_i^\mu(\rho(s, r, i, \mu), s)| \le c_{13}D \ ,$$

where $r \mapsto \rho(s, r, i, \mu)$ is a function relating mesh-points along the 0-time line with mesh-points along the t_s-time line, constructed by the rule

$$(13.7.6) \qquad\qquad\qquad \rho(0, r, i, \mu) = r \ ,$$

(13.7.7)
$$\rho(s, r, i, \mu) = \begin{cases} \rho(s-1, r, i, \mu) - 1 \ , & \text{if } \lambda_i^\mu(\rho(s-1, r, i, \mu), s-1) < a_s\lambda \ , \\ \rho(s-1, r, i, \mu) + 1 \ , & \text{if } \lambda_i^\mu(\rho(s-1, r, i, \mu), s-1) > a_s\lambda \ , \end{cases}$$

with λ denoting, as before, the ratio of space and time mesh lengths.

It is now possible to establish the following proposition, which improves Theorem 13.2.1 by removing the "randomness" hypothesis in the selection of the sequence $\wp$:

Theorem 13.7.1 *The algorithm induced by any sequence* $\wp = \{a_0, a_1, \cdots\}$, *which is equidistributed on the interval* $(-1, 1)$ *in the sense of* (13.2.6), *is consistent.*

In the proof, which may be found in the references cited in Section 13.10, one expresses the right-hand side of (13.2.2) in terms of the virtual waves that partition U_h and proceeds to show that it tends to zero, as $h \downarrow 0$, whenever the sequence $\wp$ is equidistributed. This happens for the following reason. Recall that in Section 13.2 we did verify the consistency of the algorithm, for any equidistributed sequence $\wp$, in the context of the linear conservation law $\partial_t u + a\lambda\partial_x u = 0$, by employing the property that every wave propagates with constant amplitude and at constant speed. The partitioning of waves performed above demonstrates that even nonlinear systems have this property, albeit in an approximate sense, and this allows to extend the argument for consistency to that case as well.

Though somewhat cumbersome to use, wave partitioning is an effective tool for obtaining precise information on local structure, large time behavior, and other qualitative properties of solutions; and in particular it is indispensable for deriving properties that hinge on the global wave pattern.

13.8 Inhomogeneous Systems of Balance Laws

It is relatively easy to adapt the construction scheme to the case of inhomogeneous, strictly hyperbolic systems of balance laws, in the general form (7.1.1). Inhomogeneity in the flux function and explicit dependence in t are the easiest to handle, especially when they fade away as $|x|$ and t tend to infinity. Source terms may be more troublesome. Accordingly, to avoid cumbersome notation, we shall illustrate the theory in the context of systems in the special form:

$$(13.8.1) \qquad \partial_t U(x,t) + \partial_x F(U(x,t)) + G(U(x,t),x) = 0 \ .$$

The function F satisfies the standard assumptions: It is defined on a ball $\mathcal{O}$ in $\mathbb{R}^n$, centered at the origin, takes values in $\mathbb{R}^n$, it is smooth and, for any $U \in \mathcal{O}$, $DF(U)$ has real distinct eigenvalues $\lambda_1(U) < \cdots < \lambda_n(U)$, so that the system is strictly hyperbolic.

The source function G is defined on $\mathcal{O} \times (-\infty, \infty)$, takes values in $\mathbb{R}^n$, and satisfies $G(0,x) \equiv 0$, so that $U \equiv 0$ is a solution of (13.8.1). Furthermore,

$$(13.8.2) \qquad |DG(U,x)| \le b \ , \quad U \in \mathcal{O} \ , \quad x \in (-\infty, \infty) \ ,$$

$$(13.8.3) \qquad |G_x(U,x)| \le g(x) \ , \quad U \in \mathcal{O} \ , \quad x \in (-\infty, \infty) \ ,$$

where b is a constant and g is an integrable function on $(-\infty, \infty)$, with

$$(13.8.4) \qquad \int_{-\infty}^{\infty} g(x)dx = \omega \ .$$

To (13.8.1) we assign initial conditions

$$(13.8.5) \qquad U(x,0) = U_0(x) \ , \quad -\infty < x < \infty \ .$$

We shall determine solutions of the initial-value problem (13.8.1), (13.8.5) as the $h \downarrow 0$ limit of approximate solutions U_h constructed by a simple adaptation of the algorithm developed in Sections 13.1 and 13.2. We start out again with a "random" sequence $\wp = \{a_0, a_1, \cdots\}$, where $a_s \in (-1, 1)$. We fix the space mesh-length h, with corresponding time mesh-length $\lambda^{-1}h$, and build, as before, the staggered grids of mesh-points (x_r, t_s), for $r+s$ even, and sampling points (y_s^r, t_s), for $r+s$ odd.

Assuming U_h is already known on $\{(x,t) : -\infty < x < \infty, 0 \le t < t_s\}$, we determine $U_h(\cdot, t_s)$ through (13.1.8), namely as a step function that is constant on intervals defined by neighboring mesh-points along the t_s-time line. Here, however, in order to account for the source term in our system, instead of using (13.2.5), we compute U_s^r in two steps:

$$(13.8.6) \qquad V_s^r = \lim_{t \uparrow \uparrow t_s} U_h(y_s^r -, t) , \qquad r + s \text{ odd} ,$$

$$(13.8.7) \qquad U_s^r = V_s^r - \lambda^{-1} h G(V_s^r, x_r) , \qquad r + s \text{ odd} .$$

Next we determine U_h on the strip $\{(x, t) : -\infty < x < \infty, t_s \leq t < t_{s+1}\}$ as a solution of (13.1.9), under initial condition (13.1.8), namely, by resolving the jump discontinuities at the mesh points, along the t_s-time line, according to the conservation law (13.1.9), ignoring the source term.

The algorithm is initiated, at $s = 0$, from the initial data, through (13.1.10).

In the above construction, the source term is handled by "operator splitting": To pass from $t = t_s$ to $t = t_{s+1}$, we first solve approximately the ordinary differential equation $\partial_t U + G(U, x) = 0$, on (t_s, t_{s+1}), through (13.8.7), and then solve, separately, the conservation law $\partial_t U + \partial_x F(U) = 0$. More elaborate construction schemes have also been applied, in which one resolves the jump discontinuities at the mesh-points according to the balance law itself. Needless to say, in the absence of self-similarity, the resolution of jump discontinuities is a hard problem. It is not necessary, however, to employ the exact solution; an approximation would suffice. For that purpose, approximations have been computed, which provide a better fit for the solution than crude operator splitting. Both approaches work and yield, in the $h \downarrow 0$ limit, solutions to (13.8.1), (13.8.5). One should expect that, for fixed h, the more refined approach provides a closer approximation to the solution than mere operator splitting, albeit at the expense of more complicated computation.

To establish the effectiveness of the algorithm, one should retrace our steps in Sections 13.2–13.7 and adapt the analysis to the present setting. It turns out that this is a tedious but straightforward process, not requiring major new ideas. Consequently, a brief outline will suffice here. The reader may find the details in the references cited in Section 13.10.

To begin with, the analogs of Theorems 13.2.1, 13.2.2 and 13.7.1, on consistency of the algorithm, generalize readily to the present situation. Namely, there is a null subset $\mathcal{N}$ of the set $\mathcal{A}$ of sequences $\wp = \{a_0, a_1, \cdots\}$ with the following property. When the U_s^r are evaluated through (13.8.6), (13.8.7), with $y_s^r = x_r + a_s h$, for $\wp \in \mathcal{A} \backslash \mathcal{N}$, then the limit U of any convergent sequence $\{U_{h_m}\}$ of the resulting family $\{U_h\}$ of approximate solutions, with $h_m \to 0$ as $m \to \infty$, is a weak solution of the initial-value problem (13.8.1), (13.8.5), which satisfies the entropy admissibility criterion for any entropy-entropy flux pair (η, q) with $\eta(U)$ convex. Furthermore, the class of equidistributed sequences $\wp$ is necessarily contained in $\mathcal{A} \backslash \mathcal{N}$.

The next task is to establish a priori bounds on U_h and the first step in that direction is to consider, in the present setting, the interaction of waves in the typical diamond Δ_s^r, $r + s$ even. By virtue of (13.8.7), the jump of U across the mesh-point (x_r, t_s) is

$$U_s^{r+1} - U_s^{r-1} = [I - \lambda^{-1} h B_s^r][V_s^{r+1} - V^{r-1}]$$

(13.8.8)
$$- \lambda^{-1} h \int_{x_{r-1}}^{x_r} G_x(V_s^{r-1}, x) dx - \lambda^{-1} h \int_{x_r}^{x_{r+1}} G_x(V_s^{r+1}, x) dx ,$$

where

(13.8.9)
$$B_s^r = \int_0^1 DG(\tau V_s^{r+1} + (1 - \tau) V_s^{r-1}) d\tau .$$

If α and β are the wave fans entering Δ_s^r through its "southwestern" and "southeastern" edge, respectively, and ε is the wave fan generated at the mesh-points (x_r, t_s), we use again, as in Section 13.3, the analysis of Section 9.6, together with our assumptions (13.8.2), (13.8.3) to get, in the place of (13.3.1),

(13.8.10)
$$|\varepsilon - (\alpha + \beta)| \le [c_3 + c_4(|\alpha| + |\beta|)] \mathscr{D}(\Delta_s^r) + \mu h(|\alpha| + |\beta|) + c_{14} h \int_{x_{r-1}}^{x_{r+1}} g(x) dx ,$$

where μ and c_{14} are positive constants depending solely on F and b. Note that the term $\mu h(|\alpha| + |\beta|)$ arises because G varies with U, while the last term accounts for the dependence of G on x.

We now consider, as in Section 13.4, mesh curves I, and with them we associate the functionals $\mathscr{L}(I)$ and $\mathscr{Q}(I)$, defined through (13.4.2) and (13.4.5). Assuming J is an immediate successor of I, as depicted in Fig. 13.4.1, we employ (13.8.10) to derive estimates analogous to (13.4.4) and (13.4.7):

(13.8.11) $$\mathscr{L}(J) \le \mathscr{L}(I) + \kappa \mathscr{D}(\Delta_s^r) + \mu h(|\alpha| + |\beta|) + c_{14} h \int_{x_{r-1}}^{x_{r+1}} g(x) dx ,$$

$$\mathscr{Q}(J) \le \mathscr{Q}(I) + [\kappa \mathscr{L}(I) - 1] \mathscr{D}(\Delta_s^r)$$
(13.8.12)
$$+ \left[\mu h(|\alpha| + |\beta|) + c_{14} h \int_{x_{r-1}}^{x_{r+1}} g(x) dx \right] \mathscr{L}(I) ,$$

where κ is some fixed upper bound of $c_3 + c_4(|\alpha| + |\beta|)$, independent of I and h. Consequently, considering again the Glimm functional $\mathscr{G}$ defined through (13.4.9), we deduce from (13.8.11) and (13.8.12):

$$\mathscr{G}(J) \le \mathscr{G}(I) - \kappa[1 - 2\kappa \mathscr{L}(I)] \mathscr{D}(\Delta_s^r)$$
(13.8.13)
$$+ [1 + 2\kappa \mathscr{L}(I)] \left[\mu h(|\alpha| + |\beta|) + c_{14} h \int_{x_{r-1}}^{x_{r+1}} g(x) dx \right] .$$

In particular, if $2\kappa \mathscr{L}(I) \le 1$, then

(13.8.14) $$\mathscr{G}(J) \le \mathscr{G}(I) + 2\mu h(|\alpha| + |\beta|) + 2c_{14} h \int_{x_{r-1}}^{x_{r+1}} g(x) dx .$$

For $s = 0, 1, 2, \cdots$, we let I_s denote the mesh curve, with infinite edges, which originates at $-\infty$, zig-zags between $t = t_s$ and $t = t_{s+1}$, and terminates at $+\infty$.

Thus the vertices of I_s will be the sampling points (y_s^r, t_s) and (y_{s+1}^{r+1}, t_{s+1}), for all r with $r + s$ odd. So long as $2\kappa \mathcal{L}(I) \leq 1$ holds for all mesh curves I, with $I_{s-1} < I < I_s$, we may apply (13.8.14) repeatedly to get

$$(13.8.15) \qquad \mathcal{G}(I_s) \leq (1 + 2\mu h)\mathcal{G}(I_{s-1}) + 2c_{14}\omega h ,$$

where ω is given by (13.8.4). Iterating (13.8.15) yields

$$(13.8.16) \qquad \mathcal{G}(I_s) \leq e^{2\mu t_s}\mathcal{G}(I_0) + 2c_{14}\omega[e^{2\mu t_s} - 1] .$$

In particular, it follows from (13.8.16) that, assuming $2\kappa \mathcal{G}(I_0) < 1$, one may fix h and T so small that $2\kappa \mathcal{L}(I) \leq 2\kappa \mathcal{G}(I) \leq 1$ for all I such that $I_0 < I < I_s$ and $t_s \leq T$. This closes the loop and establishes that (13.8.16) will indeed hold for $t_s \in [0, T)$.

Next we note that (13.8.16) readily implies an estimate

$$(13.8.17) \qquad TV_{(-\infty,\infty)}U_h(\cdot, t) \leq c_1 e^{2\mu t}[TV_{(-\infty,\infty)}U_0(\cdot) + \omega] ,$$

for h sufficiently small and $t \in [0, T)$. With the help of (13.8.17), one may derive, as in Sections 13.5 and 13.6, a bound on $\sup|U_h|$, over $(\infty, \infty) \times [0, T)$, as well as an estimate

$$(13.8.18) \qquad \int_{-\infty}^{\infty} |U_h(x, t) - U_h(x, \tau)|dx$$
$$\leq c_2 e^{2\mu t}(|t - \tau| + h)[TV_{(-\infty,\infty)}U_0(\cdot) + \omega] ,$$

valid for $0 \leq \tau < t < T$. Combining (13.8.17) with (13.8.18), we infer that the total variation of U_h over $(-\infty, \infty) \times [0, T)$ is bounded, uniformly in h, so that one may extract convergent sequences whose limits will be solutions of (13.8.1), (13.8.5). We have thus sketched the proof of the following

Theorem 13.8.1 *Under the assumptions* (13.8.2), (13.8.3) *on the source term, there are positive constants* δ_0 *and* δ_1 *such that if*

$$(13.8.19) \qquad \sup_{(-\infty,\infty)} |U_0(\cdot)| < \delta_0 ,$$

$$(13.8.20) \qquad TV_{(-\infty,\infty)}U_0(\cdot) < \delta_1 ,$$

then there exists a solution U of (13.8.1), (13.8.5), *which is a function of bounded variation defined on* $(-\infty, \infty) \times [0, T)$ *and taking values in* $\mathcal{O}$. *This solution satisfies the entropy admissibility criterion for any entropy-entropy flux pair* (η, q) *of the system, with* $\eta(U)$ *convex. Furthermore, for each fixed $t \in [0, T)$, $U(\cdot, t)$ is a function of bounded variation on* $(-\infty, \infty)$ *and*

$$(13.8.21) \qquad TV_{(-\infty,\infty)}U(\cdot, t) \leq c_1 e^{2\mu t}[TV_{(-\infty,\infty)}U_0(\cdot) + \omega] .$$

In particular, the life span of the solution increases to infinity as $TV_{(-\infty,\infty)}U_0(\cdot)$ *and* ω *shrink to zero.*

In view of the above proposition, it is useful to identify classes of source terms for which solutions exist globally in time. Below we consider two distinct mechanisms that may induce infinite life span.

We begin with the case where the effect of the source term is dissipative. To motivate the natural hypothesis on G, we linearize (13.8.1) about the origin, and then express U as a linear combination $\sum V_i R_i(0)$ of the right eigenvectors of $DF(0)$, in which case the principal part of the system decouples,

$$(13.8.22) \quad \partial_t V_i(x, t) + \lambda_i(0)\partial_x V_i(x, t) + \sum_{j=1}^{n} A_{ij}(x)V_j(x, t) = 0 , \quad i = 1, \cdots, n ,$$

where

$$(13.8.23) \qquad A(x) = [R_1(0), \cdots, R_n(0)]^{-1} DG(0, x)[R_1(0), \cdots, R_n(0)] .$$

The natural condition that renders (13.8.22) stable in $L^1(-\infty, \infty)$ is that the matrices $A(x)$ are uniformly, strictly diagonally dominant:

$$(13.8.24) \qquad A_{ii}(x) - \sum_{j \neq i}|A_{ij}(x)| \geq v > 0 , \quad i = 1, \cdots, n .$$

Equivalently,

$$(13.8.25) \qquad\qquad |I - \tau A(x)| \leq 1 - v\tau ,$$

for τ positive small, where on the left-hand side we have the norm of $I - \tau A(x)$ as an operator on ℓ_1^n.

It turns out that (13.8.24) is also the natural condition for stability in BV of the system (13.8.1). To see this, one has to return to (13.8.8) and apply more carefully the results of Section 9.6 to get, in the place of (13.8.10), the more precise estimate:

$$(13.8.26) \quad \begin{aligned} \varepsilon = {}&[I - \lambda^{-1}hH_s^r](\alpha + \beta) + O(1)\mathscr{D}(\Delta_s^r) \\ &+ O(h)|v_s^{r-1}|(|\alpha| + |\beta|) + O(h)\int_{x_{r-1}}^{x_{r+1}} g(x)dx , \end{aligned}$$

where H_s^r is an $n \times n$ matrix which is close to A, provided U_h takes values in a sufficiently small neighborhood of the origin. Therefore, by using (13.8.25), we now get, in the place of (13.8.11), the more precise estimate

$$(13.8.27) \quad \mathscr{L}(J) \leq \mathscr{L}(I) - \frac{1}{2}v\lambda^{-1}h(|\alpha| + |\beta|) + \kappa\mathscr{D}(\Delta_s^r) + c_{14}h\int_{x_{r-1}}^{x_{r+1}} g(x)dx .$$

Assuming, as before, that $2\kappa\mathscr{L}(I) \leq 1$ for $I_{s-1} < I < I_s$, we obtain, in lieu of (13.8.15),

$$(13.8.27) \qquad\qquad \mathscr{G}(I_s) \leq (1 - \rho h)\mathscr{G}(I_{s-1}) + 2c_{14}\omega h ,$$

for some $\rho > 0$, whence we deduce

(13.8.28) $$\mathscr{G}(I_s) \le e^{-\rho t_s}\mathscr{G}(I_0) + 2c_{14}\rho^{-1}\omega \; .$$

One may thus preserve $TV_{(-\infty,\infty)}U_h(\cdot, t)$ small, uniformly on $[0, \infty)$, provided $TV_{(-\infty,\infty)}U_0(\cdot)$ and ω are sufficiently small.

The simplest way to keep $\sup_{(-\infty,\infty)}|U_h(\cdot, t)|$ also small for $t \in [0, \infty)$, is by assuming that $U_h(\cdot, t)$ has compact support, and so the final result will be stated in that special situation:

Theorem 13.8.2 *Under the assumptions* (13.8.2), (13.8.3), (13.8.24) *on the source term, there are positive constants* δ *and* ω_0 *such that if* (13.8.4) *holds with* $\omega \le \omega_0$ *and the initial data* U_0 *have compact support and satisfy*

(13.8.29) $$TV_{(-\infty,\infty)}U_0(\cdot) < \delta \; ,$$

then there exists a solution U *of* (13.8.1), (13.8.5), *which is a function of locally bounded variation on* $(-\infty, \infty) \times [0, \infty)$, *taking values in* $\mathscr{O}$. *This solution satisfies the entropy admissibility criterion for any entropy-entropy flux pair* (η, q) *of the system, with* $\eta(U)$ *convex. Furthermore, for each fixed* $t \in [0, T)$, $U(\cdot, t)$ *is a function of bounded variation on* $(-\infty, \infty)$ *and*

(13.8.30) $$TV_{(-\infty,\infty)}U(\cdot, t) \le c_1 e^{-\rho t}TV_{(-\infty,\infty)}U_0(\cdot) + c\omega \; ,$$

for some $\rho > 0$. *In particular, when* G *does not depend explicitly on* x, $TV_{(-\infty,\infty)}U(\cdot, t)$ *decays exponentially fast to zero, as* $t \to \infty$.

We may also have global existence under an entirely different situation, namely when the characteristic speeds of the system are all bounded away from zero and the source term decays, as $|x| \to \infty$. The expectation is that under such conditions the bulk of the wave originating at $t = 0$ shall travel at nonzero speed and thus will eventually enter, and stay, in the region where the source term is small and has insignificant influence. It is clear that to verify the above conjecture, it is imperative to exhibit the global wave pattern and track the bulk of the wave. This may be effected by the method of wave partitioning discussed in Section 13.7. The precise result established in the literature reads:

Theorem 13.8.3 *Assume that*

(13.8.31) $$|\lambda_i(U)| \ge \sigma > 0 \; , \quad U \in \mathscr{O} \; , \quad i = 0, \cdots, n \; ,$$

(13.8.32) $$|G(U, x)| + |DG(U, x)| \le f(x) \; , \quad U \in \mathscr{O} \; , \quad x \in (-\infty, \infty) \; ,$$

where

(13.8.33) $$\int_{-\infty}^{\infty} f(x)dx = \omega \; .$$

There are positive constants ω_0 *and* δ *such that if* $\omega < \omega_0$ *and*

(13.8.34) $$TV_{(-\infty,\infty)}U_0(\cdot) < \delta \; ,$$

then there exists a solution U of (13.8.1), (13.8.5), *which is a function of locally bounded variation on* $(-\infty, \infty) \times [0, \infty)$, *taking values in* $\mathcal{O}$. *Furthermore, for each fixed* $t \in [0, \infty)$, $U(\cdot, t)$ *is a function of bounded variation on* $(-\infty, \infty)$ *and*

$$(13.8.35) \qquad TV_{(-\infty,\infty)}U(\cdot, t) \le c_1[TV_{(-\infty,\infty)}U_0(\cdot) + \omega] \,.$$

An interesting application of the above result is on the system (7.1.11) that governs the isentropic flow through a duct of slowly varying cross section $a(x)$. We rewrite (7.1.11) as

$$(13.8.36) \qquad \begin{cases} \partial_t v + \partial_x(\rho v) + a^{-1}(x)a'(x)\rho v = 0 \\ \partial_t(\rho v) + \partial_x[\rho v^2 + p(\rho)] + a^{-1}(x)a'(x)\rho v^2 = 0 \,, \end{cases}$$

which is in the form (13.8.1). Clearly, in order to satisfy the assumption (13.8.32), (13.8.33) of Theorem 13.8.3, one needs to assume that the cross section becomes rapidly constant as $|x| \to \infty$.

13.9 Breakdown of Weak Solutions

We have seen that wave cancellation, together with dispersion, manage to offset wave amplification, keeping the growth of the total variation of the solution in check, so long as the total variation of the initial data is sufficiently small. For genuinely nonlinear systems of two conservation laws, only the oscillation of the initial data need be small; cf. Chapter XII. It will be demonstrated here that such restrictions on the initial data are generally necessary: For systems of at least three conservation laws, wave patterns exist which resonate to drive the oscillation and/or total variation of solutions to infinity, in finite time.

Consider the system

$$(13.9.1) \qquad \begin{cases} \partial_t u + \partial_x(uv + w) = 0 \\ \partial_t v + \partial_x(\frac{1}{16}v^2) = 0 \\ \partial_t w + \partial_x(u - uv^2 - vw) = 0 \,. \end{cases}$$

The characteristic speeds are $\lambda_1 = -1, \lambda_2 = \frac{1}{8}v, \lambda_3 = 1$, so that strict hyperbolicity holds for $-8 < v < 8$. The first and third characteristic families are linearly degenerate, while the second characteristic family is genuinely nonlinear. Clearly, the system is partially decoupled: The second, Burgers-like, equation by itself determines v.

The Rankine-Hugoniot jump condition for a shock of speed s, joining the state (u_-, v_-, w_-), on the left, with the state (u_+, v_+, w_+), on the right, here read

$$(13.9.2) \qquad \begin{cases} u_+v_+ - u_-v_- + w_+ - w_- = s(u_+ - u_-) \\ \frac{1}{16}v_+^2 - \frac{1}{16}v_-^2 = s(v_+ - v_-) \\ u_+ - u_- - u_+v_+^2 + u_-v_-^2 - v_+w_+ + v_-w_- = s(w_+ - w_-) \,. \end{cases}$$

One then easily sees that 1-shocks are 1-contact discontinuities, with $s = -1$, $v_- = v_+$ and

(13.9.3)$_1$ $$w_+ - w_- = -(v_\pm + 1)(u_+ - u_-) \ .$$

Similarly, 3-shocks are 3-contact discontinuities, with $s = 1$, $v_- = v_+$ and

(13.9.3)$_3$ $$w_+ - w_- = -(v_\pm - 1)(u_+ - u_-) \ .$$

Finally, for 2-shocks, $s = \frac{1}{16}(v_- + v_+)$, and $v_+ < v_-$, in order to satisfy the Lax E-condition.

We now construct a piecewise constant, admissible solution of (13.9.1) with wave pattern depicted in Fig. 13.9.1: Two 2-shocks issue from the points $(-1, 0)$ and $(1, 0)$, with respective speeds $\frac{1}{4}$ and $-\frac{1}{4}$. On the left of the left 2-shock, $v = 4$; on the right of the right 2-shock, $v = -4$; and $v = 0$ between the two 2-shocks. A 1-shock issues from the origin $(0, 0)$ and upon colliding with the left 2-shock it is partly transmitted as a 1-shock and partly reflected as a 3-shock. This 3-shock, upon impinging on the right 2-shock, is in turn partly transmitted as a 3-shock and partly reflected as a 1-shock, and the process is repeated ad infinitum.

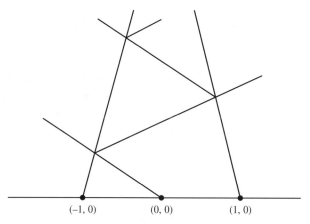

(-1, 0)   (0, 0)   (1, 0)

Fig. 13.9.1.

By checking the Rankine-Hugoniot conditions (13.9.2), one readily verifies that, for instance, initial data

(13.9.4) $\quad (u(x, 0), v(x, 0), w(x, 0)) = \begin{cases} (-65, 4, 225) \ , & -\infty < x < -1 \\ (15, 0, -15) \ , & -1 < x < 0 \\ (-15, 0, 15) \ , & 0 < x < 1 \\ (-63, -4, -225) \ , & 1 < x < \infty \end{cases}$

generate a solution with the above structure.

The aim is to demonstrate that each reflection increases the strength of the shock by a constant factor. With collisions becoming progressively more frequent as the distance between the two 2-shocks is decreasing, until they finally coalesce at $t = 4$, the conclusion will then be that the oscillation of the solution explodes as $t \uparrow 4$. It will be convenient to measure the strength of 1- and 3-shocks by the jump of u across them.

Let us first examine the interaction depicted in Fig. 13.9.2, where a 1-shock hits the left 2-shock, from the right.

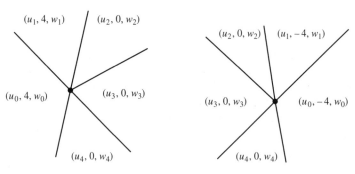

Fig. 13.9.2.   Fig. 13.9.3.

We need to compare the strength $|u_3 - u_2|$ of the reflected 3-shock with the strength $|u_3 - u_4|$ of the incident 1-shock. We write the Rankine-Hugoniot conditions, (13.9.2) or (13.9.3), as applicable, for the five shocks involved in the interaction:

(13.9.5)
$$\begin{cases} w_3 - w_4 = -(u_3 - u_4) \\ w_1 - w_0 = -5(u_1 - u_0) \\ w_3 - w_2 = u_3 - u_2 \\ -4u_0 + w_4 - w_0 = \dfrac{1}{4}(u_4 - u_0) \\ u_4 - u_0 + 16u_0 + 4w_0 = \dfrac{1}{4}(w_4 - w_0) \\ -4u_1 + w_2 - w_1 = \dfrac{1}{4}(u_2 - u_1) \\ u_2 - u_1 + 16u_1 + 4w_1 = \dfrac{1}{4}(w_2 - w_1) \, . \end{cases}$$

After elementary eliminations, one arrives at

(13.9.6)
$$u_3 - u_2 = -\frac{10}{9}(u_3 - u_4) \, ,$$

which shows that as the 1-shock is reflected into a 3-shock the strength increases by a factor 10/9.

Next we examine the interaction depicted in Fig. 13.9.3, where a 3-shock hits the right 2-shock from the left. By writing again the Rankine-Hugoniot conditions, completely analogous to (13.9.5), and after straightforward eliminations, one ends up once more with Equation (13.9.6). Thus, the strength $|u_2 - u_3|$ of the reflected 1-shock exceeds the strength $|u_4 - u_3|$ of the incident 3-shock by a factor 10/9.

We have now confirmed that the oscillation of the solution blows up as $t \uparrow 4$. The above setting, which renders the calculation particularly simple, may appear at first as a singular, isolated example. However, after some reflection one realizes that the wave resonance persists under small perturbations of the equations and/or initial data, i.e., this kind of catastrophe is sort of generic.

Catastrophes of other nature may occur as well: The total variation may blow up even though the oscillation remains bounded. This may be demonstrated in the context of the system

(13.9.7)
$$\begin{cases} \partial_t u + \partial_x (uv^2 + w) = 0 \\ \partial_t v + \partial_x \left(\tfrac{1}{16} v^2 \right) = 0 \\ \partial_t w + \partial_x (u - uv^4 - v^2 w) = 0 \ , \end{cases}$$

which has the same characteristic speeds as (13.9.1), and similarly admits piece-wise constant solutions with the wave pattern depicted in Fig. 13.9.1. It is possible to adjust the speeds of the two 2-shocks in such a manner that after any two successive reflections 1- and 3-shocks regain their original left and right states, i.e., the solution takes values in a finite set of states. On the other hand, as t approaches from below the time t^* of collision of the two 2-shocks, the number of shocks, of fixed strength, that cross the t-time line grows without bound thus driving the total variation to infinity. Details may be found in the references cited in Section 13.10.

It is not currently known whether such catastrophes may occur even in systems arising in Continuum Physics, which, as we have seen, are endowed with special features, such as convex entropy-entropy flux pairs.

13.10 Notes

The random choice method was developed in the fundamental paper of Glimm [1]. It is in that work that the ideas of consistency (Section 13.2), wave interactions (Section 13.3), and the Glimm functional (Section 13.4) were first introduced, and Theorem 13.1.1 was first established, for genuinely nonlinear systems. The Glimm functional may be defined in the context of more general solutions; cf. Schatzman [1]. The construction of solutions with large variation for the special system (13.4.21) of isothermal gas dynamics is due to Nishida [1] (see also Luskin and Temple [1], Poupaud, Rascle and Vila [1] and Ying and Wang [1]). The notions of wave cancellation, approximate conservation laws and approximate characteristics (Section 13.3), which were introduced in the important memoir by

Glimm and Lax [1], provide the vehicle for deriving properties of solutions of genuinely nonlinear systems of two conservation laws, constructed by the random choice method (see Section 12.11). There is voluminous literature on extensions and applications of the random choice method. For systems of mixed type, see Pego and Serre [1]. For initial-boundary value problems, cf. Liu [7], Luskin and Temple [1], Dubroca and Gallice [1], Sablé-Tougeron [1] and Frid [1]. Additional references are given in the books of Smoller [1] and Serre [9].

The extension of the random choice method to general systems that are not necessarily genuinely nonlinear, which was just hinted in Remark 13.4.1, is treated in great detail in Liu [11].

The derivation of bounds on the supremum, roughly outlined in Section 13.6, is taken from the thesis of R. Young [1], where the reader may find the technical details. In fact, this work introduces a new length scale for the Cauchy problem, which may be used in order to relax, under special circumstances, the requirement of small total variation on the initial data, for certain systems of more than two conservation laws. In that direction, see Temple and Young [1,2], and Cheverry [3]. Local or global solutions under initial data with large total variation are also constructed by Alber [1] and Schochet [3,4].

The method of wave partitioning is developed in Liu [4], for genuinely nonlinear systems and in Liu [11], for general systems, and is used to establish the deterministic consistency of the algorithm for equidistributed sequences (Theorem 13.7.1) as well as many important properties of solutions. Thus, for general systems with characteristic families that are either piecewise genuinely nonlinear or linearly degenerate, Liu [11] describes the local structure of solutions and shows, in particular, that any point of discontinuity of the solution is either a point of classical jump discontinuity or a point of wave interaction. Furthermore, the set of points of jump discontinuity comprise a countable family of Lipschitz curves (shocks), while the set of points of wave interaction is at most countable (compare with Theorem 12.7.1, for genuinely nonlinear systems of two conservation laws). As $t \to \infty$, solutions of (13.1.1) approach the solution of the Riemann problem with data (9.1.12), where $U_\ell = U_0(-\infty)$ and $U_r = U_0(+\infty)$; cf. Liu [6,8,11].

The details of the proof of Theorems 13.8.1 and 13.8.2 are found in Dafermos and Hsiao [1]. For an application to the system of isentropic elasticity with frictional damping, see Dafermos [20]. Spherically symmetric solutions of the Euler equations with damping are constructed in Hsiao, Tao and Yang [1]. See also Yang [1]. Theorem 13.8.3 is taken from Liu [10].

The demonstration of the breakdown of solutions, presented in Section 13.9, follows Jenssen [1]. Another example of a quasilinear hyperbolic system of three equations (not in conservation form), with linearly degenerate characteristic families, in which the supremum of (even) smooth solutions blows up in finite time is recorded in Jeffrey [2]. Systems of three conservation laws which exhibit phenomena of instability, like amplification of solutions at a very high rate, are found in R. Young [2]. Rapid magnification in the total variation of solutions of certain systems is also demonstrated in Joly, Métivier and Rauch [2], by the methodology of geometric optics (see Section 15.9). The mechanisms that induce instability differ from case to case, but they all involve wave interactions of three distinct characteristic families. No instability has been detected thus far in solutions of systems with physical interest. It is conceivable that the special features of such systems, e.g. the presence of entropy-entropy flux pairs, may offset the agents of instability. The work of R. Young [3,4] indicates that, for the system of nonisentropic gas dynamics, solutions with periodic initial data remain bounded but do not necessarily decay, as in the isentropic case.

Chapter XIV. The Front Tracking Method and Standard Riemann Semigroups

A method is described, in this chapter, for constructing solutions of the initial-value problem for hyperbolic systems of conservation laws by tracking the waves and monitoring their interactions as they collide. Interactions between shocks are easily resolved by solving Riemann problems; this is not the case, however, with interactions involving rarefaction waves. The random choice method, expounded in Chapter XIII, side-steps this difficulty by stopping the clock before the onset of wave collisions and reapproximating the solution by step functions. In contrast, the front tracking approach circumvents the obstacle by disposing of rarefaction waves altogether and resolving all Riemann problems in terms of shocks only. Such solutions generally violate the admissibility criteria. Nevertheless, considering the close local proximity between shock and rarefaction wave curves in state space, any rarefaction wave may be approximated arbitrarily close by fans of (inadmissible) shocks of very small strength. The expectation is that in the limit, as this approximation becomes finer, one recovers admissible solutions.

The implementation of the front tracking algorithm, with proof that it converges, will be presented here first for scalar conservation laws and then in the context of genuinely nonlinear strictly hyperbolic systems of conservation laws of any size.

By establishing contraction with respect to a suitably weighted L^1 distance, it will be demonstrated that solutions of genuinely nonlinear systems, constructed by the front tracking method, may be realized as orbits of the *Standard Riemann Semigroup*, which is defined on the set of functions with small total variation and is Lipschitz continuous in L^1. It will further be shown that any BV solution, which satisfies reasonable stability conditions, is also identifiable with the orbit of the Standard Riemann Semigroup issuing from its initial data. This establishes, in particular, uniqueness for the initial-value problem within a broad class of BV solutions, including those constructed by the random choice method, as well as those whose trace along space-like curves has bounded variation, encountered in earlier chapters.

The chapter will close with a discussion of the structural stability of the wave pattern under perturbations of the initial data.

14.1 The Scalar Conservation Law

This section discusses the construction of the admissible solution to the initial-value problem for scalar conservation laws by a front tracking scheme that aims at eliminating rarefaction waves. The building blocks will be wave fans composed of constant states, admissible "compressive" shocks, and inadmissible "rarefaction" shocks of small strength.

The admissible solution of the Riemann problem for the scalar conservation law $\partial_t u + \partial_x f(u) = 0$, with C^1 flux f, was constructed in Section 9.3: The left end-state u_ℓ and the right end-state u_r are joined by the wave fan

$$(14.1.1) \qquad u(x,t) = [g']^{-1}\left(\frac{x}{t}\right) ,$$

where g is the convex envelope of f over $[u_\ell, u_r]$, when $u_\ell < u_r$, or the concave envelope of f over $[u_r, u_\ell]$, when $u_\ell > u_r$. Intervals on which g' is constant yield shocks, while intervals over which g' is strictly monotone generate rarefaction waves. The same construction applies even when f is merely Lipschitz, except that now, in addition to shocks and rarefaction waves, the ensuing wave fan may contain intermediate constant states, namely, the jump points of g'. In particular, when f, and thereby g, are piecewise linear, the wave fan does not contain any rarefaction waves but is composed of shocks and constant states only (Fig. 14.1.1).

We now consider the initial-value problem

$$(14.1.2) \qquad \begin{cases} \partial_t u(x,t) + \partial_x f(u(x,t)) = 0 , & -\infty < x < \infty , \quad 0 \le t < \infty , \\ u(x,0) = u_0(x) , & -\infty < x < \infty , \end{cases}$$

for a scalar conservation law, where the flux f is Lipschitz continuous on $(-\infty, \infty)$ and the initial datum u_0 takes values in a bounded interval $[-M, M]$ and has bounded total variation over $(-\infty, \infty)$.

To solve (14.1.2), one first approximates the flux f by a sequence $\{f_m\}$ of piecewise linear functions, such that the graph of f_m is a polygonal line inscribed in the graph of f, with vertices at the points $\left(\frac{k}{m}, f\left(\frac{k}{m}\right)\right)$, $k \in \mathbb{N}$. Next, one realizes the initial datum u_0 as the a.e. limit of a sequence $\{u_{0m}\}$ of step functions, where u_{0m} takes values in the set $\mathcal{U}_m = \{\frac{k}{m} : k \in \mathbb{N}, |k| \le mM\}$, and its total variation

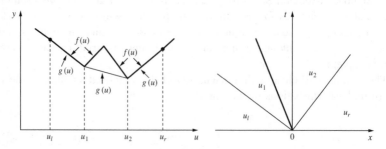

Fig. 14.1.1.

does not exceed the total variation of u_0 over $(-\infty, \infty)$. Finally, one solves the initial-value problem

$$(14.1.3) \quad \begin{cases} \partial_t u(x,t) + \partial_x f_m(u(x,t)) = 0 , & -\infty < x < \infty , \quad 0 \le t < \infty , \\ u(x,0) = u_{0m}(x) , & -\infty < x < \infty , \end{cases}$$

for $m = 1, 2, \cdots$. The aim is to show that the admissible solution u_m of (14.1.3) is a piecewise constant function, taking values in $\mathscr{U}_m$, which is constructed by solving a finite number of Riemann problems for the conservation law $(14.1.3)_1$; and that the sequence $\{u_m\}$ converges to the admissible solution u of (14.1.2).

The construction of u_m is initiated by solving the Riemann problems that resolve the jump discontinuities of u_{0m} into wave fans of shocks and constant states in $\mathscr{U}_m$. In turn, wave interactions induced by shock collisions are similarly resolved, in the order they occur, into wave fans of shocks and constant states in $\mathscr{U}_m$, resulting from the solution of Riemann problems. It should be noted that the admissible solution of the Riemann problem for $(14.1.3)_1$, with end-states in $\mathscr{U}_m$, is also a solution of $(14.1.2)_1$, albeit not necessarily an admissible one, because in that context some of the jump discontinuities may be rarefaction shocks. Thus, in addition to being the admissible solution of (14.1.3), u_m is a (generally inadmissible) solution of $(14.1.2)_1$.

We demonstrate that the number of shock collisions that may be encountered in the implementation of the above algorithm is a priori bounded, and hence u_m is constructed on the entire upper half-plane in finite steps. The reason is that each shock interaction simplifies the wave pattern by lowering either the number of shocks, measured by the number $j_m(t)$ of points of jump discontinuity of the step function $u_m(\cdot, t)$, or the number of "oscillations", counted by the lap number $\ell_m(t)$ of $u_m(\cdot, t)$.

For the case of a step function $v(\cdot)$ on $(-\infty, \infty)$ the *lap number* ℓ, previously encountered in Section 11.8, is set $\ell = 0$ when $v(\cdot)$ is monotone, while when $v(\cdot)$ is nonmonotone it is defined as the largest positive integer with the property that there exist $\ell + 2$ points $-\infty < x_0 < x_1 < \cdots < x_{\ell+1} < \infty$ of continuity of $v(\cdot)$, such that $[v(x_{i+1}) - v(x_i)][v(x_i) - v(x_{i-1})] < 0, i = 1, \cdots, \ell$.

Clearly, both $j_m(t)$ and $\ell_m(t)$ stay constant along the open time intervals between consecutive shock collisions; they may only change across $t = 0$ and as shocks collide. When k shocks, joining (left, right) states $(u_0, u_1), \cdots, (u_{k-1}, u_k)$, collide at one point, the ensuing interaction is called *monotone* if the finite sequence $\{u_0, u_1, \cdots, u_k\}$ is monotone. Such an interaction produces a single shock joining the state u_0, on the left, with the state u_k, on the right. In particular, monotone interactions leave $\ell_m(t)$ unchanged, while lowering the value of $j_m(t)$ at least by one. In contrast, across nonmonotone interactions $\ell_m(t)$ decreases at least by one, while the value of $j_m(t)$ may change in either direction, but in any case it cannot increase by more than $s_m - 1$, s_m being the number of jump points of f'_m over the interval $(-M, M)$; thus $s_m - 1 < 2Mm$. It follows that the integer-valued function $p_m(t) = j_m(t) + s_m \ell_m(t)$ stays constant along the open time intervals between consecutive shock collisions, while decreasing at least by one across any monotone or nonmonotone shock collision. Across $t = 0$, $\ell_m(0+) = \ell_m(0)$ and $j_m(0+) \leq (s_m + 1)j_m(0)$. Therefore, $(s_m + 1)[j_m(0) + \ell_m(0)]$ provides an upper bound for the total number of shock collisions involved in the construction of u_m.

As function of t, the total variation of $u_m(\cdot, t)$ over $(-\infty, \infty)$ stays constant along time intervals between consecutive shock collisions; it does not change across monotone shock collisions; and it decreases across nonmonotone shock collisions. Hence,

(14.1.4)
$$TV_{(-\infty,\infty)}u_m(\cdot, t) \leq TV_{(-\infty,\infty)}u_{m0}(\cdot) \leq TV_{(-\infty,\infty)}u_0(\cdot) , \quad 0 \leq t < \infty .$$

Since the speed of any shock of u_m cannot exceed the Lipschitz constant c of f over $[-M, M]$, (14.1.4) implies

(14.1.5)
$$\int_{-\infty}^{\infty} |u_m(x, t) - u_m(x, \tau)|dx \leq c|t - \tau|TV_{(-\infty,\infty)}u_0(\cdot) , \quad 0 \leq \tau < t < \infty .$$

By virtue of Theorem 1.7.1, (14.1.4) together with (14.1.5) yield that the total variation of u_m over any compact subset of $(-\infty, \infty) \times [0, \infty)$ is bounded, uniformly in m. Hence, by account of Theorem 1.7.2, $\{u_m\}$ contains a subsequence $\{u_{m_k}\}$ which converges a.e. to some function u of locally bounded variation on $(-\infty, \infty) \times [0, \infty)$.

As discussed in Chapter VI, since u_m is the admissible solution of (14.1.3),

(14.1.6)
$$\int_0^{\infty} \int_{-\infty}^{\infty} [\partial_t \psi \eta(u_m) + \partial_x \psi q_m(u_m)]dxdt + \int_{-\infty}^{\infty} \psi(x, 0)\eta(u_{0m}(x))dx \geq 0 ,$$

for any convex entropy η, with associated entropy flux $q_m = \int \eta'df_m$, and all nonnegative Lipschitz test functions ψ on $(-\infty, \infty) \times [0, \infty)$, with compact support. As $m \to \infty$, $\{u_{0m}\}$ converges, a.e. on $(-\infty, \infty)$, to u_0, and $\{q_m\}$ converges, uniformly on $[-M, M]$, to the function $q = \int \eta'df$, namely, the entropy flux associated with the entropy η in the conservation law $(14.1.2)_1$. Upon passing to the limit in (14.1.6), along the subsequence $\{m_k\}$, we deduce

(14.1.7) $$\int_0^{\infty} \int_{-\infty}^{\infty} [\partial_t \psi \eta(u) + \partial_x \psi q(u)]dxdt + \int_{-\infty}^{\infty} \psi(x, 0)\eta(u_0(x))dx \geq 0 ,$$

which implies that u is the admissible solution of (14.1.2). By uniqueness, we infer that the whole sequence $\{u_m\}$ converges to u.

14.2 Front Tracking for Systems of Conservation Laws

Consider a system of conservation laws, in canonical form

(14.2.1) $$\partial_t U + \partial_x F(U) = 0 ,$$

which is strictly hyperbolic (7.2.8), and each characteristic family is either genuinely nonlinear (7.6.13) or linearly degenerate (7.5.2). The object of this section is to introduce a front tracking algorithm which solves the initial-value problem (13.1.1), under initial data U_0 with small total variation, and provides, in particular, an alternative proof of the existence Theorem 13.1.1.

The instrument of the algorithm will be special Riemann solvers, which will be employed to resolve jump discontinuities into centered wave fans composed of jump discontinuities and constant states, approximating the admissible solution of the Riemann problem. In implementing the algorithm, the initial data are approximated by step functions whose jump discontinuities are then resolved into wave fans. Interactions induced by the collision of jump discontinuities are in turn resolved, in the order they occur, into similar wave fans. It will suffice to consider the generic situation, in which no more than two jump discontinuities may collide simultaneously. The expectation is that such a construction will produce an approximate solution of the initial-value problem in the class of piecewise constant functions.

The first item on the agenda is how to design suitable Riemann solvers. The experience with the scalar conservation law, in Section 14.1, suggests that one should synthesize the centered wave fans with constant states, admissible shocks, and inadmissible rarefaction shocks of small strength.

In an *admissible i-shock*, the right state U_+ lies on the i-th shock curve through the left state U_-, that is, in the notation of Section 9.3, $U_+ = \Phi_i(\tau; U_-)$, with $\tau < 0$ when the i-th characteristic family is genuinely nonlinear (compressive shock) or with $\tau \leqslant 0$ when the i-th characteristic family is linearly degenerate (contact discontinuity). The amplitude is τ, the strength is measured by $|\tau|$, and the speed s is set by the Rankine-Hugoniot jump condition (8.1.2).

Instead of actual rarefaction shocks, it is more convenient to employ "rarefaction fronts", namely jump discontinuities which join states lying on a rarefaction wave curve and propagate with characteristic speed. Thus, in an i-*rarefaction front* (which may arise only when the i-th characteristic family is genuinely nonlinear) the right state U_+ lies on the i-th rarefaction wave curve through the left state U_-, i.e., $U_+ = \Phi_i(\tau; U_-)$, with $\tau > 0$. Both, amplitude and strength are measured by τ, and the speed is set equal to $\lambda_i(U_+)$. Clearly, these fronts violate not only the entropy admissibility criterion but even the Rankine-Hugoniot jump condition, albeit only slightly when their strength is small.

Centered rarefaction waves may be approximated by centered wave fans composed of constant states and rarefaction fronts with strength not exceeding some prescribed magnitude $\delta > 0$. Consider some i-rarefaction wave, centered, for definiteness, at the origin, which joins the state U_-, on the left, with the state U_+, on the right. Thus, U_+ lies on the i-rarefaction curve through U_-, say $U_+ = \Phi_i(\tau; U_-)$, for some $\tau > 0$. If ν is the smallest integer which is larger than τ/δ, we set $U^0 = U_-$, $U^\nu = U_+$, $U^\mu = \Phi_i(\mu\delta; U_-)$, $\mu = 1, \cdots, \nu - 1$, and approximate the rarefaction wave, inside the sector $\lambda_i(U_-) < \frac{x}{t} < \lambda_i(U_+)$, by the wave fan

$$(14.2.2) \qquad U(x,t) = U^\mu, \qquad \lambda_i(U^{\mu-1}) < \frac{x}{t} < \lambda_i(U^\mu), \qquad \mu = 1, \cdots, \nu.$$

We are thus naturally lead to an *Approximate Riemann Solver*, which resolves the jump discontinuity between a state U_ℓ, on the left, and U_r, on the right, into a wave fan composed of constant states, admissible shocks, and rarefaction fronts, by the following procedure: The starting point is the admissible solution of the Riemann problem, consisting of $n + 1$ constant states $U_\ell = U_0, U_1, \cdots, U_n = U_r$, where U_{i-1} is joined to U_i by an admissible i-shock or an i-rarefaction wave. To pass to the approximation, the domain and values of the constant states, and thereby all shocks, are retained, whereas, as described above, any rarefaction wave is replaced, within its sector, by a fan of constant states and rarefaction fronts of the same family, with strength not exceeding δ (Fig. 14.2.1).

Our earlier success with the scalar case may raise expectations that a front tracking algorithm in which all shock interactions are resolved via the above approximate, though relatively accurate, Riemann solver will produce an approximate solution of our system, converging to an admissible solution of the initial-value problem as the allowable strength δ of rarefaction fronts shrinks to zero. Unfortunately, such an approach would generally fail, for the following reason: By contrast to the case for scalar conservation laws, wave interactions in systems tend to increase the complexity of the wave pattern so that collisions become progressively more frequent and the algorithm may grind to a stop in finite time. As a remedy, in order to prevent the proliferation of waves, only shocks and rarefaction fronts of substantial strength shall be tracked with relative accuracy. The rest shall not be totally disregarded but shall be treated with less accuracy: They will be lumped together to form jump discontinuities, dubbed "pseudoshocks", which propagate with artificial, supersonic speed.

A *pseudoshock* is allowed to join arbitrary states U_- and U_+. Its strength is measured by $|U_+ - U_-|$ and its assigned speed is a fixed upper bound λ_{n+1} of $\lambda_n(U)$, for U in the range of the solution. Clearly, pseudoshocks are more serious violators of the Rankine-Hugoniot jump condition than rarefaction fronts, and may thus wreak havoc in the approximate solution, unless their total strength is kept very small.

To streamline the exposition, i-rarefaction fronts and i-shocks (compression or contact discontinuities) together will be dubbed *i-fronts*. Fronts and pseudoshocks

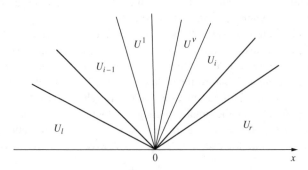

Fig. 14.2.1.

will be called collectively *waves*. Thus an i-front will be an i-*wave* and a pseudoshock will be termed $(n + 1)$-*wave*. For notational convenience, the strength of waves of any type will be denoted by $|\alpha|, |\beta|, |\gamma|, \cdots$, notwithstanding that, as postulated above, $\alpha, \beta, \gamma, \cdots$ could represent scalar amplitudes in the case of fronts but vector amplitudes in the case of pseudoshocks.

Under circumstances to be specified below, the jump discontinuity generated by the collision of two waves shall be resolved via s *Simplified Riemann Solver*, which allows fronts to pass through the point of interaction without affecting their strength, while introducing an outgoing pseudoshock in order to bridge the resulting mismatch in the states. The following cases may arise.

Suppose that, for $i < j$, a j-front, joining the states U_ℓ and U_m, collides with an i-front, joining the states U_m and U_r; see Fig. 14.2.2. Thus $U_m = \Phi_j(\tau_\ell; U_\ell)$ and $U_r = \Phi_i(\tau_r; U_m)$. To implement the Simplified Riemann Solver, one determines the state $U_p = \Phi_i(\tau_r; U_\ell)$ and $U_q = \Phi_j(\tau_\ell; U_p)$. Then, the outgoing wave fan will be composed of the i-front, joining the states U_ℓ and U_p, the j-front, joining the states U_p and U_q, plus the pseudoshock that joins U_q with U_r.

Suppose next that an i-front joining the states U_ℓ and U_m, collides with another i-front, joining the states U_m and U_r (no such collision may occur unless at least one of these fronts is a compressive shock); see Fig. 14.2.3. Thus $U_m = \Phi_i(\tau_\ell; U_\ell)$ and $U_r = \Phi_i(\tau_r; U_m)$. Upon setting $U_q = \Phi_i(\tau_\ell + \tau_r; U_\ell)$, the outgoing wave fan will be composed of the i-front, joining the states U_ℓ and U_q, plus the pseudoshock that joins U_q with U_r.

Finally, suppose a pseudoshock, joining the states U_ℓ and U_m, collides with an i-front, joining the states U_m and U_r; see Fig. 14.2.4. Hence, $U_r = \Phi_i(\tau; U_m)$. We determine $U_q = \Phi_i(\tau; U_\ell)$. The outgoing wave fan will be composed of the i-front, joining the states U_ℓ and U_q, plus the pseudoshock that joins U_q with U_r.

In implementing the front tracking algorithm, one fixes, at the outset, the supersonic speed λ_{n+1} of pseudoshocks, sets the delimiter δ for the strength of rarefaction fronts, and also specifies a third parameter $\sigma > 0$, which rules how jump discontinuities are to be resolved:

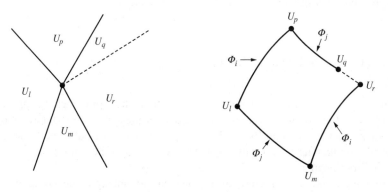

Fig. 14.2.2.

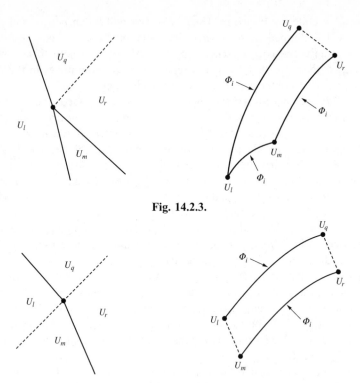

Fig. 14.2.3.

Fig. 14.2.4.

- Jump discontinuities resulting from the collision of two fronts, with respective amplitudes α and β, must be resolved via the Approximate Riemann Solver if $|\alpha||\beta| > \sigma$, or via the Simplified Riemann Solver if $|\alpha||\beta| \leq \sigma$.
- Jump discontinuities resulting from the collision of a pseudoshock with any front must be resolved via the Simplified Riemann Solver.
- Jump discontinuities of the step function approximating the initial data are to be resolved via the Approximate Riemann Solver.

14.3 The Global Wave Pattern

Starting out from some fixed initial step function, the front tracking algorithm, described in the previous section, will produce a piecewise constant function U on a maximal time interval $[0, T)$. In principle, T may turn out to be finite, if the number of collisions grows without bound as $t \uparrow T$, so the onus is to show that this shall not happen.

To understand the structure of U, one has to untangle the complex wave pattern. Towards that end, waves have to be tracked not just between consecutive collisions but globally, from birth to extinction or in perpetuity. The waves are granted global identity through the following convention: An i-wave involved in a collision does not necessarily terminate there, but generally continues on as the outgoing i-wave from that point of wave interaction. Any ambiguities that may arise in applying the above rule will be addressed and resolved below.

Pseudoshocks are generated by the collision of two fronts, resolved via the Simplified Riemann Solver, as depicted in Figs. 14.2.2 or 14.2.3. On the other hand, i-fronts may be generated either at $t = 0$, from the resolution of some jump discontinuity of the initial step function, or at $t > 0$, by the collision of a j-front with a k-front, where $j \neq i \neq k$, which is resolved via the Approximate Riemann Solver.

Every wave is carrying throughout its life span a number μ, identifying its *generation order*, that is the maximum number of collisions predating its birth. Thus, fronts originating at $t = 0$ are assigned generation order $\mu = 0$. Any other new wave, which is necessarily generated by the collision of two waves, with respective generation orders say μ_1 and μ_2, is assigned generation order $\mu = \max\{\mu_1, \mu_2\} + 1$.

As postulated above, waves retain their generation order as they traverse points of interaction. Ambiguity may arise when, in a collision of an i-rarefaction front with a j-front, resolved via the Approximate Riemann Solver, the outgoing i-wave fan contains two i-rarefaction fronts. In that case, the slower of these fronts, with strength δ, is designated as the prolongation of the incoming i-front, while the other i-front, with strength $< \delta$, is regarded as a new front and is assigned a higher generation order, in accordance to the standard rule. Ambiguity may also arise when two fronts of the same family collide, since the outgoing wave fan may include (at most) one front of that family. In that situation, the convention is that the front with the lower generation order is designated to get through, while the other one is terminated. In case both fronts are of the same generation order, any one of them, arbitrarily, may be designated as the survivor. Of course, both fronts may be terminated upon colliding, as depicted in Fig. 14.2.3, in the (nongeneric) case where one of them is a compression shock, the other is a rarefaction front of the same characteristic family and both have the same strength. Pseudoshocks may also be extinguished in finite time by colliding with a front, as depicted in Fig. 14.2.4, in the (nongeneric) case $U_q = U_r$.

We now introduce the following notions, which will establish a connection with the approach pursued in Chapters X–XII.

For $i = 1, \cdots, n$, an i-*characteristic* associated with U is a Lipschitz, polygonal line $x = \xi(t)$ which traverses constant states, say $\overline{U}$, at classical i-characteristic speed, $\dot{\xi} = \lambda_i(\overline{U})$, but upon impinging on an i-front, or a generation point thereof, it adheres to that front, following it throughout its lifespan. Thus, in particular, any i-front is an i-characteristic. By analogy, $(n + 1)$-*characteristics* are defined as straight lines with slope λ_{n+1}. Thus, pseudoshocks are $(n + 1)$-characteristics.

Consider now an oriented Lipschitz curve with graph $\mathscr{C}$, which divides the upper half-plane into its "positive" and "negative" side. We say $\mathscr{C}$ is *nonresonant* if the set $\{1, \cdots, n, n + 1\}$ can be partitioned into three, pairwise disjoint, possibly empty, subsets $\mathcal{N}_-$, $\mathcal{N}_0$ and $\mathcal{N}_+$, with the following properties: $\mathcal{N}_-$ and $\mathcal{N}_+$ each consists of up to $n + 1$ consecutive integers, while $\mathcal{N}_0$ may contain at most one member. For $i \in \mathcal{N}_-$ (or $i \in \mathcal{N}_+$) any i-characteristic impinging on $\mathscr{C}$ crosses from the positive to the negative (or from the negative to the positive) side. On the other hand, if $i \in \mathcal{N}_0$, any i-characteristic impinging upon $\mathscr{C}$, from either its positive or its negative side, is absorbed by $\mathscr{C}$, i.e., $\mathscr{C}$ itself is an i-characteristic.

Noteworthy examples of nonresonant curves include:

(a) Any i-characteristic, in particular any i-wave. In that case $\mathcal{N}_- = \{1, \cdots, i-1\}$, $\mathcal{N}_0 = \{i\}$ and $\mathcal{N}_+ = \{i + 1, \cdots, n + 1\}$.

(b) Any *space-like curve*. Assuming $\lambda_1(U) < 0 < \lambda_{n+1}$, these may be represented by Lipschitz functions $t = \hat{t}(x)$, $-\infty < x < \infty$, with $1/\lambda_1 < d\hat{t}/dx < 1/\lambda_{n+1}$, a.e. on $(-\infty, \infty)$. In that case, $\mathcal{N}_+ = \{1, \cdots, n + 1\}$ while both $\mathcal{N}_-$ and $\mathcal{N}_0$ are empty.

The relevance of the above shall become clear in the next section.

14.4 Approximate Solutions

The following definition collects all the requirements on a piecewise constant function, of the type produced by the front tracking algorithm, so as to qualify as a reasonable approximation to the solution of our initial-value problem:

Definition 14.4.1 For $\delta > 0$, a δ-*approximate solution* of the hyperbolic system of conservation laws (14.2.1) is a piecewise constant function U, defined on $(-\infty, \infty) \times [0, \infty)$ and satisfying the following conditions: The domains of the constant states are bordered by jump discontinuities, called waves, each propagating with constant speed along a straight line segment $x = y(t)$. Any wave may originate either at a point of the x-axis, $t = 0$, or at a point of collision of other waves, and generally terminates upon colliding with another wave, unless no such collision occurs in which case it propagates all the way to infinity. Only two incoming waves may collide simultaneously, but any (finite) number of outgoing waves may originate at a point of collision. There is a finite number of points of collision, waves and constant states. The waves are of three types:

(a) *Shocks.* An (approximate) i-shock $x = y(t)$ borders constant states U_-, on the left, and U_+, on the right, which can be joined by an admissible i-shock, i.e., $U_+ = W_i(\tau; U_-)$, with $\tau < 0$ when the i-characteristic family is genuinely nonlinear or $\tau \lessgtr 0$ when the i-characteristic family is linearly degenerate, and propagates approximately at the shock speed $s = s_i(\tau; U_-)$:

(14.4.1) $$|\dot{y}(\cdot) - s| \leq \delta .$$

(b) *Rarefaction Fronts.* An (approximate) i-rarefaction front $x = y(t)$ borders constant states U_-, on the left, and U_+, on the right, which can be joined by an i-rarefaction wave with strength $\leq \delta$, i.e., $U_+ = V_i(\tau; U_-)$, with $0 < \tau \leq \delta$, and propagates approximately at characteristic speed:

(14.4.2)
$$|\dot{y}(\cdot) - \lambda_i(U_+)| \leq \delta .$$

(c) *Pseudoshocks.* A pseudoshock $x = y(t)$ may border arbitrary states U_- and U_+ and propagates at the specified supersonic speed:

(14.4.3)
$$\dot{y}(\cdot) = \lambda_{n+1} .$$

The combined strength of pseudoshocks does not exceed δ:

(14.4.4)
$$\sum |U(y(t)+, t) - U(y(t)-, t)| \leq \delta , \quad 0 < t < \infty ,$$

where for each t the summation runs over all pseudoshocks $x = y(\cdot)$ which cross the t-time line.

If, in addition, the step function $U(\cdot, 0)$ approximates the initial data U_0 in L^1, within distance δ,

(14.4.5)
$$\int_{-\infty}^{\infty} |U(x, 0) - U_0(x)| dx \leq \delta ,$$

then U is called a *δ-approximate solution of the initial-value problem* (13.1.1).

The extra latitude afforded by the above definition in allowing the speed of (approximate) shocks and rarefaction fronts to (slightly) deviate from their more accurate values granted by the front tracking algorithm provides some flexibility that may be put to good use for securing that no more than two fronts may collide simultaneously.

The effectiveness of front tracking will be demonstrated through the following

Theorem 14.4.1 *Assume $U_0 \in BV(-\infty, \infty)$, with $TV_{(-\infty,\infty)}U_0(\cdot) \leq a \ll 1$. Fix any small positive δ, and approximate U_0 by some step function $U_{0\delta}$ such that $TV_{(-\infty,\infty)}U_{0\delta}(\cdot) \leq TV_{(-\infty,\infty)}U_0(\cdot)$ and $\|U_{0\delta}(\cdot) - U_0(\cdot)\|_{L^1(-\infty,\infty)} \leq \delta$. Then the front tracking algorithm with initial data $U_{0\delta}$, fixed supersonic speed λ_{n+1} for pseudoshocks, delimiter δ for the strength of rarefaction fronts, and sufficiently small parameter σ (depending on δ and on the number of jump points of $U_{0\delta}$) generates a δ-approximate solution U_δ of the initial-value problem (13.1.1). Any sequence of δ's converging to zero contains a subsequence $\{\delta_k\}$ such that $\{U_{\delta_k}\}$ converges, a.e. on $(-\infty, \infty) \times [0, \infty)$, to a BV solution U of (13.1.1), which satisfies the entropy admissibility condition for any convex entropy-entropy flux pair (η, q) of the system (14.2.1), together with the estimates (13.1.5) and (13.1.6). Furthermore, the trace of U on any Lipschitz graph on the upper half-plane which is nonresonant relative to all U_δ, has bounded variation.*

The above proposition reestablishes the assertions of Theorem 13.1.1. The property that the trace of U along nonresonant curves has bounded variation establishes a connection with the class of solutions discussed in Chapter XII.

The demonstration of Theorem 14.4.1 is quite lengthy and will be presented, in installments, in the next three sections. However, the following road map may prove useful at this juncture.

As already noted in Section 14.3, once the step function $U_{0\delta}$ has been designated, the front tracking algorithm will produce U_δ, at least on a time interval $[0, T)$, which as we shall see later is $[0, \infty)$. We will be assuming throughout that the range of U_δ is contained in a ball of small radius in state space, a condition that must be verified a posteriori. The constants $c_1, c_2, \cdots, \kappa, \cdots$ which will appear in the course of the proof, all depend solely on bounds of F and its derivatives in that ball.

The first step will be to establish an estimate

$$(14.4.6) \qquad TV_{(-\infty,\infty)}U_\delta(\cdot, t) \le c_1 TV_{(-\infty,\infty)}U_0(\cdot) , \quad 0 \le t < T ,$$

on the total variation, together with a bound on the total amount of wave interaction. By account of the construction of U_δ, (14.4.6) will immediately imply

$(14.4.7)$

$$\int_{-\infty}^{\infty} |U_\delta(x, t) - U_\delta(x, \tau)| dx \le c_2 |t - \tau| TV_{(-\infty,\infty)}U_0(\cdot) , \quad 0 \le \tau < t < T ,$$

with $c_2 = cc_1$, where c is any upper bound of the wave speeds; for instance c is the maximum of λ_{n+1} and $-\inf \lambda_1(U)$. The usefulness of these estimates is twofold: First, they will assist in the task of verifying that U_δ meets the requirements set by Definition 14.4.1. Secondly, they will induce compactness that allows to pass to the $\delta \downarrow 0$ limit.

In verifying that U_δ is a δ-approximate solution, the requirements (14.4.1), (14.4.2) and (14.4.3), on the speed of shocks, rarefaction fronts and pseudoshocks, are patently met, due to the specifications of the construction. Moreover, the selection of the delimiter entails that the strengths of rarefaction fronts will be bounded by δ. The remaining requirements, namely that the combined strength of pseudoshocks is also bounded by δ, as in (14.4.4), and that the number of collisions is finite, will be established by insightful analysis of the wave pattern. In particular, this will furnish the warranty that U_δ is generated, in finite steps, on the entire upper half-plane, i.e., $T = \infty$.

The final step in the proof will complete the construction of the solution to (13.1.1) by passing to the $\delta \downarrow 0$ limit in U_δ, via a compactness argument relying on the estimates (14.4.6) and (14.4.7).

14.5 Bounds on the Total Variation

As in Section 13.4, $TV_{(-\infty,\infty)}U_\delta(\cdot, t)$ will be measured through

$$(14.5.1) \qquad\qquad L(t) = \sum |\gamma| ,$$

namely the sum of the strengths of all jump discontinuities that cross the t-time line. Clearly, $L(\cdot)$ stays constant along time intervals between consecutive collisions of fronts and only changes across points of wave interaction. To estimate these changes, we have to investigate the various types of collisions.

Suppose a j-front of amplitude α collides with an i-front of amplitude β. When $|\alpha||\beta| \geq \sigma$, so that the resulting jump discontinuity is resolved, via the Approximate Riemann Solver, into a full wave fan $\varepsilon = (\varepsilon_1, \cdots, \varepsilon_n)$, then, by virtue of Theorem 9.6.1[1],

$$(14.5.2) \qquad |\varepsilon_j - \alpha| + |\varepsilon_i - \beta| + \sum_{k \neq i,j} |\varepsilon_k| = O(1)|\alpha||\beta| ,$$

if $i < j$, or

$$(14.5.3) \qquad |\varepsilon_i - \alpha - \beta| + \sum_{k \neq i} |\varepsilon_k| = O(1)|\alpha||\beta| ,$$

if $i = j$. On the other hand, when $|\alpha||\beta| < \sigma$, in which case the resulting jump discontinuity is resolved, via the Simplified Riemann Solver, as shown in Fig. 14.2.2 or 14.2.3, the amplitude of the colliding fronts is conserved. The strength of the generated outgoing pseudoshock is easily estimated from the wave diagrams in state space:

$$(14.5.4) \qquad |U_r - U_q| = O(1)|\alpha||\beta| .$$

Consider next the case depicted in Fig. 14.2.4, where a pseudoshock collides with an i-front of amplitude β. Since the amplitude of the i-front is conserved across the collision, analysis of the wave diagram in state space, Fig. 14.2.4, yields that the strength of the outgoing pseudoshock is related to the strength of the incoming pseudoshock by

$$(14.5.5) \qquad |U_r - U_q| = |U_m - U_\ell| + O(1)|\beta||U_m - U_\ell| .$$

Let I denote the set of $t \in (0, T)$ where collisions occur. We let Δ denote the "jump" operator from $t-$ to $t+$, for $t \in I$. By account of the analysis of wave interactions, above, we infer

$$(14.5.6) \qquad \Delta L(t) \leq \kappa |\alpha||\beta| , \quad t \in I ,$$

where $|\alpha|$ and $|\beta|$ are the strengths of the waves which collide at t.

Our strategy for keeping $TV_{(-\infty,\infty)}U_\delta(\cdot, t)$ under control is to show that any increase of $L(\cdot)$ allowed by (14.5.6) is offset by the simultaneous decrease in the amount of potential wave interaction.

A j-wave and an i-wave, with the former crossing the t-time line to the left of the latter, are called *approaching* when either $i < j$, or $i = j$ and at least one of these waves is a compression shock.

---

[1] If the outgoing k-wave is a fan of k-rarefaction fronts, ε_k denotes the cumulative amplitude and $|\varepsilon_k|$ stands for the cumulative strength of these fronts.

The potential for wave interaction at $t \in (0, \tau) \backslash I$ will be measured by

$$(14.5.7) \qquad Q(t) = \sum |\zeta||\xi| , \quad t \in (0, T) \backslash I ,$$

where the summation runs over all pairs of approaching waves, with strengths say $|\zeta|$ and $|\xi|$, which cross the t-time line. In particular,

$$(14.5.8) \qquad Q(t) \leq \frac{1}{2} L(t)^2 , \quad t \in (0, T) \backslash I .$$

Clearly, $Q(\cdot)$ stays constant along time intervals between consecutive collisions. On the other hand, at any $t \in I$ where waves with strength $|\alpha|$ and $|\beta|$ collide, our analysis of wave interactions implies

$$(14.5.9) \qquad \Delta Q(t) \leq -|\alpha||\beta| + \kappa |\alpha||\beta| L(t-) , \quad t \in I .$$

In analogy to the Glimm functional (13.4.9), we set

$$(14.5.10) \qquad G(t) = L(t) + 2\kappa Q(t) , \quad t \in (0, T) \backslash I .$$

Combining (14.5.10) with (14.5.6) and (14.5.9), yields

$$(14.5.11) \qquad \Delta G(t) \leq \kappa [2\kappa G(t-) - 1] |\alpha||\beta|, \quad t \in (0, T) \backslash I.$$

Assume now the total variation of the initial data is so small that $4\kappa L(0+) \leq 1$. Then, by account of (14.5.10) and (14.5.8), $G(0+) \leq 2L(0+) \leq (2\kappa)^{-1}$. This together with (14.5.11) and a simple induction argument yields $\Delta G(t) \leq 0, t \in I$, i.e., $G(\cdot)$ is nonincreasing. Hence

$$(14.5.12) \qquad L(t) \leq G(t) \leq G(0+) \leq 2L(0+) , \quad t \in (0, T) \backslash I ,$$

which establishes the desired estimate (14.4.6).

Next we estimate the total amount of wave interaction. Since $\kappa L(t-) \leq \frac{1}{2}$, (14.5.9) yields

$$(14.5.13) \qquad \Delta Q(t) \leq -\frac{1}{2} |\alpha||\beta| , \quad t \in I .$$

By summing (14.5.13) over all $t \in I$, and upon using (14.5.8),

$$(14.5.14) \qquad \sum |\alpha||\beta| \leq L(0+)^2 ,$$

where the summation runs over the set of collisions in $(-\infty, \infty) \times (0, T)$.

Let us now consider any Lipschitz graph $\mathscr{C}$ in $(-\infty, \infty) \times [0, T)$, which is nonresonant relative to U_δ, as defined in Section 14.3. The aim is to estimate the total variation of the trace of U_δ on $\mathscr{C}$, measured by the sum $L_{\mathscr{C}} = \sum |\gamma|$ of the strengths of all waves that impinge on $\mathscr{C}$.

Let J stand for the set of $t \in (0, T)$ where some wave impinges on $\mathscr{C}$. For $t \in (0, T) \backslash (I \bigcup J)$ we set

$$(14.5.15) \qquad M(t) = \Sigma_- |\gamma| + \Sigma_+ |\gamma| + \Sigma_0 |\gamma| ,$$

where the summation Σ_- (or Σ_+) runs over the i-waves, with $i \in \mathcal{N}_-$ (or $\mathcal{N}_+$), which cross the t-time line on the positive (or negative) side of $\mathscr{C}$; while Σ_0 runs over all i-waves, with $i \in \mathcal{N}_0$, which cross the t-time line on either side of $\mathscr{C}$. Clearly,

$$(14.5.16) \qquad \Delta M(t) = -|\gamma| , \quad t \in J \backslash I ,$$

$$(14.5.17) \qquad \Delta M(t) \le \kappa |\alpha||\beta| , \quad t \in I \backslash J ,$$

$$(14.5.18) \qquad \Delta M(t) \le -|\gamma| + \kappa |\alpha||\beta| , \quad t \in I \cap J ,$$

where $|\alpha|$ and $|\beta|$ are the strengths of the waves colliding at $t \in I$ and $|\gamma|$ is the strength of the wave that impinges on $\mathscr{C}$ at $t \in J$. Summing the above inequalities over all $t \in I \bigcup J$ and using (14.5.14) together with $4\kappa L(0+) \le 1$, we conclude

$$(14.5.19) \qquad L_G \le M(0+) + \kappa \sum |\alpha||\beta| \le 2L(0+) .$$

Another important implication of the boundedness of the amount of wave inter-action is that the total number of collisions is finite and bounded, independently of T. Indeed, recall that the Approximate Riemann Solver is employed to resolve col-lisions only when the product of the strengths of the two incoming fronts exceeds σ. By virtue of (14.5.14), the number of such collisions is bounded by $L(0+)^2/\sigma$. Fronts are generated exclusively by the application of the Approximate Riemann Solver to resolve jump discontinuities of $U_{0\delta}$ or collisions of fronts. Therefore, the number of fronts is bounded. Any two fronts may collide at most once in their lifetime, so the number of collisions between fronts is also bounded. Since all pseudoshocks are generated by collisions of fronts, the number of pseudoshocks is likewise bounded. But then, even the number of collisions between fronts and pseudoshocks must be bounded. To summarize, the total number of collisions is finite, bounded solely in terms of δ, σ, and the number of jump points of $U_{0\delta}$. Consequently, the front tracking algorithm generates U_δ, in finite steps, on the entire upper half-plane. In particular, the estimates (14.4.6) and (14.4.7) will hold for $0 \le t < \infty$ and $0 \le \tau < t < \infty$, respectively.

14.6 Bounds on the Combined Strength of Pseudoshocks

The final task for verifying that U_δ is a δ-approximate solution of (14.2.1) is to establish requirement (14.4.4). The notion of generation order was introduced in Section 14.3. Waves of high generation order are produced after a large number of collisions and so it should be expected that their strength is small. Indeed, the first step in our argument is to show that the combined strength of all waves, and thus in particular of all pseudoshocks, of sufficiently high generation order is arbitrarily small. To that end, the analysis of Section 14.5 shall be refined by sorting out and monitoring separately the waves according to their generation order.

We know by now that the total number of collisions is bounded, and hence the generation order of all waves lies in a finite range, $0 \le \mu \le \nu$. Note, however, that the magnitude of v depends penultimately on δ, and should be expected to grow without bounds as $\delta \downarrow 0$. For $\mu = 0, 1, \cdots, v$ and $t \in [0, \infty)\backslash I$, we let $L_\mu(t)$ denote the sum of the strengths of all waves with generation order $\geq \mu$ which cross the t-time line; and $Q_\mu(t)$ stand for the sum of the products of the strengths of all couples of approaching waves that cross the t-time line and have generation order μ_1, μ_2 with $\max\{\mu_1, \mu_2\} \geq \mu$. Thus, in particular, $L_0(t) = L(t)$ and $Q_0(t) = Q(t)$. Finally, we identify the set I_μ of times $t \in I$ in which a wave of generation order μ collides with a wave of generation order $\leq \mu$.

Collisions between waves of generation order $\leq \mu - 2$ cannot affect waves of generation order $\geq \mu$, and so

$$(14.6.1) \qquad \Delta L_\mu(t) = 0 , \quad t \in I_0 \cup \cdots \cup I_{\mu-2} .$$

Any change in $L_\mu(\cdot)$ at $t \in I$ must be induced by the collision of two waves, of which at least one is of generation order $\geq \mu - 1$. These colliding waves, with strengths say $|\alpha|$ and $|\beta|$, are contributing $|\alpha||\beta|$ to $Q_{\mu-1}(t-)$ but nothing to $Q_{\mu-1}(t+)$. As in Section 14.5, the resulting drop in $Q_{\mu-1}(\cdot)$ can be used to offset the potential increment of $L_\mu(\cdot)$, which is bounded by $\kappa|\alpha||\beta|$:

$$(14.6.2) \qquad \Delta L_\mu(t) + 2\kappa \Delta Q_{\mu-1}(t) \leq 0 , \quad t \in I_{\mu-1} \cup \cdots \cup I_v .$$

By similar arguments one verifies the inequalities

$$(14.6.3) \qquad \Delta Q_\mu(t) + 2\kappa \Delta Q(t) L_\mu(t-) \leq 0 , \quad t \in I_0 \cup \cdots \cup I_{\mu-2} ,$$

$$(14.6.4) \qquad \Delta Q_\mu(t) + 2\kappa \Delta Q_{\mu-1}(t) L(t-) \leq 0 , \quad t \in I_{\mu-1} ,$$

$$(14.6.5) \qquad \Delta Q_\mu(t) \leq 0 , \quad t \in I_\mu \cup \cdots \cup I_v ,$$

which govern the change of $Q_\mu(\cdot)$ across collisions of various orders.

A superscript $+$ or $-$ will be employed below to indicate "positive" or "negative" part: $w^+ = \max\{w, 0\}$, $w^- = \max\{-w, 0\}$. The aim is to monitor the quantities

$$(14.6.6) \qquad \hat{L}_\mu = \sup_t L_\mu(t) , \quad \hat{Q}_\mu = \sum_{t \in I}[\Delta Q_\mu(t)]^+ ,$$

for $\mu = 1, \cdots, v$, and show

$$(14.6.7) \qquad \hat{L}_\mu \leq 2^{-\mu} c_3 a , \quad \hat{Q}_\mu \leq 2^{-\mu+3} c_3^2 a^2 ,$$

where a is the bound on $TV_{(-\infty,\infty)}U_0(\cdot)$.

From (14.6.1), (14.6.2) and the "initial condition" $L_\mu(0+) = 0, \mu = 1, \cdots, v$, follows

$$(14.6.8) \qquad \hat{L}_\mu \leq 2\kappa \sum_{t \in I}[\Delta Q_{\mu-1}(t)]^- , \quad \mu = 1, \cdots, v .$$

Next we focus on (14.6.3), (14.6.4) and (14.6.5), with "initial condition" $Q_\mu(0+) = 0$. Recalling (14.5.8), (14.5.12) and using

(14.6.9) $$\sum_{t \in I} [\Delta Q(t)]^- = Q(0+) - Q(\infty) \leq \frac{1}{2} L(0+)^2 ,$$

we deduce

(14.6.10) $\hat{Q}_\mu \leq \kappa L(0+)^2 \hat{L}_\mu + 4\kappa L(0+) \sum_{t \in I} [\Delta Q_{\mu-1}(t)]^- , \quad \mu = 1, \cdots, \nu .$

We combine (14.6.8) with (14.6.10). Assuming the total variation of the initial data is so small that $10 \kappa L(0+) \leq 1$, we deduce

(14.6.11) $$\hat{Q}_\mu \leq \frac{1}{2} \sum_{t \in I} [\Delta Q_{\mu-1}(t)]^- , \quad \mu = 1, \cdots, \nu .$$

In particular, for $\mu = 1$ and by account of (14.6.9), $\hat{Q}_1 \leq \frac{1}{4} L(0+)^2$.

We finally notice that, for $\mu = 1, \cdots, \nu$, since $Q_\mu(0+) = 0$,

(14.6.12) $$\sum_{t \in I} [\Delta Q_\mu(t)]^- = \sum_{t \in I} [\Delta Q_\mu(t)]^+ - Q_\mu(\infty) \leq \hat{Q}_\mu .$$

Therefore, (14.6.11) yields $\hat{Q}_\mu \leq \frac{1}{2} \hat{Q}_{\mu-1}, \mu = 2, \cdots, \nu$, which in turn implies $\hat{Q}_\mu \leq 2^{-\mu-1} L(0+)^2$. This together with (14.6.9) and (14.6.10) gives $\hat{L}_\mu \leq 2^{-\mu-2} L(0+)$. We have thus established (14.6.7).

It is now clear that one can fix μ_0 sufficiently large that the combined strength of all waves of generation order $\geq \mu_0$, which is majorized by $\hat{L}_{\mu_0}$, does not exceed $\frac{1}{2} \delta$.

In order to estimate the combined strength of pseudoshocks of generation order $< \mu_0$, the first step is to estimate their number. For $\mu = 0, \cdots, \nu$, let K_μ denote the number of waves of generation order $\leq \mu$. A crude upper bound for K_μ may be derived by the following argument. The number of outgoing waves produced by resolving a jump discontinuity, via any of the two Riemann solvers, is bounded by a number b/δ. Thus, $K_0 \leq \frac{b}{\delta} N$, where N is the number of jump points of $U_{0\delta}$. Since any two waves may collide at most once in their lifetime, the number of collisions that may generate waves of generation order μ is bounded by $\frac{1}{2} K_{\mu-1}^2$. Therefore,

(14.6.13) $$K_\mu \leq K_{\mu-1} + \frac{b}{2\delta} K_{\mu-1}^2 \leq \frac{b}{\delta} K_{\mu-1}^2 ,$$

whence one readily deduces

(14.6.14) $$K_\mu \leq \left(\frac{b}{\delta} \right)^{2^{\mu+1}} N^{2^\mu} .$$

Next we estimate the strength of individual pseudoshocks. Any pseudoshock is generated by the collision of two fronts, with strengths $|\alpha|$ and $|\beta|$ such that $|\alpha||\beta| \leq \sigma$, which is thus resolved via the Simplified Riemann Solver, as depicted in Figs. 14.2.2 and 14.2.3. It then follows from the corresponding interaction estimate (14.5.4) that the strength of any pseudoshock at birth does not exceed $c_4 \sigma$. By account of (14.5.5), the collision of a pseudoshock with a front of strength $|\beta|$, as depicted in Fig. 14.2.4, may increase its strength at most by a factor $1+\kappa|\beta|$. Consequently, the strength of a pseudoshock may ultimately grow at most by the factor $\Pi(1+\kappa|\gamma|)$, where the product runs over all fronts with which the pseudoshock collides during its life span. Since pseudoshocks are nonresonant, the estimate (14.5.19) here applies and implies $\sum|\gamma| \leq 2L(0+)$. Assuming $2\kappa L(0+) \leq 1$, we thus conclude that the strength of each pseudoshock, at any time, does not exceed $3c_4\sigma$.

It is now clear that by employing the upper bound for K_{μ_0-1} provided by (14.6.14), and upon selecting σ sufficiently small, one guarantees that the combined strength of pseudoshocks of generation order $< \mu_0$ is bounded by $\frac{1}{2}\delta$. In conjunction with our earlier estimate on the total strength of pseudoshocks of generation order $\geq \mu_0$, this establishes (14.4.4).

14.7 Compactness and Consistency

In this section, the proof of Theorem 14.4.1 will be completed by passing to the $\delta \downarrow 0$ limit. Here we will just be assuming that $\{U_\delta\}$ is any family of δ-approximate solutions, in the sense of Definition 14.4.1, with δ positive and small, which satisfy estimates (14.4.6) and (14.4.7). Thus, we shall not require the special features of the particular δ-approximate solutions constructed via the front tracking algorithm, for instance that shocks propagate with the correct shock speed.

Let us fix any test function ϕ, with compact support in $(-\infty, \infty) \times [0, T)$. By applying Green's theorem,

$$
\begin{aligned}
(14.7.1) \quad & \int_0^\infty \int_{-\infty}^\infty [\partial_t\phi U_\delta + \partial_x\phi F(U_\delta)]dxdt + \int_{-\infty}^\infty \phi(x,0)U_\delta(x,0)dx \\
& = -\int_0^\infty \sum \phi(y(t),t)\{F(U_\delta(y(t)+,t)) - F(U_\delta(y(t)-,t)) \\
& \qquad\qquad - \dot{y}(t)[U_\delta(y(t)+,t) - U_\delta(y(t)-,t)]\}dt ,
\end{aligned}
$$

where for each t the summation runs over all jump discontinuities $x = y(\cdot)$ which cross the t-time line.

When the jump discontinuity $x = y(\cdot)$ is an (approximate) shock, then by virtue of (14.4.1),

(14.7.2)
$$
\begin{aligned}
& |F(U_\delta(y(t)+,t)) - F(U_\delta(y(t)-,t)) - \dot{y}(t)[U_\delta(y(t)+,t) - U_\delta|y(t)-,t)]| \\
& \leq \delta|U_\delta(y(t)+,t) - U_\delta(y(t)-,t)| .
\end{aligned}
$$

Similarly, when $x = y(\cdot)$ is an (approximate) rarefaction front, with strength $\leq \delta$, then by account of the proximity between shock and rarefaction wave curves, and (14.4.2),

(14.7.3)

$$|F(U_\delta(y(t)+,t)) - F(U_\delta(y(t)-,t)) - \dot{y}(t)[U_\delta(y(t)+,t) - U_\delta(y(t)-,t)]|$$
$$\le c_5\delta|U_\delta(y(t)+,t) - U_\delta(y(t)-,t)| \ .$$

Finally, when $x = y(\cdot)$ is a pseudoshock,

(14.7.4)
$$|F(U_\delta(y(t)+,t)) - F(U_\delta(y(t)-,t))|$$
$$\le c_6|U_\delta(y(t)+,t) - U_\delta(y(t)-,t)| \ .$$

By combining (14.7.2), (14.7.3), (14.7.4) with (14.4.6) and (14.4.4), we deduce that, for any fixed test function ϕ, the right-hand side of (14.7.1) is bounded by $C_\phi[TV_{(-\infty,\infty)}U_0(\cdot) + 1]\delta$ and thus tends to zero as $\delta \downarrow 0$.

By virtue of (14.4.6), (14.4.7) and Theorem 1.7.1, any sequence of δ's converging to zero contains a subsequence $\{\delta_k\}$ such that $\{U_{\delta_k}\}$ converges a.e. to some $U \in BV_{\text{loc}}$. Passing to the limit in (14.7.1) along the sequence $\{\delta_k\}$, and using (14.4.5), we conclude that U is indeed a weak solution of (13.1.1).

By passing to the $\delta \downarrow 0$ limit in (14.4.6) and (14.4.7), one verifies that U satisfies (13.1.5) and (13.1.6). Furthermore, if $\mathscr{C}$ is any Lipschitz graph which is nonresonant relative to U_δ, for all δ, then, as shown in Section 14.5, the trace of U_δ on $\mathscr{C}$ has bounded variation, uniformly in δ, and thus, passing to the $\delta \downarrow 0$ limit, yields that the trace of U on $\mathscr{C}$ will have the same property.

To conclude the proof, assume (η, q) is an entropy-entropy flux pair for the system (14.2.1), with $\eta(U)$ convex. Let ϕ be any nonnegative test function, with compact support in $(-\infty, \infty) \times [0, T)$. By Green's theorem,

(14.7.5)
$$\int_0^\infty \int_{-\infty}^\infty [\partial_t\phi\eta(U_\delta) + \partial_x\phi q(U_\delta)]dxdt + \int_{-\infty}^\infty \phi(x,0)\eta(U_\delta(x,0))dx$$
$$= -\int_0^\infty \sum \phi(y(t),t)\{q(U_\delta(y(t)+,t)) - q(U_\delta(y(t)-,t))$$
$$- \dot{y}(t)[\eta(U_\delta(y(t)+,t)) - \eta(U_\delta(y(t)-,t))]\}dt \ ,$$

where, as in (14.7.1), for each t the summation runs over all jump discontinuities $x = y(\cdot)$ which cross the t-time line.

When $x = y(\cdot)$ is an (approximate) shock, the entropy inequality (8.5.1) together with (14.4.1) imply

(14.7.6)

$$q(U_\delta(y(t)+,t)) - q(U_\delta(y(t)-,t)) - \dot{y}(t)[\eta(U_\delta(y(t)+,t)) - \eta(U_\delta(y(t)-,t))]$$
$$\le c_7\delta|U_\delta(y(t)+,t) - U_\delta(y(t)-,t)| \ .$$

When $x = y(\cdot)$ is an (approximate) rarefaction front, with strength $\le \delta$, Theorem 8.5.1 together with (14.4.2) yield

(14.7.7)

$$|q(U_\delta(y(t)+,t)) - q(U_\delta(y(t)-,t)) - \dot{y}(t)[\eta(U_\delta(y(t)+,t)) - \eta(U_\delta(y(t)-,t))]|$$
$$\le c_8\delta|U_\delta(y(t)+,t) - U_\delta(y(t)-,t)| \ .$$

Finally, when $x = y(\cdot)$ is a pseudoshock,

(14.7.8)

$$|q(U_\delta(y(t)+,t)) - q(U_\delta(y(t)-,t)) - \dot{y}(t)[\eta(U_\delta(y(t)+,t) - \eta(U_\delta(y(t)-,t))]|$$
$$\leq c_9|U_\delta(y(t)+,t) - U_\delta(y(t)-,t)| .$$

By combining (14.7.6), (14.7.7), (14.7.8) with (14.4.6) and (14.4.4), we deduce that, for fixed test function ϕ, the right-hand side of (14.7.5) is bounded from below by $-C_\phi[TV_{(-\infty,\infty)}U_0(\cdot) + 1]\delta$. Therefore, passing to the limit along the $\{\delta_k\}$ sequence, we conclude that the solution U satisfies the inequality (13.2.17), which expresses the entropy admissibility condition. The proof of Theorem 14.4.1 is now complete.

14.8 Continuous Dependence on Initial Data

The remainder of this chapter will address the issue of uniqueness and stability of solutions to the initial-value problem (13.1.1). The existence proofs via Theorems 13.1.1 and 14.4.1, which rely on compactness arguments, offer no clue on that question. We will approach the subject via the approximate solutions generated by the front tracking algorithm. By monitoring the time evolution of a certain functional, we will demonstrate that δ-approximate solutions depend continuously on their initial data, modulo corrections of order δ. This will induce stability for solutions obtained by passing to the $\delta \downarrow 0$ limit.

Our earlier experiences with the scalar conservation law strongly suggest that the L^1 topology should provide the proper setting for continuous dependence. However, the L^1 distance shall not be measured via the standard L^1 metric but through a functional ρ, specially designed for the task at hand.

Let us consider two δ-approximate solutions U and $\overline{U}$ of (14.2.1). Fixing any point (x, t) of continuity for both U and $\overline{U}$, we shall measure the distance between the vectors $U(x, t)$ and $\overline{U}(x, t)$ in the special curvilinear coordinate system whose coordinate curves are the shock curves, with both the admissible and the nonadmissible branches retained. To that end, the vector $\overline{U}(x, t) - U(x, t)$ is represented by curvilinear "coordinates" $p_1(x, t), \cdots, p_n(x, t)$, obtained by means of the following process: One envisages a "virtual" jump discontinuity with left state $U(x, t)$ and right state $\overline{U}(x, t)$, and resolves it into a wave fan composed of $n + 1$ constant states joined exclusively by (admissible or nonadmissible) virtual shocks. For $|U(x, t) - \overline{U}(x, t)|$ sufficiently small, this resolution is unique and can be effected, via the implicit function theorem, by retracing the steps of the admissible solution to the Riemann problem, in Section 9.3, with the wave fan curves Φ_i here replaced by the shock curves W_i. We denote the amplitude of the resulting virtual i-shock by $p_i(x, t)$ and its speed by $s_i(x, t)$. The distance between $U(x, t)$ and $\overline{U}(x, t)$ will now be measured by the suitably weighted sum $\sum g_i(x, t)|p_i(x, t)|$ of the strengths of the n virtual shocks, and accordingly the distance between the two approximate solutions at time t will be measured through the functional

$$(14.8.1) \qquad \rho(U(\cdot,t),\overline{U}(\cdot,t)) = \sum_{i=1}^{n}\int_{-\infty}^{\infty} g_i(x,t)|p_i(x,t)|dx \ .$$

We proceed to introduce suitable weights g_i. Let I and $\overline{I}$ denote the sets of collision times for U and $\overline{U}$, and consider the corresponding potentials for wave interaction $Q(t)$ and $\overline{Q}(t)$, defined through (14.5.7), for $t \in (0,\infty)\backslash I$ and $t \in (0,\infty)\backslash\overline{I}$, respectively. For $t \in (0,\infty)\backslash(I\bigcup\overline{I})$ and any point of continuity x of both $U(\cdot,t)$ and $\overline{U}(\cdot,t)$, we define

$$(14.8.2) \qquad g_i(x,t) = 1 + \kappa[Q(t)+\overline{Q}(t)] + \nu A_i(x,t) \ ,$$

where κ and ν are sufficiently large positive constants, to be fixed later, and

$$(14.8.3) \qquad A_i(x,t) = \Sigma_-|\gamma| + \overline{\Sigma}_-|\gamma| + \Sigma_+|\gamma| + \overline{\Sigma}_+|\gamma| + \Sigma_0|\gamma| + \overline{\Sigma}_0|\gamma| \ .$$

In (14.8.3), Σ_- (or $\overline{\Sigma}_-$) sums the strengths of all j-fronts of U (or $\overline{U}$), with $j = i+1,\cdots,n$, which cross the t-time line to the left of the point x; Σ_+ (or $\overline{\Sigma}_+$) sums the strengths of all j-fronts of U (or $\overline{U}$), with $j = 1,\cdots,i-1$, which cross the t-time line to the right of the point x; Σ_0 (or $\overline{\Sigma}_0$) sums the strengths of all i-fronts of U (or $\overline{U}$) which cross the t-time line to the left (or right) of the point x, when $p_i(x,t) < 0$, or to the right (or left) of the point x, when $p_i(x,t) > 0$. Thus, one may justifiably say that $A_i(x,t)$ represents the total strength of the fronts of U and $\overline{U}$ that cross the t-time line and approach the virtual i-shock at (x,t).

Once κ and ν have been fixed, the total variation of the initial data shall be restricted to be so small that $\frac{1}{2} \le g_i(x,t) \le 2$. Then, $\rho(U(\cdot,t),\overline{U}(\cdot,t))$ will be equivalent to the L^1 distance of $U(\cdot,t)$ and $\overline{U}(\cdot,t)$:

$$(14.8.4) \qquad \begin{aligned} \frac{1}{C}\|U(\cdot,t)-\overline{U}(\cdot,t)\|_{L^1(-\infty,\infty)} &\le \rho(U(\cdot,t),\overline{U}(\cdot,t)) \\ &\le C\|U(\cdot,t)-\overline{U}(\cdot,t)\|_{L^1(-\infty,\infty)} \ . \end{aligned}$$

It is easily seen that in the scalar case, $n = 1$, the functional ρ introduced by (14.8.1) is closely related to the functional ρ, defined by (11.8.11), when the latter is restricted to step functions.

The aim is to show that $\rho(U(\cdot,t),\overline{U}(\cdot,t))$ is nonincreasing, modulo corrections of order δ:

$$(14.8.5)$$
$$\rho(U(\cdot,t),\overline{U}(\cdot,t)) - \rho(U(\cdot,\tau),\overline{U}(\cdot,\tau)) \le \omega\delta(t-\tau) \ , \qquad 0 < \tau < t < \infty \ .$$

Notice that across points of I or $\overline{I}$, $Q(t)$ or $\overline{Q}(t)$ decreases by an amount approximately equal to the product of the strengths of the two colliding waves, while $A_i(x,t)$ may increase at most by a quantity of the same order of magnitude. Therefore, upon fixing κ/ν sufficiently large, $\rho(U(\cdot,t),\overline{U}(\cdot,t))$ will be decreasing across points of I or $\overline{I}$. Between consecutive points of $I\bigcup\overline{I}$, $\rho(U(\cdot,t),\overline{U}(\cdot,t))$ is continuously differentiable, hence to establish (14.8.5) it will suffice to show

$$(14.8.6) \qquad \frac{d}{dt}\rho(U(\cdot,t),\overline{U}(\cdot,t)) \le \omega\delta \ .$$

From (14.8.1),

$$(14.8.7) \qquad \frac{d}{dt}\rho(U(\cdot,t),\overline{U}(\cdot,t)) = \sum_y \sum_{i=1}^n \{g_i^-|p_i^-| - g_i^+|p_i^+|\}\dot{y} ,$$

where $\sum_y$ runs over all waves $x = y(\cdot)$ of U and $\overline{U}$ which cross the t-time line, and $g_i^\pm$, $p_i^\pm$ and $\dot{y}$ stand for $g_i(y(t)\pm, t)$, $p_i(y(t)\pm, t)$ and $\dot{y}(t)$. By adding and subtracting, appropriately, the speed $s_i^\pm = s_i(y(t)\pm, t)$ of the virtual i-shocks, one may recast (14.8.7) in the form

$$(14.8.8) \qquad \frac{d}{dt}\rho(U(\cdot,t),\overline{U}(\cdot,t)) = \sum_y \sum_{i=1}^n E_i(y(\cdot),t) ,$$

where

$$(14.8.9) \qquad \begin{aligned} E_i(y(\cdot),t) &= g_i^+(s_i^+ - \dot{y})|p_i^+| - g_i^-(s_i^- - \dot{y})|p_i^-| \\ &= (g_i^+ - g_i^-)(s_i^+ - \dot{y})|p_i^-| + g_i^-(s_i^+ - s_i^-)|p_i^-| \\ &\quad + g_i^+(s_i^+ - \dot{y})(|p_i^+| - |p_i^-|) . \end{aligned}$$

Suppose first $x = y(\cdot)$ is a pseudoshock, say of U. Then $g_i^+ = g_i^-$ and (14.8.9) yields

$$(14.8.10) \qquad \sum_{i=1}^n E_i(y(\cdot),t) \leq c_{10}|U(y(t)+,t) - U(y(t)-,t)| .$$

Thus, by virtue of (14.4.4), the portion of the sum on the right-hand side of (14.8.8) that runs over all pseudoshocks of U is bounded by $c_{10}\delta$. Of course, this equally applies to the portion of the sum that runs over all pseudoshocks of $\overline{U}$.

We now turn to the case $x = y(\cdot)$ is a j-front of U or $\overline{U}$, with amplitude say γ. To complete the proof of (14.8.6), one has to show that

$$(14.8.11) \qquad \sum_{i=1}^n E_i(y(\cdot),t) \leq c_{11}\delta|\gamma| .$$

What follows, is a road map to the proof of (14.8.11), which will expose the main ideas and will explain, in particular, why the weight function $g_i(x,t)$ was designed according to (14.8.2). The detailed proof, which is quite laborious, is found in the references cited in Section 14.12.

Let us first examine the three terms on the right-hand side of (14.8.9) for $i \neq j$. By virtue of (14.8.2), $g_i^+ - g_i^-$ equals $\nu|\gamma|$ when $j > i$, or $-\nu|\gamma|$ when $j < i$. In either case, the first term

$$(14.8.12) \qquad (g_i^+ - g_i^-)(s_i^+ - \dot{y})|p_i^-| \cong -\nu|\lambda_i - \lambda_j||p_i^-||\gamma|$$

is strongly negative. The idea is that this term dominates the other two terms, rendering the desired inequality (14.8.6). Indeed, the second term is majorized by $c_{12}|p_i^-||\gamma|$, which is clearly dominated by (14.8.12), when ν is sufficiently large.

One estimates the remaining term by the following argument. The amplitudes $(p_1^-, \cdots, p_n^-)$ or $(p_1^+, \cdots, p_n^+)$ of the virtual shocks result respectively from the resolution of the jump discontinuity between U^- and $\overline{U}^-$ or U^+ and $\overline{U}^+$, where $U^\pm = U(y(t)\pm, t)$ and $\overline{U}^\pm = \overline{U}(y(t)\pm, t)$.

Assuming, for definiteness, that $x = y(\cdot)$ is a front of U, we have $\overline{U}^- = \overline{U}^+$, while the states U^- and U^+ are connected, in state space, by a j-wave curve. Consequently, to leading order, $p_j^+ \cong p_j^- - \gamma$ while, for any $k \neq j$, $p_k^+ \cong p_k^-$. Indeed, a study of the wave curves easily yields the estimate

(14.8.13)

$$|p_j^+ - p_j^- + \gamma| + \sum_{k \neq j} |p_k^+ - p_k^-| = O(1)[\delta + |p_j^-|(|p_j^-| + |\gamma|) + \sum_{k \neq j} |p_k^-|]|\gamma| \,,$$

which in turn implies

(14.8.14)
$$E_i(y(\cdot), t) \leq -av|p_i^-||\gamma|$$
$$+ c_{12}\left[\delta + |p_j^-|(|p_j^-| + |\gamma|) + \sum_{k \neq j} |p_k^-|\right]|\gamma| \,,$$

with $a > 0$.

For $i = j$, the estimation of $E_i(y(\cdot), t)$ is more delicate, as the j-front may resonate with the virtual i-shock. The same difficulty naturally arises, and has to be addressed, even for the scalar conservation law. In fact, the scalar case was already treated, in Section 11.8, albeit in a different guise. For the system, one has to examine separately a number of cases, depending on whether $x = y(\cdot)$ is a shock or a rarefaction front, in conjunction with the signs of p_j^- and p_j^+. The resulting estimates, which slightly vary from case to case but are essentially equivalent, are derived in the references. For example, when either $x = y(\cdot)$ is a j-rarefaction front and $0 < p_j^- < p_j^+$ or $x = y(\cdot)$ is a j-shock and $p_j^+ < p_j^- < 0$,

(14.8.15)
$$E_j(y(\cdot), t) \leq -bv|p_j^-||\gamma|(|p_j^-| + |\gamma|)$$
$$+ c_{13}\left[\delta + |p_j^-|(|p_j^-| + |\gamma|) + \sum_{k \neq j} |p_k^-|\right]|\gamma| \,,$$

where $b > 0$.

We now sum the inequalities (14.8.14), for $i \neq j$, together with the inequality (14.8.15). Upon selecting v sufficiently large to offset the possibly positive terms, we arrive at (14.8.11). As noted earlier, this implies (14.8.6), which in turn yields (14.8.5). Recalling (14.8.4), we conclude

(14.8.16) $\|U(\cdot, t) - \overline{U}(\cdot, t)\|_{L^1(-\infty, \infty)} \leq C^2 \|U(\cdot, 0) - \overline{U}(\cdot, 0)\|_{L^1(-\infty, \infty)} + C\omega\delta t \,,$

which establishes that δ-approximate solutions depend continuously on their initial data, modulo δ. The implications on actual solutions, obtained as $\delta \downarrow 0$, will be discussed in the following section.

14.9 The Standard Riemann Semigroup

As a corollary of the stability properties of approximate solutions, established in the previous section, it will be shown here that any solution to our system constructed as the $\delta \downarrow 0$ limit of some sequence of δ-approximate solutions is uniquely determined by its initial data and may be identified with a trajectory of a L^1-Lipschitz semigroup, defined on a closed subset of $L^1(-\infty, \infty)$.

The first step in our investigation is to locate the domain of the semigroup. This must be a set which is positively invariant for solutions. Motivated by the analysis in Section 14.5, with any step function $V(\cdot)$, of compact support and small total variation over $(-\infty, \infty)$, we associate a number $H(V(\cdot))$ determined by the following procedure. The jump discontinuities of $V(\cdot)$ are resolved into fans of admissible shocks and rarefaction waves, by solving classical Riemann problems. Before any wave collisions may occur, one measures the total strength L and the potential for wave interaction Q of these outgoing waves and then sets $H(V(\cdot)) = L + 2\kappa Q$, where κ is a sufficiently large positive constant. Suppose a δ-approximate solution U, with initial data V, is constructed by the front tracking algorithm of Section 14.2. By the rules of the construction, all jump discontinuities of V will be resolved via the Approximate Riemann Solver and so, for any $\delta > 0, H(V(\cdot))$ will coincide with the initial value $G(0+)$ of the Glimm-type function $G(t)$ defined through (14.5.10). At a later time, as the Simplified Riemann Solver comes into play, $G(t)$ and $H(U(\cdot, t))$ may part from each other. In particular, by contrast to $G(t)$, $H(U(\cdot, t))$ will not necessarily be nonincreasing with t. Nevertheless, when κ is sufficiently large, $H(U(\cdot, t)) \le H(U(\cdot, t-))$ and $H(U(\cdot, t+)) \le H(U(\cdot, t-))$. Hence $H(U(\cdot, t)) \le H(V(\cdot))$ for any $t \ge 0$ and so sets of step functions $\{V(\cdot) : H(V(\cdot)) < r\}$ are positively invariant for δ-approximate solutions constructed by the front tracking algorithm. Following this preparation, we define the set that will serve as the domain of the semigroup by

(14.9.1) $\mathcal{D} = c\ell\{$step functions $V(\cdot)$ with compact support $: H(V(\cdot)) < r\}$,

where $c\ell$ denotes closure in $L^1(-\infty, \infty)$. By virtue of Theorem 1.7.2, the members of $\mathcal{D}$ are functions of bounded variation over $(-\infty, \infty)$, with total variation bounded by cr. The main result is

Theorem 14.9.1 *For r sufficiently small, there is a family of maps $S_t : \mathcal{D} \mapsto \mathcal{D}$, $t \in [0, \infty)$, with the following properties.*

(a) L^1-*Lipschitz continuity on* $\mathcal{D} \times [0, \infty)$: *For any* $V, \overline{V} \in \mathcal{D}$ *and* $t, \tau \in [0, \infty)$,

(14.9.2) $\|S_t \circ V(\cdot) - S_\tau \circ \overline{V}(\cdot)\|_{L^1(-\infty,\infty)} \le \kappa\{\|V(\cdot) - \overline{V}(\cdot)\|_{L^1(-\infty,\infty)} + |t - \tau|\}$.

(b) $\{S_t : t \in [0, \infty)\}$ *has the semigroup property, namely*

(14.9.3) $S_0 = $ identity ,

(14.9.4) $S_{t+\tau} = S_t S_\tau$, $t, \tau \in [0, \infty)$.

(c) *If U is any solution of* (13.1.1), *with initial data $U_0 \in \mathscr{D}$, which is the $\delta \downarrow 0$ limit of some sequence of δ-approximate solutions, then*

$$(14.9.5) \qquad U(\cdot, t) = S_t \circ U_0(\cdot) , \quad t \in [0, \infty) .$$

Proof. Let U and $\overline{U}$ be two solutions of (13.1.1), with initial data U_0 and $\overline{U}_0$, which are $\delta \downarrow 0$ limits of sequences of δ-approximate solutions $\{U_{\delta_n}\}$ and $\{\overline{U}_{\overline{\delta}_n}\}$, respectively. No assumption is made that these approximate solutions have necessarily been constructed by the front tracking algorithm. So long as the total variation is sufficiently small to meet the requirements of Section 14.8, we may apply (14.8.16) to get

$$(14.9.6) \qquad \begin{aligned} \|U_{\delta_n}(\cdot, t) - \overline{U}_{\overline{\delta}_n}(\cdot, t)\|_{L^1(-\infty,\infty)} &\leq C^2 \|U_{\delta_n}(\cdot, 0) - \overline{U}_{\overline{\delta}_n}(\cdot, 0)\|_{L^1(-\infty,\infty)} \\ &\quad + C\omega \max\{\delta_n, \overline{\delta}_n\} t . \end{aligned}$$

Passing to the limit, $n \to \infty$, we deduce

$$(14.9.7) \qquad \|U(\cdot, t) - \overline{U}(\cdot, t)\|_{L^1(-\infty,\infty)} \leq C^2 \|U_0(\cdot) - \overline{U}_0(\cdot)\|_{L^1(-\infty,\infty)} .$$

When r is sufficiently small, Theorem 14.4.1 asserts that for any $U_0 \in \mathscr{D}$ one can generate solutions U of (13.1.1) as limits of sequences $\{U_{\delta_n}\}$ of δ-approximate solutions constructed by the front tracking algorithm. Moreover, the initial values of U_δ may be selected so that $H(U_\delta(\cdot, 0)) < r$, in which case, as noted above, $H(U_\delta(\cdot, t)) < r$ and thereby $U(\cdot, t) \in \mathscr{D}$, for any $t \in [0, \infty)$. By virtue of (14.9.7), all these solutions must coincide so that U is uniquely defined. In fact, (14.9.7) further implies that U must even coincide with any solution, with initial data U_0, which is derived as the $\delta \downarrow 0$ limit of any sequence of δ-approximate solutions, regardless of whether they were constructed by the front tracking algorithm.

Once U has thus been identified, we define S_t through (14.9.5). The Lipschitz continuity property (14.9.2) follows by combining (14.9.7) with (13.1.6), and (14.9.3) is obvious. To verify (14.9.4), it suffices to notice that for any fixed $\tau > 0$, $U(\cdot, \tau + \cdot)$ is a solution of (13.1.1), with initial data $U(\cdot, \tau)$, which is derived as the $\delta \downarrow 0$ limit of δ-approximate solutions and thus, by uniqueness, must coincide with $S_t \circ U(\cdot, \tau)$. The proof is complete.

The term Standard Riemann Semigroup is commonly used for S_t, as a reminder that its building block is the solution of the Riemann problem. The question whether this semigroup also encompasses solutions derived via alternative methods will be addressed in the next section.

14.10 Uniqueness of Solutions

Uniqueness for the initial-value problem (13.1.1) shall be established here by demonstrating that any solution in a reasonable function class can be identified with the trajectory of the Standard Riemann Semigroup which emanates from the initial data. As shown in Section 14.9, this is indeed the case for solutions constructed by front tracking.

For fair comparison one should limit, at the outset, the investigation to solutions U for which $U(\cdot, t)$ resides in the domain $\mathscr{D}$ of the Standard Riemann Semigroup, defined through (14.9.1). As noted earlier, this implies, in particular, that $U(\cdot, t)$ has bounded variation over $(-\infty, \infty)$:

$$(14.10.1) \qquad\qquad TV_{(-\infty,\infty)}U(\cdot, t) \le cr .$$

It then follows from Theorem 4.1.2 that $t \mapsto U(\cdot, t)$ is L^1-Lipschitz,

$$(14.10.2) \qquad \int_{-\infty}^{\infty} |U(x, t) - U(x, \tau)|dx \le c'r|t - \tau| , \qquad 0 \le \tau < t < \infty ,$$

and U is in BV_{loc} on $(-\infty, \infty) \times [0, \infty)$. Hence, as pointed out in Section 10.1, there is $\mathscr{N} \subset [0, \infty)$, of measure zero, such that any (x, t) with $t \notin \mathscr{N}$ and $U(x-, t) = U(x+, t)$ is a point of approximate continuity of U while any (x, t) with $t \notin \mathscr{N}$ and $U(x-, t) \ne U(x+, t)$ is a point of approximate jump discontinuity of U, with one-sided approximate limits $U_{\pm} = U(x\pm, t)$ and associated shock speed determined through the Rankine-Hugoniot jump condition (8.1.2).

It is currently unknown whether uniqueness prevails within the above class of solutions. Accordingly, one should endow solutions with additional structure. Here we will experiment with the

Tame Oscillation Condition: there are positive constants λ and β such that

$$(14.10.3) \qquad |U(x\pm, t + h) - U(x\pm, t)| \le \beta TV_{[x-\lambda h, x+\lambda h]}U(\cdot, t) ,$$

for all $x \in (-\infty, \infty), t \in [0, \infty)$ and any $h > 0$.

Clearly, solutions constructed by either the random choice method or the front tracking algorithm satisfy this condition, and so do also the solutions to systems of two conservation laws considered in Chapter XII.

The Tame Oscillation Condition induces uniqueness:

Theorem 14.10.1 *Any solution* U *of the initial-value problem* (13.1.1), *with* $U(\cdot, t) \in \mathscr{D}$, *for all* $t \in [0, \infty)$, *which satisfies the Tame Oscillation Condition* (14.10.3), *coincides with the trajectory of the Standard Riemann Semigroup* S_t, *emanating from the initial data:*

$$(14.10.4) \qquad U(\cdot, t) = S_t \circ U_0(\cdot) , \qquad t \in [0, \infty) .$$

In particular, U *is uniquely determined by its initial data.*

Proof. The demonstration will be quite lengthy. The first step is to show that at every $\tau \notin \mathscr{N}, U(\cdot, t)$ is tangential to the trajectory of S_t emanating from $U(\cdot, \tau)$:

$$(14.10.5) \qquad \limsup_{h\downarrow 0} \frac{1}{h} \|U(\cdot, \tau + h) - S_h \circ U(\cdot, \tau)\|_{L^1(-\infty,\infty)} = 0 .$$

Then we shall verify that (14.10.5), in turn, implies (14.10.4).

Fixing $\tau \notin \mathcal{N}$, we will establish (14.10.5) by the following procedure. For any fixed bounded interval $[a, b]$ and $\varepsilon > 0$, arbitrarily small, we will construct some function U^* on a strip $[a, b] \times [\tau, \tau + \delta]$ such that

(14.10.6)
$$\limsup_{h \downarrow 0} \frac{1}{h} \|U(\cdot, \tau + h) - U^*(\cdot, h)\|_{L^1(a,b)} \le c_{14} r \varepsilon ,$$

(14.10.7)
$$\limsup_{h \downarrow 0} \frac{1}{h} \|S_h \circ U(\cdot, \tau) - U^*(\cdot, h)\|_{L^1(a,b)} \le c_{14} r \varepsilon .$$

Naturally, such a U^* shall provide a local approximation to the solution of (13.1.1) with initial data $U_0(\cdot) = U(\cdot, \tau)$, and will be constructed accordingly by patching together local approximate solutions of two types, one fitting to points of strong jump discontinuity, the other suitable for regions with small local oscillation.

We begin by fixing λ which is larger than the absolute value of all characteristic speeds and also sufficiently large for the Tame Oscillation Condition (14.10.3) to apply.

With any point (y, τ) of approximate jump discontinuity for U, with approximate limits $U_\pm = U(y\pm, \tau)$ and shock speed s, we associate the sector $\mathcal{K} = \{(x, \sigma) : \sigma > 0, |x - y| \le \lambda\sigma\}$, on which we consider the solution $U^\sharp = U^\sharp_{(y,\tau)}$ defined by

(14.10.8)
$$U^\sharp(x, \sigma) = \begin{cases} U_-, & \text{for } x < y + s\sigma \\ U_+, & \text{for } x > y + s\sigma . \end{cases}$$

We prove that

(14.10.9)
$$\lim_{h \downarrow 0} \frac{1}{h} \int_{y-\lambda h}^{y+\lambda h} |U(x, \tau + h) - U^\sharp(x, h)| dx = 0 .$$

Indeed, for $0 \le \sigma \le h$, let us set

(14.10.10)
$$\phi_h(\sigma) = \frac{1}{h} \int_{y-\lambda h}^{y+\lambda h} |U(x, \tau + \sigma) - U^\sharp(x, \sigma)| dx .$$

Suppose $\phi_h(h) > 0$. Since $\sigma \mapsto U(\cdot, \tau + \sigma) - U^\sharp(\cdot, \sigma)$ is L^1-Lipschitz, with constant say γ, we infer that, for $h \ll 1, \phi_h(h) < 2\gamma$ and $\phi_h(\sigma) \ge \frac{1}{2}\phi_h(h)$, for any σ will $h - \sigma \le \frac{h}{2\gamma}\phi_h(h)$. Then

(14.10.11)
$$\frac{1}{h^2} \int_0^h \int_{y-\lambda h}^{y+\lambda h} |U(x, \tau + \sigma) - U^\sharp(x, \sigma)| dx d\sigma = \frac{1}{h} \int_0^h \phi_h(\sigma) d\sigma \ge \frac{1}{4\gamma}\phi_h^2(h) .$$

As $h \downarrow 0$, the left-hand side of (14.10.11) tends to zero, by virtue of Theorem 1.7.3, and this verifies (14.10.9).

To handle small oscillations, let us fix any interval (ζ, ξ), with midpoint say z. On the triangle $\mathcal{T} = \{(x, \sigma) : \sigma > 0, \zeta + \lambda\sigma < x < \xi - \lambda\sigma\}$, we construct the solution $U^\flat = U^\flat_{(z,\tau)}$ of the linear Cauchy problem

(14.10.12) $\partial_t U^\flat + A^\flat \partial_x U^\flat = 0$,

(14.10.13) $U^\flat(x, 0) = U(x, \tau)$,

where $A^\flat$ is the constant matrix $DF(U(z, \tau))$. The aim is to establish the estimate

(14.10.14)
$$\int_{\zeta+\lambda h}^{\xi-\lambda h} |U(x, \tau + h) - U^\flat(x, h)|dx$$
$$\leq c_{15}[TV_{(\zeta,\xi)}U(\cdot, \tau)] \int_0^h TV_{(\zeta+\lambda\sigma,\xi-\lambda\sigma)}U(\cdot, \tau + \sigma)d\sigma .$$

Integrating (14.10.12) along characteristic directions and using (14.10.13) yields

(14.10.15) $L_i^\flat U^\flat(x, h) = L_i^\flat U(x - \lambda_i^\flat h, \tau)$, $i = 1, \cdots, n$,

where $L_i^\flat = L_i(U(z, \tau))$ is a left eigenvector of $A^\flat$ associated with the eigenvalue $\lambda_i^\flat = \lambda_i(U(z, \tau))$. For fixed i, we may assume without loss of generality that $\lambda_i^\flat = 0$, since this may be achieved by the change of variables $x \mapsto x - \lambda_i^\flat t$, $F(U) \mapsto F(U) - \lambda_i^\flat U$. In that case, since U satisfies (14.2.1) in the sense of distributions,

(14.10.16)
$$\int_{\zeta+\lambda h}^{\xi-\lambda h} \phi(x)L_i^\flat[U(x, \tau + h) - U^\flat(x, h)]dx$$
$$= \int_{\zeta+\lambda h}^{\xi-\lambda h} \phi(x)L_i^\flat[U(x, \tau + h) - U(x, \tau)]dx$$
$$= \int_0^h \int_{\zeta+\lambda h}^{\xi-\lambda h} \partial_x\phi(x)L_i^\flat F(U(x, \tau + \sigma))dxd\sigma ,$$

for any test function $\phi \in C_0^\infty(\zeta + \lambda h, \xi - \lambda h)$. Taking the supremum over all such ϕ with $|\phi(x)| \leq 1$, yields

(14.10.17)
$$\int_{\zeta+\lambda h}^{\xi-\lambda h} |L_i^\flat[U(x, \tau + h) - U^\flat(x, h)]|dx$$
$$\leq \int_0^h TV_{(\zeta+\lambda h,\xi-\lambda h)}L_i^\flat F(U(\cdot, \tau + \sigma))d\sigma .$$

Given $\zeta + \lambda h < x < y < \xi - \lambda h$, let us set, for brevity, $V = U(x, \tau + \sigma)$ and $W = U(y, \tau + \sigma)$. Recalling the notation (8.1.4), one may write

(14.10.18)
$$F(V) - F(W) = A(V, W)(V - W)$$
$$= A^\flat(V - W) + [A(V, W) - A^\flat](V - W) .$$

We now note that $L_i^\flat A^\flat = 0$. Furthermore, $A(V, W) - A^\flat$ is bounded in terms of the oscillation of U inside the triangle $\mathcal{T}$, which is in turn bounded in terms of the total variation of $U(\cdot, \tau)$ over (ζ, ξ), by virtue of the Tame Oscillation Condition (14.10.3). Therefore, (14.10.17) yields the estimate

$$(14.10.19) \qquad \int_{\zeta+\lambda h}^{\xi-\lambda h} |L_i^\flat[U(x,\tau+h)-U^\flat(x,h)]|dx$$

$$\leq c_{16}[TV_{(\zeta,\xi)}U(\cdot,\tau)] \int_0^h TV_{(\zeta+\lambda\sigma,\xi-\lambda\sigma)}U(\cdot,\tau+\sigma)d\sigma \ .$$

Since (14.10.19) holds for $i = 1, \cdots, n$, (14.10.14) readily follows.

We have now laid the preparation for synthesizing a function U^* that satisfies (14.10.6). We begin by identifying a finite collection of open intervals (ζ_j, ξ_j), $j = 1, \cdots, J$, with the following properties:

(i) $[a,b] \subset \bigcup_{j=1}^{J} [\zeta_j, \xi_j]$.

(ii) The intersection of any three of these intervals is empty.

(iii) $TV_{(\zeta_j,\xi_j)}U(\cdot,\tau) < \varepsilon$, for $j = 1, \cdots, J$.

With each (ζ_j, ξ_j), we associate, as above, the triangle $\mathscr{T}_j$ and the approximate solution $U^\flat_{(z_j,\tau)}$ relative to the midpoint z_j. We also consider $[a,b]\backslash \bigcup_{j=1}^{J}(\zeta_j,\xi_j)$, which is a finite set $\{y_1, \cdots, y_K\}$ containing the points where strong shocks cross the τ-time line between a and b. With each y_k we associate the sector $\mathscr{H}_k$ and the corresponding approximate solution $U^\sharp_{(y_k,\tau)}$ (see Fig. 14.10.1). We then set

$$(14.10.20) \qquad U^*(x,h) = \begin{cases} U^\sharp_{(y_k,\tau)}(x,h), & \text{for } (x,h) \in \mathscr{H}_k \backslash \bigcup_{\ell=1}^{k-1} \mathscr{H}_\ell \\[2ex] U^\flat_{(z_j,\tau)}(x,h), & \text{for } (x,h) \in \mathscr{T}_j \backslash \bigcup_{\ell=1}^{j-1} \mathscr{T}_\ell . \end{cases}$$

Clearly, for h sufficiently small $U^*(\cdot,h)$ is defined for all $x \in [a,b]$ and

$$\int_a^b |U(x,\tau+h)-U^*(x,h)|dx$$

$$(14.10.21) \qquad \leq \sum_{k=1}^{K} \int_{y_k-\lambda h}^{y_k+\lambda h} |U(x,\tau+h)-U^\sharp_{(y_k,\tau)}(x,h)|dx$$

$$+ \sum_{j=1}^{J} \int_{\zeta_j+\lambda h}^{\xi_j-\lambda h} |U(x,\tau+h)-U^\flat_{(z_j,\tau)}(x,h)|dx \ .$$

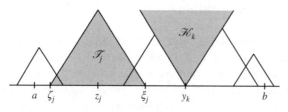

Fig. 14.10.1.

Upon combining (14.10.21), (14.10.9), (14.10.14) and (14.10.1), we arrive at (14.10.6), with $c_{14} = 2cc_{15}$.

We now note that $S_{t-\tau} \circ U(\cdot, \tau)$ defines, for $t \geq \tau$, another solution of (14.2.1) which has the same properties, complies with the same bounds, and has identical restriction to $t = \tau$ with U. Therefore, this solution must equally satisfy the analog of (14.10.6), namely (14.10.7). Finally, (14.10.6) and (14.10.7) together yield (14.10.5).

It remains to show that (14.10.5) implies (14.10.4). To that end, we fix $t > 0$ and any, arbitrarily small, $\varepsilon > 0$. By virtue of (14.10.5) and the Vitali covering lemma, there is a collection of pairwise disjoint closed subintervals $[\tau_k, \tau_k + h_k]$, $k = 1, \cdots, K$, of $[0, t]$, with $0 \leq \tau_1 < \cdots < \tau_K < t$, such that $\tau_k \notin \mathcal{N}$ and

$$(14.10.22) \qquad 0 \leq t - \sum_{k=1}^{K} h_k < \varepsilon ,$$

$$(14.10.23) \quad \|U(\cdot, \tau_k + h_k) - S_{h_k} \circ U(\cdot, \tau_k)\|_{L^1(-\infty, \infty)} < \varepsilon h_k , \quad k = 1, \cdots, K .$$

By the triangle inequality,

$$
(14.10.24) \quad
\begin{aligned}
\|U(\cdot, t) &- S_t \circ U_0(\cdot)\|_{L^1(-\infty, \infty)} \\
&\leq \sum_{k=0}^{K} \|S_{t-\tau_{k+1}} \circ U(\cdot, \tau_{k+1}) - S_{t-\tau_k - h_k} \circ U(\cdot, \tau_k + h_k)\|_{L^1(-\infty, \infty)} \\
&\quad + \sum_{k=1}^{K} \|S_{t-\tau_k - h_k} \circ U(\cdot, \tau_k + h_k) - S_{t-\tau_k} \circ U(\cdot, \tau_k)\|_{L^1(-\infty, \infty)} .
\end{aligned}
$$

In the first summation on the right-hand side of (14.10.24), $\tau_0 + h_0$ is to be interpreted as 0, and τ_{K+1} is to be interpreted as t. The general term in this summation is bounded by $\kappa(1 + c'r)(\tau_{k+1} - \tau_k - h_k)$, on account of (14.9.2) and (14.10.2). Hence the first sum is bounded by $\kappa(1 + c'r)\varepsilon$, because of (14.10.22). Turning now to the second summation, since $S_{t-\tau_k} = S_{t-\tau_k - h_k} S_{h_k}$,

$$(14.10.25) \quad \|S_{t-\tau_k - h_k} \circ U(\cdot, \tau_k + h_k) - S_{t-\tau_k} \circ U(\cdot, \tau_k)\|_{L^1(-\infty, \infty)} \leq \kappa \varepsilon h_k ,$$

by virtue of (14.9.2) and (14.10.23). Therefore, the second sum is bounded by $\kappa t \varepsilon$. Thus the right-hand side of (14.10.24) can be made arbitrarily small and this establishes (14.10.4). The proof is complete.

14.11 Structure of Solutions

In earlier chapters we studied in great detail the structure of BV solutions for scalar conservation laws as well as for systems of two conservation laws. The front tracking method, with its simplicity and explicitness, provides an appropriate vehicle for extending the investigation to genuinely nonlinear systems of arbitrary size. The aim of the study is to determine what features of piecewise constant solutions are being inherited by the BV solutions generated via the limit process.

In addition to providing a fairly detailed picture of local structure and regularity, this approach uncovers various stability characteristics of solutions and elucidates the issue of structural stability of the wave pattern. A sample of results will be stated below, without proofs. The reader may find a detailed exposition in the literature cited in the next section.

Approximation schemes, like random choice or front tracking, which employ wave discretization, are effective because they expose the wave pattern thus allowing us to measure the key quantities of wave strength and interaction potential. As a first step towards developing a qualitative theory, one should devise an intrinsic realization of the wave pattern, within the framework of general BV solutions. This is done in the following way.

Consider any solution U of (14.2.1), in the same function class as in the previous sections. For fixed t, the function $U(\cdot, t)$, which is of bounded variation, induces in the standard manner a signed vector-valued measure μ on $(-\infty, \infty)$, with continuous part μ^c and atomic part μ^a. Then one associates with each characteristic family i a (scalar-valued) measure μ_i, with continuous part μ_i^c defined through

$$(14.11.1) \qquad \int_{\mathbb{R}} \phi d\mu_i^c = \int_{\mathbb{R}} \phi L_i(U) d\mu^c ,$$

and atomic part μ_i^a which is supported in the countable set of points $\bar{x}$ of jump discontinuity of $U(\cdot, t)$ and is evaluated according to the following prescription: $\mu_i^a(\{\bar{x}\})$ is the amplitude of the i-wave in the admissible wave fan that resolves the jump discontinuity with left state $U(\bar{x}-, t)$ and right state $U(\bar{x}+, t)$. Clearly, μ_i encodes the pattern of i-waves that cross the t-time line. In particular, its negative part μ_i^- is associated with compression waves, including i-shocks, while its positive part μ_i^+ is associated with rarefaction waves. The total strength of i-waves is naturally measured by the total variation $|\mu_i|(\mathbb{R})$ of μ_i, and so

$$(14.11.2) \qquad L = \sum_{i=1}^{n} |\mu_i|(\mathbb{R})$$

will represent the combined strength of waves of all characteristic families that cross the t-time line. The potential for wave interactions is then defined as

$$(14.11.3) \qquad \begin{aligned} Q = &\sum_{1 \le i < j \le n} (|\mu_j| \otimes |\mu_i|)(\{(x, y) : x < y\}) \\ &+ \sum_{1 \le i \le n} (\mu_i^- \otimes |\mu_i|)(\{(x, y) : x \ne y\}) . \end{aligned}$$

Finally, $H = L + 2\kappa Q$ plays the role of the Glimm-type functional H introduced in Section 14.9. It is nonincreasing with time and thus controls the growth of L.

The following proposition describes the local structure of solutions.

Theorem 14.11.1 *Let U be a solution of (14.2.1). Fix a point $(\bar{x}, \bar{t})$ on the upper half-plane and consider the rescaled function*

(14.11.4) $U_\alpha(x, t) = U(\overline{x} + \alpha x, \overline{t} + \alpha t)$, $\alpha > 0$.

Then, for any $t \in (-\infty, \infty)$, *as* $\alpha \downarrow 0$, $U_\alpha(\cdot, t)$ *converges in* L^1_{loc} *to* $\overline{U}(\cdot, t)$, *where* $\overline{U}$ *is a self-similar solution of* (14.2.1). *On the upper half-plane,* $t \geq 0$, $\overline{U}$ *coincides with the admissible solution of the Riemann problem for* (14.2.1), *with left state* $U(\overline{x}-, \overline{t})$ *and right state* $U(\overline{x}+, \overline{t})$. *On the lower half-plane,* $t \leq 0$, $\overline{U}$ *contains only admissible shocks and/or compression waves. Furthermore, as* $\alpha \downarrow 0$, *the measures* $\mu^\pm_i$ *of rarefaction and compression i-waves for* $U_\alpha(\cdot, t)$ *converge, in the weak topology of measures, to the corresponding measures* $\overline{\mu}^\pm_i$ *for* $\overline{U}(\cdot, t)$.

The next proposition states that jump discontinuities of sizable strength arrange themselves along a finite collection of shocks.

Theorem 14.11.2 *Let* U *be a solution of* (14.2.1). *For each* $\varepsilon > 0$, *the set of points of the upper half-plane where the local oscillation of* U *is larger than* ε *is contained in a set* $\mathscr{F}_\varepsilon$ *which is the union of a finite set* $\mathscr{I}_\varepsilon$ *of points, and the graphs of a finite collection* $\{x = y_k(t), t \in [a_k, b_k] : k = 1, \cdots, K\}$ *of Lipschitz curves. With each* $k = 1, \cdots, K$ *is associated a (possibly empty) countable subset* $\mathscr{T}_k$ *of* $[a_k, b_k]$ *such that* $y_k(\cdot)$ *is differentiable on* $(a_k, b_k)\backslash \mathscr{T}_k$ *and* $t \mapsto \dot{y}_k(t)$ *has bounded variation over* (a_k, b_k). *For any* $t \in (a_k, b_k)\backslash \mathscr{T}_k$, *as one approaches the point* $(y_k(t), t)$ *from either side of the graph of* $x = y_k(\cdot)$, U *attains one-sided classical limits* $U_\pm$, *where* U_- *and* U_+ *are joined by a shock of speed* $\dot{y}_k(t)$ *and strength* $> c\varepsilon$.

Applying the above proposition for a sequence of ε's that tends to zero, one draws the following corollary which establishes regularity for solutions:

Theorem 14.11.3 *Any solution* U *of* (14.2.1) *on the upper half-plane is continuous, except on a set* $\mathscr{F}$ *which is the union of a countable set* $\mathscr{I}$ *of points and the graphs of a countable collection* $\{x = y_k(t), t \in [a_k, b_k], k = 1, 2, \cdots\}$ *of Lipschitz curves. The derivative* $\dot{y}_k(t)$ *exists at any* $t \in (a_k, b_k)$ *with* $(y_k(t), t) \notin \mathscr{I}$, *and the function* $t \mapsto \dot{y}_k(t)$ *has bounded variation on* (a_k, b_k). *Furthermore, as one approaches the point* $(y_k(t), t) \notin \mathscr{I}$ *from either side of the graph of* $x = y_k(\cdot)$, U *attains distinct one-sided limits* $U_\pm$, *where* U_- *and* U_+ *are joined by a shock of speed* $\dot{y}_k(t)$.

14.12 Notes

The front tracking method for scalar conservation laws was introduced by Dafermos [2] and is developed in Hedstrom [1], Holden, Holden and Høegh-Krohn [1], Holden and Holden [1], Holden and Risebro [1], Risebro and Tveito [2], Gimse and Risebro [1], Gimse [1], and Pan and Lin [1]. It has been employed, especially by the Norwegian School, as a computational tool. In fact, a similar approach had already been used for computations in the 1960's, by Barker [1]. For a survey, see the lecture notes of Holden and Risebro [2]. The method was extended to genuinely nonlinear systems of two conservation laws by DiPerna [3] and then to systems of larger size, independently, by Bressan [2] and Risebro [1]. For early applications to special systems see Alber [2], L.-W. Lin [1], Risebro and Tveito [1] and Wendroff [2]. The presentation here, Sections 14.2–14.7, follows the approach of Bressan and employs a technical simplification due to Baiti and Jenssen [2]. The notion of nonresonant curve is introduced here for the first time.

The Standard Riemann Semigroup was originally constructed by Bressan [1], for linearly degenerate systems, Bressan and Colombo [1], for genuinely nonlinear systems of two conservation laws, and finally by Bressan, Crasta and Piccoli [1], for systems of n conservation laws, based on estimates derived in Bressan [3] and Bressan and Colombo [3]. The presentation in Sections 14.8–14.9 follows the simplified approach of Bressan, Liu and Yang [1], in which the basic estimate is derived by means of the functional ρ introduced by Liu and Yang [1,2]. Actually, L^1 stability has now been established, by Liu and Yang [4], even via the Glimm scheme. It should be noted that, in contrast to the scalar case, there is no standard L^1-contractive metric for systems (Temple [2]).

Uniqueness under the Tame Oscillation Condition was established by Bressan and Goatin [1], improving an earlier theorem by Bressan and LeFloch [1] which required a Tame Variation Condition. Uniqueness also prevails when the Tame Oscillation Condition is substituted by the assumption that the trace of solutions along space-like curves has bounded variation; see Bressan and Lewitska [1]. For an alternative approach, based on Haar's method, see Hu and LeFloch [1]. Uniqueness is also discussed in Oleinik [3], Liu [3], DiPerna [5], Dafermos and Geng [2], Heibig [1], LeFloch and Xin [1], Bressan [3] and Chen and Frid [7].

A detailed study of the regularity of solutions and the structural stability of the wave pattern is found in Bressan and LeFloch [2].

The front tracking method is nicely presented in the lecture notes of Bressan [4]. For further developments of this approach, for handling large initial data, initial-boundary value problems, systems that are merely piecewise genuinely nonlinear and inhomogeneous systems of balance laws, see Bressan [5], Bressan and Colombo [2], Baiti and Jenssen [1], Colombo and Risebro [1], Amadori [1], Amadori and Colombo [1], Amadori and Guerra [1], Crasta and Piccoli [1] and Ancona and Marson [1]. Estimates on the rate of convergence are derived by Lucier [1], in the scalar case, and Bressan and Marson [1], for systems.

Chapter XV. Compensated Compactness

Approximate solutions to hyperbolic systems of conservation laws may be generated in a variety of ways: by the method of vanishing viscosity, through difference approximations, by relaxation schemes, etc. The topic for discussion in this chapter is whether solutions may be constructed as limits of sequences of approximate solutions that are only bounded in some L^p space. Since the systems are nonlinear, the difficulty lies in that the construction schemes are generally consistent only when the sequence of approximating solutions converges strongly, whereas the assumed L^p bounds only guarantee weak convergence: Approximate solutions may develop high frequency oscillations of finite amplitude which play havoc with consistency. The aim is to demonstrate that entropy inequalities may save the day by quenching rapid oscillations thus enforcing strong convergence of the approximating solutions. Some indication of this effect was alluded in Section 1.9.

The principal tools in the investigation will be the notion of Young measure and the functional analytic method of compensated compactness. The former naturally induces the very general class of measure-valued solutions and the latter is employed to verify that nonlinearity reduces measure-valued solutions to traditional ones. Due to its heavy reliance on entropy dissipation, the method appears to be applicable mainly to systems endowed with a rich family of entropy-entropy flux pairs, most notably the scalar conservation law and systems of just two conservation laws. Despite this limitation, the approach is quite fruitful, not only due to the wealth of important systems with such structure, but also because it provides valuable insight in the stabilizing role of entropy dissipation as well as in the "schizophrenic" stabilizing-destabilizing behavior of nonlinearity. Different manifestations of these factors were already encountered in earlier chapters.

Out of a host of known applications of the method, only the simplest shall be presented here, pertaining to the scalar conservation law, genuinely nonlinear systems of two conservation laws, and the system of isentropic elasticity and gas dynamics.

15.1 The Young Measure

The stumbling block for establishing consistency of construction schemes that generate weakly convergent sequences of approximate solutions lies in that it is generally impossible to pass weak limits under nonlinear functions. Suppose Ω is an open subset of $\mathbb{R}^m$ and $\{U_k\}$ is a sequence in $L^\infty(\Omega; \mathbb{R}^n)$ which converges in L^∞ weak* to some limit $\overline{U}$. If g is any continuous real-valued function on $\mathbb{R}^n$, the sequence $\{g(U_k)\}$ will contain subsequences which converge in L^∞ weak*, say to $\overline{g}$, but in general $\overline{g} \neq g(\overline{U})$. It turns out that the limit behavior of such sequences, for all continuous g, is encoded in a family $\{\nu_X : X \in \Omega\}$ of probability measures on $\mathbb{R}^n$, which is constructed by the following procedure.

Let $M(\mathbb{R}^n)$ denote the space of bounded Radon measures on $\mathbb{R}^n$, which is isometrically isomorphic to the dual of the space $C(\mathbb{R}^n)$ of bounded continuous functions. With $k = 1, 2, \cdots$ and any $X \in \Omega$, we associate the Dirac mass $\delta_{U_k(X)} \in M(\mathbb{R}^n)$, positioned at the point $U_k(X)$, and realize the family $\{\delta_{U_k(X)} : X \in \Omega\}$ as an element ν_k of the space $L_w^\infty(\Omega; M(\mathbb{R}^n))$, which is isometrically isomorphic to the dual of $L^1(\Omega; C(\mathbb{R}^n))$. By virtue of standard weak compactness and separability theorems, there is a subsequence $\{\nu_j\}$ of $\{\nu_k\}$ which converges weakly* to some $\nu \in L_w^\infty(\Omega; M(\mathbb{R}^n))$. Thus, $\nu = \{\nu_X : X \in \Omega\}$ and, as $j \to \infty$,

$$(15.1.1) \quad \int_\Omega \psi(X, U_j(X))dX = \int_\Omega \langle \delta_{U_j(X)}, \psi(X, \cdot) \rangle dX \to \int_\Omega \langle \nu_X, \psi(X, \cdot) \rangle dX ,$$

for any $\psi \in C(\Omega \times \mathbb{R}^n)$. The supports of the $\delta_{U_j(X)}$ are uniformly bounded and hence the ν_X must have compact support. Furthermore, since the $\delta_{U_j(X)}$ are probability measures, so are the ν_X. In particular, applying (15.1.1) for $\psi(X, U) = \phi(X)g(U)$, where $\phi \in C(\Omega)$ and $g \in C(\mathbb{R}^n)$, we arrive at the following

Theorem 15.1.1 *Let Ω be any open subset of $\mathbb{R}^m$. Then any bounded sequence $\{U_k\}$ in $L^\infty(\Omega; \mathbb{R}^n)$ contains a subsequence $\{U_j\}$, together with a measurable family $\{\nu_X : X \in \Omega\}$ of probability measures with compact support, such that, for any $g \in C(\mathbb{R}^n)$,*

$$(15.1.2) \qquad\qquad g(U_j) \rightharpoonup \overline{g} , \quad \text{as } j \to \infty ,$$

in L^∞ weak, where*

$$(15.1.3) \qquad\qquad \overline{g}(X) = \langle \nu_X, g \rangle = \int_{\mathbb{R}^n} g(U)d\nu_X(U) .$$

The collection $\{\nu_X : X \in \Omega\}$ constitutes the family of *Young measures* associated with the subsequence $\{U_j\}$. To gain some insight, let us consider the ball B_r in Ω, with center at some $X \in \Omega$, radius r and measure $|B_r|$. On account of our construction of ν_X, it is easy to see that

$$(15.1.4) \qquad \nu_X = \lim_{r\downarrow 0} \lim_{j\uparrow\infty} \frac{1}{|B_r|} \int_{B_r} \delta_{U_j(Y)}dY , \quad \text{a.e. on } \Omega ,$$

where the limits are to be understood in the weak* sense. Notice that the averaged integral on the right-hand side of (15.1.4) may be interpreted as the probability distribution of the values of $U_j(Y)$ as Y is selected uniformly at random from B_r. Thus, according to (15.1.4), ν_X represents the limiting probability distribution of the values of U_j near X.

By virtue of (15.1.2) and (15.1.3), the subsequence $\{U_j\}$ converges, in L^∞ weak*, to the mean $\overline{U} = \langle v_X, U \rangle$ of the Young measures. The limit $\overline{g}$ of $\{g(U_j)\}$ will satisfy $\overline{g} = g(\overline{U})$, for all $g \in C(\mathbb{R}^n)$, if and only if v_X reduces to the Dirac mass $\delta_{\overline{U}(X)}$ positioned at $\overline{U}(X)$. In that case, $\{|U_j|\}$ will converge to $|\overline{U}|$, which implies that $\{U_j\}$ will converge to $\overline{U}$ strongly in $L^p_{\text{loc}}(\Omega)$, for any $1 \leq p < \infty$, and some subsequence of $\{U_j\}$ will converge to $\overline{U}$ a.e. on Ω. Hence, to establish strong convergence of $\{U_j\}$, one needs to verify that the support of the Young measure is confined to a point.

Certain applications require more general versions of Theorem 15.1.1. Young measures v_X are defined even when the sequence $\{U_k\}$ is merely bounded in some $L^p(\Omega; \mathbb{R}^n)$, with $1 < p < \infty$. If Ω is bounded, the v_X are still probability measures and (15.1.2), (15.1.3) hold for all continuous functions g which satisfy a growth condition $|g(U)| \leq c(1 + |U|^q)$, for some $0 < q < p$. In that case, convergence in (15.1.2) is weakly in $L^r(\Omega)$, for $1 < r < p/q$. By contrast, when Ω is unbounded, the v_X may have mass less than one, because in the process of constructing them, as one passes to the $j \to \infty$ limit, part of the mass may leak out at infinity.

15.2 Compensated Compactness and the div-curl Lemma

The theory of compensated compactness strives to classify bounded (weakly compact) sets in L^p space endowed with additional structure that falls short of (strong) compactness but still manages to render certain nonlinear functions weakly continuous. This is nicely illustrated by means of the following proposition, the celebrated *div-curl lemma*, which commands a surprisingly broad gamut of applications.

Theorem 15.2.1 *Given an open subset Ω of $\mathbb{R}^m$, let $\{G_j\}$ and $\{H_j\}$ be sequences of vector fields in $L^2(\Omega; \mathbb{R}^m)$ converging weakly to respective limits $\overline{G}$ and $\overline{H}$, as $j \to \infty$. Assume both $\{\text{div } G_j\}$ and $\{\text{curl } H_j\}$ lie in compact subsets of $W^{-1,2}(\Omega)$. Then*

$$(15.2.1) \qquad G_j \cdot H_j \to \overline{G} \cdot \overline{H} , \quad \text{as } j \to \infty ,$$

in the sense of distributions.

Proof. What follows is a demonstration in the special case $m = 2$, which is all that will be needed for the intended applications in this book. The reader may find alternative proofs, for general m, in the references cited in Section 15.9.

The first remark is that, for any $\chi \in C_0^\infty(\Omega)$, the sequences $\{\chi G_j\}$ and $\{\chi H_j\}$ satisfy the same assumptions as $\{G_j\}$ and $\{H_j\}$, so we may assume, without loss of generality, that $\{G_j\}$ and $\{H_j\}$ have compact support in Ω. We enclose the support in a square S, then extend the G_j and the H_j, first to $S \backslash \Omega$ by zero and thence to all of $\mathbb{R}^2$ by periodicity.

Upon subtracting $\overline{G}$ from G_j, we are reduced to $\overline{G} = 0$. Moreover, since one may subtract from H_j its average value over S, we need only deal with the case $\int_S H_j \, dX = 0$.

We now perform a Helmholtz decomposition of H_j into a gradient and a solenoidal field:

$$(15.2.2) \qquad H_{j1} = \partial_1 \phi_j + \partial_2 \psi_j \, , \qquad H_{j2} = \partial_2 \phi_j - \partial_1 \psi_j \, .$$

The functions ϕ_j and ψ_j are determined as periodic solutions of the equations

$$(15.2.3) \qquad \Delta \phi_j = \operatorname{div} H_j \, , \qquad \Delta \psi_j = \operatorname{curl} H_j \, ,$$

whence it follows that $\{\phi_j\}$ lies in a bounded set of $W^{1,2}(S)$, while $\{\psi_j\}$ lies in a compact set of $W^{1,2}(S)$.

Finally, we note the simple identity

$$(15.2.4) \qquad \begin{aligned} G_j \cdot H_j &= \partial_1 (\phi_j G_{j1}) + \partial_2 (\phi_j G_{j2}) - \phi_j \operatorname{div} G_j \\ &\quad - \partial_1 (\psi_j G_{j2}) + \partial_2 (\psi_j G_{j1}) - \psi_j \operatorname{curl} G_j \, . \end{aligned}$$

Each term on the right-hand side of (15.2.4) tends to zero, in the sense of distributions, as $j \to \infty$, and this establishes (15.2.1). The proof is complete.

In the applications, the following technical result is often helpful for verifying the hypotheses of Theorem 15.2.1.

Lemma 15.2.1 *Let Ω be an open subset of $\mathbb{R}^m$ and $\{\phi_j\}$ a bounded sequence in $W^{-1,p}(\Omega)$, for some $p > 2$. Furthermore, let $\phi_j = \chi_j + \psi_j$, where $\{\chi_j\}$ lies in a compact set of $W^{-1,2}(\Omega)$, while $\{\psi_j\}$ lies in a bounded set of the space of measures $M(\Omega)$. Then $\{\phi_j\}$ lies in a compact set of $W^{-1,2}(\Omega)$.*

Proof. Consider the (unique) functions g_j and h_j in $W_0^{1,2}(\Omega)$ which solve the equations

$$(15.2.5) \qquad \Delta g_j = \chi_j \, , \qquad \Delta h_j = \psi_j \, .$$

By standard elliptic theory, $\{g_j\}$ lies in a compact set of $W_0^{1,2}(\Omega)$ while $\{h_j\}$ lies in a compact set of $W_0^{1,q}(\Omega)$, for $1 < q < \frac{m}{m-1}$. Since $\phi_j = \Delta(g_j + h_j)$, $\{\phi_j\}$ is contained in a compact set of $W^{-1,q}(\Omega)$. But $\{\phi_j\}$ is bounded in $W^{-1,p}(\Omega)$, with $p > 2$; hence, by interpolation between $W^{-1,q}$ and $W^{-1,p}$, it follows that $\{\phi_j\}$ lies in a compact set of $W^{-1,2}(\Omega)$. The proof is complete.

15.3 Measure-Valued Solutions for Systems of Conservation Laws and Compensated Compactness

Consider a system of conservation laws,

$$(15.3.1) \qquad \partial_t U + \partial_x F(U) = 0 \ ,$$

and suppose $\{U_k\}$ is a sequence of approximate solutions in an open subset Ω of $\mathbb{R}^2$, namely

$$(15.3.2) \qquad \partial_t U_k + \partial_x F(U_k) \to 0 \ , \quad \text{as } k \to \infty \ ,$$

in the sense of distributions on Ω. For example, $\{U_k\}$ may have been derived via the vanishing viscosity approach, that is $U_k = U_{\mu_k}$, with $\mu_k \downarrow 0$ as $k \to \infty$, where U_μ is the solution of the parabolic system

$$(15.3.3) \qquad \partial_t U_\mu + \partial_x F(U_\mu) = \mu \partial_x^2 U_\mu \ .$$

When $\{U_k\}$ lies in a bounded set of $L^\infty(\Omega; \mathbb{R}^n)$, following the discussion in Section 15.1, one may extract a subsequence $\{U_j\}$, associated with a family of Young probability measures $\{v_{x,t} : (x, t) \in \Omega\}$ such that $h(U_j) \rightharpoonup \langle v, h \rangle$, as $j \to \infty$, in L^∞ weak*, for any continuous h. In particular, by account of (15.3.2),

$$(15.3.4) \qquad \partial_t \langle v_{x,t}, U \rangle + \partial_x \langle v_{x,t}, F(U) \rangle = 0 \ .$$

One may thus interpret $v_{x,t}$ as a new type of weak solution for (15.3.1):

Definition 15.3.1 A *measure-valued solution* for the system of conservation laws (15.3.1), in an open subset Ω of $\mathbb{R}^2$, is a measurable family $\{v_{x,t} : (x, t) \in \Omega\}$ of probability measures which satisfies (15.3.4) in the sense of distributions on Ω.

Clearly, any traditional weak solution $U \in L^\infty(\Omega; \mathbb{R}^n)$ of (15.3.1) may be identified with the measure-valued solution $v_{x,t} = \delta_{U(x,t)}$. However, the class of measure-valued solutions is definitely broader than the class of traditional solutions. For instance, if U and V are any two traditional solutions of (15.3.1) in $L^\infty(\Omega; \mathbb{R}^n)$, then for any fixed $\alpha \in (0, 1)$,

$$(15.3.5) \qquad v_{x,t} = \alpha \delta_{U(x,t)} + (1 - \alpha) \delta_{V(x,t)}$$

defines a measure-valued solution, which is nontraditional.

At first glance, the notion of measure-valued solution may appear too broad to be relevant. However, abandoning the premise that solutions should assign at each point (x, t) a specific value to the state vector provides the means for describing effectively a class of physical phenomena, like phase transitions, where at the macroscopic level a mixture of phases may occupy the same point in space-time. We shall not develop these ideas here, but rather regard measure-valued solutions as stepping stones towards constructing traditional solutions.

The notion of admissibility naturally extends from traditional to measure-valued solutions. The measure-valued solution $v_{x,t}$ on Ω is said to satisfy the *entropy admissibility condition*, relative to the entropy-entropy flux pair (η, q) of (15.3.1), if

$$(15.3.6) \qquad \partial_t \langle v_{x,t}, \eta(U) \rangle + \partial_x \langle v_{x,t}, q(U) \rangle \leq 0 \,,$$

in the sense of distributions on Ω.

Returning to our earlier example, suppose $v_{x,t}$ is generated through a sequence $\{U_{\mu_j}\}$ of solutions to the parabolic system (15.3.3). If (η, q) is any entropy-entropy flux pair for (15.3.1), multiplying (15.3.3) by $D\eta(U_\mu)$ and using (7.4.1) yields the identity

$$(15.3.7) \qquad \partial_t \eta(U_\mu) + \partial_x q(U_\mu) = \mu \partial_x^2 \eta(U_\mu) - \mu \partial_x U_\mu^T D^2 \eta(U_\mu) \partial_x U_\mu \,.$$

In particular, when η is convex the last term on the right-hand side of (15.3.7) is nonpositive. We thus conclude that any measure-valued solution $v_{x,t}$ of (15.3.1), constructed by the vanishing viscosity approach relative to (15.3.3), satisfies the entropy admissibility condition (15.3.6), for any entropy-entropy flux pair (η, q) with η convex.

Lest it be thought that admissibility suffices to reduce measure-valued solutions to traditional ones, it should be noted that when two traditional solutions U and V satisfy the entropy admissibility condition for an entropy-entropy flux pair (η, q), then so does also the nontraditional measure-valued solution $v_{x,t}$ defined by (15.3.5). On the other hand, admissibility may be an agent for uniqueness and stability in the framework of measure-valued solutions as well. In that direction, it has been shown (references in Section 15.9) that any measure-valued solution $v_{x,t}$ of a scalar conservation law, on the upper half-plane, which satisfies the entropy admissibility condition for all convex entropy-entropy flux pairs, and whose initial values are Dirac masses, $v_{x,0} = \delta_{u_0(x)}$ for some $u_0 \in L^\infty(-\infty, \infty)$, necessarily reduces to a traditional solution, i.e., $v_{x,t} = \delta_{u(x,t)}$, where u is the unique admissible solution of the conservation law, with initial data $u(x, 0) = u_0(x)$. In particular, this implies that for scalar conservation laws any measure-valued solution constructed by the vanishing viscosity approach, with traditional initial data, reduces to a traditional solution.

Returning to the system (15.3.1), a program will be outlined for verifying that the measure-valued solution induced by the family of Young measures $\{v_{x,t} : (x, t) \in \Omega\}$ associated with a sequence $\{U_j\}$ of approximate solutions reduces to a traditional solution. This program will then be implemented for special systems. As already noted in Section 1.9, when (15.3.1) is hyperbolic, approximate solutions may develop sustained rapid oscillations, which prevent strong convergence of the sequence $\{U_j\}$. Thus, our enterprise is destined to fail, unless the approximate solutions somehow embody a mechanism that quenches oscillations. From the standpoint of the theory of compensated compactness, such a mechanism is manifested in the condition

(15.3.8) $\partial_t \eta(U_j) + \partial_x q(U_j) \subset$ compact set in $W_{\mathrm{loc}}^{-1,2}(\Omega)$,

for any entropy-entropy flux pair (η, q) of (15.3.1).

To see the implications of (15.3.8), consider any two entropy-entropy flux pairs (η_1, q_1) and (η_2, q_2). As $j \to \infty$, $\{\eta_1(U_j)\}$, $\{\eta_2(U_j)\}$, $\{q_1(U_j)\}$ and $\{q_2(U_j)\}$ converge to $\bar\eta_1 = \langle \nu, \eta_1 \rangle$, $\bar\eta_2 = \langle \nu, \eta_2 \rangle$, $\bar q_1 = \langle \nu, q_1 \rangle$ and $\bar q = \langle \nu, q_2 \rangle$, respectively, where for brevity we set $\nu_{x,t} = \nu$. By (15.3.8), both div $(q_2(U_j), \eta_2(U_j))$ and curl $(\eta_1(U_j), -q_1(U_j))$ lie in compact sets of $W_{\mathrm{loc}}^{-1,2}(\Omega)$. Hence, on account of Theorem 15.2.1,

(15.3.9) $\eta_1(U_j)q_2(U_j) - \eta_2(U_j)q_1(U_j) \rightharpoonup \bar\eta_1 \bar q_2 - \bar\eta_2 \bar q_1$,   as $j \to \infty$,

in $L^\infty(\Omega)$ weak*, or equivalently

(15.3.10) $\langle \nu, \eta_1 \rangle \langle \nu, q_2 \rangle - \langle \nu, \eta_2 \rangle \langle \nu, q_1 \rangle = \langle \nu, \eta_1 q_2 - \eta_2 q_1 \rangle$.

The plan is to use (15.3.10), for strategically selected entropy-entropy flux pairs, in order to demonstrate that the support of the Young measure ν is confined to a single point. Clearly, such a program may have a fair chance for success only when there is flexibility to construct entropy-entropy flux pairs with prescribed specifications. For all practical purposes, this requirement limits the applicability of the method to scalar conservation laws, systems of two conservation laws, and the special class of systems of more than two conservation laws that are endowed with a rich family of entropies (see Section 7.4). On the other hand, the method offers considerable flexibility in regard to construction scheme, as it only requires that the approximate solutions satisfy (15.3.8).

For illustration, let us verify (15.3.8) for the case the system (15.3.1) is endowed with a uniformly convex entropy, Ω is the upper half-plane, and the sequence $\{U_j\}$ of approximate solutions is generated by the vanishing viscosity approach, $U_j = U_{\mu_j}$, where U_μ is the solution of (15.3.3) on the upper half-plane, with initial data

(15.3.11) $U_\mu(x, 0) = U_{0\mu}(x)$, $-\infty < x < \infty$,

lying in a bounded set of $L^\infty(-\infty, \infty) \cap L^2(-\infty, \infty)$.

Let η be a uniformly convex entropy, so that $D^2\eta(U)$ is positive definite. We can assume $0 \le \eta(U) \le c|U|^2$, since otherwise we simply substitute η with the entropy $\eta^*(U) = \eta(U) - \eta(0) - D\eta(0)U$. Upon integrating (15.3.7) over the upper half-plane, we obtain the estimate

(15.3.12) $\mu \displaystyle\int_0^\infty \int_{-\infty}^\infty |\partial_x U_\mu(x, t)|^2 dx dt \le C$,

where C is independent of μ.

Consider now any, not necessarily convex, entropy-entropy flux pair (η, q), and fix some open bounded subset Ω of the upper half-plane. In reference to (15.3.7), the left-hand side is bounded in $W^{-1,p}(\Omega)$, for any $1 \le p < \infty$. The right-hand side is the sum of two terms: By virtue of (15.3.12), the first one tends to zero, as $\mu \downarrow 0$, in $W^{-1,2}(\Omega)$, and thus lies in a compact set of $W^{-1,2}(\Omega)$. The second lies in a bounded set of $M(\Omega)$, again by account of (15.3.12). Therefore, (15.3.8) follows from Lemma 15.2.1.

15.4 Scalar Conservation Laws

Here we shall see how the program outlined in the previous section may be realized in the case of the scalar conversation law

$$(15.4.1) \qquad \partial_t u + \partial_x f(u) = 0.$$

Theorem 15.4.1 *Let Ω be an open subset of $\mathbb{R}^2$ and $\{u_k(x, t)\}$ a bounded sequence in $L^\infty(\Omega)$ with*

$$(15.4.2) \qquad \partial_t \eta(u_k) + \partial_x q(u_k) \subset \text{compact set in } W_{\text{loc}}^{-1,2}(\Omega) ,$$

for any entropy-entropy flux pair of (15.4.1). Then there is a subsequence $\{u_j\}$ such that

$$(15.4.3) \qquad u_j \rightharpoonup \overline{u} , \quad f(u_j) \rightharpoonup f(\overline{u}) , \quad \text{as } j \to \infty ,$$

in L^∞ weak*. Furthermore, if the set of u with $f''(u) \neq 0$ in dense in $\mathbb{R}$, then $\{u_j\}$, or a subsequence thereof, converges almost everywhere to $\overline{u}$ on Ω.

Proof. By applying Theorem 15.1.1, we extract the subsequence $\{u_j\}$ and the associated family of Young measures $\nu = \nu_{x,t}$ so that $h(u_j) \rightharpoonup \langle \nu, h \rangle$, for any continuous function h. In particular, $u_j \rightharpoonup \overline{u} = \langle \nu, u \rangle$ and $f(u_j) \rightharpoonup \langle \nu, f \rangle$. We thus have to show $\langle \nu, f \rangle = f(\overline{u})$; and that ν reduces to the Dirac mass when there is no interval on which $f'(u)$ is constant.

We employ (15.3.10) for the two entropy-entropy flux pairs $(u, f(u))$ and $(f(u), g(u))$, where

$$(15.4.4) \qquad g(u) = \int_0^u [f'(v)]^2 dv ,$$

to get

$$(15.4.5) \qquad \langle \nu, u \rangle \langle \nu, g \rangle - \langle \nu, f \rangle \langle \nu, f \rangle = \langle \nu, ug - f^2 \rangle .$$

From Schwarz's inequality,

$$(15.4.6) \qquad [f(u) - f(\overline{u})]^2 \leq (u - \overline{u})[g(u) - g(\overline{u})] ,$$

we deduce

$$(15.4.7) \qquad \langle \nu, [f(u) - f(\overline{u}]^2 - (u - \overline{u})[g(u) - g(\overline{u})] \rangle \leq 0 .$$

Upon using (15.4.5), (15.4.7) reduces to

$$(15.4.8) \qquad\qquad [\langle v, f \rangle - f(\bar{u})]^2 \leq 0 \,,$$

whence $\langle v, f \rangle = f(\bar{u})$. In particular, the left-hand side of (15.4.7) will vanish. Hence, (15.4.6) must hold as an equality for u in the support of v. However, Schwarz's inequality (15.4.6) may hold as equality only if f' is constant on the interval with endpoints $\bar{u}$ and u. When no such interval exists, the support of v collapses to a single point and v reduces to the Dirac mass $\delta_{\bar{u}}$. The proof is complete.

As indicated in the previous section, one may generate a sequence $\{u_k\}$ that satisfies the assumptions of Theorem 15.4.1 by the method of vanishing viscosity, setting $u_k = u_{\mu_k}$, $\mu_k \to 0$ as $k \to \infty$, where u_μ is the solution of

$$(15.4.9) \qquad\qquad \partial_t u_\mu + \partial_x f(u_\mu) = \mu \partial_x^2 u_\mu \,,$$

on the upper half-plane, with initial data

$$(15.4.10) \qquad\qquad u_\mu(x, 0) = u_{0\mu}(x) \,, \quad -\infty < x < \infty \,,$$

that are uniformly bounded in $L^\infty(-\infty, \infty) \cap L^2(-\infty, \infty)$. Indeed, the resulting $\{u_k\}$ will be bounded in L^∞, since $\|u_\mu\|_{L^\infty} \leq \|u_{0\mu}\|_{L^\infty}$ by virtue of the maximum principle. Moreover, (15.4.2) will hold for all entropy-entropy flux pairs (η, q), by the general argument of Section 15.3, which applies here, in particular, because (15.4.1) possesses the uniformly convex entropy u^2. Finally, $\mu \partial_x^2 u_\mu \to 0$, as $\mu \downarrow 0$, in the sense of distributions. We thus arrive at the following

Theorem 15.4.2 *Suppose $u_{0\mu} \rightharpoonup u_0$, as $\mu \downarrow 0$, in $L^\infty(-\infty, \infty)$ weak*. Then there is a sequence $\{\mu_j\}$, $\mu_j \to 0$ as $j \to \infty$, such that the sequence $\{u_{\mu_j}\}$ of solutions of (15.4.9), (15.4.10) converges in L^∞ weak* to some function $\bar{u}$, which is a solution of (15.4.1), on the upper half-plane, with initial data $\bar{u}(x, 0) = u_0(x)$ on $(-\infty, \infty)$. Furthermore, if the set of u with $f''(u) \neq 0$ is dense in $\mathbb{R}$, then $\{u_{\mu_j}\}$, or a subsequence thereof, converges almost everywhere to $\bar{u}$ on the upper half-plane.*

15.5 A Relaxation Scheme for Scalar Conservation Laws

The aim here is to construct solutions to the scalar conservation law (15.4.1) by passing to the weak limit in a relaxation scheme, with the help of the theory of compensated compactness. The present scheme will be different from the scheme discussed in Section 6.7.

We consider the system

$$(15.5.1) \qquad \begin{cases} \partial_t u_\mu + \partial_x v_\mu = 0 \\[2mm] \partial_t v_\mu + a^2 \partial_x u_\mu = \dfrac{1}{\mu}[f(u_\mu) - v_\mu] \,, \end{cases}$$

where a is a positive constant, to be fixed later, and μ is a positive relaxation parameter. One might expect that as $\mu \downarrow 0$ the stiff term on the right-hand side will force v_μ to "relax" to its equilibrium value $f(u_\mu)$, in which case the first equation of (15.5.1) shall reduce to (15.4.1). Such expectation, however, shall not materialize unless the effect of the stiff term is dissipative, preventing the v_μ from "escaping" as $\mu \downarrow 0$.

To motivate the appropriate condition for dissipativeness, we set

$$(15.5.2) \qquad\qquad v_\mu = f(u_\mu) + \mu w_\mu$$

and substitute into (15.5.1). Dropping, formally, all terms of order μ and then eliminating $\partial_t u_\mu$ between the resulting two equations yields

$$(15.5.3) \qquad\qquad w_\mu = [f'(u_\mu)^2 - a^2]\partial_x u_\mu \ .$$

Upon combining (15.5.1)$_1$ with (15.5.2) and (15.5.3), we conclude that, formally, to leading order, u_μ satisfies the equation

$$(15.5.4) \qquad\qquad \partial_t u_\mu + \partial_x f(u_\mu) = \mu \partial_x\{[a^2 - f'(u_\mu)^2]\partial_x u_\mu\} \ .$$

Clearly, for dissipativeness, the "viscosity" on the right-hand side of (15.5.4) must be positive, and this motivates the so called *subcharacteristic condition*

$$(15.5.5) \qquad\qquad -a < f'(u) < a \ ,$$

stipulating that the characteristic speed of the relaxed equation must be confined between the minimum and the maximum characteristic speed of the system.

We now examine the implications of the subcharacteristic condition on the entropy-entropy flux pairs of (15.5.1). Since this is a system of balance (rather than conservation) laws, the companion balance laws will generally include a source term, like in (3.2.1):

$$(15.5.6) \qquad\qquad \partial_t \phi(u_\mu, v_\mu) + \partial_x \psi(u_\mu, v_\mu) = -\frac{1}{\mu}h(u_\mu, v_\mu) \ .$$

The integrability conditions, relating the entropy ϕ, entropy flux ψ and entropy production h read as follows:

$$(15.5.7) \qquad \begin{cases} \psi_u(u, v) = a^2 \phi_v(u, v) \ , \\ \psi_v(u, v) = \phi_u(u, v) \ , \end{cases}$$

$$(15.5.8) \qquad\qquad h(u, v) = \phi_v(u, v)[v - f(u)] \ .$$

We are interested in entropy-entropy flux pairs (ϕ, ψ) which reduce on the equilibrium curve $v = f(u)$ to entropy-entropy flux pairs (η, q) of the relaxed equation (15.4.1), that is

$$(15.5.9) \qquad\qquad \phi(u, f(u)) = \eta(u) \ , \quad \psi(u, f(u)) = q(u) \ .$$

If one regards u as the "space" and v as the "time" variable, the subcharacteristic condition (15.5.5) guarantees that the curve $v = f(u)$ is space-like for the system (15.5.7), and hence unique (ϕ, ψ) may be determined from arbitrary Cauchy data (15.5.9). In fact, (ϕ, ψ) may be computed explicitly as follows. The general solution of (15.5.7) is

(15.5.10)
$$\begin{cases} \phi(u, v) = r(au + v) + s(au - v) \,, \\ \psi(u, v) = ar(au + v) - as(au - v) \,. \end{cases}$$

One readily verifies that ϕ is (strictly) convex if and only if both r and s are (strictly) convex. After a straightforward calculation, it is seen that (ϕ, ψ) from (15.5.10) will satisfy (15.5.9), with $q'(u) = \eta'(u)f'(u)$, if and only if

(15.5.11)
$$r'(au + f(u)) = s'(au - f(u)) = \frac{1}{2a}\eta'(u) \,.$$

In particular, η convex yields ϕ convex.

Another implication of (15.5.11) is that $\phi_v(u, f(u)) = 0$. Combined with convexity of ϕ, this implies $\phi_v(u, v) > 0$ for $v > f(u)$ and $\phi_v(u, v) < 0$ for $v < f(u)$, thus rendering the stiff term dissipative:

(15.5.12)
$$h(u, v) \begin{cases} = 0 \ \ \text{if} \ \ v = f(u) \,, \\ > 0 \ \ \text{if} \ \ v \neq f(u) \,. \end{cases}$$

Theorem 15.5.1 *Under the subcharacteristic condition* (15.5.5), *the Cauchy problem for the system* (15.5.1), *with initial data*

(15.5.13)
$$(u_\mu(x, 0), v_\mu(x, 0)) = (u_{0\mu}(x), v_{0\mu}(x)) \,, \quad -\infty < x < \infty \,,$$

in $L^\infty(-\infty, \infty) \cap L^2(-\infty, \infty)$, *possesses a bounded (weak) solution* (u_μ, v_μ) *on the upper half-plane. Furthermore,*

(15.5.14)
$$\frac{1}{\mu} \int_0^\infty \int_{-\infty}^\infty [v_\mu - f(u_\mu)]^2 dx dt \le c \int_{-\infty}^\infty [u_{0\mu}^2(x) + v_{0\mu}^2(x)]dx \,,$$

where c is independent of μ.

Proof. The system (15.5.1) being semilinear hyperbolic, a local solution exists and may be continued for as long as it remains bounded in L^∞. Thus, for global existence it will suffice to establish L^∞ bounds. We normalize v and $f(u)$ by $f(0) = 0$. We construct the entropy-entropy flux pair (ϕ_m, ψ_m) induced by (15.5.11), for $\eta(u) = \frac{a}{m}u^{2m}$, and normalized by $\phi_m(0, 0) = 0$, $\psi_m(0, 0) = 0$. Notice that the first derivatives of $\phi_m(u, v)$ vanish at $(0, 0)$. We now integrate (15.5.6) over $(-\infty, \infty) \times [0, t]$ to get

(15.5.15)
$$\int_{-\infty}^\infty \phi_m(u_\mu(x, t), v_\mu(x, t))dx \le \int_{-\infty}^\infty \phi_m(u_{0\mu}(x), v_{0\mu}(x))dx \,.$$

By (15.5.11), $(cw)^{2m} \leq r_m(w) \leq (Cw)^{2m}$ and $(cw)^{2m} \leq s_m(w) \leq (Cw)^{2m}$. Therefore, raising (15.5.15) to the power $\frac{1}{2m}$ and letting $m \to \infty$, we obtain bounds for $\|u_\mu(\cdot, t)\|_{L^\infty(-\infty,\infty)}$ and $\|v_\mu(\cdot, t)\|_{L^\infty(-\infty,\infty)}$, in terms of $\|u_{0\mu}(\cdot)\|_{L^\infty(-\infty,\infty)}$ and $\|v_{0\mu}(\cdot)\|_{L^\infty(-\infty,\infty)}$, and otherwise independent of t and μ. This establishes global existence of bounded solutions to (15.5.1), (15.5.13).

Finally, we integrate (15.5.6) over $(-\infty, \infty) \times [0, \infty)$ to get

$$(15.5.16) \qquad \frac{1}{\mu} \int_0^\infty \int_{-\infty}^\infty h(u_\mu, v_\mu)dxdt \leq \int_\infty^\infty \phi(u_{0\mu}(x), v_{0\mu}(x))dx .$$

We apply (15.5.16) for the special entropy $\phi_2(u, v)$. Using the properties of $h(u, v)$, from (15.5.8), we deduce (15.5.14). The proof is complete.

We have now laid the preparation for passing to the relaxation limit $\mu \downarrow 0$:

Theorem 15.5.2 *Consider the family* $\{(u_\mu, v_\mu)\}$ *of solutions of the initial-value problem* (15.5.1), (15.5.13), *with initial data* $\{(u_{0\mu}, v_{0\mu})\}$ *that are bounded in* $L^\infty(-\infty, \infty) \cap L^2(-\infty, \infty)$ *and* $u_{0\mu} \to u_0$, *as* $\mu \downarrow 0$, *in* L^∞ *weak*. Then there is a sequence* $\{\mu_j\}$, *with* $\mu_j \to 0$ *as* $j \to \infty$, *such that* $\{(u_{\mu_j}, v_{\mu_j})\}$ *converges, in* L^∞ *weak*, to* $(\overline{u}, f(\overline{u}))$, *where* $\overline{u}$ *is a solution of* (15.4.1)), *on the upper half-plane, with initial data* $\overline{u}(x, 0) = u_0(x)$ *on* $(-\infty, \infty)$. *Furthermore, if the set of u with* $f''(u) \neq 0$ *is dense in* $\mathbb{R}$, *then* $\{(u_{\mu_j}, v_{\mu_j})\}$, *or a subsequence thereof, converges almost everywhere to* $(\overline{u}, f(\overline{u}))$ *on the upper half-plane.*

Proof. By Theorem 15.5.1, $\{(u_\mu, v_\mu)\}$ lies in a bounded set of L^∞.

We fix any entropy-entropy flux pair (η, q) of (15.4.1), consider the entropy-entropy flux pair (ϕ, ψ) of (15.5.1) generated by solving the Cauchy problem (15.5.7), (15.5.9), and use (15.5.6), (15.5.8) to write

$$
\begin{aligned}
& \partial_t \eta(u_\mu) + \partial_x q(u_\mu) \\
(15.5.17) \quad & = \partial_t[\phi(u_\mu, f(u_\mu)) - \phi(u_\mu, v_\mu)] + \partial_x[\psi(u_\mu, f(u_\mu)) - \psi(u_\mu, v_\mu)] \\
& \quad - \frac{1}{\mu}\phi_v(u_\mu, v_\mu)[v_\mu - f(u_\mu)] .
\end{aligned}
$$

Both $\phi(u_\mu, f(u_\mu)) - \phi(u_\mu, v_\mu)$ and $\psi(u_\mu, f(u_\mu)) - \psi(u_\mu, v_\mu)$ tend to zero in L^2, as $\mu \downarrow 0$, by virtue of (15.5.14). Therefore, the first two terms on the right-hand side of (15.5.17) tend to zero in $W^{-1,2}$, as $\mu \downarrow 0$. On the other hand, the third term lies in a bounded set of L^1, again by account of (15.5.14), upon recalling that $\phi_v(u, f(u)) = 0$.

We now fix any sequence $\{\mu_k\}$, with $\mu_k \to 0$ as $k \to \infty$, and set $(u_k, v_k) = (u_{\mu_k}, v_{\mu_k})$. In virtue of the above, Lemma 15.2.1 implies that (15.4.2) holds for any entropy-entropy flux pair (η, q) of (15.4.1), where Ω is the upper half-plane. Theorem 15.4.1 then yields (15.4.3), for some subsequence $\{u_j\}$. In turn, (15.4.3) together with (15.5.14) imply $v_j \rightharpoonup f(\overline{u})$, in L^∞ weak*. In particular, $\overline{u}$ is a solution of (15.4.1), with initial data u_0, because of (15.5.1)$_1$.

When the set of u with $f''(u) \neq 0$ is dense in $\mathbb{R}$, $\{u_j\}$, or a subsequence thereof, converges to $\bar{u}$ almost everywhere, by account of Theorem 15.4.1. It then follows from (15.5.14) that, likewise, $\{v_j\}$ converges to $f(\bar{u})$ almost everywhere. The proof is complete.

By combining (15.5.6), (15.5.12), (15.5.14) and (15.5.9), we infer that, at least in the case where $\{u_j\}$ converges almost everywhere, the limit $\bar{u}$ will satisfy the entropy admissibility condition, for all convex entropy-entropy flux pairs.

Notice that Theorem 15.5.2 places no restriction on the initial values $v_{0\mu}$ of v_μ, beyond the requirement that they be bounded. In particular, $v_{0\mu}$ may lie far apart from the relaxed value $f(u_{0\mu})$. In that situation, a boundary layer must form along the 0-time line, because according to the theorem v will have to relax instantaneously to $f(u)$, at $t > 0$.

15.6 Genuinely Nonlinear Systems of Two Conservation Laws

The program outlined in Section 15.3 will here be implemented for genuinely nonlinear systems (15.3.1) of two conservation laws. In particular, our system will be endowed with a coordinate system of Riemann invariants (z, w), normalized as in (12.1.2), and the condition of genuine nonlinearity will be expressed by (12.1.3), namely $\lambda_z < 0$ and $\mu_w > 0$. Moreover, the system will be equipped with a rich family of entropy-entropy flux pairs, including the Lax pairs constructed in Section 12.2, which will play a pivotal role in the analysis.

We show that the entropy conditions, in conjunction with genuine nonlinearity, quench rapid oscillations:

Theorem 15.6.1 *Let Ω be an open subset of $\mathbb{R}^2$ and $\{U_k(x, t)\}$ a bounded sequence in $L^\infty(\Omega; \mathbb{R}^2)$ with*

$$(15.6.1) \qquad \partial_t \eta(U_k) + \partial_x q(U_k) \subset \text{compact set in } W_{\text{loc}}^{-1,2}(\Omega) ,$$

for any entropy-entropy flux pair (η, q) of (15.3.1). Then there is a subsequence $\{U_j\}$ which converges almost everywhere on Ω.

Proof. By applying Theorem 15.1.1, we extract a subsequence $\{U_j\}$ and identify the associated family of Young measures $\nu_{x,t}$. We have to show that, for almost all (x, t), the support of $\nu_{x,t}$ is confined to a single point and so this measure reduces to the Dirac mass. It will be expedient to monitor the Young measure on the plane of the Riemann invariants (z, w), rather than in the original state space.

We thus let ν denote the Young measure at any fixed point $(x, t) \in \Omega$, relative to the (z, w) variables, and let $\mathscr{R} = [z^-, z^+] \times [w^-, w^+]$ be the smallest rectangle that contains the support of ν. We need to show $z^- = z^+$ and $w^- = w^+$. Arguing by contradiction, assume $z^- < z^+$.

We consider the Lax entropy-entropy flux pairs (12.2.5), which will be here labeled (η_k, q_k), so as to display explicitly the dependence on the parameter k. We shall use the η_k as weights for redistributing the mass of ν, reallocating it near the boundary of $\mathscr{R}$. To that end, with each large positive integer k we associate probability measures $\nu_k^\pm$ on $\mathscr{R}$, defined through their action on continuous functions $h(z, w)$:

$$(15.6.2) \qquad \langle \nu_k^\pm, h \rangle = \frac{\langle \nu, h\eta_{\pm k} \rangle}{\langle \nu, \eta_{\pm k} \rangle} .$$

Because of the factor e^{kz} in the definition of η_k, the measure ν_k^- (or ν_k^+) is concentrated near the left (or right) side of $\mathscr{R}$. As $k \to \infty$, the sequences $\{\nu_k^-\}$ and $\{\nu_k^+\}$, or subsequences thereof, will converge, weakly* in the space of measures, to probability measures ν^- and ν^+ which are respectively supported by the left side $[z^-] \times [w^-, w^+]$ and the right side $[z^+] \times [w^-, w^+]$ of $\mathscr{R}$.

We apply (15.3.10) for any fixed entropy-entropy flux pair (η, q) and the Lax pairs $(\eta_{\pm k}, q_{\pm k})$ to get

$$(15.6.3) \qquad \langle \nu, q \rangle - \frac{\langle \nu, q_{\pm k} \rangle}{\langle \nu, \eta_{\pm k} \rangle} \langle \nu, \eta \rangle = \frac{\langle \nu, \eta_{\pm k} q - \eta q_{\pm k} \rangle}{\langle \nu, \eta_{\pm k} \rangle} .$$

From (12.2.5) and (12.2.7) we infer

$$(15.6.4) \qquad q_{\pm k} = \left[\lambda + O\left(\frac{1}{k}\right) \right] \eta_{\pm k} .$$

Therefore, letting $k \to \infty$ in (15.6.3) yields

$$(15.6.5) \qquad \langle \nu, q \rangle - \langle \nu^\pm, \lambda \rangle \langle \nu, \eta \rangle = \langle \nu^\pm, q - \lambda \eta \rangle .$$

Next, we apply (15.3.10) for the Lax pairs (η_{-k}, q_{-k}) and (η_k, q_k), thus obtaining

$$(15.6.6) \qquad \frac{\langle \nu, q_k \rangle}{\langle \nu, \eta_k \rangle} - \frac{\langle \nu, q_{-k} \rangle}{\langle \nu, \eta_{-k} \rangle} = \frac{\langle \nu, \eta_{-k} q_k - \eta_k q_{-k} \rangle}{\langle \nu, \eta_{-k} \rangle \langle \nu, \eta_k \rangle} .$$

By virtue of (15.6.4), the left-hand side of (15.6.6) tends to $\langle \nu^+, \lambda \rangle - \langle \nu^-, \lambda \rangle$, as $k \to \infty$. On the other hand, the right-hand side tends to zero, because the numerator is $O(k^{-1})$ while

$$(15.6.7) \qquad \langle \nu, \eta_{\pm k} \rangle \geq c \exp[\pm \tfrac{1}{2}(z^- + z^+)] .$$

Hence,

$$(15.6.8) \qquad \langle \nu^-, \lambda \rangle = \langle \nu^+, \lambda \rangle .$$

Combining (15.6.5) with (15.6.8),

$$(15.6.9) \qquad \langle \nu^-, q - \lambda \eta \rangle = \langle \nu^+, q - \lambda \eta \rangle .$$

We apply (15.6.9) for $(\eta, q) = (\eta_k, q_k)$. On account of (12.2.12), for k large,

$$(15.6.10) \qquad \begin{cases} \langle \nu^-, q_k - \lambda \eta_k \rangle \leq C\dfrac{1}{k} \exp(kz^-) \\[2mm] \langle \nu^+, q_k - \lambda \eta_k \rangle \geq c\dfrac{1}{k} \exp(kz^+) , \end{cases}$$

which yields the desired contradiction to $z^- < z^+$. Similarly one shows $w^- = w^+$, so that $\mathscr{R}$ collapses to a single point. The proof is complete.

The stumbling block in employing the above theorem for constructing solutions to our system (15.3.1) is that, at the time of this writing, it has not been established that sequences of approximate solutions, produced by any of the available schemes, are bounded in L^∞. Thus, boundedness has to be imposed as an extraneous (and annoying) assumption. On the other hand, once boundedness is taken for granted, it is not difficult to verify the other requirement of Theorem 15.6.1, namely (15.6.1). In particular, when the sequence of U_k is generated via the vanishing viscosity approach, as solutions of the parabolic system (15.3.3), condition (15.6.1) follows directly from the discussion in Section 15.3, because genuinely nonlinear systems of two conservation laws are always endowed with uniformly convex entropies. For example, as shown in Section 12.2, under the normalization condition (12.1.4), the Lax entropy η_k is convex, for k sufficiently large. We thus have

Theorem 15.6.2 *For $\mu > 0$, let U_μ denote the solution on the upper half-plane of the genuinely nonlinear parabolic system of two conservation laws (15.3.3) with initial data (15.3.11), where $U_{0\mu} \rightharpoonup U_0$ in $L^\infty(-\infty, \infty)$ weak*, as $\mu \downarrow 0$. Suppose the family $\{U_\mu\}$ lies in a bounded subset of L^∞. Then, there is a sequence $\{\mu_j\}$, $\mu_j \to 0$ as $j \to \infty$, such that $\{U_{\mu_j}\}$ converges, almost everywhere on the upper half-plane, to a solution $\overline{U}$ of (15.3.1) with initial values $\overline{U}(x, 0) = U_0(x)$, for $-\infty < x < \infty$.*

One obtains entirely analogous results for sequences of approximate solutions generated by a class of one-step difference schemes with a three-point domain of dependence:

(15.6.11)
$$U(x, t + \Delta t) - U(x, t)$$
$$= \frac{\alpha}{2} G(U(x, t), U(x + \Delta x, t)) - \frac{\alpha}{2} G(U(x - \Delta x, t), U(x, t)) ,$$

where $\alpha = \Delta t / \Delta x$ is the ratio of mesh-lengths and G, possibly depending on α, is a function which satisfies the consistency condition $G(U, U) = F(U)$. The class includes the *Lax-Friedrichs scheme*, with

(15.6.12)
$$G(V, W) = \frac{1}{2}[F(V) + F(W)] + \frac{1}{\alpha}(V - W) ,$$

and also the *Godunov scheme*, where $G(V, W)$ denotes the state in the wake of the solution to the Riemann problem for (15.3.1), with left state V and right state W. The condition of uniform boundedness on L^∞ of the approximate solutions has to be extraneously imposed in these cases as well.

15.7 The System of Isentropic Elasticity

The assertion of Theorem 15.6.1 is obviously false when the system (15.3.1) is linear. On the other hand, genuine nonlinearity is far too strong a restriction: It may be allowed to fail along a finite collection of curves in state space, so long as these curves intersect transversely the level curves of the Riemann invariants. This will be demonstrated here in the context of the system (7.1.6) of conservation laws of one-dimensional, isentropic thermoelasticity,

(15.7.1) $\bullet$ $\begin{cases} \partial_t u - \partial_x v = 0 \\ \partial_t v - \partial_x \sigma(u) = 0 \ , \end{cases}$

under the assumption $\sigma''(u) \neq 0$ for $u \neq 0$, but $\sigma''(0) = 0$, so that genuine nonlinearity fails along the line $u = 0$ in state space. Nevertheless, the analog of Theorem 15.6.1 still holds:

Theorem 15.7.1 *Let Ω be an open subset of $\mathbb{R}^2$ and $\{(u_k, v_k)\}$ a bounded sequence in $L^\infty(\Omega; \mathbb{R}^2)$ with*

(15.7.2) $\partial_t \eta(u_k, v_k) + \partial_x q(u_k, v_k) \subset$ compact set in $W_{loc}^{-1,2}(\Omega)$,

for any entropy-entropy flux pair (η, q) of (15.7.1). Then there is a subsequence $\{(u_j, v_j)\}$ which converges almost everywhere on Ω.

Proof. As in the proof of Theorem 15.6.1, we extract a subsequence $\{(u_j, v_j)\}$ and identify the associated family of Young measures $\nu_{x,t}$. We fix (x, t) in Ω and monitor the Young measure ν at (x, t) relative to the Riemann invariants

(15.7.3) $z = \int_0^u [\sigma'(\omega)]^{\frac{1}{2}} d\omega + v \ , \quad w = -\int_0^u [\sigma'(\omega)]^{\frac{1}{2}} d\omega + v \ .$

We need to show that the smallest rectangle $\mathscr{R} = [z^-, z^+] \times [w^-, w^+]$ which contains the support of ν collapses to a single point.

By retracing the steps in the proof of Theorem 15.6.1, which do not depend on the genuine nonlinearity of the system, we rederive (15.6.9). The remainder of the argument will depend on the relative positions of $\mathscr{R}$ and the straight line $z = w$ along which genuine nonlinearity fails.

Suppose first the line $z = w$ does not intersect the right side of $\mathscr{R}$, i.e., $z^+ \notin [w^-, w^+]$. In that case, (15.6.10) are still in force, yielding $z^- = z^+$. Hence $\mathscr{R}$ collapses to $[z^+] \times [w^-, w^+]$, which, according to our assumption, lies entirely in the genuinely nonlinear region and so the familiar argument implies $w^- = w^+$, verifying the assertion of the theorem. Similar arguments apply when the line $z = w$ misses any one of the other three sides of $\mathscr{R}$.

It thus remains to examine the case where the line $z = w$ intersects all four sides of $\mathscr{R}$, i.e., $z^- = w^-$ and $z^+ = w^+$. Even in that situation, by virtue of (12.2.12), $q_k - \lambda \eta_k$ does not change sign on segments $[z^-] \times [w^- + \varepsilon, w^+]$ and

$[z^+] \times [w^-, w^+ - \varepsilon]$, so that the familiar argument will still go through, showing $z^- = z^+$, unless the measures ν^- and ν^+ are respectively concentrated in the vertices (z^-, w^-) and (z^+, w^+). When that happens, (15.6.9) reduces to

$$(15.7.4) \quad q(z^-, w^-) - \lambda(z^-, w^-)\eta(z^-, w^-) = q(z^+, w^+) - \lambda(z^+, w^+)\eta(z^+, w^+) .$$

In particular, let us apply (15.7.4) for the trivial entropy-entropy flux pair $(u, -v)$. At the "southwestern" vertex, $u^- = 0$ and $v^- = z^- = w^-$, while at the "northeastern" vertex, $u^+ = 0$ and $v^+ = z^+ = w^+$. Therefore, (15.7.4) yields $z^- = z^+ = w^- = w^+$. The proof is complete.

Smoothness of $\sigma(u)$ cannot be generally relaxed as examples indicate that the assertion of the above proposition may be false when $\sigma''(u)$ is discontinuous at $u = 0$.

In particular, Theorem 15.7.1 applies when the elastic medium responds like a "hard spring", that is, σ is concave at $u < 0$ and convex at $u > 0$:

$$(15.7.5) \qquad\qquad u\sigma''(u) > 0 , \quad u \neq 0 .$$

For that case, it is possible to establish L^∞ bounds on the approximate solutions constructed by the vanishing viscosity method, namely, as solutions to a Cauchy problem

$$(15.7.6) \qquad \begin{cases} \partial_t u_\mu - \partial_x v_\mu = \mu \partial_x^2 u_\mu \\ \partial_t v_\mu - \partial_x \sigma(u_\mu) = \mu \partial_x^2 v_\mu , \end{cases}$$

$$(15.7.7) \qquad (u_\mu(x, 0), v_\mu(x, 0)) = (u_{0\mu}(x), v_{0\mu}(x)) , \quad -\infty < x < \infty .$$

Theorem 15.7.2 *Under the assumption* (15.7.5), *for any* $M > 0$, *the set* $\mathcal{U}_M$, *defined by*

$$(15.7.8) \qquad \mathcal{U}_M = \{(u, v) : -M \leq z(u, v) \leq M, \ -M \leq w(u, v) \leq M\} ,$$

where z *and* w *are the Riemann invariants* (15.7.3) *of* (15.7.1), *is a (positively) invariant region for solutions of* (15.7.6), (15.7.7).

Proof. The standard proof is based on the maximum principle. An alternative proof will be presented here, which relies on entropies and thus is closer to the spirit of the hyperbolic theory. It has the advantage of requiring less regularity for solutions of (15.7.6). Moreover, it readily extends to any other approximation scheme, which, like (15.7.6), is dissipative under convex entropies of (15.7.1).

For the system (15.7.1), the equations (7.4.1) that determine entropy-entropy flux pairs (η, q) reduce to

$$(15.7.9) \qquad \begin{cases} q_u(u, v) = -\sigma'(u)\eta_v(u, v) \\ q_v(u, v) = -\eta_u(u, v) . \end{cases}$$

Notice that (15.7.9) admits the family of solutions

(15.7.10) $\eta_m(u, v) = Y_m(u)\cosh(mv) - 1$,

(15.7.11) $q_m(u, v) = -\dfrac{1}{m}Y_m'(u)\sinh(mv)$,

where $m = 1, 2, \cdots$ and Y_m is the solution of the ordinary differential equation

(15.7.12) $Y_m''(u) = m^2\sigma'(u)Y_m(u)$, $-\infty < u < \infty$,

with initial conditions

(15.7.13) $Y_m(0) = 1$, $Y_m'(0) = 0$.

A simple calculation gives

(15.7.14) $\eta_{muu}\eta_{mvv} - \eta_{muv}^2 \geq m^2[m^2\sigma'Y_m^2 - Y_m'^2]$.

Moreover, by virtue of (15.7.12),

(15.7.15) $[m^2\sigma'Y_m^2 - Y_m'^2]' = m^2\sigma''Y_m^2$.

Consequently, (15.7.5) implies that the right-hand side of (15.7.14) is positive and hence $\eta_m(u, v)$ is a convex function on $\mathbb{R}^2$. Furthermore, $\eta_m(0, 0) = 0$ and $\eta_{mu}(0, 0) = \eta_{mv}(0, 0) = 0$, so that $\eta_m(u, v)$ is positive definite.

Next we examine the asymptotics of $\eta_m(u, v)$ as $m \to \infty$. The change of variables $(u, Y_m) \mapsto (\xi, X_m)$:

(15.7.16) $\xi = \displaystyle\int_0^u [\sigma'(\omega)]^{\frac{1}{2}}d\omega$,

(15.7.17) $X_m = (\sigma')^{\frac{1}{4}}Y_m$,

transforms (15.7.12) into

(15.7.18) $\ddot{X}_m = m^2 X_m + [\tfrac{1}{4}(\sigma')^{-2}\sigma''' - \tfrac{5}{16}(\sigma')^{-3}(\sigma'')^2]X_m$,

with asymptotics

(15.7.19) $X_m(\xi) = [A + O(\dfrac{1}{m})]e^{m|\xi|}$,

as $m \to \infty$.

Upon combining (15.7.10) with (15.7.17), (15.7.19), (15.7.16) and (15.7.3), we deduce

(15.7.20) $\displaystyle\lim_{m\to\infty} \eta_m(u, v)^{\frac{1}{m}} = \begin{cases} \exp[z(u, v)] , & \text{if } u > 0, \quad v > 0, \\ \exp[w(u, v)] , & \text{if } u < 0, \quad v > 0, \\ \exp[-w(u, v)] , & \text{if } u > 0, \quad v < 0, \\ \exp[-z(u, v)] , & \text{if } u < 0, \quad v < 0. \end{cases}$

We now consider the solution (u_μ, v_μ) of (15.7.6), (15.7.7), where $(u_{0\mu}, v_{0\mu})$ lie in $L^2(-\infty, \infty)$ and take values in the region $\mathcal{U}_M$, defined by (15.7.8). We write (15.3.7), with $U_\mu = (u_\mu, v_\mu)$, $\eta = \eta_m$, $q = q_m$, and integrate it over $(-\infty, \infty) \times [0, t]$, to get

$$(15.7.21) \qquad \int_{-\infty}^{\infty} \eta_m(u_\mu(x, t), v_\mu(x, t))dx \leq \int_{-\infty}^{\infty} \eta_m(u_{0\mu}(x), v_{0\mu}(x))dx \ .$$

Raising (15.7.21) to the power $1/m$, letting $m \to \infty$ and using (15.7.20), we conclude that $(u_\mu(\cdot, t), v_\mu(\cdot, t))$ takes values in the region $\mathcal{U}_M$. The proof is complete.

The above proposition, in conjunction with Theorem 15.7.1, yields an existence theorem for the system (15.7.1), which is free from extraneous assumptions:

Theorem 15.7.3 *Let (u_μ, v_μ) be the solution of the initial-value problem (15.7.6), (15.7.7), on the upper half-plane, where $(u_{0\mu}, v_{0\mu}) \rightharpoonup (u_0, v_0)$ in $L^\infty(-\infty, \infty)$ weak*. Under the assumption (15.7.5), there is a sequence $\{\mu_j\}$, $\mu_j \to 0$ as $j \to \infty$, such that $\{(u_{\mu_j}, v_{\mu_j})\}$ converges almost everywhere on the upper half-plane to a solution $(\bar{u}, \bar{v})$ of (15.7.1) with initial values $(\bar{u}(x, 0), \bar{v}(x, 0)) = (u_0(x), v_0(x))$, $-\infty < x < \infty$.*

The assumption (15.7.5) and the use of the special, artifical viscosity (15.7.7) are essential in the proof of Theorem 15.7.3, because they appear to be indispensable for establishing uniform L^∞ bounds on approximate solutions. At the same time, it is interesting to know whether one may construct solutions to (15.7.1) by passing to the zero viscosity limit in the system (8.6.3) of viscoelasticity, or at least in the model system

$$(15.7.22) \qquad \begin{cases} \partial_t u_\mu - \partial_x v_\mu = 0 \\ \partial_t v_\mu - \partial_x \sigma(u_\mu) = \mu \partial_x^2 v_\mu \ , \end{cases}$$

which is close to it.

Even though we do not have uniform L^∞ estimates for solutions of (15.7.22), as this system is not dissipative with respect to all convex entropies of (15.7.1), we still have a number of estimates of L^p type, the most prominent among them being the "energy inequality" induced by the physical entropy-entropy flux pair (7.4.10). It is thus natural to inquire whether the method of compensated compactness is applicable in conjunction to such estimates. Of course, this would force us to abandon L^∞ and consider Young measures in the framework of L^p, a possibility already raised in Section 15.1. It turns out that this approach is effective for the problem at hand, albeit at the expense of elaborate analysis, so just the conclusion shall be recorded here. The proof is found in the references cited in Section 15.9.

Theorem 15.7.4 *Consider the system (15.7.22), where (a) $\sigma'(u) \geq \sigma_0 > 0$, for $-\infty < u < \infty$; (b) σ'' may vanish at most at one point on $(-\infty, \infty)$; (c) $\sigma'(u)$*

grows like $|u|^\alpha$, *as* $|u| \to \infty$, *for some* $\alpha \geq 0$; *and* (d) $\sigma''(u)$ *and* $\sigma'''(u)$ *grow no faster than* $|u|^{\alpha-1}$, *as* $|u| \to \infty$. *Let* (u_μ, v_μ) *be the solution of the Cauchy problem* (15.7.22), (15.7.7), *where* $\{(u_{0\mu}, v_{0\mu})\}$ *are functions in* $W^{1,2}(-\infty, \infty)$, *which have uniformly bounded total energy,*

$$(15.7.23) \qquad \int_{-\infty}^{\infty} \left[\frac{1}{2} v_{0\mu}^2(x) + \Sigma(u_{0\mu}) \right] dx \leq C ,$$

have relatively tame oscillations,

$$(15.7.24) \qquad \mu \int_{-\infty}^{\infty} [v_{0\mu}'(x)]^2 dx \to 0 , \quad \text{as } \mu \to 0 ,$$

and converge, $u_{0\mu} \to u_0$, $v_{0\mu} \to v_0$, *as* $\mu \to 0$, *in the sense of distributions. Then there is a sequence* $\{\mu_j\}$, $\mu_j \to 0$ *as* $j \to \infty$, *such that* $\{(u_{\mu_j}, v_{\mu_j})\}$ *converges in* L_{loc}^p, *for any* $1 < p < 2$, *to a solution* $(\bar{u}, \bar{v})$ *of* (15.7.1) *with initial values* $(\bar{u}(x, 0), \bar{v}(x, 0)) = (u_0(x), v_0(x))$, $-\infty < x < \infty$.

15.8 The System of Isentropic Gas Dynamics

The system (7.1.8) of isentropic gas dynamics, for a polytropic gas, in Eulerian coordinates, the first hyperbolic system of conservation laws ever to be derived, has served over the past two centuries as proving ground for testing the theory. It is thus fitting to conclude this work with the application of the method of compensated compactness to that system.

It is instructive to monitor the system simultaneously in its original form (7.1.8), with state variables density ρ and velocity v, as well as in its canonical form

$$(15.8.1) \qquad \begin{cases} \partial_t \rho + \partial_x m = 0 \\ \partial_t m + \partial_x \left[\dfrac{m^2}{\rho} + \kappa \rho^\gamma \right] = 0 , \end{cases}$$

with state variables density ρ and momentum $m = \rho v$. The physical range for density is $0 \leq \rho < \infty$, while v and m may take any values in $(-\infty, \infty)$.

For convenience, we scale the state variables so that $\kappa = (\gamma - 1)^2/4\gamma$, and set $\theta = \frac{1}{2}(\gamma - 1)$, in which case the characteristic speeds (7.2.10) and the Riemann invariants (7.3.3) assume the form

$$(15.8.2) \qquad \lambda = -\theta \rho^\theta + v = -\theta \rho^\theta + \frac{m}{\rho} , \qquad \mu = \theta \rho^\theta + v = \theta \rho^\theta + \frac{m}{\rho} ,$$

$$(15.8.3) \qquad z = -\rho^\theta + v = -\rho^\theta + \frac{m}{\rho} , \qquad w = \rho^\theta + v = \rho^\theta + \frac{m}{\rho} .$$

It is not difficult to construct sequences of approximate solutions taking values in compact sets of the state space $[0, \infty) \times (-\infty, \infty)$. For example, one may follow the vanishing viscosity approach relative to the system

$$(15.8.4) \quad \begin{cases} \partial_t \rho_\mu + \partial_x m_\mu = \mu \partial_x^2 \rho_\mu \\ \partial_t m_\mu + \partial_x \left[\dfrac{m_\mu^2}{\rho_\mu} + \kappa \rho_\mu^\gamma \right] = \mu \partial_x^2 m_\mu \, , \end{cases}$$

which admits the family of (positively) invariant regions

$$(15.8.5) \qquad \mathscr{U}_M = \{ (\rho, m) : \rho \geq 0, -M \leq z(\rho, m) \leq w(\rho, m) \leq M \} \, .$$

Furthermore, solutions of (15.8.4) on the upper half-plane, with initial data that are bounded in $L^\infty(-\infty, \infty) \cap L^2(-\infty, \infty)$, satisfy

$$(15.8.6) \qquad \partial_t \eta(\rho_\mu, m_\mu) + \partial_x q(\rho_\mu, m_\mu) \subset \text{compact set in } W_{loc}^{-1,2} \, ,$$

for any entropy-entropy flux pair (η, q) of (15.8.1). Approximate solutions with analogous properties are also constructed by finite difference schemes, like the Lax-Friedrichs scheme and the Godunov scheme. They all lead to the following existence theorem:

Theorem 15.8.1 *For any $\gamma > 1$, there exists a bounded solution (ρ, v) of the system (7.1.8) on the upper half-plane, with assigned initial data*

$$(15.8.7) \qquad (\rho(x, 0), v(x, 0)) = (\rho_0(x), v_0(x)) \, , \quad -\infty < x < \infty \, ,$$

where $(\rho_0, v_0) \in L^\infty(-\infty, \infty)$ and $\rho_0(x) \geq 0$, for $-\infty < x < \infty$. Furthermore, the solution satisfies the entropy admissibility condition

$$(15.8.8) \qquad \partial_t \eta(\rho, m) + \partial_x q(\rho, m) \leq 0 \, ,$$

for any entropy-entropy flux pair (η, q) of (15.8.1), with $\eta(\rho, m)$ convex.

The proof employs (15.3.10) to establish that the support of the Young measure, associated with a sequence of approximate solutions, either reduces to a single point in state space or is confined to the axis $\rho = 0$ (vacuum state).

As function of (ρ, v), any entropy η of (7.1.8) satisfies the integrability condition

$$(15.8.9) \qquad \eta_{\rho\rho} = \theta^2 \rho^{\gamma-3} \eta_{vv} \, .$$

The above equation is singular along the axis $\rho = 0$, and the nature of the singularity changes as one crosses the threshold $\gamma = 3$. Accordingly, different arguments have to be used for treating the cases $\gamma < 3$ and $\gamma > 3$.

Of relevance here are the so called *weak entropies*, which vanish at $\rho = 0$. Upon setting $\eta_\rho(0, v) = g(v)$, they admit the representation

$$(15.8.10) \qquad \eta(\rho, v) = \int_{-\infty}^{\infty} \chi(\rho, \xi - v) g(\xi) d\xi \, ,$$

where

$$(15.8.11) \qquad \chi(\rho, v) = \begin{cases} (\rho^{2\theta} - v^2)^s , & \text{if } \rho^{2\theta} > v^2 , \\ 0, & \text{if } \rho^{2\theta} \le v^2 , \end{cases}$$

with $s = \frac{1}{2}\frac{3-\gamma}{\gamma-1}$. Thus χ is the fundamental solution of (15.8.9) under initial conditions $\eta(0, v) = 0$, $\eta_\rho(0, v) = \delta_0(v)$.

The classical kinetic theory predicts the value $\gamma = 1 + \frac{2}{n}$ for the adiabatic exponent of a gas with n degrees of freedom. When the number of degrees of freedom is odd, $n = 2\ell + 1$, the exponent s in (15.8.11) is the integer ℓ. In this special situation the analysis of weak entropies and thereby the reduction of the Young measure is substantially simplified. However, even in that simpler case the proof is quite technical and shall be delegated to the references cited in Section 15.9. Only the degenerate case $\gamma = 3$ will be presented here.

For $\gamma = 3$, i.e. $\theta = 1$, (15.8.2) and (15.8.3) yield $\lambda = z$ and $\mu = w$, in which case the two characteristic families totally decouple. In particular, (12.2.1) reduce to $q_z = z\eta_z$, $q_w = w\eta_w$, so that there are entropy-entropy flux pairs (η, q) which depend solely on z, for example $(2z, z^2)$ and $(3z^2, 2z^3)$.

Suppose now a sequence $\{(\rho_{\mu_k}, m_{\mu_k})\}$ of solutions of (15.8.4), with $\mu_k \to 0$ as $k \to \infty$, induces a weakly convergent subsequence $\{(z_j, w_j)\}$ of Riemann invariants with associated family $\nu_{x,t}$ of Young measures. We fix (x, t), set $\nu_{x,t} = \nu$ and apply (15.3.10) for the two entropy-entropy flux pairs $(2z, z^2)$ and $(3z^2, 2z^3)$ to get

$$(15.8.12) \qquad 4\langle \nu, z \rangle \langle \nu, z^3 \rangle - 3\langle \nu, z^2 \rangle \langle \nu, z^2 \rangle = \langle \nu, z^4 \rangle .$$

Next we consider the inequality

$$(15.8.13) \qquad z^4 - 4z^3\bar{z} + 6z^2\bar{z}^2 - 4z\bar{z}^3 + \bar{z}^4 = (z - \bar{z})^4 \ge 0 ,$$

where $\bar{z} = \langle \nu, z \rangle$, and apply the measure ν to it, thus obtaining

$$(15.8.14) \qquad \langle \nu, z^4 \rangle - 4\langle \nu, z^3 \rangle \langle \nu, z \rangle + 6\langle \nu, z^2 \rangle \langle \nu, z \rangle^2 - 3\langle \nu, z \rangle^4 \ge 0 .$$

Combining (15.8.14) with (15.8.12) yields

$$(15.8.15) \qquad -3[\langle \nu, z^2 \rangle - \langle \nu, z \rangle^2]^2 \ge 0 ,$$

whence $\langle \nu, z^2 \rangle = \langle \nu, z \rangle^2$. Therefore, $\{z_j\}$ converges strongly to $\bar{z} = \langle \nu, z \rangle$. Similarly one shows that $\{w_j\}$ converges strongly to $\bar{w} = \langle \nu, w \rangle$. In particular, $(\bar{z}, \bar{w})$ induces a solution $(\bar{\rho}, \bar{v})$ of (7.1.8) by $\bar{\rho} = \frac{1}{2}(\bar{w} - \bar{z})$ and $\bar{v} = \frac{1}{2}(\bar{w} + \bar{z})$.

15.9 Notes

The method of compensated compactness was introduced by Murat [1] and Tartar [1,2]. The program of employing the method for constructing solutions to hyperbolic conservation laws was designed by Tartar [2,3], who laid down the fundamental condition (15.3.10) and demonstrated its use in the context of the scalar case. The first application to systems, due to DiPerna [6], provided the impetus for intensive development of these ideas, which has produced a substantial body of research. The presentation here only scratches the surface. A clear introduction is also found in the lecture notes of Evans [1] and Hörmander [1], the monograph of Malek, Neças and Rokyta [1], as well as the treatise of Taylor [1]. For more detailed, and deeper development of the subject the reader is referred to the book of Serre [9] and the forthcoming monograph of G.-Q. Chen.

The Young measure was introduced in L.C. Young [1]. The presentation here follows Ball [2], where the reader may find generalizations beyond the L^∞ framework, as well as commentary and references to alternative constructions.

For an introduction to the theory of compensated compactness, see the lecture notes of Tartar [1,2,3]. The div-curl lemma is due to Murat and Tartar and Lemma 15.2.1 is generally known as Murat's lemma (Murat [2]).

The notion of a measure-valued solution is due to DiPerna [9]. For further developments of the theory and applications to the construction of solutions to systems of conservation laws, including those of mixed type modeling phase transitions, see Chen and Frid [2], Coquel and LeFloch [1], Demengel and Serre [1], Frid [1], Poupaud and Rascle [1], Roytburd and Slemrod [1], Schochet [2] and Szepessy [1,2].

The scalar conservation law was first treated via the method of compensated compactness by Tartar [2]. The clever argument employed in the proof of Theorem 15.4.1 was communicated to the author by Luc Tartar, in May 1986. See also Vecchi [1].

The competition between viscosity and dispersion is investigated in Schonbek [1].

The program of constructing solutions to hyperbolic conservation laws via relaxation was initiated by Liu [13], who was motivated by the ideas of Whitham [2]. The discussion of the scheme in Section 15.5 is adapted from Chen, Levermore and Liu [1] and Jin and Xin [1]. Other studies of relaxation schemes include Chen and Liu [1], Collet and Rascle [1], Coquel and Perthame [1], Klingenberg and Lu [1], Lattanzio and Marcatti [1], Lu and Klingenberg [1], Marcati and Natalini [1], Marcati and Rubino [1], Natalini [1], Tveito and Winther [1] and Yong [2]. A survey is found in Natalini [3]. Tzavaras [3] discusses the interpretation of relaxation schemes in the context of Continuum Mechanics.

Shonbek [1] considers a balance law with singular source.

The scalar conservation law is treated in the L^p framework by Yang, Zhu and Zhao [1].

The treatment of the genuinely nonlinear system of two conservation laws, in Section 15.6, and the system of isentropic elasticity with a single inflection point, in Section 15.7, follows the pioneering paper of DiPerna [6]. Counterexamples to Theorem 15.7.1, when $\sigma''(u)$ is discontinuous at $u = 0$, are exhibited in Greenberg [3] and Greenberg and Rascle [1].

A very efficient approach, due to Serre [2,9], has rendered the method of compensated compactness sufficiently flexible to treat systems of two conservation laws even when characteristic families are linearly degenerate, strict hyperbolicity fails, etc. The construction of solutions to many interesting systems of this type is effected in Chen [6], Chen and Glimm [1,2], Chen and Kan [1], Heidrich [1], Kan, Santos and Xin [1], Lu [1], Marcati and Natalini [1], Rubino [1] and Zhao [1]. Since the analysis relies heavily on the availability of a rich family of entropies, the application of the method to systems of more than two conservation laws is currently limited to special systems in which the shock and rarefaction wave curves coincide for all but at most two characteristic families (Benzoni-Gavage and Serre [1]) and to the system of nonisentropic gas dynamics for a very special equation of state (Chen and Dafermos [1]). For a variety of systems, the large time behavior of solutions with initial values that are either periodic or L^1 perturbations of Riemann data is established in Chen and Frid [1,3,4,5,7], by combining scale invariance with compactness. See also Serre and Xiao [1].

The system of isentropic elasticity was treated in the L^p framework by J.W. Shearer [1], P. Lin [1] and Serre and Shearer [1]. The theory of invariant regions via the maximum principle is due to Chueh, Conley and Smoller [1] (see also Hoff [2]). A systematic discussion, with several examples, is found in Serre [9]. The connection between stability of relaxation schemes and existence of invariant regions is discussed in Serre [13]. The proof of Theorem 15.7.2 is taken from Dafermos [13]. See also Serre [3] and Venttsel' [1].

The system of isentropic gas dynamics was first treated by the method of compensated compactness in DiPerna [7], for the special values $\gamma = 1 + \frac{2}{n}$, $n = 2\ell + 1$, of the adiabatic exponent. Subsequently, G.-Q. Chen [1] and Ding, Chen and Luo [1] extended the analysis to any γ within the range $(1, \frac{5}{3}]$. For a survey, see G.-Q. Chen [2]. The case $\gamma \geq 3$ was solved by Lions, Perthame and Tadmor [1], and finally the full range $1 < \gamma < \infty$ is covered in Lions, Perthame and Souganidis [1]. The argument presented here, for the special case $\gamma = 3$ was communicated to the author by G.-Q. Chen. The more general, genuinely nonlinear system (7.1.7), for a nonpolytropic gas, was treated by Chen and LeFloch [1] under the assumption that near the vacuum state the pressure function $p(\rho)$, together with its first four derivatives, behave like $p(\rho) \sim \kappa \rho^\gamma$.

The method of compensated compactness is the only vehicle that is currently available for establishing weak solutions to hyperbolic systems of conservation laws via the method of vanishing viscosity; however, a more traditional approach is applicable in the special case where the solution happens to be smooth or piecewise smooth (Goodman and Xin [1], Lin and Yang [1] and Yu [1]).

An important test on systems of conservation laws is how the solution operator interacts with highly oscillatory initial data, say $U_{0\varepsilon}(x) = V(x, x/\varepsilon)$, where $V(x, \cdot)$ is periodic and ε is a small positive parameter. When the system is linear, the rapid oscillations are transported along characteristics and their frequency and amplitude persist for $t > 0$. On the opposite extreme, when the system is strictly hyperbolic and genuinely nonlinear, the results of Sections 15.4 and 15.6 indicate that, as $\varepsilon \to 0$, the resulting family of solutions $U_\varepsilon(x, t)$ contains sequences which converge strongly to solutions with initial value the weak limit of $\{U_{0\varepsilon}\}$, that is for $t > 0$ the solution operator quenches the rapid oscillations of the initial data. It is interesting to investigate intermediate situations, where some characteristic families may be linearly degenerate, strict hyperbolicity fails, etc. Following the study of many particular examples (cf. Bonnefille [1], Chen [3,4,5], E [1], Heibig [1], Rascle [1] and Serre [5,8]), a coherent theory of *propagation of oscillations* seems to be emerging (Serre [9]).

An alternative approach to propagation of oscillations is provided by the method of *weakly nonlinear geometric optics* which derives asymptotic expansions for solutions of hyperbolic systems under initial data oscillating with high frequency and small amplitude. Following the pioneering work of Landau [1], Lighthill [1] and Whitham [1], extensive literature has emerged, of purely formal, semirigorous or rigorous nature, dealing with the cases of a single phase, or possibly resonating multiphases, etc. See, for example, Choquet-Bruhat [1], Hunter and Keller [1,2], Majda and Rosales [1], Majda, Rosales and Schonbek [1], Pego [1], Hunter [1], Joly, Métivier and Rauch [1,3] and Cheverry [1]. It is remarkable that the asymptotic expansions remain valid even after shocks develop in the solution; see DiPerna and Majda [1], Schochet [5] and Cheverry [2]. A survey is found in Majda [4] and a systematic presentation is given in Serre [9].

Bibliography

Abeyaratne, R. and J.K. Knowles

1. Kinetic relations and the propagation of phase boundaries in solids. Arch. Rational Mech. Anal. **114** (1991), 119–154.
2. On the propagation of maximally dissipative phase boundaries in solids. Quart. Appl. Math. **50** (1992), 149–172.

Alber, H.D.

1. Local existence of weak solutions to the quasilinear wave equation for large initial values. Math. Z. **190** (1985), 249–276.
2. Global existence and large time behaviour of solutions for the equations of nonisentropic gas dynamics to initial values with unbounded support. Preprint No. 15, Sonderforschungsbereich 256, Bonn, 1988.

Alekseyevskaya, T.V.

1. Study of the system of quasilinear equations for isotachophoresis. Adv. Appl. Math. **11** (1990), 63–107.

Alinhac, S.

1. *Blowup for Nonlinear Hyperbolic Equations.* Boston: Birkhäuser, 1995.

Amadori, D.

1. Initial-boundary value problems for nonlinear systems of conservation laws. Nonl. Diff. Eqs. Appl. **4** (1997), 1–42.

Amadori, D., Baiti, P., LeFloch, P.G. and B. Piccoli

1. Nonclassical shocks and the Cauchy problem for nonconvex conservation laws. J. Diff. Eqs. **151** (1999), 345–372.

Amadori, D. and M. Colombo

1. Continuous dependence for 2×2 systems of conservation laws with boundary. J. Diff. Eqs. **138** (1997), 229–266.

Amadori, D. and G. Guerra

1. Global BV solutions and relaxation limit for a system of conservation laws. (Preprint).

Ancona, F. and A. Marson

1. Well-posedness for general 2×2 systems of conservation laws. (Preprint).

Antman, S.S.

1. *The Theory of Rods.* Handbuch der Physik, Vol. VIa/2. Berlin: Springer, 1972.
2. The equations for the large vibrations of strings. Amer. Math. Monthly **87** (1980), 359–370.
3. *Nonlinear Problems of Elasticity.* New York: Springer, 1995.

398   Bibliography

Antman, S.S. and R. Malek-Madani

1. Traveling waves in nonlinearly viscoelastic media and shock structure in elastic media. Quart. Appl. Math. **46** (1988), 77–93.

Anzellotti, G.

1. Pairings between measures and bounded functions and compensated compactness. Ann. Mat. Pura Appl. **135** (1983), 293–318.

Azevedo, A.V., Marchesin, D., Plohr, B.J. and K. Zumbrun

1. Nonuniqueness of solutions of Riemann problems. ZAMP **47** (1996), 977–998.

Baiti, P. and H.K. Jenssen

1. Well-posedness for a class of 2×2 conservation laws with L^∞ data. J. Diff. Eqs. **140** (1997), 161–185.
2. On the front tracking algorithm. J. Math. Anal. Appl. **217** (1998), 395–404.

Ball, J.M.

1. Convexity conditions and existence theorems in nonlinear elasticity. Arch. Rational Mech. Anal. **63** (1977), 337–403.
2. A version of the fundamental theorem for Young measures. *Partial Differential Equations and Continuum Models of Phase Transitions*, pp. 241–259, ed. M. Rascle, D. Serre and M. Slemrod. Lecture Notes in Physics No. **344**. Berlin: Springer, 1989.

Ballou, D.

1. Solutions to nonlinear hyperbolic Cauchy problems without convexity conditions. Trans. AMS **152** (1970), 441–460.
2. Weak solutions with a dense set of discontinuities. J. Diff. Eqs. **10** (1971), 270–280.

Bardos, C., Leroux, A.-Y. and J.-C. Nédélec

1. First order quasilinear equations with boundary conditions. Comm. PDE **4** (1979), 1017–1034.

Barker, L.M.

1. A computer program for shock wave analysis. Sandia National Labs. Albuquerque, 1963.

Bateman, H.

1. Some recent researches on the motion of fluids. Monthly Weather Review **43** (1915), 163–170.

Bauman, P. and D. Phillips

1. Large time behavior of solutions to a scalar conservation law in several space dimensions. Trans. AMS **298** (1986), 401–419.

Benabdallah, A. and D. Serre

1. Problèmes aux limites pour des systèmes hyperboliques nonlinéaires de deux equations à une dimension d'espace. C. R. Acad. Sci. Paris, Série I, **305** (1987), 677–680.

Bénilan, Ph. and M.G. Crandall

1. Regularizing effects of homogeneous evolution equations. Am. J. Math. Supplement dedicated to P. Hartman (1981), 23–39.

Bénilan, Ph. and S. Kruzkov

1. Conservation laws with continuous flux functions. No DEA Nonlinear Differential Equations Appl. **3** (1996), 395–419.

Benzoni-Gavage, S.

1. On a representation formula for B. Temple systems. SIAM J. Math. Anal. **27** (1996), 1503–1519.
2. Stability of subsonic planar phase boundaries in a van der Waals fluid. Arch. Rational Mech. Anal. (To appear).

Benzoni-Gavage, S. and D. Serre

1. Compacité par compensation pour une classe de systèmes hyperboliques de p lois de conservation ($p \geq 3$). Rev. Math. Iberoamericana **10** (1994), 557–579.

Bethe, H.

1. Report on the theory of shock waves for an arbitrary equation of state. Report No. PB-32189. Clearinghouse for Federal Scientific and Technical Information, U.S. Dept. of Commerce, Washington DC, 1942.

Bloom, F.

1. *Mathematical Problems of Classical Nonlinear Electromagnetic Theory.* Harlow: Longman, 1993.

Boillat, G.

1. *La Propagation des Ondes.* Paris: Gauthier-Villars, 1965.
2. Chocs characteristiques. C. R. Acad. Sci. Paris, Série I, **274** (1972), 1018–1021.
3. Involutions des systèmes conservatifs. C. R. Acad. Sci. Paris, Série I, **307** (1988), 891–894.

Boillat, G. and T. Ruggeri

1. Hyperbolic principal subsystems: entropy convexity and subcharacteristic conditions. Arch. Rational Mech. Anal. **137** (1997), 305–320.

Bonnefille, M.

1. Propagation des oscillations dans deux classes de systèmes hyperboliques (2×2 et 3×3). Comm. PDE **13** (1988), 905–925.

Bouchut, F. and F. James

1. Duality solutions for pressureless gases. C. R. Acad. Sci. Paris, Série I, **326** (1998), 1073–1078.

Bouchut, F. and B. Perthame

1. Kruzkov's estimates for scalar conservation laws revisited. Trans. AMS, **350** (1998), 2847–2870.

Brenier, Y.

1. Résolution d'équations d'évolutions quasilinéaires en dimension n d'espace à l'aide d'équations linéaires en dimension $n + 1$. J. Diff. Eqs. **50** (1983), 375–390.

Brenier, Y. and L. Corrias

1. A kinetic formulation for multi-branch entropy solutions of scalar conservation laws. Ann. Inst. Henri Poincaré **15** (1998), 169–190.

Brenier Y. and E. Grenier

1. Sticky particles and scalar conservation laws. SIAM J. Num. Anal. **35** (1998), 2317–2328.

Brenner, P.

1. The Cauchy problem for the symmetric hyperbolic systems in L_p. Math. Scand. **19** (1966), 27–37.

400 Bibliography

Bressan, A.

1. Contractive metrics for nonlinear hyperbolic systems. Indiana U. Math. J. **37** (1988), 409–421.
2. Global solutions of systems of conservation laws by wave-front tracking. J. Math. Anal. Appl. **170** (1992), 414–432.
3. The unique limit of the Glimm scheme. Arch. Rational Mech. Anal. **130** (1995), 205–230.
4. *Lecture Notes on Systems of Conservation Laws.* Trieste: SISSA, 1995.
5. The semigroup approach to systems of conservation laws. Math. Contemp. **10** (1996), 21–74.

Bressan, A. and R. M. Colombo

1. The semigroup generated by 2×2 conservation laws. Arch. Rational Mech. Anal. **133** (1995), 1–75.
2. Unique solutions of 2×2 conservation laws with large data. Indiana U. Math. J. **44** (1995), 677–725.
3. Decay of positive waves in nonlinear systems of conservation laws. Ann. Scu. Norm. Sup. Pisa **IV-26** (1998), 133–160.

Bressan, A., Crasta, G. and B. Piccoli

1. Well posedness of the Cauchy problem for $n \times n$ systems of conservation laws. Memoirs AMS. (To appear).

Bressan, A. and P. Goatin

1. Oleinik type estimates and uniqueness for $n \times n$ conservation laws. J. Diff. Eqs. **156** (1999), 26–49.

Bressan, A. and P.G. LeFloch

1. Uniqueness of weak solutions to hyperbolic systems of conservation laws. Arch. Rational Mech. Anal. **140** (1997), 301–317.
2. Structural stability and regularity of entropy solutions to hyperbolic systems of conservation laws. Indiana U. Math. J. (To appear).

Bressan, A. and M. Lewicka

1. A uniqueness condition for hyperbolic systems of conservation laws. (Preprint).

Bressan, A, Liu, T.-P. and T. Yang

1. L^1 stability estimates for $n \times n$ conservation laws. Arch. Rational Mech. Anal. (To appear).

Bressan, A. and A. Marson

1. Error bounds for a deterministic version of the Glimm scheme. Arch. Rational Mech. Anal. **142** (1998), 155–176.

Brio, M. and J.K. Hunter

1. Rotationally invariant hyperbolic waves. Comm. Pure Appl. Math. **43** (1990), 1037–1053.

Burgers, J.

1. Application of a model system to illustrate some points of the statistical theory of free turbulence. Neder. Akad. Wefensh. Proc. **43** (1940), 2–12.

Burton, C.V.

1. On plane and spherical sound-waves of finite amplitude. Philos. Magazine, Ser. 5, **35** (1893), 317–333.

Cabannes, H.

1. *Theoretical Magnetofluiddynamics.* New York: Academic Press, 1970.

Caginalp, G.

1. Nonlinear equations with coefficients of bounded variation in two space variables. J. Diff. Eqs. **43** (1982), 134–155.

Čanić, S.

1. On the influence of viscosity on Riemann solutions. J. Dyn. Diff. Eqs. **10** (1998), 109–149.

Čanić, S. and B.L. Keyfitz

1. Quasi-one-dimensional Riemann problems and their role in self-similar two-dimensional problems. Arch. Rational Mech. Anal. **144** (1998), 233–258.
2. Riemann problems for the two-dimensional unsteady transonic small disturbance equation. SIAM J. Appl. Math. **58** (1998), 636–665.

Čanić, S. and G.R. Peters

1. Nonexistence of Riemann solutions and Majda-Pego instability. (Preprint).

Čanić, S. and B.J. Plohr

1. Shock wave admissibility for quadratic conservation laws. J. Diff. Eqs. **118** (1995), 293–335.

Cauchy, A.-L.

1. Recherches sur l'équilibre et le mouvement intérieur des corps solides ou fluides, élastiques ou non élastiques. Bull. Soc. Philomathique (1823), 9–13.
2. De la pression ou tension dans un corps solide. *Exercises de Mathématiques* **2** (1827), 42–56.
3. Sur les relations qui existent dans l'état d'équilibre d'un corps solide ou fluide, entre les pressions ou tensions et les forces accélératrices. *Exercises de Mathématiques* **2** (1827), 108–111.
4. Sur l'équilibre et le mouvement intérieur des corps considérés comme des masses continues. *Exercises de Mathématiques* **4** (1829), 293–319.

Challis, J.

1. On the velocity of sound. Philos. Magazine **32** (1848), 494–499.

Chang, T. and L. Hsiao

1. Riemann problem and discontinuous initial value problem for typical quasilinear hyperbolic system without convexity. Acta Math. Sinica **20** (1977), 229–231.
2. A Riemann problem for the system of conservation laws of aerodynamics without convexity. Acta Math. Sinica **22** (1979), 719–732.
3. *The Riemann Problem and Interaction of Waves in Gas Dynamics.* Harlow: Longman, 1989.

Chen, G.-Q.

1. Convergence of the Lax-Friedrichs scheme for isentropic gas dynamics (III). Acta Math. Scientia **6** (1986), 75–120.
2. The compensated compactness method and the system of isentropic gas dynamics. Berkeley: Math. Sci. Res. Inst. Preprint #00527-91, 1990.
3. Propagation and cancellation of oscillations for hyperbolic systems of conservation laws. Comm. Pure Appl. Math. **44** (1991), 121–139.
4. Hyperbolic systems of conservation laws with a symmetry. Comm. PDE **16** (1991), 1461–1487.

5. The method of quasidecoupling for discontinuous solutions to conservation laws. Arch. Rational Mech. Anal. **121** (1992), 131–185.
6. Remarks on global solutions to the compressible Euler equations with spherical symmetry. Proc. Royal Soc. Edinburgh **127A** (1997), 243–259.

Chen, G.-Q. and C.M. Dafermos

1. The vanishing viscosity method in one-dimensional thermoelasticity. Trans. AMS **347** (1995), 531–541.

Chen, G.-Q., Du, Q. and E. Tadmor

1. Spectral viscosity approximations to multidimensional scalar conservation laws. Math. Comp. **61** (1993), 629–643.

Chen, G.-Q. and H. Frid

1. *Asymptotic Stability and Decay of Solutions of Conservation Laws.* Lecture Notes, Northwestern U., 1996.
2. Existence and asymptotic behavior of the measure-valued solutions for degenerate conservation laws. J. Diff. Eqs. **127** (1996), 197–224.
3. Asymptotic stability of Riemann waves for conservation laws. ZAMP **48** (1997), 30–44.
4. Large time behavior of entropy solutions of conservation laws. J. Diff. Eqs. **152** (1999), 308–357.
5. Decay of entropy solutions of nonlinear conservation laws. Arch. Rational Mech. Anal. **146** (1999), 95–127.
6. Divergence measure fields and conservation laws. Arch. Rational Mech. Anal. **147** (1999), 89–118.
7. Uniqueness and asymptotic stability of entropy solutions for the compressible Euler equations. Trans. AMS. (To appear).

Chen, G.-Q. and J. Glimm

1. Global solutions to the compressible Euler equations with geometric structure. Comm. Math. Phys. **180** (1996), 153–193.
2. Global solutions to the cylindrically symmetric rotating motion of isentropic gas. ZAMP **47** (1996), 353–372.

Chen, G.-Q. and P.-T. Kan

1. Hyperbolic conservation laws with umbilic degeneracy, I. Arch. Rational Mech. Anal. **130** (1995), 231–276.

Chen, G.-Q. and P.G. LeFloch

1. Compressible Euler equations with general pressure law and related equations. Arch. Rational Mech. Anal. (To appear).

Chen, G.-Q., Levermore, C.D. and T.-P. Liu

1. Hyperbolic conservation laws with stiff relaxation terms and entropy. Comm. Pure Appl. Math. **47** (1994), 787–830.

Chen, G.-Q., Li, D. and D. Tan

1. Structure of Riemann solutions for two-dimensional scalar conservation laws. J. Diff. Eqs. **127** (1996), 124–147.

Chen, G.-Q. and T.-P. Liu

1. Zero relaxation and dissipation limits for hyperbolic conservation laws. Comm. Pure Appl. Math. **46** (1993), 755–781.

Chen, J.

1. Conservation laws for the relativistic *p*-system. Comm. PDE **20** (1995), 1605–1646.

Cheng, K.-S.

1. Asymptotic behavior of solutions of a conservation law without convexity conditions. J. Diff. Eqs. **40** (1981), 343–376.
2. Decay rate of periodic solutions for a conservation law. J. Diff. Eqs. **42** (1981), 390–399.
3. A regularity theorem for a nonconvex scalar conservation law. J. Diff. Eqs. **61** (1986), 79–127.

Cheverry, C.

1. The modulation equations of nonlinear geometric optics. Comm. PDE **21** (1996), 1119–1140.
2. Justification de l'optique géométrique non linéaire pour un système de lois de conservations. Duke Math. J. **87** (1997), 213–263.
3. Système de lois de conservations et stabilité BV. Mémoires Soc. Math. France No. **75** (1998).
4. Regularizing effects for multidimensional scalar conservation laws. Ann. Inst. Henri Poincaré. (To appear).

Choksi, R.

1. The conservation law $\partial_y u + \partial_x \sqrt{1 - u^2} = 0$ and deformations of fibre reinforced materials. SIAM J. Appl. Math. **56** (1996), 1539–1560.

Choquet-Bruhat, V.

1. Ondes asymptotiques et approchées pour systèmes d'équations aux dérivées paratielles nonlinéaires. J. Math. Pures Appl. **48** (1969), 117–158.

Chorin, A.J.

1. Random choice solution of hyperbolic systems. J. Comp. Physics **22** (1976), 517–533.

Christoffel, E.B.

1. Untersuchungen über die mit der Fortbestehen linearer partieller Differentialgleichungen verträglichen Unstetigkeiten. Ann. Mat. Pura Appl. **8** (1877), 81–113.

Chueh, K.N., Conley, C.C. and J.A. Smoller

1. Positively invariant regions for systems of nonlinear diffusion equations. Indiana U. Math. J. **26** (1977), 372–411.

Ciarlet, P.G.

1. *Mathematical Elasticity.* Amsterdam: North-Holland, 1988.

Clausius, R.

1. Über einer veränderte Form des zweiten Hauptsatzes der mechanischen Wärmetheorie. Ann. Physik **93** (1854), 481–506.

Coleman, B.D. and E.H. Dill

1. Thermodynamic restrictions on the constitutive equations of electromagnetic theory. ZAMP **22** (1971), 691–702.

Coleman, B.D. and V.J. Mizel

1. Existence of caloric equations of state in thermodynamics. J. Chem. Phys. **40** (1964), 1116–1125.

Coleman, B.D. and W. Noll

1. The thermodynamics of elastic materials with heat conduction and viscosity. Arch. Rational Mech. Anal. **13** (1963), 167–178.

Collet, J.F. and M. Rascle

1. Convergence of the relaxation approximation to a scalar nonlinear hyperbolic equation arising in chromatography. ZAMP **47** (1996), 400–409.

Colombo, R.M. and N.H. Risebro

1. Continuous dependence in the large for some equations of gas dynamics. Comm. PDE **23** (1998), 1693–1718.

Conley, C. and J. Smoller

1. Viscosity matrices for two-dimensional nonlinear hyperbolic systems. Comm. Pure Appl. Math. **23** (1970), 867–884.
2. Viscosity matrices for two-dimensional nonlinear hyperbolic systems, II. Amer. J. Math. **94** (1972), 631–650.

Conlon, J.G.

1. Asymptotic behavior for a hyperbolic conservation law with periodic initial data. Comm. Pure Appl. Math. **32** (1979), 99–112.

Conlon, J. and T.-P. Liu

1. Admissibility criteria for hyperbolic conservation laws. Indiana U. Math. J. **30** (1981), 641–652.

Conway, E.

1. The formation and decay of shocks of a conservation law in several dimensions. Arch. Rational Mech. Anal. **64** (1977), 47–57.

Conway, E.D. and J.A. Smoller

1. Global solutions of the Cauchy problem for quasi-linear first order equations in several space variables. Comm. Pure Appl. Math. **19** (1966), 95–105.

Coquel, F. and P.G. LeFloch

1. Convergence of finite difference schemes for conservation laws in several space variables: a general theory. SIAM J. Num. Anal. **30** (1993), 675–700.

Coquel, F. and B. Perthame

1. Relaxation of energy and approximate Riemann solvers for general pressure laws in fluid dynamics. SIAM J. Num. Anal. **35** (1998), 2223–2249.

Corli, A. and M. Sablé-Tougeron

1. Perturbations of bounded variation of a strong shock wave. J. Diff. Eqs. **138** (1997), 195–228.

Cosserat, E. and F.

1. *Théorie des Corps Déformables.* Paris: Hermann, 1909.

Courant, R. and K.O. Friedrichs

1. *Supersonic Flow and Shock Waves.* New York: Wiley-Interscience, 1948.

Crandall, M.G.

1. The semigroup approach to first-order quasilinear equations in several space varibles. Israel J. Math. **12** (1972), 108–132.

Crandall, M.G. and T.M. Liggett

1. Generation of semi-groups of nonlinear transformations of general Banach spaces. Amer. J. Math. **93** (1971), 265–298.

Crandall, M.G. and A. Majda

1. The method of fractional steps for conservation laws. Math. Comput. **34** (1980), 285–314.

Crasta G. and B. Piccoli

1. Viscosity solutions and uniqueness for systems of inhomogeneous balance laws. Discr. Cont. Dyn. Syst. **3** (1997), 477–502.

Dacorogna, B.

1. *Weak Continuity and Weak Lower Semicontinuity of Non-Linear Functionals.* Lecture Notes in Math. **922** (1982). Berlin: Springer.

Dafermos, C.M.

1. Asymptotic behavior of solutions of a hyperbolic conservation law. J. Diff. Eqs. **11** (1972), 416–424.
2. Polygonal approximations of solutions of the initial value problem for a conservation law. J. Math. Anal. Appl. **38** (1972), 33–41.
3. The entropy rate admissibility criterion for solutions of hyperbolic conservation laws. J. Diff. Eqs. **14** (1973), 202–212.
4. Solution of the Riemann problem for a class of hyperbolic systems of conservation laws by the viscosity method. Arch. Rational Mech. Anal. **52** (1973), 1–9.
5. Structure of solutions of the Riemann problem for hyperbolic systems of conservation laws. Arch. Rational Mech. Anal. **53** (1974), 203–217.
6. Quasilinear hyperbolic systems that result from conservation laws. *Nonlinear Waves*, pp. 82–102, ed. S. Leibovich and A. R. Seebass. Ithaca: Cornell U. Press, 1974.
7. Characteristics in hyperbolic conservation laws. *Nonlinear Analysis and Mechanics: Heriot-Watt Symposium*, Vol. I, pp. 1–58, ed. R.J. Knops. London: Pitman, 1977.
8. Generalized characteristics and the structure of solutions of hyperbolic conservation laws. Indiana U. Math. J. **26** (1977), 1097–1119.
9. The second law of thermodynamics and stability. Arch. Rational Mech. Anal. **70** (1979), 167–179.
10. Hyperbolic systems of conservation laws. *Systems of Nonlinear Partial Differential Equations*, pp. 25–70, ed. J.M. Ball. Dordrecht: D. Reidel 1983.
11. Regularity and large time behavior of solutions of a conservation law without convexity. Proc. Royal Soc. Edinburgh **99A** (1985), 201–239.
12. Quasilinear hyperbolic systems with involutions. Arch. Rational Mech. Anal. **94** (1986), 373–389.
13. Estimates for conservation laws with little viscosity. SIAM J. Math. Anal. **18** (1987), 409–421.
14. Trend to steady state in a conservation law with spatial inhomogeneity. Quart. Appl. Math. **45** (1987), 313–319.
15. Admissible wave fans in nonlinear hyperbolic systems. Arch. Rational Mech. Anal. **106** (1989), 243–260.
16. Generalized characteristics in hyperbolic systems of conservation laws. Arch. Rational Mech. Anal. **107** (1989), 127–155.
17. Equivalence of referential and spatial field equations in continuum physics. Notes Num. Fluid Mech. **43** (1993), 179–183.
18. Large time behavior of solutions of hyperbolic systems of conservation laws with periodic initial data. J. Diff. Eqs. **121** (1995), 183–202.
19. Stability for systems of conservation laws in several space dimensions. SIAM J. Math. Anal. **26** (1995), 1403–1414.
20. A system of hyperbolic conservation laws with frictional damping. ZAMP, Special Issue, **46** (1995), S294–S307.
21. Entropy and the stability of classical solutions of hyperbolic systems of conservation laws. Lecture Notes in Math. **1640** (1996), 48–69. Berlin: Springer.

Dafermos, C.M. and R.J. DiPerna

1. The Riemann problem for certain classes of hyperbolic systems of conservation laws. J. Diff. Eqs. **20** (1976), 90–114.

Dafermos, C.M. and X. Geng

1. Generalized characteristics in hyperbolic systems of conservation laws with special coupling. Proc. Royal Soc. Edinburgh **116A** (1990), 245–278.
2. Generalized characteristics, uniqueness and regularity of solutions in a hyperbolic system of conservation laws. Ann. Inst. Henri Poincaré **8** (1991), 231–269.

Dafermos, C.M. and W.J. Hrusa

1. Energy methods for quasilinear hyperbolic initial-boundary value problems. Applications to elastodynamics. Arch. Rational Mech. Anal. **87** (1985), 267–292.

Dafermos, C.M. and L. Hsiao

1. Hyperbolic systems of balance laws with inhomogeneity and dissipation. Indiana U. Math. J. **31** (1982), 471–491.

Dal Masso, G., LeFloch, P. and F. Murat

1. Definition and weak stability of nonconservative products. J. Math. Pures Appl. **74** (1995), 483–548.

Demengel, F. and D. Serre

1. Nonvanishing singular parts of measure-valued solutions for scalar hyperbolic equations. Comm. PDE **16** (1991), 221–254.

DeVore, R.A. and B.J. Lucier

1. On the size and smoothness of solutions to nonlinear hyperbolic conservation laws. SIAM J. Math. Anal. **27** (1996), 684–707.

Diehl, S.

1. A conservation law with point source and discontinuous flux function modelling continuous sedimentation. SIAM J. Appl. Math. **56** (1996), 388–419.

Ding, X., Chen, G.-Q. and P. Luo

1. Convergence of the Lax-Friedrichs scheme for the isentropic gas dynamics (I)–(II). Acta Math. Scientia **5** (1985), 483–500, 501–540; **7** (1987), 467–480.

DiPerna, R.J.

1. Singularities of solutions of nonlinear hyperbolic systems of conservation laws. Arch. Rational Mech. Anal. **60** (1975), 75–100.
2. Decay and asymptotic behavior of solutions to nonlinear hyperbolic systems of conservation laws. Indiana U. Math. J. **24** (1975), 1047–1071.
3. Global existence of solutions to nonlinear hyperbolic systems of conservation laws. J. Diff. Eqs. **20** (1976), 187–212.
4. Decay of solutions of hyperbolic systems of conservation laws with a convex extension. Arch. Rational Mech. Anal. **64** (1977), 1–46.
5. Uniqueness of solutions to hyperbolic conservation laws. Indiana U. Math. J. **28** (1979), 137–188.
6. Convergence of approximate solutions to conservation laws. Arch. Rational Mech. Anal. **82** (1983), 27–70.
7. Convergence of the viscosity method for isentropic gas dynamics. Comm. Math. Phys. **91** (1983), 1–30.
8. Compensated compactness and general systems of conservation laws. Trans. A.M.S. **292** (1985), 283–420.
9. Measure-valued solutions to conservation laws. Arch. Rational Mech. Anal. **88** (1985), 223–270.

DiPerna, R.J. and P.-L. Lions

1. On the Cauchy problem for Boltzmann equations: Global existence and weak stability. Ann. Math. **130** (1989), 321–366.

DiPerna, R. and A. Majda

1. The validity of nonlinear geometric optics for weak solutions of conservation laws. Comm. Math. Phys. **98** (1985), 313–347.

Douglis, A.

1. Layering methods for nonlinear partial differential equations of first order. Ann. Inst. Fourier, Grenoble **22** (1972), 141–227.

DuBois, F. and P.G. LeFloch

1. Boundary conditions for nonlinear hyperbolic systems of conservation laws. J. Diff. Eqs. **71** (1988), 93–122.

Dubroca, B. and G. Gallice

1. Résultats d'existence et d'unicité du problème mixte pour des systèmes hyperbolique de lois de conservation monodimensionels. Comm. PDE **15** (1990), 59–80.

Duhem, P.

1. Recherches sur l'hydrodynamique. Ann. Toulouse **3** (1901), 315–377.

E, Weinan

1. Propagation of oscillations in the solutions of $1 - d$ compressible fluid equations. Comm. PDE **17** (1992), 347–370.
2. Homogenization of scalar conservation laws with oscillatory forcing terms. SIAM J. Appl. Math. **52** (1992), 959–972.
3. Aubry-Mather theory and periodic solutions of the forced Burgers equation. Comm. Pure Appl. Math. **52** (1999), 811–828.

E, Weinan, Khanin, K., Mazel, A. and Ya Sinai

1. Invariant measures for Burgers equation with stochastic forcing. (Preprint).

E, Weinan, Rykov, Yu. and Ya. G. Sinai

1. Generalized variational principles, global existence of weak solutions and behavior with random initial data for systems of conservation laws arising in adhesion particle dynamics. Comm. Math. Phys. **177** (1996), 349–380.

E, Weinan and D. Serre

1. Correctors for the homogenization of conservation laws with oscillatory forcing terms. Asymptotic Analysis **5** (1992), 311–316.

Earnshaw, S.

1. On the mathematical theory of sound. Trans. Royal Soc. London **150** (1860), 133–148.

Engquist, B. and Weinan E

1. Large time behavior and homogenization of solutions of two-dimensional conservation laws. Comm. Pure Appl. Math. **46** (1993), 1–26.

Ercole, G.

1. Delta-shock waves as self-similar viscosity limits. Quart. Appl. Math. (To appear).

Euler, L.

1. Principes généraux du mouvement des fluides. Mém. Acad. Sci. Berlin **11** (1755), 274–315.

2. Supplément aux recherches sur la propagation du son. Mém. Acad. Sci. Berlin **15** (1759), 210–240.

Evans, L.C.

1. *Weak Convergence Methods for Nonlinear Partial Differential Equations.* CBMS Regional Conference Series in Mathematics No. **74**. Providence: American Mathematical Society, 1990.

Evans, L.C. and R.F. Gariepy

1. *Measure Theory and Fine Properties of Functions.* Boca Raton: CRC Press, 1992.

Fan, H.-T.,

1. A limiting "viscosity" approach to the Riemann problem for materials exhibiting change of phase. Arch. Rational Mech. Anal. **116** (1992), 317–338.
2. One phase Riemann problems and wave interactions in systems of conservation laws of mixed type. SIAM J. Math. Anal. **24** (1993), 840–865.

Fan H.-T. and J.K. Hale

1. Large time behavior in inhomogeneous conservation laws. Arch. Rational Mech. Anal. **125** (1993), 201–216.
2. Attractors in inhomogeneous conservation laws and parabolic regularizations. Trans. AMS **347** (1995), 1239–1254.

Fan, H., Jin S. and Z.-H. Teng

1. Zero reaction limit for hyperbolic conservation laws with source terms. J. Diff. Eqs. (To appear).

Federer, H.

1. *Geometric Measure Theory.* New York: Springer, 1969.

Feireisl, E. and H. Petzeltová

1. Long-time behaviour for multidimensional scalar conservation laws. J. Reine Angew. Math. (To appear).

Ferziger, J.H. and H.G. Kaper

1. *Mathematical Theory of Transport Processes in Gases.* Amsterdam: North-Holland, 1972.

Fife, P.C. and X. Geng

1. Mathematical aspects of electrophoresis. *Reaction-Diffusion Equations*, pp. 139–172, eds. K.J. Brown and A.A. Lacey. Oxford: Clarendon Press, 1990.

Filippov, A.F.

1. *Differential Equations with Discontinuous Righthand Sides.* Dordrecht: Kluwer, 1988.

Foy, R.L.

1. Steady state solutions of hyperbolic systems of conservation laws with viscosity terms. Comm. Pure Appl. Math. **17** (1964), 177–188.

Freistühler, H.

1. Instability of vanishing viscosity approximation to hyperbolic systems of conservation laws with rotational invariance. J. Diff. Eqs. **87** (1990), 205–226.
2. Rotational degeneracy of hyperbolic systems of conservation laws. Arch. Rational Mech. Anal. **113** (1991), 39–64.
3. Dynamical stability and vanishing viscosity. A case study of a non-strictly hyperbolic system. Comm. Pure Appl. Math. **45** (1992), 561–582.

Freistühler, H. and T.-P. Liu

1. Nonlinear stability of overcompressive shock waves in a rotationally invariant system of viscous conservation laws. Comm. Math. Phys. **153** (1993), 147–158.

Freistühler, H. and D. Serre

1. L^1 stability of shock waves in scalar viscous conservation laws. Comm. Pure Appl. Math. **51** (1998), 291–301.

Freistühler, H. and P. Szmolyan

1. Existence and bifurcation of viscous profiles for all intermediate magnetohydrodynamic shock waves. SIAM J. Math. Anal. **26** (1995), 112–128.

Frid, H.

1. Initial-boundary value problems for conservation laws. J. Diff. Eqs. **128** (1996), 1–45.
2. Measure-valued solutions to initial-boundary value problems for certain systems of conservation laws: Existence and dynamics. Trans. AMS **348** (1996), 51–76.

Frid H. and I.-S. Liu

1. Oscillation waves in Riemann problems for phase transitons. Quart. Appl. Math. **56** (1998), 115–135.

Friedrichs, K.O.

1. Nonlinear hyperbolic differential equations for functions of two independent variables. Am. J. Math. **70** (1948), 555–589.
2. On the laws of relativistic electro-magneto-fluid dynamics. Comm. Pure Appl. Math. **27** (1974), 749–808.

Friedrichs, K.O. and P.D. Lax

1. Systems of conservation equations with a convex extension. Proc. Natl. Acad. Sci. USA **68** (1971), 1686–1688.

Fries, C.

1. Nonlinear asymptotic stability of general small-amplitude viscous Laxian shock waves. J. Diff. Eqs. **146** (1998), 185–202.
2. Stability of viscous shock waves associated with non-convex modes. Arch. Rational Mech. Anal. (To appear).

Gardner, R.A. and K. Zumbrun

1. The gap lemma and geometric criteria for instability of viscous shock profiles. Comm. Pure Appl. Math. **51** (1998), 797–855.

Gelfand, I.

1. Some problems in the theory of quasilinear equations. Usp. Mat. Nauk **14** (1959), 87–158. English translation: AMS Translations, Ser. II, **29**, 295–381.

Gilbarg, D.

1. The existence and limit behavior of the one-dimensional shock layer. Am. J. Math. **73** (1951), 256–274.

Gimse, T.

1. Conservation laws with discontinuous flux functions. SIAM J. Math. Anal. **24** (1993), 279–289.

Gimse, T. and N.H. Risebro

1. Solution of the Cauchy problem for a conservation law with a discontinuous flux function. SIAM J. Math. Anal. **23** (1992), 635–648.

Gisclon, M.

1. Etude des conditions aux limites pour un système strictement hyperbolique via l'approximation parabolique. J. Math. Pures Appl. **75** (1996), 485–508.

Gisclon, M. and D. Serre

1. Etude des conditions aux limites pour un système strictement hyperbolique via l'approximation parabolique. C. R. Acad. Sci. Paris, Série I, **319** (1994), 377–382.

Giusti, E.

1. *Minimal Surfaces and Functions of Bounded Variation.* Boston: Birkhäuser, 1984.

Glimm, J.

1. Solutions in the large for nonlinear hyperbolic systems of equations. Comm. Pure Appl. Math. **18** (1965), 697–715.
2. The interaction of nonlinear hyperbolic waves. Comm. Pure Appl. Math. **41** (1988), 569–590.

Glimm, J. and P.D. Lax

1. Decay of solutions of systems of nonlinear hyperbolic conservation laws. Memoirs AMS, No. **101** (1970).

Godin, P.

1. Global shock waves in some domains for the isentropic irrotational potential flow equations. Comm. PDE **22** (1997), 1929–1997.

Godlewski, E. and P.-A. Raviart

1. *Hyperbolic Systems of Conservation Laws.* Paris: Ellipses, 1991.
2. *Numerical Approximation of Hyperbolic Systems of Conservation Laws.* New York: Springer, 1996.

Godunov, S.K.

1. An interesting class of quasilinear systems. Dokl. Akad. Nauk SSSR **139** (1961), 521–523. English translation: Soviet Math. **2** (1961), 947–949.
2. *Elements of Continuum Mechanics.* Moscow: Nauka, 1978.
3. Lois de conservation et integrales d'énergie des équations hyperboliques. Lecture Notes in Math. **1270** (1987), 135–149. Berlin: Springer.

Goodman, J.

1. Nonlinear asymptotic stability of viscous shock profiles for conservation laws. Arch. Rational Mech. Anal. **95** (1986), 325–344.

Goodman, J., Szepessy A., and K. Zumbrun

1. A remark on stability of viscous waves. SIAM J. Math. Anal. **25** (1994), 1463–1467.

Goodman J. and Z.P. Xin

1. Viscous limits for piecewise smooth solutions to systems of conservation laws. Arch. Rational Mech. Anal. **121** (1992), 235–265.

Grassin, M. and D. Serre

1. Existence de solutions globales et régulières aux équations d'Euler pour un gaz parfait isentropique. C. R. Acad. Sci. Paris, Série I, **325** (1997), 721–726.

Greenberg, J.M.

1. On the elementary interactions for the quasilinear wave equation. Arch. Rational Mech. Anal. **43** (1971), 325–349.
2. On the interaction of shocks and simple waves of the same family, Parts I and II. Arch. Rational Mech. Anal. **37** (1970), 136–160; **51** (1973), 209–217.

3. Smooth and time periodic solutions to the quasilinear wave equation. Arch. Rational Mech. Anal. **60** (1975), 29–50.

Greenberg, J.M. and M. Rascle

1. Time-periodic solutions to systems of conservation laws. Arch. Rational Mech. Anal. **115** (1991), 395–407.

Greenberg, J.M. and D.D.M. Tong

1. Decay of periodic solutions of $\partial_t u + \partial_x f(u) = 0$. J. Math. Anal. Appl. **43** (1973), 56–71.

Grenier, E.

1. Boundary layers for viscous perturbations of noncharacteristic quasilinear hyperbolic problems. J. Diff. Eqs. **143** (1998), 110–146.

Grot, R.A.

1. Relativistic continuum physics: electromagnetic interactions. *Continuum Physics*, Vol. III, pp. 129–219, ed. A.C. Eringen. New York: Academic Press, 1976.

Guckenheimer, J.

1. Solving a single conservaton law. Lecture Notes in Math. **468** (1975), 108–134. Berlin: Springer.
2. Shocks and rarefactions in two space dimensions. Arch. Rational Mech. Anal. **59** (1975), 281–291.

Gurtin, M.E.

1. *An Introduction to Continuum Mechanics*. New York: Academic Press, 1981.

Hadamard, J.

1. *Leçons sur la Propagation des Ondes et les Equations de l'Hydrodynamique*. Paris: Hermann, 1903.

Hagan, R. and M. Slemrod

1. The viscosity-capillarity criterion for shocks and phase transitions. Arch. Rational Mech. Anal. **83** (1983), 333–361.

Hanyga, A.

1. *Mathematical Theory of Non-Linear Elasticity*. Warszawa: PWN, 1985.

Härterich, J.

1. Heteroclinic orbits between rotating waves in hyperbolic balance laws. Proc. Royal Soc. Edinburgh **129A** (1999), 519–538.

Hattori, H.

1. The Riemann problem for a van der Waals fluid with entropy rate admissibility criterion. Isothermal case. Arch. Rational Mech. Anal. **92** (1986), 247–263.
2. The Riemann problem for a van der Waals fluid with entropy rate admissibility criterion. Nonisothermal case. J. Diff. Eqs. **65** (1986), 158–174.
3. The entropy rate admissibility criterion and the double phase boundary problem. Contemp. Math. **60** (1987), 51–65.
4. The Riemann problem and the existence of weak solutions to a system of mixed-type in dynamic phase transitions. J. Diff. Eqs. **146** (1998), 287–319.

Hayes, B. and P.G. LeFloch

1. Nonclassical shocks and kinetic relations: Scalar conservaton laws. Arch. Rational Mech. Anal. **139** (1997), 1–56.
2. Nonclassical shocks and kinetic relations: Strictly hyperbolic systems. SIAM J. Math. Anal. (To appear).

Hedstrom, G.W.

1. Some numerical experiments with Dafermos's method for nonlinear hyperbolic equations. Lecture Notes in Math. **267** (1972), 117–138. Berlin: Springer.

Heibig, A.

1. Error estimates for solutions to hyperbolic systems of conservation laws. Comm. PDE **18** (1993), 281–304.
2. Existence and uniqueness of solutions for some hyperbolic systems of conservation laws. Arch. Rational Mech. Anal. **126** (1994), 79–101.

Heibig, A. and A. Sahel

1. A method of characteristics for some systems of conservation laws. SIAM J. Math. Anal. **29** (1998), 1467–1480.

Heibig, A. and D. Serre

1. Etude variationnelle du problème de Riemann. J. Diff. Eqs. **96** (1992), 56–88.

Heidrich, A.

1. Global weak solutions to initial-boundary value problems for the one-dimensional quasi-linear wave equation with large data. Arch. Rational Mech. Anal. **126** (1994), 333–368.

Hoff, D.

1. The sharp form of Oleinik's entropy condition in several space variables. Trans. AMS **276** (1983), 707–714.
2. Invariant regions for systems of conservation laws. Trans. AMS **289** (1985), 591–610.

Holden, H.

1. On the Riemann problem for a prototype of a mixed type conservation law. Comm. Pure Appl. Math. **40** (1987), 229–264.

Holden, H. and L. Holden

1. First order nonlinear scalar hyperbolic conservation laws in one dimension. *Ideas and Methods in Mathematical Analysis, Stochastics and Applications*, pp. 480–510, eds. S. Albeveiro, J.E. Fenstad, H. Holden and T. Lindstrøm. Cambridge: Cambridge University Press, 1992.

Holden, H., Holden, L. and R. Høegh-Krohn

1. A numerical method for first order nonlinear scalar hyperbolic conservation laws in one dimension. Computers and Maths. with Appl. **15** (1988), 595–602.

Holden, H. and N.H. Risebro

1. A method of fractional steps for scalar conservation laws without the CFL condition. Math. in Comp. **60** (1993), 221–232.
2. *Front Tracking for Conservation Laws*. Lecture Notes, Norwegian University of Science and Technology, 1999.

Hölder, E.

1. Historischer Überblick zur mathematischen Theorie von Unstetigkeitswellen seit Riemann und Christoffel. *E.B. Christoffel*, pp. 412–434, ed. P.L. Butzer and F. Fehér. Basel: Birkhäuser, 1981.

Hopf, E.

1. The partial differential equation $u_t + uu_x = \mu u_{xx}$. Comm. Pure Appl. Math. **3** (1950), 201–230.

Hörmander, L.

1. *Lectures on Nonlinear Hyperbolic Differential Equations.* Paris: Springer, 1997.

Hsiao, L.

1. The entropy rate admissibility criterion in gas dynamics. J. Diff. Eqs. **38** (1980), 226–238.
2. *Quasilinear Hyperbolic Systems and Dissipative Mechanisms.* Singapore: World Scientific, 1997.

Hsiao, L. and T. Chang

1. Perturbations of the Riemann problem in gas dynamics. J. Math. Anal. Appl. **79** (1981), 436–460.

Hsiao, L. and P. DeMottoni

1. Existence and uniqueness of the Riemann problem for a nonlinear system of conservation laws of mixed type. Trans. AMS **322** (1990), 121–158.

Hsiao, L., Tao, L. and T. Yang

1. Global BV solutions of compressible Euler equations with spherical symmetry and damping. J. Diff. Eqs. **146** (1998), 203–225.

Hsiao, L. and Zhang Tung

1. Riemann problem for 2×2 quasilinear hyperbolic system without convexity. Ke Xue Tong Bao **8** (1978), 465–469.

Hu, J. and P.G. LeFloch

1. L^1 continuous dependence property for systems of conservation laws. Arch. Rational Mech. Anal. (To appear).

Huang, Feimin

1. Existence and uniqueness of discontinuous solutions for a hyperbolic system. Proc. Royal Soc. Edinburgh, **127A** (1997), 1193–1205.

Hubert, F. and D. Serre

1. Fast-slow dynamics for parabolic perturbations of conservation laws. Comm. PDE **21** (1996), 1587–1608.

Hughes, T.J.R., Kato, T. and J.E. Marsden

1. Well-posed quasi-linear second-order hyperbolic systems with applications to nonlinear elastodynamics and general relativity. Arch. Rational Mech. Anal. **63** (1977), 273–294.

Hugoniot, H.

1. Sur la propagation du movement dans les corps et spécialement dans les gaz parfaits. J. Ecole Polytechnique **58** (1889), 1–125.

Hunter, J.

1. Interaction of elastic waves. Stud. Appl. Math. **86** (1992), 281–314.

Hunter, J.K., and J.B. Keller

1. Weakly nonlinear high frequency waves. Comm. Pure Appl. Math. **36** (1983), 547–569.
2. Nonlinear hyperbolic waves. Proc. Royal Soc. London **417A** (1988), 299–308.

Ilin, A.M. and O.A. Oleinik

1. Behavior of the solutions of the Cauchy problem for certain quasilinear equations for unbounded increase of the time. Dokl. Akad. Nauk SSSR **120** (1958), 25–28. English translation: AMS Translations, Ser. II, **42**, 19–23.

Isaacson, E.L., Marchesin, D. and B. Plohr

1. Transitional waves for conservation laws. SIAM J. Math. Anal. **21** (1990), 837–866.

Isaacson, E.L., Marchesin, D., Plohr, B. and J.B. Temple

1. The Riemann problem near a hyperbolic singularity: The classification of quadratic Riemann problems I. SIAM J. Appl. Math. **48** (1988), 1009–1032.

Isaacson, E. and J.B. Temple

1. The Riemann problem near a hyperbolic singularity. SIAM J. Appl. Math. **48** (1988), 1287–1318.

Izumiya, S. and G.T. Kossioris

1. Geometric singularities for solutions of single conservation laws. Arch. Rational Mech. Anal. **139** (1997), 255–290.

James, F., Peng, Y.J. and B. Perthame

1. Kinetic formulation for chromatography and some other hyperbolic systems. J. Math. Pures Appl. **74** (1995), 367–385.

James, R.D.

1. The propagation of phase boundaries in elastic bars. Arch. Rational Mech. Anal. **73** (1980), 125–158.

Jeffrey, A.

1. *Magnetohydrodynamics.* Edinburgh: Oliver and Boyd, 1966.
2. *Quasilinear Hyperbolic Systems and Waves.* London: Pitman, 1976.

Jenssen, H.K.

1. Blowup for systems of conservation laws. SIAM J. Appl. Math. (To appear).
2. A note on the spreading of characteristics for nonconvex conservation laws. (Preprint).

Jin, S. and Z. Xin

1. The relaxation schemes for systems of conservation laws in arbitrary space dimensions. Comm. Pure Appl. Math. **48** (1995), 235–276.

John, F.

1. Formation of singularities in one-dimensional nonlinear wave propagation. Comm. Pure Appl. Math. **27** (1974), 377–405.
2. Blow-up for quasilinear wave equations in three space dimensions. Comm. Pure Appl. Math. **34** (1981), 29–53.

Johnson, J.N. and R. Chéret

1. *Classic Papers in Shock Compression Science.* New York: Springer, 1998.

Joly, J.-L., Métivier, G. and J. Rauch

1. Resonant one-dimensional nonlinear geometric optics. J. Funct. Anal. **114** (1993), 106–231.
2. A nonlinear instability for 3×3 systems of conservation laws. Comm. Math. Phys. **162** (1994), 47–59.
3. Coherent and focusing multi-dimensional nonlinear geometric optics. Ann. Sci. ENS **28** (1995), 51–113.

Joseph, K.T. and P.G. LeFloch

1. Boundary layers in weak solutions of hyperbolic conservation laws. Arch. Rational Mech. Anal. **147** (1999), 47–88.

Jouguet, E.

1. Sur la propagation des discontinuités dans les fluides. C. R. Acad. Sci. Paris **132** (1901), 673–676.

Kalašnikov, A.S.

1. Construction of generalized solutions of quasi-linear equations of first order without convexity conditions as limits of solutions of parabolic equations with a small parameter. Dokl. Akad. Nauk SSSR **127** (1959), 27–30.

Kan, P.T., Santos, M.M. and Z. Xin

1. Initial-boundary value problem for conservation laws. Comm. Math. Phys. **186** (1997), 701–730.

Kato, T.

1. The Cauchy problem for quasi-linear symmetric hyperbolic systems. Arch. Rational Mech. Anal. **58** (1975), 181–205.

Katsoulakis, M.A. and A.E. Tzavaras

1. Contractive relaxation systems and the scalar multidimensional conservation law. Comm. PDE **22** (1997), 195–233.

Kawashima, S. and A. Matsumura

1. Stability of shock profiles in viscoelasticity with non-convex constitutive relations. Comm. Pure Appl. Math. **47** (1994), 1547–1569.

Keyfitz, B.L.

1. Change of type in three-phase flow: A simple analogue. J. Diff. Eqs. **80** (1989), 280–305.
2. Admissibility conditions for shocks in systems that change type. SIAM J. Math. Anal. **22** (1991), 1284–1292.

Keyfitz, B.L. and H.C. Kranzer

1. A system of nonstrictly hyperbolic conservation laws arising in elasticity theory. Arch. Rational Mech. Anal. **72** (1980), 219–241.
2. A viscosity approximation to a system of conservation laws with no classical Riemann solution. Lecture Notes in Math. **1402** (1989), 185–197. Berlin: Springer.
3. Spaces of weighted measures for conservation laws with singular shock solutions. J. Diff. Eqs. **118** (1995), 420–451.

Kim, Y.J.

1. A self-similar viscosity approach for the Riemann problem in isentropic gas dynamics and the structure of the solutions. (Preprint).

Kirchhoff, G.

1. Ueber den Einfluss der Wärmeleitung in einem Gase auf die Schallbewegung. Ann. Physik **134** (1868), 177–193.

Klainerman, S. and A. Majda

1. Formation of singularities for wave equations including the nonlinar vibrating string. Comm. Pure Appl. Math. **33** (1980), 241–263.

Klingenberg, C. and Y. Lu

1. Cauchy problem for hyperbolic conservation laws with a relaxation term. Proc. Royal Soc. Edinburgh, **126A** (1996), 821–828.

Klingenberg, C. and N.H. Risebro

1. Convex conservation law with discontinuous coefficients. Comm. PDE **20** (1995), 1959–1990.

Kohler, M.

1. Behandlung von Nichtgleichgewichtsvorgängen mit Hilfe eines Extremalprinzipes. Zeit. Physik **124** (1948), 772–789.

Kröner, D.

1. *Numerical Schemes for Conservation Laws.* Chichester: John Wiley, 1997.

Kruzkov, S.

1. First-order quasilinear equations with several space variables. Mat. Sbornik **123** (1970), 228–255. English translation: Math. USSR Sbornik **10** (1970), 217–273.

Kuznetsov, N.

1. Weak solutions of the Cauchy problem for a multi-dimensional quasilinear equation. Mat. Zam. **2** (1967), 401–410. English translation: Math. Notes Acad. USSR **2** (1967), 733–739.

Landau, L.D.

1. On shock waves at large distances from their place of origin. J. Phys. USSR **9** (1945), 495–500.

Lattanzio, C. and P. Marcati

1. The zero relaxation limit for the hydrodynamic Whitham traffic flow model. J. Diff. Eqs. **141** (1997), 150–178.

Lax, P.D.

1. Weak solutions of nonlinear hyperbolic equations and their numerical computation. Comm. Pure Appl. Math. **7** (1954), 159–193.
2. Hyperbolic systems of conservation laws. Comm. Pure Appl. Math. **10** (1957), 537–566.
3. Development of singularities of solutions of nonlinear hyperbolic partial differential equations. J. Math. Phys. **5** (1964), 611–613.
4. Shock waves and entropy. *Contributions to Functional Analysis* pp. 603–634, ed. E.A. Zarantonello. New York: Academic Press, 1971.
5. *Hyperbolic Systems of Conservation Laws and the Mathematical Theory of Shock Waves.* CBMS Regional Conference Series in Mathematics No. **11**. Philadelphia: SIAM, 1973.
6. The multiplicity of eigenvalues. Bull. AMS (New Series) **6** (1982), 213–214.

LeFloch, P.G.

1. Explicit formula for scalar non-linear conservation laws with boundary conditions. Math. Meth. Appl. Sci. **10** (1988), 265–287.
2. Entropy weak solutions to nonlinear hyperbolic systems in nonconservative form. Comm. PDE **13** (1988), 669–727.
3. Propagating phase boundaries: formulation of the problem and existence via the Glimm scheme. Arch. Rational Mech. Anal. **123** (1993), 153–197.
4. An introduction to nonclassical shocks of systems of conservation laws. *An Introduction to Recent Developments in Theory and Numerics for Conservation Laws*, pp. 28–72, eds. D. Kröner, N. Ohlberger and C. Rohde. Berlin: Springer, 1999.

LeFloch, P.G. and A.E. Tzavaras

1. Representation of weak limits and definition of nonconservative products. SIAM J. Math. Anal. **30** (1999), 1309–1342.

LeFloch, P.G. and Z.P. Xin

1. Uniqueness via the adjoint problems for systems of conservation laws. Comm. Pure Appl. Math. **46** (1993), 1499–1533.

Leibovich, L.

1. Solutions of the Riemann problem for hyperbolic systems of quasilinear equations without convexity conditions. J. Math. Anal. Appl. **45** (1974), 81–90.

LeVeque, R.J.

1. *Numerical Methods for Conservation Laws.* Basel: Birkhäuser, 1990.

LeVeque, R.J. and B. Temple

1. Convergence of Godunov's method for a class of 2×2 systems of conservation laws. Trans. AMS **288** (1985), 115–123.

Li, J., Zhang, T. and S. Yang

1. *The Two-Dimensional Riemann Problem in Gas Dynamics.* Harlow: Longman, 1998.

Li, Ta-tsien

1. *Global Classical Solutions for Quasilinear Hyperbolic Systems.* New York: Wiley, 1994.

Li, Ta-tsien and Wen-ci Yu

1. *Boundary Value Problems for Quasilinear Hyperbolic Systems.* Durham: Duke University Math. Series V, 1985.

Li, Ta-tsien, Zhou, Yi and De-xing Kong

1. Weak linear degeneracy and global classical solutions for general quasilinear hyperbolic systems. Comm. PDE, **19** (1994), 1263–1317.

Lighthill, M.J.

1. A method for rendering approximate solutions to physical problems uniformly valid. Philos. Magazine **40** (1949), 1179–1201.

Lin, Long-Wei

1. On the vacuum state for the equations of isentropic gas dynamics. J. Math. Anal. Appl. **121** (1987), 406–425.

Lin, Long-Wei and Tong Yang

1. Convergence of the viscosity method for the system of isentropic gas dynamics in Lagrangian coordinates. J. Diff. Eqs. **102** (1993), 330–341.

Lin, Peixiong

1. Young measures and an application of compensated compactness to one-dimensional nonlinear elastodynamics. Trans. AMS **329** (1992), 377–413.

Lindquist, W.B.

1. The scalar Riemann problem in two spatial dimensions: Piecewise smoothness of solutions. SIAM J. Math. Anal. **17** (1986), 1178–1197.

Lions, P.-L.

1. *Generalized Solutions of Hamilton-Jacobi Equations.* London: Pitman, 1982.
2. *Mathematical Topics in Fluid Mechanics* Vols. I–II. Oxford: Oxford University Press, 1996–1998.

Lions, P.-L., Perthame, B. and P.E. Souganidis

1. Existence and stability of entropy solutions for the hyperbolic systems of isentropic gas dynamics in Eulerian and Lagrangian coordinates. Comm. Pure Appl. Math. **49** (1996), 599–638.

Lions, P.-L., Perthame, B. and E. Tadmor

1. Kinetic formulation for the isentropic gas dynamics and p-systems. Comm. Math. Phys. **163** (1994), 415–431.
2. A kinetic formulation of multidimensional scalar conservation laws and related equations. J. AMS **7** (1994), 169–191.

Liu, T.-P.

1. The Riemann problem for general system of conservation laws. J. Diff. Eqs. **18** (1975), 218–234.
2. The entropy condition and the admissibility of shocks. J. Math. Anal. Appl. **53** (1976), 78–88.
3. Uniqueness of weak solutions of the Cauchy problem for general 2×2 conservation laws. J. Diff. Eqs. **20** (1976), 369–388.
4. The deterministic version of the Glimm scheme. Comm. Math. Phys. **57** (1977), 135–148.
5. Decay to N-waves of solutions of general systems of nonlinear hyperbolic conservation laws. Comm. Pure Appl. Math. **30** (1977), 585–610.
6. Linear and nonlinear large-time behavior of solutions of hyperbolic conservation laws. Comm. Pure Appl. Math. **30** (1977), 767–796.
7. Initial-boundary value problems for gas dynamics. Arch. Rational Mech. Anal. **64** (1977), 137–168.
8. Asymptotic behavior of solutions of general systems of nonlinear hyperbolic conservation laws. Indiana U. Math. J. **27** (1978), 211–253.
9. Development of singularities in the nonlinear waves for quasilinear hyperbolic partial differential equations. J. Diff. Eqs. **33** (1979), 92–111.
10. Quasilinear hyperbolic systems. Comm. Math. Phys. **68** (1979), 141–172.
11. Admissible solutions of hyperbolic conservation laws. Memoirs AMS **30** (1981), No. 240.
12. Nonlinear stability of shock waves for viscous conservation laws. Memoirs AMS **56** (1985), No. 328.
13. Hyperbolic conservation laws with relaxation. Comm. Math. Phys. **108** (1987), 153–175.
14. Pointwise convergence to N-waves for solutions of hyperbolic conservation laws. Bull. Inst. Math. Acad. Sinica **15** (1987), 1–17.
15. Nonlinear resonance for quasilinear hyperbolic equation. J. Math. Phys. **28** (1987), 2593–2602.
16. On the viscosity criterion for hyperbolic conservation laws. *Viscous Profiles and Numerical Methods for Shock Waves*, pp. 105–114, ed. M. Shearer. Philadelphia: SIAM, 1991.
17. Pointwise convergence to shock waves for viscous conservation laws. Comm. Pure Appl. Math. **50** (1997), 1113–1182.

Liu, T.-P., Matsumura, A. and K. Nishihara

1. Behavior of solutions for the Burgers equation with boundary corresponding to rarefaction waves. SIAM J. Math. Anal. **29** (1998), 293–308.

Liu, T.-P. and K. Nishihara

1. Asymptotic behavior of scalar viscous conservation laws with boundary effect. J. Diff. Eqs. **133** (1997), 296–320.

Liu, T.-P. and M. Pierre

1. Source-solutions and asymptotic behavior in conservation laws. J. Diff. Eqs. **51** (1984), 419–441.

Liu, T.-P. and Z. Xin

1. Nonlinear stability of rarefaction waves for compressible Navier-Stokes equations. Comm. Math. Phys. **118** (1986), 451–465.
2. Stability of viscous shock waves associated with a system of nonstrictly hyperbolic conservation laws. Comm. Pure Appl. Math. **45** (1992), 361–388.
3. Pointwise decay to contact discontinuities for systems of viscous conservation laws. Asian J. Math. **1** (1997), 34–84.

Liu, T.-P. and T. Yang

1. L_1 stability of conservation laws with coinciding Hugoniot and characteristic curves. Indiana U. Math. J. (To appear).
2. L_1 stability of weak solutions for 2×2 systems of hyperbolic conservation laws. J. AMS **12** (1999), 729–774.
3. A new entropy functional for a scalar conservation law. Comm. Pure Appl. Math. **52** (1999), 1427–1442.
4. Well-posedness theory for hyperbolic conservation laws. Comm. Pure Appl. Math. **52** (1999), 1553–1586.

Liu, T.-P. and S.-H. Yu

1. Propagation of a stationary shock layer in the presence of a boundary. Arch. Rational Mech. Anal. **139** (1997), 57–82.
2. Viscous Riemann problems. (In preparation).

Liu, T.-P. and Y. Zeng

1. Large time behavior of solutions for general quasilinear hyperbolic-parabolic systems of conservaton laws. Memoirs AMS **125** (1997), No. 549.
2. Compressible Navier-Stokes equations with zero heat conductivity. J. Diff. Eqs. **153** (1999), 225–291.

Liu, T.-P. and K. Zumbrun

1. On nonlinear stability of general undercompressive viscous shock waves. Comm. Math. Phys. **174** (1995), 319–345.

Lu, Yun-Guang

1. Convergence of the viscosity method for some nonlinear hyperbolic systems. Proc. Royal Soc. Edinburgh **124A** (1994), 341–352.

Lu, Y.-G. and C. Klingenberg

1. The Cauchy problem for hyperbolic conservation laws with three equations. J. Math. Anal. Appl. **202** (1996), 206–216.

Lucier, B.J.

1. A moving mesh numerical method for hyperbolic conservation laws. Math. Comput. **46** (1986), 59–69.
2. Regularity through approximation for scalar conservation laws. SIAM J. Math. Anal. **19** (1988), 763–773.

Luo, T. and D. Serre

1. Linear stability of shock profiles for a rate-type viscoelastic system with relaxation. Quart. Appl. Math. **56** (1998), 569–586.

Luskin, M. and B. Temple

1. The existence of a global weak solution to the nonlinear waterhammer problem. Comm. Pure Appl. Math. **35** (1982), 697–735.

Lyapidevskii, V. Yu

1. The continuous dependence on the initial conditions of the generalized solutions of the gas-dynamic system of equations. Zh. Vychisl. Mat. Mat. Fiz. **14** (1974), 982–991.

Lyberopoulos, A.N.

1. Large-time structure of solutions of scalar conservation laws without convexity in the presence of a linear source field. J. Diff. Eqs. **99** (1992), 342–380.
2. A Poincaré-Bendixson theorem for scalar balance laws. Proc. Royal Soc. Edinburgh **124A** (1994), 589–607.

Lyons, W.K.

1. Conservation laws with sharp inhomogeneities. Quart. Appl. Math. **40** (1983), 385–393.

Majda, A.

1. The stability of multi-dimensional shock fronts. Memoirs AMS **41** (1983), No. 275.
2. The existence of multi-dimensional shock fronts. Memoirs AMS **43** (1983), No. 281.
3. *Compressible Fluid Flow and Systems of Conservation Laws in Several Space Variables.* New York: Springer, 1984.
4. Nonlinear geometric optics for hyperbolic systems of conservation laws. *Oscillation Theory, Computation and Methods of Compensated Compactness*, pp. 115–165, eds. C. Dafermos, J.L. Ericksen, D. Kinderlehrer and M. Slemrod. New York: Springer, 1986.

Majda, A. and R. Pego

1. Stable viscosity matrices for systems of conservation laws. J. Diff. Eqs. **56** (1985), 229–262.

Majda, A. and R.R. Rosales

1. Resonantly interacting weakly nonlinear hyperbolic waves, I. A single space variable. Studies Appl. Math. **71** (1984), 149–179.

Majda, A., Rosales, R. and M. Schonbek

1. A canonical system of integrodifferential equations arising in resonant nonlinear acoustics. Studies Appl. Math. **79** (1988), 205–262.

Málek, J., Nečas, J. and M. Rokyta

1. *Weak and Measure-Valued Solutions to Evolutionary PDEs.* London: Chapman & Hall, 1996.

Marcati, P. and R. Natalini

1. Weak solutions to a hydrodynamic model for semiconductors and relaxation to the drift-diffusion equation. Arch. Rational Mech. Anal. **129** (1995), 129–145.
2. Weak solutions to a hydrodynamic model for semiconductors: the Cauchy problem. Proc. Royal Soc. Edinburgh **125A** (1995), 115–131.

Marcati, P. and B. Rubino

1. Hyperbolic to parabolic relaxation theory for quasilinear first order systems. (Preprint).

Marsden, J.E. and T.J.R. Hughes

1. *Mathematical Foundations of Elasticity.* Englewood Cliffs: Prentice-Hall, 1983.

Mascia, C.

1. Qualitative behavior of conservation laws with reaction term and nonconvex flux. Quart. Appl. Math. (To appear).

Mascia, C. and C. Sinestrari

1. The perturbed Riemann problem for a balance law. Adv. Diff. Eqs. **2** (1997), 779–810.

Matano, H.

1. Nonincrease of the lap-number of a solution for a one-dimensional semilinear parabolic equation. J. Fac. Sci. Univ. Tokyo, Sect. 1A **29** (1982), 401–441.

Mock, M.S.

1. A topological degree for orbits connecting critical points of autonomous systems. J. Diff. Eqs. **38** (1980), 176–191.

Morrey, C.B.

1. Quasiconvexity and the lower semicontinuity of multiple integrals. Pacific J. Math. **2** (1952), 25–53.

Müller, I. and T. Ruggeri

1. *Rational Extended Thermodynamics*. (Second Edition). New York: Springer, 1998.

Müller, S. and I. Fonseca

1. A-quasiconvexity, lower semicontinuity and Young measures. Max-Planck-Institut für Mathematik, Leipzig, Preprint No. 18, 1998.

Murat, F.

1. Compacité par compensation. Ann. Scuola Norm. Sup. Pisa Sci. Fis. Mat. **5** (1978), 489–507.
2. L' injection du cône positif de H^{-1} dans $W^{-1,q}$ est compacte pour tout $q < 2$. J. Math. Pures Appl. **60** (1981), 309–322.

Natalini, R.

1. Convergence to equilibrium for the relaxation approximations of conservation laws. Comm. Pure Appl. Math. **49** (1996), 795–823.
2. A discrete kinetic approximation of entropy solutions to multidimensional scalar conservation laws. J. Diff. Eqs. **148** (1998), 292–317.
3. Recent results on hyperbolic relaxation problems. *Analysis of Systems of Conservation Laws*, pp. 128–198, ed. H. Freistühler. London: Chapman and Hall/CRC, 1998.

Natalini, R., Sinestrari C. and A. Tesei

1. Incomplete blowup of solutions of quasilinear hyperbolic balance laws. Arch. Rational Mech. Anal. **135** (1996), 259–296.

Nessyahu, H. and E. Tadmor

1. The convergence rate of approximate solutions for nonlinear scalar conservation laws. SIAM J. Num. Anal. **29** (1992), 1505–1519.

von Neumann, J.

1. Theory of shock waves. *Collected Works*, Vol. VI, pp. 178–202. Oxford: Pergamon Press, 1963.
2. Oblique reflection of shocks. *Collected Works*, Vol. VI, pp. 238–299. Oxford: Pergamon Press, 1963.
3. Refraction, intersection and reflection of shock waves. *Collected Works*, Vol. VI, pp. 300–308. Oxford: Pergamon Press, 1963.

Nickel, K.

1. Gestaltaussagen über Lösungen parabolischer Differentialgleichungen. J. Reine Angew. Math. **211** (1962), 78–94.

Nishida, T.

1. Global solution for an initial boundary value problem of a quasilinear hyperbolic system. Proc. Japan Acad. **44** (1968), 642–646.

Noelle, S.

1. Development of singularities for the complex Burgers equation. Nonl. Anal. **26** (1986), 1313–1321.
2. Radially symmetric solutions for a class of hyperbolic systems of conservation laws. ZAMP **48** (1997), 676–679.

Noll, W.

1. A mathematical theory of the mechanical behavior of continuous media. Arch. Rational Mech. Anal. **2** (1958), 197–226.
2. The foundations of classical mechanics in the light of recent advances in continuum mechanics. *The Axiomatic Method*, pp. 266–281. Amsterdam: North Holland, 1959.

Oleinik, O.A.

1. The Cauchy problem for nonlinear equations in a class of discontinuous functions. Dokl. Akad. Nauk SSSR **95** (1954), 451–454. English translation: AMS Translations, Ser. II, **42**, 7–12.
2. Discontinuous solutions of non-linear differential equations. Usp. Mat. Nauk **12** (1957), 3–73. English translation: AMS Translations, Ser. II, **26**, 95–172.
3. On the uniqueness of the generalized solution of the Cauchy problem for a nonlinear system of equations occurring in mechanics. Usp. Mat. Nauk **12** (1957), 169–176.
4. Uniqueness and stability of the generalized solution of the Cauchy problem for quasi-linear equation. Usp. Mat. Nauk **14** (1959), 165–170. English translation: AMS Translations, Ser. II, **33**, 285–290.

Osher, S. and E. Tadmor

1. On the convergence of difference approximations to scalar conservation laws. Math. Comp. **50** (1988), 19–51.

Ostrov, D.N.

1. Asymptotic behavior of two interreacting chemicals in a chromatography reactor. SIAM J. Math. Anal. **27** (1996), 1559–1596.

Otto, F.

1. Initial-boundary value problem for a scalar conservation law. C. R. Acad. Sci. Paris, Série I, **322** (1996), 729–734.
2. A regularizing effect of nonlinear transport equations. Quart. Appl. Math. **56** (1998), 355–375.

Pan, Tao and Longwei Lin

1. The global solution of the scalar nonconvex conservation law with boundary condition. J. PDE **8** (1995), 371–383; **11** (1998), 1–8.

Pant, V.

1. Global entropy solutions for isentropic relativistic fluid dynamics. Comm. PDE **21** (1996), 1609–1641.

Pego, R.L.

1. Stable viscosities and shock profiles for systems of conservation laws. Trans. AMS **282** (1984), 749–763.
2. Nonexistence of a shock layer in gas dynamics with a nonconvex equation of state. Arch. Rational Mech. Anal. **94** (1986), 165–178.
3. Some explicit resonating waves in weakly nonlinear gas dynamics. Studies Appl. Math. **79** (1988), 263–270.

Pego, R.L. and D. Serre

1. Instabilities in Glimm's scheme for two systems of mixed type. SIAM J. Num. Anal. **25** (1988), 965–988.

Pence, T.J.

1. On the mechanical dissipation of solutions to the Riemann problem for impact involving a two-phase elastic material. Arch. Rational Mech. Anal. **117** (1992), 1–52.

Pericak-Spector, K.A. and S.J. Spector

1. Nonuniqueness for a hyperbolic system: cavitation in nonlinear elastodynamics. Arch. Rational Mech. Anal. **101** (1988), 293–317.
2. On dynamic cavitation with shocks in nonlinear elasticity. Proc. Royal Soc. Edinburgh **127A** (1997), 837–857.

Perthame, B.

1. Uniqueness and error estimates in first order quasilinear conservation laws via the kinetic entropy defect measure. J. Math. Pures Appl. **77** (1998), 1055–1064.

Perthame, B. and M. Pulvirenti

1. On some large systems of random particles which approximate scalar conservation laws. Asympt. Anal. **10** (1995), 263–278.

Perthame, B. and E. Tadmor

1. A kinetic equation with kinetic entropy functions for scalar conservation laws. Comm. Math. Phys. **136** (1991), 501–517.

Peters, G.R. and S. Čanić

1. On the oscillatory solutions in hyperbolic conservation laws. (Preprint).

Poisson, S.D.

1. Mémoire sur la théorie du son. J. Ecole Polytechnique, **7** (1808), 319–392.
2. Mémoire sur les équations générales de l'équilibre et du mouvement des corps élastiques et des fluides. J. Ecole Polytechnique, **13** (1831), 1–174.

Poupaud, F. and M. Rascle

1. Measure solutions to the linear multi-dimensional transport equation with non-smooth coefficients. Comm. PDE **22** (1997), 337–358.

Poupaud, F., Rascle, M. and J.P. Vila

1. Global solutions to the isothermal Euler-Poisson system with arbitrarily large data. J. Diff. Eqs. **123** (1995), 93–121.

Qin, Tiehu

1. Symmetrizing the nonlinear elastodynamic system. J. Elasticity. **50** (1998), 245–252.

Quinn, B. (B.L. Keyfitz)

1. Solutions with shocks: an example of an L_1-contraction semi-group. Comm. Pure Appl. Math. **24** (1971), 125–132.

Rankine, W.J.M.

1. On the thermodynamic theory of waves of finite longitudinal disturbance. Phil. Trans. Royal Soc. London **160** (1870), 277–288.

Rascle, M.

1. On the static and dynamic study of oscillations for some nonlinear hyperbolic systems of conservation laws. Ann. Inst. Henri Poincaré **8** (1991), 333–350.

Rauch, J.

1. *BV* estimates fail for most quasilinear hyperbolic systems in dimension greater than one. Comm. Math. Phys. **106** (1986), 481–484.

Rayleigh, Lord (J.W. Strutt)

1. *The Theory of Sound*, Vol II. London: Macmillan, 1878.
2. Note on tidal bores. Proc. Royal Soc. London **81A** (1908), 448–449.
3. Aerial plane waves of finite amplitude. Proc. Royal Soc. London **84A** (1910), 247–284.

Rezakhanlou, F.

1. Microscopic structure of shocks in one-conservation laws. Ann. Inst. Henri Poincaré **12** (1995), 119–153.

Rhee, H.-K., Aris, R. and N.R. Amundson

1. *First-Order Partial Differential Equations*, Vols. I–II. Englewood Cliffs: Prentice-Hall, 1986–1989.

Riemann, B.

1. Ueber die Fortpflanzung ebener Luftwellen von endlicher Schwingungsweite. Gött. Abh. Math. Cl. **8** (1860), 43–65.

Risebro, N.H.

1. A front-tracking alternative to the random choice method. Proc. AMS **117** (1993), 1125–1139.

Risebro, N.H. and A. Tveito

1. Front tracking applied to a non-strictly hyperbolic system of conservation laws. SIAM J. Sci. Statist. Comput. **12** (1991), 1401–1419.
2. A front tracking method for conservation laws in one dimension. J. Comput. Phys. **101** (1992), 130–139.

Rivlin, R.S. and J.L. Ericksen

1. Stress-deformation relations for isotropic materials. J. Rational Mech. Anal. **4** (1955), 323–425.

Rosakis, P.

1. An equal area rule for dissipative kinetics of propagating strain discontinuities. SIAM J. Appl. Math. **55** (1995), 100–123.

Roytburd V. and M. Slemrod

1. Positively invariant regions for a problem in phase transitions. Arch. Rational Mech. Anal. **93** (1986), 61–79.

Roždestvenskii, B.L.

1. A new method of solving the Cauchy problem in the large for quasilinear equations. Dokl. Akad. Nauk SSSR **138** (1961), 309–312.

Roždestvenskii, B.L. and N.N. Janenko

1. *Systems of Quasilinear Equations and Their Applications to Gas Dynamics*. Moscow: Nauka, 1978. English translation: Providence: American Mathematical Society, 1983.

Rubino, B.

1. On the vanishing viscosity approximation to the Cauchy problem for a 2×2 system of conservation laws. Ann. Inst. Henri Poincaré **10** (1993), 627–656.

Ruggeri, T.

1. Galilean invariance and entropy principle for systems of balance laws. Cont. Mech. Therm. **1** (1989), 3–20.
2. Convexity and symmetrization in relativistic theories. Cont. Mech. Therm. **2** (1990), 163–177.

Ruggeri, T. and A. Strumia

1. Main field and convex covariant density for quasilinear hyperbolic systems. Ann. Inst. Henri Poincaré, Section A, **34** (1981), 65–84.

Sablé-Tougeron, M.

1. Méthode de Glimm et problème mixte. Ann. Inst. Henri Poincaré **10** (1993), 423–443.

Schaeffer, D.

1. A regularity theorem for conservation laws. Adv. in Math. **11** (1973), 368–386.

Schaeffer D. and M. Shearer

1. The classification of 2 × 2 nonstrictly hyperbolic conservation laws, with application to oil recovery. Comm. Pure Appl. Math. **40** (1987), 141–178.

Schatzman, M.

1. Continuous Glimm functionals and uniqueness of solutions of the Riemann problem. Indiana U. Math. J. **34** (1985), 533–589.

Schauder, J.

1. Cauchy'sches Problem für partielle Differentialgleichungen erster Ordnung. Comment. Math. Helvetici **9** (1937), 263–283.

Schecter, S., Marchesin, D. and B.J. Plohr

1. Structurally stable Riemann solutions. J. Diff. Eqs. **126** (1996), 303–354.
2. Classification of codimension-one Riemann solutions. (Preprint).

Schochet, S.

1. The compressible Euler equations in a bounded domain. Comm. Math. Phys. **104** (1986), 49–75.
2. Examples of measure-valued solutions. Comm. PDE **14** (1989), 545–575.
3. Glimm scheme for systems with almost planar interactions. Comm. PDE **16** (1991), 1423–1440.
4. Sufficient conditions for local existence via Glimm's scheme for large BV data. J. Diff. Eqs. **89** (1991), 317–354.
5. Resonant nonlinear geometric optics for weak solutions of conservation laws. J. Diff. Eqs. **113** (1994), 473–504.

Shochet, S. and E. Tadmor

1. The regularized Chapman-Enskog expansion for scalar conservation laws. Arch. Rational Mech. Anal. **119** (1992), 95–107.

Schonbek, M.E.

1. Convergence of solutions to nonlinear dispersive equations. Comm. PDE **7** (1982), 959–1000.
2. Existence of solutions to singular conservation laws. SIAM J. Math. Anal. **15** (1984), 1125–1139.

Serre, D.

1. Solutions à variation bornée pour certains systèmes hyperboliques de lois de conservation. J. Diff. Eqs. **67** (1987), 137–168.

2. La compacité par compensation pour les systèmes non linéaires de deux équations a une dimension d'espace. J. Math. Pures Appl. **65** (1987), 423–468.
3. Domaines invariants pour les systèmes hyperboliques de lois de conservation. J. Diff. Eqs. **69** (1987), 46–62.
4. Les ondes planes en électromagnétisme non linéaire. Physica D **31** (1988), 227–251.
5. Oscillations non linéaires des systèmes hyperboliques: méthodes et résultats qualitatif. Ann. Inst. Henri Poincaré **8** (1991), 351–417.
6. Systèmes hyperboliques riches de lois de conservation. *Nonlinear PDE's and their Applications*, ed. H. Brézis and J.-L. Lions, Harlow: Longman, 1992.
7. Integrability of a class of systems of conservation laws. Forum Math. **4** (1992), 607–623.
8. Oscillations non-linéaires de haute fréquence. Dim $\geq$ 2. Nonlinear Variational Problems and Partial Differential Equations, ed. A. Marino and M.K.V. Murthy, Harlow: Longman, 1995.
9. *Systèmes de Lois de Conservation*, Vols. I–II. Paris: Diderot, 1996. English translation: *Systems of Conservation Laws*, Vols. 1–2. Cambridge: Cambridge University Press, 1999.
10. Stabilité L^1 pour les lois de conservation scalaires visqueses. C. R. Acad. Sci. Paris, Série I, **323** (1996), 359–363.
11. Solutions classiques globales des équations d'Euler pour un fluide parfait compressible. Ann. Inst. Fourier, Grenoble **47** (1997), 139–153.
12. Solutions globales des systèmes paraboliques de lois de conservations. Ann. Inst. Fourier, Grenoble **48** (1998), 1069–1091.
13. Relaxation semi-linéaire et cinétique des lois de conservation. Ann. Inst. Henri Poincaré. (To appear).

Serre, D. and J. Shearer

1. Convergence with physical viscosity for nonlinear elasticity. (Preprint).

Serre, D. and L. Xiao

1. Asymptotic behavior of large weak entropy solutions of the damped p-system. J. PDE **10** (1997), 355–368.

Serre D. and K. Zumbrun

1. Viscous and inviscid stability of multidimensional planar shock fronts. (Preprint).

Sevennec, B.

1. Geometry of hyperbolic systems of conservation laws. Bull. Soc. Math. France **122** (1994), Suppl. **56**.

Sever, M.

1. The rate of total entropy generation for Riemann problems. J. Diff. Eqs. **87** (1990), 115–143.

Shearer, J.W.

1. Global existence and compactness in L^p for the quasi-linear wave equation. Comm. PDE **19** (1994), 1829–1877.

Shearer, M.

1. Nonuniqueness of admissible solutions of Riemann initial value problems for a system of conservation laws of mixed type. Arch. Rational Mech. Anal. **93** (1986), 45–59.
2. The Riemann problem for the planar motion of an elastic string. J. Diff. Eqs. **61** (1986), 149–163.
3. The Riemann problem for 2×2 systems of hyperbolic conservation laws with case I quadratic nonlinearities. J. Diff. Eqs. **80** (1989), 343–363.

Shearer, M. and D. Schaeffer

1. Riemann problem for nonstrictly hyperbolic 2×2 systems of conservation laws. Trans. AMS **304** (1987), 267–306.

Shearer, M., Schaeffer, D., Marchesin, D. and P. Paes-Leme

1. Solution of the Riemann problem for a prototype 2×2 system of non-strictly hyperbolic conservation laws. Arch. Rational Mech. Anal. **97** (1987), 299–329.

Shu, C.W.

1. *TVB* uniformly high order schemes for conservation laws. Math. Comp. **49** (1987), 105–121.

Sideris, T.

1. Formation of singularities in three-dimensional compressible fluids. Comm. Math. Phys. **101** (1985), 475–485.

Šilhavý, M.

1. *The Mechanics and Thermodynamics of Continuous Media.* Berlin: Springer, 1997.

Sinestrari, C.

1. Instability of discontinuous traveling waves for hyperbolic balance laws. J. Diff. Eqs. **134** (1997), 269–285.
2. The Riemann problem for an inhomogeneous conservation law without convexity. SIAM J. Math. Anal. **28** (1997), 109–135.

Slemrod, M.

1. Admissibility criteria for propagating phase boundaries in a van der Waals fluid. Arch. Rational Mech. Anal. **81** (1983), 301–315.
2. Dynamic phase transitions in a van der Waals fluid. J. Diff. Eqs. **52** (1984), 1–23.
3. A limiting "viscosity" approach to the Riemann problem for materials exhibiting change of phase. Arch. Rational Mech. Anal. **105** (1989), 327–365.
4. Resolution of the spherical piston problem for compressible isotropic gas dynamics via a self-similar viscous limit. Proc. Royal Soc. Edinburgh **126A** (1996), 1309–1340.

Slemrod, M. and A.E. Tzavaras

1. A limiting viscosity approach for the Riemann problem in isentropic gas dynamics. Indiana U. Math. J. **38** (1989), 1047–1074.
2. Shock profiles and self-similar fluid dynamics limits. J. Transport Th. Stat. Phys. **25** (1996), 531–542.

Smith, R.G.

1. The Riemann problem in gas dynamics. Trans. AMS **249** (1979), 1–50.

Smoller, J.

1. *Shock Waves and Reaction-Diffusion Equations* (Second Edition). New York: Springer, 1994.

Smoller J. and B. Temple

1. Shock wave solutions of the Einstein equations. The Oppenheimer-Snyder model of gravitational collapse extended to the case of nonzero pressure. Arch. Rational Mech. Anal. **128** (1994), 249–297.
2. General relativistic shock waves that extend the Oppenheimer-Snyder model. Arch. Rational Mech. Anal. **138** (1997), 239–277.

Smoller, J.A., Temple, J.B. and Z.-P. Xin

1. Instability of rarefaction shocks in systems of conservation laws. Arch. Rational Mech. Anal. **112** (1990), 63–81.

Sod, G.A.

1. *Numerical Methods in Fluid Dynamics.* Cambridge: Cambridge University Press, 1985.

Stokes, G.G.

1. On a difficulty in the theory of sound. Philos. Magazine **33** (1848), 349–356.
2. On a difficulty in the theory of sound. *Mathematical and Physical Papers*, Vol II, pp. 51–55. Cambridge: Cambridge University Press, 1883.

Šverak, V.

1. Rank-one convexity does not imply quasiconvexity. Proc. Royal Soc. Edinburgh **120A** (1992), 185–189.

Szepessy, A.

1. Measure-valued solutions of scalar conservation laws with boundary conditions. Arch. Rational Mech. Anal. **107** (1989), 181–193.
2. An existence result for scalar conservation laws using measure valued solutions. Comm. PDE **14** (1989), 1329–1350.

Szepessy, A. and Z. Xin

1. Nonlinear stability of viscous shock waves. Arch. Rational Mech. Anal. **122** (1993), 53–103.

Szepessy A. and K. Zumbrun

1. Stability of rarefaction waves in viscous media. Arch. Rational Mech. Anal. **133** (1996), 249–298.

Tadmor, E.

1. *Approximate Solutions of Nonlinear Conservation Laws.* Lecture Notes in Math. **1697** (1998), 1–149. Berlin: Springer.

Tadmor, E. and T. Tassa

1. On the piecewise smoothness of entropy solutions to scalar conservation laws. Comm. PDE **18** (1993), 1631–1652.

Tan, De Chun

1. Riemann problems for hyperbolic systems of conservation laws with no classical wave solutions. Quart. Appl. Math. **51** (1993), 765–776.

Tan, D. and T. Zhang

1. Two-dimensional Riemann problems for a hyperbolic system of nonlinear conservation laws. I; II. J. Diff. Eqs. **111** (1994), 203–254; 255–282.

Tan, D., Zhang T. and Y. Zheng

1. Delta shock waves as limits of vanishing viscosity for hyperbolic systems of conservation laws. J. Diff. Eqs. **112** (1994), 1–32.

Tang, Z.-J. and T.C.T. Ting

1. Wave curves for the Riemann problem of plane waves in isotropic elastic solids. Int. J. Eng. Sci. **25** (1987), 1343–1381.

Tartar, L.C.

1. Cours Peccot, Collège de France 1977.

2. Compensated compactness and applications to partial differential equations. *Nonlinear Analysis and Mechanics: Heriot-Watt Symposium*, Vol IV, pp. 136–212, ed. R.J. Knops. London: Pitman, 1979.
3. The compensated compactness method applied to systems of conservation laws. *Systems of Nonlinear Partial Differential Equations*, pp. 263–285, ed. J.M. Ball. Dordrecht: D. Reidel, 1983.

Taub, A.H.

1. Relativistic Rankine-Hugoniot equations. Phys. Rev. **74** (1948), 328–334.

Taylor, M.E.

1. *Partial Differential Equations III*. New York: Springer, 1996.

Temple, B.

1. Systems of conservation laws with invariant submanifolds. Trans. AMS **280** (1983), 781–795.
2. No L^1-contractive metric for systems of conservation laws. Trans. AMS **288** (1985), 471–480.
3. Decay with a rate for noncompactly supported solutions of conservation laws. Trans. AMS **298** (1986), 43–82.

Temple, B. and R. Young

1. The large time existence of periodic solutions for the compressible Euler equations. Mat. Contemp. **11** (1996), 171–190.
2. The large time stability of sound waves. Comm. Math. Phys. **179** (1996), 417–465.

Trivisa, K.

1. A priori estimates in hyperbolic systems of conservation laws via generalized characteristics. Comm. PDE **22** (1997), 235–267.

Truesdell, C.A. and W. Noll

1. *The Non-Linear Field Theories of Mechanics*. Handbuch der Physik, Vol. III/3. Berlin: Springer, 1965.

Truesdell, C.A. and R.A. Toupin

1. *The Classical Field Theories*. Handbuch der Physik, Vol. III/1. Berlin: Springer, 1960.

Truskinovsky, L.

1. Structure of an isothermal phase discontinuity. Soviet Physics Doklady **30** (1985), 945–948.
2. Transition to detonation in dynamic phase changes. Arch. Rational Mech. Anal. **125** (1994), 375–397.

Tsarev, S.P.

1. On Poisson brackets and one-dimensional systems of hydrodynamic type. Dokl. Akad. Nauk SSSR **282** (1985), 534–537.

Tupciev, V.A.

1. The problem of decomposition of an arbitrary discontinuity for a system of quasilinear equations without the convexity condition. Ž. Vyčisl Mat. i Mat. Fiz. **6** (1966), 527–547. English translation: USSR Comp. Math. Math. Phys. **6** (1966), 161–190.
2. On the method for introducing viscosity in the study of problems involving the decay of a discontinuity. Dokl. Akad. Nauk SSSR **211** (1973), 55–58. English translation: Soviet Math. **14** (1973), 978–982.

Tveito A. and R. Winther

1. On the rate of convergence to equilibrium for a system of conservation laws with a relaxation term. SIAM J. Math. Anal. **28** (1997), 136–161.

Tzavaras, A.E.

1. Elastic as limit of viscoelastic response in a context of self-similar viscous limits. J. Diff. Eqs. **123** (1995), 305–341.
2. Wave interactions and variation estimates for self-similar zero-viscosity limits in systems of conservation laws. Arch. Rational Mech. Anal. **135** (1996), 1–60.
3. Materials with internal variables and relaxation to conservation laws. Arch. Rational Mech. Anal. **146** (1999), 129–155.

Vecchi, I.

1. A note on entropy compactification for scalar conservation laws. Nonlinear Analysis **15** (1990), 693–695.

Venttsel', T.D.

1. Estimates of solutions of the one-dimensional system of equations of gas dynamics with "viscosity" nondepending on "viscosity". Soviet Math. J. **31** (1985), 3148–3153.

Volpert, A.I.

1. The spaces BV and quasilinear equations. Mat. Sbornik **73** (1967), 255–302. English translation: Math. USSR Sbornik **2** (1967), 225–267.

Wagner, D.H.

1. The Riemann problem in two space dimensions for a single conservation law. SIAM J. Math. Anal. **14** (1983), 534–559.
2. Equivalence of the Euler and Lagrangian equations of gas dynamics for weak solutions. J. Diff. Eqs. **68** (1987), 118–136.
3. Conservation laws, coordinate transformations, and differential forms. *Hyperbolic Problems: Theory, Numerics, Applications*, pp. 471–477, eds. J. Glimm, M.J. Graham, J.W. Grove and B.J. Plohr. Singapore: World Scientific, 1996.

Wang, C.C. and C. Truesdell

1. *Introduction to Rational Elasticity*. Leyden: Noordhoff, 1973.

Wang, Dehua

1. Global solutions and stability for self-gravitating isentropic gases. J. Math. Anal. Appl. **229** (1999), 530–542.

Wang, Z. and X. Ding

1. Uniqueness of generalized solution for the Cauchy problem of transportation equations. Acta Math. Scientia **17** (1997), 341–352.

Wang, Z., Huang, F. and X. Ding

1. On the Cauchy problem of transportation equations. Acta Math. Appl. Sinica **13** (1997), 113–122.

Weber, H.

1. *Die Partiellen Differential-Gleichungen der Mathematischen Physik*, Zweiter Band, Vierte Auflage. Braunschweig: Friedrich Vieweg und Sohn, 1901.

Weinberger, H.

1. Long-time behavior for a regularized scalar conservation law in the absence of genuine nonlinearity. Ann. Inst. Henri Poincaré **7** (1990), 407–425.

Wendroff, B.
1. The Riemann problem for materials with nonconvex equation of state. J. Math. Anal. Appl. **38** (1972), 454–466, 640–658.
2. An analysis of front tracking for chromatography. Acta Appl. Math. **30** (1993), 265–285.

Weyl, H.
1. Shock waves in arbitrary fluids. Comm. Pure Appl. Math. **2** (1949), 103–122.

Whitham, G.B.
1. The flow pattern of a supersonic projectile. Comm. Pure Appl. Math. **5** (1952), 301–348.
2. *Linear and Nonlinear Waves.* New York: Wiley-Interscience, 1974.

Wu, Z.-Q.
1. The ordinary differential equation with discontinuous right-hand members and the discontinuous solutions of the quasilinear partial differential equations. Acta Math. Sinica **13** (1963), 515–530. English translation: Scientia Sinica **13** (1964), 1901–1907.

Xin, Z.
1. On the linearized stability of viscous shock profiles for systems of conservation laws. J. Diff. Eqs. **100** (1992), 119–136.
2. Zero dissipation limit to rarefaction waves for the one-dimensional Navier-Stokes equations of compressible isentropic gases. Comm. Pure Appl. Math. **46** (1993), 621–665.
3. On nonlinear stability of contact discontinuities. *Hyperbolic Problems: Theory, Numerics, Applications*, pp. 249–257, eds. J. Glimm, M.J. Graham, J.W. Grove and B.J. Plohr. Singapore: World Scientific, 1996.
4. Viscous boundary layers and their stability. J. PDE **11** (1998), 97–124.

Xu, Xiangsheng
1. Asymptotic behavior of solutions of hyperbolic conservation laws $u_t + (u^m)_x = 0$ as $m \to \infty$ with inconsistent initial values. Proc. Royal Soc. Edinburgh **113A** (1989), 61–71.

Yan, Baisheng
1. Cavitation solutions to homogeneous van der Waals type fluids involving phase transitions. Quart. Appl. Math. **53** (1995), 721–730.

Yang, Tong
1. A functional integral approach to shock wave solutions of the Euler equations with spherical symmetry, I. Comm. Math. Phys. **171** (1995), 607–638. II. J. Diff. Eqs. **130** (1996), 162–178.

Yang, T., Zhu, C. and H. Zhao
1. Compactness framework of L^p approximate solutions for scalar conservation laws. J. Math. Anal. Appl. **220** (1998), 164–186.

Ying, L.A. and C.H. Wang
1. Global solutions of the Cauchy problem for a nonhomogeneous quasilinear hyperbolic system. Comm. Pure Appl. Math. **33** (1980), 579–597.

Yong, W.-A.
1. A simple approach to Glimm's interaction estimates. Appl. Math. Letters **12** (1999), 29–34.
2. Boundary conditions for hyperbolic systems with stiff source terms. Arch. Rational Mech. Anal. (To appear).

Young, L.C.

1. Generalized curves and the existence of an attained absolute minimum in the calculus of variations. Comptes Rendus de la Société des Sciences et des Lettres de Varsovie, Classe III, **30** (1937), 212–234.

Young, R.

1. Sup-norm stability for Glimm's scheme. Comm. Pure Appl. Math. **46** (1993), 903–948.
2. Exact solutions to degenerate conservation laws. SIAM J. Math. Anal. **30** (1999), 537–558.
3. Sustained solutions for gas dynamics. Comm. PDE. (To appear).
4. Periodic solutions to conservation laws. Contemp. Math. (To appear).

Yu, Shih-Hsien

1. Zero dissipation limit to solutions with shocks for systems of hyperbolic conservation laws. Arch. Rational Mech. Anal. **146** (1999), 275–370.

Zeldovich, Ya. and Yu. Raizer

1. *Physics of Shock Waves and High-Temperature Hydrodynamic Phenomena*, Vols. I–II. New York: Academic Press, 1966–1967.
2. *Elements of Gas Dynamics and the Classical Theory of Shock Waves.* New York: Academic Press, 1968.

Zeng, Yanni

1. Convergence to diffusion waves of solutions to nonlinear viscoelastic model with fading memory. Comm. Math. Phys. **146** (1992), 585–609.
2. L^1 asymptotic behavior of compressible isentropic viscous $1 - D$ flow. Comm. Pure Appl. Math. **47** (1994), 1053–1082.
3. L^p asymptotic behavior of solutions to hyperbolic-parabolic systems of conservation laws. Arch. Math. **66** (1996), 310–319.

Zhang, Peng and Tong Zhang

1. Generalized characteristic analysis and Guckenheimer structure. J. Diff. Eqs. **152** (1999), 409–430.

Zhang, T. and Y. Zheng

1. Two-dimensional Riemann problem for a single conservation law. Trans. AMS **312** (1989), 589–619.
2. Axisymmetric solutions of the Euler equations for polytropic gases. Arch. Rational Mech. Anal. **142** (1998), 253–279.

Zhao, Huijiang

1. Global existence in L^4 for a nonstrictly hyperbolic conservation law. (Preprint).

Zheng, Songmu

1. *Nonlinear Parabolic Equations and Hyperbolic-Parabolic Coupled Systems.* Harlow: Longman, 1995.

Zheng, Yuxi

1. *Two-Dimensional Riemann Problems for Systems of Conservation Laws.* (Monograph, to appear).

Zhu, G.S. and T.C.T. Ting

1. Classification of 2×2 non-strictly hyperbolic systems for plane waves in isotropic elastic solids. Int. J. Eng. Sci. **27** (1989), 1621–1638.

Ziemer, W.P.

1. Cauchy flux and sets of finite perimenter. Arch. Rational Mech. Anal. **84** (1983), 189–201.
2. *Weakly Differentiable Functions.* New York: Springer, 1989.

Zumbrun, K.

1. *N*-waves in elasticity. Comm. Pure Appl. Math. **46** (1993), 75–95.
2. Decay rates for nonconvex systems of conservation laws. Comm. Pure Appl. Math. **46** (1993), 353–386.

Zumbrun K. and P. Howard

1. Pointwise semigroup methods and stability of viscous shock waves. Indiana U. Math. J. **47** (1998), 63–85.

Author Index

Subject Index

Grundlehren der mathematischen Wissenschaften

A Series of Comprehensive Studies in Mathematics

A Selection

Printing: Mercedes-Druck, Berlin
Binding: Stürtz AG, Würzburg

1-MONTH